Management of Periprosthetic Joint Infections (PJIs)

Woodhead Publishing Series in Biomaterials

Management of Periprosthetic Joint Infections (PJIs)

Edited By

J.J. Chris Arts and Jan Geurts

ELSEVIER

AMSTERDAM • BOSTON • CAMBRIDGE • HEIDELBERG
LONDON • NEW YORK • OXFORD • PARIS • SAN DIEGO
SAN FRANCISCO • SINGAPORE • SYDNEY • TOKYO
Woodhead Publishing is an imprint of Elsevier

WP
WOODHEAD PUBLISHING

Woodhead Publishing is an imprint of Elsevier
The Officers' Mess Business Centre, Royston Road, Duxford, CB22 4QH, United Kingdom
50 Hampshire Street, 5th Floor, Cambridge, MA 02139, United States
The Boulevard, Langford Lane, Kidlington, OX5 1GB, United Kingdom

Notices
Knowledge and best practice in this field are constantly changing. As new research and experience broaden our understanding, changes in research methods, professional practices, or medical treatment may become necessary.

Practitioners and researchers must always rely on their own experience and knowledge in evaluating and using any information, methods, compounds, or experiments described herein. In using such information or methods they should be mindful of their own safety and the safety of others, including parties for whom they have a professional responsibility.

To the fullest extent of the law, neither the Publisher nor the authors, contributors, or editors, assume any liability for any injury and/or damage to persons or property as a matter of products liability, negligence or otherwise, or from any use or operation of any methods, products, instructions, or ideas contained in the material herein.

Library of Congress Cataloging-in-Publication Data
A catalog record for this book is available from the Library of Congress

British Library Cataloguing-in-Publication Data
A catalogue record for this book is available from the British Library

ISBN: 978-0-08-100205-6 (print)
ISBN: 978-0-08-100242-1 (online)

Publisher: Matthew Deans
Acquisition Editor: Laura Overend
Editorial Project Manager: Lucy Beg
Production Project Manager: Poulouse Joseph
Cover Designer: Victoria Pearson

Contents

List of contributors

S. Alazzawi The Royal London Hospital, London, United Kingdom

M.H. Amini Cleveland Clinic, Orthopaedic and Rheumatologic Institute, Cleveland, OH, United States

J.J. Arts Maastricht University Medical Centre, Maastricht; Eindhoven University of Technology, Eindhoven, The Netherlands

B.D. Brause Hospital for Special Surgery; Weill Cornell Medical Center, New York, NY, United States

L. Drago IRCCS Galeazzi Institute; University of Milan, Milan, Italy

M.B. Ekkelenkamp University Medical Center Utrecht, Utrecht, The Netherlands

P.J. Evans Cleveland Clinic, Orthopaedic and Rheumatologic Institute, Cleveland, OH, United States

D. George University College London Hospitals, London, United Kingdom

J.A.P. Geurts Maastricht University Medical Center, Maastricht, The Netherlands

F.S. Haddad Institute of Sport, Exercise and Health; University College Hospital, London, United Kingdom

M.W. Henry Hospital for Special Surgery; Weill Cornell Medical Center, New York, NY, United States

D.J. Hulsen Maastricht University Medical Centre, Maastricht, The Netherlands

O. Khan Guy's and St. Thomas' NHS Foundation Trust, London, United Kingdom

N.C. Lindfors University of Helsinki, Helsinki, Finland

N. Logoluso Orthopaedic Research Institute Galeazzi, Milan, Italy

K.N. Malizos University of Thessaly, Medical School, Biopolis, Larissa, Greece

G. Manzi CHU Rennes, Rennes, France

A.O. Miller Hospital for Special Surgery; Weill Cornell Medical Center, New York, NY, United States

J.C.E. Odekerken Maastricht University Medical Centre, Maastricht; Zuyd University of Applied Sciences, Heerlen, The Netherlands

L.L. Reubsaet University Medical Center Utrecht, Utrecht, The Netherlands

E.T. Ricchetti Cleveland Clinic, Orthopaedic and Rheumatologic Institute; Cleveland Clinic Lerner College of Medicine of Case Western Reserve University, Cleveland, OH, United States

C.L. Romanò Orthopaedic Research Institute Galeazzi, Milan, Italy

S. Scarponi Orthopaedic Research Institute Galeazzi, Milan, Italy

O.P.P. Temmerman NWZ, Centre for Orthopaedic Research (CORAL), Alkmaar, The Netherlands

M. Toscano IRCCS Galeazzi Institute; University of Milan, Milan, Italy

N.A. van Gestel Eindhoven University of Technology, Eindhoven, The Netherlands

T.A. van Vugt Maastricht University Medical Centre, Maastricht, The Netherlands

S.E. Varitimidis University of Thessaly, Medical School, Biopolis, Larissa, Greece

S.J. Verberne NWZ, Centre for Orthopaedic Research (CORAL), Alkmaar, The Netherlands

G.H.I.M. Walenkamp Maastricht University Medical Center, Maastricht, The Netherlands

T.J. Welting Maastricht University Medical Centre, Maastricht, The Netherlands

Preface

This book summarizes Management of Periprosthetic Joint Infections (PJIs) from fundamental aspects on prevention and diagnostics to surgical and pathogen-directed antibiotic treatment concepts. Furthermore, attention is placed on the biofilm biological response, biomaterials for treatment of orthopedic infections, and on experimental models in orthopedic infection research.

The primary reason for the compilation of this book is the fact that there is high and multidisciplinary interest in PJIs and osteomyelitis treatment. As clinical lecturers, we have been confronted with a lot of interest in this topic over the years, but unable to find an adequate summary of these topics directed toward clinical implementation for a multidisciplinary audience.

This book is by no means intended as a comprehensive overview but aims to raise awareness and stimulation of discussion regarding Management of Periprosthetic Joint Infections in clinical practice.

We believe it will present a useful and practical addition to your clinical practice and education.

Sincerely yours,
Chris Arts and Jan Geurts, editors

Part One

Fundamentals of periprosthetic joint infections

Prevention of joint infections

1

A.O. Miller*,†, M.W. Henry*,†, B.D. Brause*,†
*Hospital for Special Surgery, New York, NY, United States, †Weill Cornell Medical Center, New York, NY, United States

1.1 Introduction

Modern arthroplasty techniques improve the mobility and the quality of life of more than one million inhabitants of the United States of America each year: in 2012, 452,000 hip and 671,000 knee arthroplasties were performed [1]. Prosthetic joint infection (PJI) is a feared complication of arthroplasty. Population incidence of PJI in developed countries has been estimated in the range of 1–2% in the first two post-operative years, with lower rates thereafter [2–5]. Incidence varies substantially across medical facilities [6]. PJIs are costly in terms of individual patients' health and expenses, legal considerations, and costs to health care systems and third parties. Economically, an individual hip PJI costs the average patient $30,000–120,000 [7–9], with projected current cumulative US national costs up to $1 billion. Infection can lead to losses in mobility, medical complications of the infection and its therapy, decreased quality of life, and loss of life. Given the annual increases in arthroplasty volume due to improved access to surgery and marked growth in target populations, strategies to prevent PJI are critical. Aiming to eliminate this dreaded complication is worthwhile, even if we cannot achieve a zero-percent infection rate.

This chapter reviews many presently available approaches to preventing PJI and some promising future avenues. Much of our understanding of PJI prevention (and management) has been made in the absence of randomized prospective studies. Indeed, there is substantial practice variability between and within medical centers worldwide. We attempt to highlight areas where well-controlled studies show significant advantages of one strategy over others, and other areas where significant controversy remains. While guidelines and consensus documents addressing PJI generally reflect available science and expertise [10,11], the management of these difficult cases ultimately depends on clinicians and patients coming to individualized decisions based on the specifics of each individual case.

1.2 Preoperative considerations

1.2.1 Preoperative patient selection and risk factors for infection

Over decades of technical improvement, arthroplasty has become available to increasingly complex groups of patients. Comorbidities powerfully alter infection risk, so careful preoperative selection and optimization of arthroplasty candidates is critical. Comorbidities are grossly categorized into modifiable and nonmodifiable risks.

Management of Periprosthetic Joint Infections (PJIs). http://dx.doi.org/10.1016/B978-0-08-100205-6.00001-X

Modifiable risks include active infection, active tobacco use, and uncontrolled diabetes. Nonmodifiable risks cannot be altered preoperatively and include prior same-site surgery, prior infection, preexisting foreign material, and age. In this categorization scheme, a substantial middle ground exists, as in the cases of diabetes and autoimmune disease, where the underlying disease may elicit a nonmodifiable increase in infection risk, which is further increased by the intensity of therapy given (for autoimmune diseases) or not (for diabetes). Nevertheless, this organizational structure helps in stratifying risk and in optimizing suitable patients for arthroplasty.

1.2.2 Nonmodifiable comorbidities

Revision arthroplasty, compared to primary arthroplasty, has a higher rate of PJI [12,13]. Likewise, prior nonarthroplasty surgery at the site of primary arthroplasty increases the risk of PJI; this observation is likely linked to the fact that posttraumatic arthritis is associated with PJI [14–16]. Prior same site PJI and increased surgical duration also increase the risk of PJI. Likely etiologic factors include changes in the local wound matrix (slow healing, impaired vascular flow, abnormal regulation of surface microbe ecology), underlying host immune factors predisposing to infection, and subclinical (unrecognized) persistent infection.

1.2.3 Modifiable comorbidities

1.2.3.1 Active pyogenic infections

Generally, treatable active infections should prompt delay in elective arthroplasty until resolved. Elective arthroplasty is contraindicated in the presence of active intraarticular infection. Distantly (>10 years) treated native infection may not confer risk for PJI [17]. Early experience in the revision of infected joint arthroplasties noted a high rate of recurrent infection requiring arthrodesis or excision of the joint [18–20]. In the present era, routine elective arthroplasty is not performed in the setting of known intraarticular infection, and when evidence of infection is incidentally encountered intraoperatively, individualized surgical decisions (often postponing implantation of the prosthesis to a later date) must be made.

The presence of skin and soft tissue infections at or near the site of surgery is likely to increase the risk of PJI, via impairment in wound healing of the infected integument and via the potential for contiguous seeding of the prosthesis. The presence of bacteremia, and infections associated with bacteremia (i.e., dental abscesses, endovascular infections, pyelonephritis), is a risk factor for PJI [21,22]. Because vascular supply to the healing bone after arthroplasty is enhanced for approximately 2 years, the risk of bacteremic seeding appears to be to be inversely proportional to the age of the prosthesis. This risk is likely highest immediately after elective arthroplasty.

Acute nonorthopedic infections requiring antimicrobials (pharyngitis, sinus infections, pneumonia, symptomatic urinary tract infection) and significant acute viral infections (including herpes zoster and influenza) usually lead to short-term postponement of surgery, pending resolution.

1.2.3.2 Colonization with Staphylococcus aureus

At any given moment, one-third of the global population is colonized with *Staphylococcus aureus*, whether resistant (MRSA, 2–4%) or sensitive (MSSA, 30–40%) to methicillin [23,24]. This colonization state is not associated with high likelihood of illness in the typical human host. In the surgical patient, however, alterations in the skin integrity can lead to inoculation of *S. aureus* (and other colonizing skin flora) into vulnerable zones. Some evidence suggests that patients colonized with *S. aureus* are likelier to have surgical site infections (SSIs) and nosocomial infections (e.g., [25–27]). There may be an important threshold quantity of colonizing staphylococci above which risk is increased. The number of organisms needed to establish infection is substantially decreased by the presence of indwelling metallic foreign bodies.

Multiple studies have evaluated decolonization strategies. Blanket decolonization of ICU patients led to decreased nosocomial infection rates in one large study [28], but not all studies have shown a positive association (i.e., [29]), and resistance to topical antiseptics in the setting of routine use remains a concern [30]. Furthermore, the cost-effectiveness of universal screening is not certain [31]. Whether or not the benefits of screening patients for *S. aureus* prior to arthroplasty outweigh the costs and inconveniences remains unclear. To date, no major expert group recommends universal screening and decolonization.

The strategy used at our institution involves selective screening of high-risk patients, including patients with prior history of staphylococcal infection. Swabs of the anterior nares, pharynx, axillae, and groin are obtained. Patients found to have *S. aureus* skin carriage undergoes decolonization with 10 applications of mupirocin to the nostrils and five applications of chlorhexidine 4% liquid solution from chin to feet, over the 5 days preceding surgery. Oral antibiotics (usually trimethoprim-sulfamethoxazole plus rifampin) are sometimes added during this time, as reported in other literature [32]. Decolonization is neither necessarily absolute nor long lasting: the goal is to decrease the burden of potential pathogens on the skin and the nose during the intra- and postoperative periods during which the patient is at maximal risk of infection.

Controversy regarding proper management of arthroplasty patients preoperatively colonized with *S. aureus* continues. Whether or not targeted or universal screening is efficacious, which screening test is optimal, and what treatment strategies (if any) work best remain important unresolved questions.

1.2.3.3 Hyperglycemia and diabetes

Hyperglycemia decreases phagocyte function and is a risk factor for nosocomial complications in many medical arenas. In various nonorthopedic areas of acute medical care, strict control of hyperglycemia has been variably associated with improved outcomes, and protocols vary greatly [33]. Preoperative hyperglycemia (generally defined as fasting glucose >180 mg/dL) is associated with PJI [34], and postoperative morning hyperglycemia may be an even better predictor of PJI [34,35]. Diabetes mellitus may be an important (nonmodifiable) risk factor, independent of the degree of disease control [36]. At our institution, a dedicated team headed by an endocrinologist helps optimize perioperative hyperglycemia, while elective surgeries are postponed for patients with glycosylated hemoglobin A1c (HbA1c) >9 mg/dL with the goal of setting 8% as the new threshold in the near future.

1.2.3.4 Dental disease

Odontogenic bacteremia is commonplace, particularly in the elderly with periodontal disease, but dentogingival bacteria are only occasional causes of late hematogenous PJI. There has been long-standing debate regarding the need for antimicrobial prophylaxis when arthroplasty patients undergo dental procedures (briefly discussed later). There is little data to support the efficacy of preoperative "dental clearance" in routine arthroplasty [22,37]. In our institution, the majority of patients are screened for severe dental issues during preoperative medical clearance, but formal clearance from a dentist is not generally required. Most importantly, the presence of focal dentogingival pain or infection must be resolved prior to elective arthroplasty.

1.2.3.5 Immunosuppression and autoimmune/rheumatologic disease

Patient with autoimmune diseases have an increased incidence of joint disorders requiring arthroplasty and frequently received immunomodulatory medications. Both the underlying diseases and their therapies can further predispose to anatomic abnormalities increasing the risk of infection. It is unclear which factors are most important, but the risk of surgical infection in such patients is increased [15,38,39].

In many patients with immunodeficiency, medications can be modified to reduce infection risk in the perioperative period, with input from the rheumatologist, transplant specialist, or hematologist-oncologist. At our institution, perioperative protocols guiding the management of a variety of immunosuppressive medications have been developed, although they typically require individualization based on the severity and specifics of the underlying illness. Generally, disease-modifying agents are withheld prior to elective arthroplasty (at a time related to their half-life) and restarted when wound healing has progressed satisfactorily.

1.2.3.6 Human immunodeficiency virus (HIV)

The majority of current arthroplasty patients with HIV are successfully treated with highly active antiretroviral medications (HAART); a reasonable expectation for these patients is an undetectable HIV viral load and a stable CD4 count substantially above 200. HIV infection does not appear to be a very strong risk factor for PJI. In HIV-positive hemophiliac populations prior to the HAART era, HIV was not associated with PJI risk. More recent studies confirm that HIV-positive arthroplasty patients with undetectable viral load and CD4 count greater than 200 have a risk profile similar to HIV-negative patients [40,41]. In one recent population-level study, patients with HIV were at slightly higher risk of postoperative wound infection, attributable to HIV's associated comorbidities; HIV itself was not an independent risk factor [42]. In our institution, HIV-positive candidates are generally cleared for elective arthroplasty when they have optimized their immunologic and virologic control on highly active antiretroviral therapy, generally in cooperation with their primary HIV provider.

1.2.3.7 Obesity and malnutrition

Obesity is associated with increased risk of PJI, via multiple mechanisms including concurrent medical comorbidities, impaired wound healing, and increased surgical duration [43,44]. Obesity is also a major driver of joint pain and arthritis. Weight loss associated with bariatric surgery has been reported to significantly improve knee complaints [45]. Short-term weight loss before arthroplasty may not measurably modify risk [46]. The morbidly obese patient needs to be counseled about the increased risks of perioperative complications including infection: a BMI >40 appears to be a cutoff above which the risk of perioperative complications is markedly elevated [47,48].

Malnutrition, a broad term intended to capture the presence of any and all nutritional deficiencies, is a separate phenomenon which is prevalent throughout the population. Variable clinical definitions of malnutrition, and the high frequency of confounding comorbidities in malnourished patients, complicate the study of its effects on surgical outcomes. There may be no better general test of nutritional status than careful clinical interview and physical examination. Various studies confirm the plausible hypothesis that malnourishment is associated with adverse surgical outcomes in arthroplasty (for instance, Refs. [49,50]). Hypoalbuminemia may be a more useful predictor of PJI and other perioperative complications than morbid obesity [51], but no single preoperative test has been found to be sensitive and specific for malnourishment. Postoperative monitoring (or replacement) of albumin is not routine practice in our institution. We believe that accurate weight measurements should be taken at every visit in an attempt to heighten the importance of nutrition and body weight as predictors of surgical, and overall health, outcomes.

Optimal strategies for improving preoperative nutritional status remain unclear. Referrals for outpatient nutrition counseling prior to elective arthroplasty may be variably associated with measurable short-term outcomes, but the surgical period may be a moment of opportunity during which positive overall health practices can be adopted, reducing the long-term health risks of malnutrition and obesity. The efficacy of carbohydrate loading immediately before surgery remains in question [52].

1.2.3.8 Integumentary alterations

There is little literature to guide the surgeon regarding the safety of arthroplasty when there is an active ulceration, lesion, or infection in the vicinity of the operative site. However, it is well documented that skin conditions such as chronic venous ulceration, psoriasis, and eczema develop altered microbiological flora, with increased colonization with more pathogenic bacteria such as *Pseudomonas aeruginosa* and *S. aureus* [53–55]. Prospective studies of nonorthopedic patients reveal skin ulceration to be a serious risk factor for SSI [56]; elective arthroplasty should not be performed in this setting. Small retrospective studies demonstrate increased rates of both postoperative wound infections and PJI in patient with active psoriasis when compared to healthy historical controls [57,58]. Active skin disease or ulcerations near the planned operative site should be treated aggressively to mitigate the risk of postoperative infection.

Venous stasis, lymphedema, and dermatologic disorders variably increase PJI risk by altering the wound healing potential, as well as the microbiology, of the surgical wound. Venous stasis and lymphedema are commonplace in the arthroplasty population and are never wholly reversible. However, compression stockings, elevation of the affected extremity, and careful preoperative fluid management can optimize symptom severity.

1.2.3.9 Tobacco, Ethanol, and intravenous drug use

Perioperative smoking cessation improves surgical outcomes (e.g., Refs. [59,60]). In addition, there is a dose-dependent effect of tobacco smoke on surgical outcomes; among smokers, consumption of more than one pack of cigarettes daily may confer a higher risk of hip arthroplasty surgical complications [61]. Although progressively shorter periods of preoperative tobacco cessation are associated with rising rates of surgical complications, it is likely that quitting at any time during the preoperative period is always better than continued tobacco use. Preoperative smoking cessation programs have been shown to lower risk of surgical complications including infection [62] but are not common at most institutions.

Although moderate alcohol use has been associated with improved survival in population studies, excessive alcohol use obviously confers health risks, not surprisingly including surgical infections [63,64]. The mechanisms involved are likely multifactorial, including social and behavioral comorbidities, preexisting nutritional deficits, the stresses of inpatient alcohol withdrawal, and direct effects of alcohol on neutrophils and other immune effectors. Evidence-based cutoffs for excessive alcohol use do not exist, but it is likely that screening and recommendations to reduce or cease consumption in patients consuming larger amounts of alcohol decrease PJI risk.

Intravenous drug use (IVDA) is strongly associated with development of PJI; incidence rates exceeding 25% have been consistently observed [65–67]. Intravenous drug use is frequently associated with bacteremia directly related to contaminated injection techniques, and users are prone to a broad range of bacterial and fungal infections, most commonly *S. aureus*. We believe IVDA to be an absolute contraindication to elective arthroplasty.

1.2.3.10 Preoperative urinary screening and management of bacteriuria

The literature regarding the contribution of asymptomatic bacteriuria (ASB) to the risk of infectious complications following surgery is neither voluminous nor robust. Catheterization of the bladder is routine in most anesthesia protocols and has been associated with transient bacteremia in the setting of asymptomatic bacteriuria [68]. Of the available published studies, however, no correlation between ASB and SSI or PJI has ever clearly been demonstrated [69,70]. These studies, however, are limited by their lack of power and by their primarily retrospective design. A recent prospective multicenter observational cohort study of nearly 2500 patients found that although ASB was an independent risk factor for PJI, there was no correlation between the microbiological causes of the ASB and the PJI [71]. In addition, preoperative treatment of ASB was not associated with a decrease in PJI. These findings suggest that ASB may be a marker of PJI risk, but not a direct cause of PJI. Currently, there are no formal guidelines on the management of ASB prior to orthopedic surgery although several experts in the field recommend proceeding with arthroplasty in the setting of ASB as long as routine prophylactic antibiotics are given (e.g., [72]). Given this lack of correlation between ASB and postoperative infection, the practice of obtaining routine urinalyses in asymptomatic patients undergoing arthroplasty is likely unwarranted [73].

1.3 Prevention of infection in the perioperative period

1.3.1 Body cleansing at home in anticipation of surgery

Patient-directed antiseptic treatment starting the night before surgery is a commonly employed practice to reduce the incidence of SSI. Routine preoperative showering with chlorhexidine gluconate (CHG) solution or antiseptic soap is part of the preoperative instructions for many types of elective surgery, including arthroplasty. Like many such measures that intuitively appear to be rational and medically sound interventions, there is very little published literature showing any benefit. There has also been no standardization of the cleansing regimen recommended. A large metaanalysis of data from multiple randomized controlled trials of patients undergoing a wide range of surgery by the Cochrane Group did not find the incidence of SSI to be lower in patients undergoing showering with CHG vs patients not showering or in patients showering with placebo [74]. Several observational prospective studies of patients specifically undergoing arthroplasty, however, did report a reduction in the incidence of SSI in patients using CHG wipes, though the results from these studies did not achieve statistical significance [75,76]. Regardless of these findings, preoperative showering or self-cleansing by patients is a common practice at many institutions.

1.3.2 Preoperative hair management

Current SHEA guidelines for the prevention of SSI recommend against elective hair removal except when the presence of hair will interfere with the operation [77]. When hair removal is considered necessary, clippers or a depilatory agent is recommended. Use of a razor to remove hair is associated with an increased risk of SSI. Patients should be advised not to shave the legs or other operative sites for several days prior to and following prosthesis implantation. The timing of hair removal is also critical to lower the risks of infection; the lowest rates of infection are observed when hair removal is done just prior to surgery.

1.3.3 Perioperative antibiotics

The use of perioperative antibiotics is an integral part of the measures taken to reduce the incidence of SSI in almost all fields of surgery, and orthopedic surgery is no exception. In addition to the extended length of stay and cost associated with treatment of orthopedic postoperative infections, SSI is also a significant risk factor for the development of prosthesis-related infections and can lead to considerable morbidity [78]. The use of intravenous antibiotics to prevent SSI is recommended for spinal surgery, repair of closed fractures, implantation of any orthopedic fixation device, and arthroplasty [79]. Antibiotic prophylaxis is not indicated in clean procedures involving the hand, knee, or foot with no implantation of foreign material.

In order to be effective, the appropriate antibiotic must be given with careful attention to timing and dose, as well as patient factors, including allergies, renal function, and weight [80]. Cefazolin is the recommended first line agent for orthopedic surgery. In the setting of allergy to cephalosporins, vancomycin or clindamycin can be used as alternatives. Although the frequency of specific bacteria causing PJI varies by the site of arthroplasty [81], there are no recommendations to adjust the antibiotic selection based on the surgical site.

Broader-spectrum preoperative antibiotic prophylaxis is not indicated in patients with poorly controlled diabetes, or who are otherwise immunosuppressed, despite their increased risk for postoperative infectious complications. However, there are certain comorbidities that should prompt adjustments to the antibiotic prophylaxis chosen for a given patient. Vancomycin should be used in patients known to be colonized with MRSA or when the patient is at high risk for MRSA in the absence of microbiologic data [79]. There is a growing body of evidence to suggest that the dual use of cefazolin and vancomycin in patients colonized or at risk of colonization with MRSA is more effective that with MRSA alone, but this has not been included in any formal recommendations [82,83]. In addition, vancomycin can be used if a true outbreak of MRSA or methicillin-resistant coagulase-negative staphylococci has been detected within the institution. Although the management of SSI prevention in patients known to be colonized with multidrug-resistant organisms (e.g., vancomycin-resistant enterococci or carbapenem-resistant Enterobacteriaceae) is not well studied, it seems reasonable to expand the regimen used to target the organism of concern (in addition to covering the usual causes of PJI).

The timing and dosing of perioperative antibiotics are both critical in allowing adequate tissue concentrations to be achieved prior to the start of surgery. The American Association of Orthopedic Surgeons [84], the Center for Disease Control and Prevention [80], and the National Surgical Infection Prevention Project [85] suggest that doses be fully infused within 1 h of incision and prior to the application of any tourniquet. It is important to note that as the dose of vancomycin increases, the time of infusion correspondingly increases. When a proximal tourniquet is used, antibiotics must be fully administered before the tourniquet is inflated [86].

In cases exceeding two half-lives of the antibiotic or in cases with excessive blood loss (>1500 ml; [79]) or fluid resuscitation (>2000 ml; [72]), redosing of the antibiotic is required to maintain the concentration above the minimal inhibitory concentrations of the common causes of SSI.

The efficacy of a properly administered first dose of antibiotics has been shown in multiple well-conducted studies [87,88]. Likewise, these studies also demonstrate the decreasing utility of these antibiotics the longer they are continued following wound closure. Numerous guidelines, including those from the Surgical Infection Society, SCIP, AAOS, and SHEA/IDSA, all recommend the discontinuation of prophylactic antibiotics within 24h of the end of surgery. Postoperative antibiotic prophylaxis in patients with indwelling drains or persistent wound drainage is not recommended.

1.3.4 Surgical skin preparation

Preoperative skin preparation with an antiseptic is a critical intervention reducing incidence of postoperative wound infections. Contamination of the prosthesis at the time of surgery is a significant cause of PJI. The most common agents are chlorhexidine gluconate, iodophors, and alcohol-containing combination products; the current CDC guidelines for the prevention of SSI do not recommend any of these agents above any other [80]. Studies comparing these agents have been equivocal as to which is superior. The largest multicenter prospective trial to date comparing methods of surgical site antisepsis in 849 patients undergoing clean-contaminated cases, primarily intraabdominal surgery, found the use of CHG with alcohol to be superior to povidone-iodine (without alcohol; [89]). A large prospective single-center sequential implementation trial compared three preoperative skin preparations in 3209 patients; chlorhexidine/isopropyl alcohol, iodine povacrylex in isopropyl alcohol, and two applications of povidone-iodine separated by an application of isopropyl alcohol. The lowest infection rate was seen in the iodine povacrylex in the isopropyl alcohol arm, but both iodophor groups achieved better outcomes than the chlorhexidine arm, with statistical significance [90].

There have been no prospective trials of surgical site skin preparation in patients undergoing arthroplasty. However, based on the available literature, there may be an advantage to alcohol-based skin antisepsis, with chlorhexidine or iodophors. One disadvantage of alcohol solutions is the risk of operating room fires [91]. Antiseptic skin preparation is never perfect because bacteria also reside within skin structures including hair follicles, sebaceous cysts, and epidermis [92–94].

1.3.5 Antibiotic-loaded bone cement (ALBC) prophylaxis in primary and revision arthroplasty

The benefit of antibiotic-loaded bone cement (ALBC) prophylaxis in primary arthroplasty remains uncertain. There is considerable concern regarding the use of prophylaxis with ALBC in the current era of very low infection rates, especially when the disadvantages accompanying routine ALBC use may outweigh any potential benefit. In prospective randomized studies, there was no difference between the use of cefuroxime in ALBC and IV cephalosporin administration [95]. In addition, cephalosporins are too potentially allergic for routine use in ALBC. Using gentamicin in ALBC compared with IV antibiotic prophylaxis, there was a lower infection rate with ALBC with 2 years follow-up, but at 10 years follow-up, there was no significant difference between the two groups [96].

Retrospective studies most often cited in support of the use of ALBC prophylaxis in primary arthroplasty have used historical controls [97–99]. Using historical controls is likely to be artifactual in a population with increasing ALBC usage and universal infection rate reduction because patients who did not receive ALBC prophylaxis underwent arthroplasty when infection rates were universally higher. Therefore it is an unequal comparison. This is particularly true for the registry data from Sweden and Norway.

There are additional problems with the use of ALBC in primary arthroplasty. Mechanical strength of the cement may be diminished if a high antibiotic concentration is used. A relatively low-dose ALBC is used to avoid adverse mechanical effects (≤1 g aminoglycoside antibiotic/40 g methyl methacrylate cement.). Aminoglycoside antibiotics may result in nephrotoxicity when mixed with some of the newer, high-viscosity cements. Although allergic reactions to aminoglycoside antibiotics or vancomycin would be extremely rare, other antibiotics, such as cephalosporins, would be much more prone to allergic reactions. Some of the more severe, immunologic reactions to antibiotics might require removal of all the ALBC which may not be feasible. The increased healthcare cost associated with the use of ALBC should be balanced with potential cost savings associated with a reduction in the rate of infection which may not be realized.

The most important disadvantage associated with the use of ALBC prophylaxis is the potential development of antimicrobial resistance with such antibiotic use. Bacteria growing on ALBC are exposed to subinhibitory levels of antibiotics which can induce mutation resistance, representing a clear reason for caution regarding widespread use of ALBC for prophylactic purposes [100]. When coagulase-negative staphylococcal PJI develops, gentamicin-resistant infection was found in 88% of patients who had gentamicin-loaded cement used in the primary arthroplasty, as compared with only 16% of patients who had plain cement used in their primary arthroplasty [101]. Bacteria recovered from revision surgery in which gentamicin-loaded cement had been used in the prior implant surgery were resistant to gentamicin in over 50% of patients [102]. The difficulty is in the balancing of the potential decrease in the prevalence of deep periprosthetic infection with the potential increase in drug-resistant organisms. Therefore many people believe that ALBC should not be used routinely for prophylaxis. It has been suggested that possibly ALBC prophylaxis should be reserved for use in patients with a higher risk of infection than ordinary, such as patients with inflammatory arthropathies, immunosuppression, diabetes mellitus, previous joint infection, malnutrition, malignancy, or hemophilia. This would be problematic as these infection-prone patients would then be at greater risk for developing antibiotic-resistant infections which would be more difficult to treat successfully.

The use of ALBC in revision arthroplasty has been accompanied by decreased prosthesis infection rates in two studies [103,104]. The disadvantages of ALBC in these patients are the same as described above for use in primary arthroplasties. More data would be helpful, but at this time the decision regarding the use of ALBC in revision arthroplasties should be made on an individual basis.

The most difficult problem in infectious diseases at the present time is the every-increasing amounts of antimicrobial resistance being encountered. We now have more untreatable microorganisms than at any time since the advent of the antibiotic era in the 1940s. We should be extremely cautious about reducing infection rates using a technique (such as ALBC) which appears to increase antimicrobial resistance.

1.3.6 The role of laminar air flow and body suits in operating rooms during arthroplasty

Laminar air flow (LAF) ventilation systems in operating rooms were popularized by Sir John Charnley in the 1960s for use during joint replacement implantations after a series of trials showed a significant decrease in infections attributed to laminar flow use. However these studies were not controlled for the variable use of prophylactic antibiotics which had been introduced at about the same time [105]. Since then the validity of investigations regarding the clinical effectiveness of LAF in reducing SSIs has been plagued by the presence of a variety of uncontrolled variables. At the present time, there are no adequately controlled, prospective studies comparing SSI rates with and without LAF [106].

Standard operating room ventilation filters air with the removal of 80–97% of particles >5 μm. Laminar air flow systems equipped with HEPA (High-Efficiency Particulate Air) filters remove 99.97% of particles >0.3 μm. In addition, LAF creates a homogeneous flow of air in the operating room with very little turbulence. LAF is expensive to install and maintain, has significant energy requirements, and requires continuous technical maintenance. However if LAF is effective in the prevention of prosthetic joint infections, it is likely to be cost-effective [107]. Bacterial counts have been found to be at least 20 times lower in the air of operating rooms employing LAF than in those with conventional ventilation systems [108]. LAF systems are designed for use in conjunction with body exhaust suits to decrease bacterial and skin particle (squames) shedding from operative personnel into the air plenum [109]. The air in operating rooms should be maintained at higher pressure than in the adjacent areas to prevent the flow of air from less sterile areas into the operating room [106]. Improper positioning of personnel and operating room can create areas of air turbulence and increases the risk of infection [106,110].

The recommended minimum ceiling area producing LAF is 320 × 320 cm; however often hospitals use smaller dimensions, such as 120 × 240 cm. This can create turbulent air flow at the fringe areas of the operating area and contamination of surgical instruments if they are not under the LAF canopy. Most of the investigations evaluating LAF systems have not considered the impact of LAF ceiling size [107].

Even though bacterial and air particulate is reduced by LAF systems, there is no conclusive effect of a reduction of postoperative wound infections [105]. This discrepancy is explainable as a result of one or more of a variety of counter-productive behaviors and conditions, including: variable use of antibiotic prophylaxis; variable use of proper body exhaust suits and helmets; inadequate plenum/canopy size produced by undersized area of ceiling-producing LAF; inconsistent maintenance of positive pressure in the operating room; improper positioning of personnel and/or equipment (heating units, lighting) within the area of LAF, allowing too many people or unnecessary activity in the operating room; incorrect positioning of the instruments table [which needs to be entirely under the LAF canopy and which incorrect positioning was noted in 94% of 829 procedures in one series] [107]; and variable cooling of the operating room air causing local wound area hypothermia which is a known risk factor for SSI [105]. In addition, the use of LAF with body exhaust suits may give surgeons a false sense of sterility security, leading to unnoticed wound contamination during prosthesis implantation [111].

The Royal College of Anesthetists has recommended that all joint replacement surgery be carried out in a laminar flow theater [105], and LAF may play an important role in reducing the incidence of SSIs in operations involving arterial bypass using grafts [112]. In 2003, the National Institutes of Health (NIH) office of research services, division of engineering services, studied operating room ventilation systems as to their effect on protecting the surgical site from contamination. Their conclusion was that LAF operating room designs represented the best options in terms of contamination control because they result in the smallest percentage of particles arriving at the surgical site [109].

Although the role of LAF in reducing SSI remains undetermined at the present time, the lack of conclusive evidence from a randomized trial is not proof of ineffectiveness [109]. Demonstrating a significant reduction in SSI due to LAF is a difficult task as a result of the prevailing low rates of SSI and PJI [107]. It has been estimated that such a study would require 10,000–70,000 patients [109]. Any strategies to reduce the number of particles in the operating room are likely to result in smaller bacterial dropout on the incision and possibly positively influence the incidence of SSI. Thus ventilation systems that reduce the number of aerosolized bacteria in the operating room are likely to be of high importance in prevention of SSI [113]. However, in what manner these technologies are used and their limitations need to be explored further in order to implement them to good effect because they can be harmful if used improperly [109]. If LAF is to be used, there must be complete and comprehensive compliance with best behavioral and environmental practices in the operating room to enhance the likelihood of the desired beneficial effect.

1.4 Prevention of infection after arthroplasty

1.4.1 Early postoperative period

1.4.1.1 Wound issues

In general, superiority of any particular wound closure technique after arthroplasty has been guided more by local experience than by evidence, despite multiple studies. Empiric treatment of draining wounds with antibiotics is rarely advisable. Superficial cellulitis in the absence of deep infection is unlikely to present as a draining surgical wound, and the ability to subsequently diagnose deep infection is diminished. Furthermore, antibiotics alter the wound microbiome, potentially predisposing toward infections with more resistant and more pathogenic colonizing bacteria.

1.4.1.2 Cellulitis and superficial wound infection

Periincisional cellulitis and suprafascial wound infections following arthroplasty are distinct from PJI and, in general, are managed quite differently. At our institution, we seek to treat patients with a finite course of antibiotics for a nonPJI arthroplasty wound infection (typically 2 weeks, either empirically toward common causes of cellulitis or directed against the results of microbiologic cultures of the wound obtained during debridement) and then observe off antibiotics for evidence of infection. If there is development of symptoms or signs consistent with SSI, the patient should then be evaluated for possible PJI.

1.4.1.3 Postoperative fever

Multiple studies have concluded that the incidence of postoperative fever is frequent, but the likelihood of PJI is very low in any individual postoperative patient with isolated fever. When evaluating a postoperative patient with fever within 72 h of surgery, an appropriate history and physical should be done. No additional workup is typically needed if there are no localizing signs of infection and there is no evidence of very high fever or sepsis [114].

1.4.2 Late considerations

Monitoring for all of the infection predispositions identified preoperatively needs to continue postoperatively. Infectious processes which could result in prosthesis infection by contiguous spread from surrounding tissues or by hematogenous seeding should be prevented or promptly treated. Patients who have chronic or recurrent dermatitis (i.e., psoriasis, eczema, dermatophytosis) need to maintain good control over these processes by themselves or with the assistance of a dermatologist. Any process which exposes the dermis or underlying structures represents a significant portal of entry for the development of a soft tissue infection and subsequent bacteremia or contiguous spread to the prosthesis. Patients need to maintain appropriate, periodic dental care and dental cleaning to reduce the frequency and magnitude of dentogingival bacteremias. Genitourinary, gastrointestinal and respiratory infections need to be treated promptly to reduce the risk of bacteremia.

Situations that are likely to cause bacteremia should be avoided, if possible. Use of prophylactic antibiotics in anticipation of bacteremic events (e.g., dental surgery, cystoscopy, surgical procedures on infected or contaminated tissues) has been suggested on the same basis on which endocarditis prophylaxis has been recommended. This approach to prevention is controversial, and no data are available with which to determine the adequacy or the cost-effectiveness of such measures. In 1997, the American Dental Association and the American Academy of Orthopedic Surgeons jointly advised that a single dose of prophylactic antibiotic be given for selected patients undergoing dental procedures associated with significant bleeding, including periodontal scaling [115]. These selected patient populations include those with inflammatory arthropathies, immunosuppression, diabetes, malnutrition, hemophilia, or previous PJI, as well as all patients undergoing these procedures within 2 years following joint replacement. This recommendation was tempered in a 2013 edition of the same guideline [116], and controversy continues [117]. The American Urologic Association and American Academy of Orthopedic Surgeons have jointly advised that prophylactic antibiotics be considered for similar selected patient populations undergoing urologic procedures associated with higher bacteremic risk [118]. Clinical decisions regarding prophylactic antibiotics for expected bacteremias in patients with prosthetic joints should be made on an individual basis.

1.5 Conclusions and future trends

Arthroplasty has improved the lives of millions of patients, but PJI causes substantial morbidity and is costly to stakeholders, from the individual patient to the public at large. Preoperative consideration of risk factors, both modifiable and fixed, is likely to decrease risk of PJI. Work remains to be done to examine the efficacy, practical considerations, and magnitude of effect of the interventions discussed in this chapter.

In the future, it is likely that more patients will be the recipients of safer devices, implanted more efficiently and quickly. It is important that infection reduction measures continue to improve commensurately as arthroplasty volume increases. The role of universal or targeted decolonization of *S. aureus* skin carriage remains to be definitively evaluated. The possibility of intrinsically infection-resistant prostheses with antibacterial or antibiofilm coatings remains intriguing but unproven in the clinical arena. The process of assessing PJI risk begins immediately as the arthritic patient first meets his orthopedist, and we expect that this will be the case in the future; we need to understand the tools to best measure risk and the interventions that are most important in its mitigation. In conclusion, while the science of PJI prevention remains imperfect and many questions remain, the medical literature clearly shows that PJIs can be prevented by careful patient selection, with preoperative risk reduction strategies and with periincisional antibiotic prophylaxis.

References

[1] Agency for Healthcare Research and Quality, HCUPnet: a tool for identifying, tracking, and analyzing national hospital statistics, Available at: http://hcupnet.ahrq.gov, 2015 (accessed 30.11.15).

[2] K. Huotari, O. Lyytikäinen, J. Ollgren, M. Virtanen, S. Seitsalo, R. Palonen, P. Rantanen, Disease burden of prosthetic joint infections after hip and knee joint replacement in Finland during 1999–2004: capture–recapture estimation, J. Hosp. Infect. 75 (3) (2010) 205–208.

[3] S. Kurtz, K. Ong, E. Lau, K. Bozic, D. Berry, J. Parvizi, Prosthetic joint infection risk after TKA in the medicare population, Clin. Orthop. Relat. Res. 468 (1) (2010) 52–56.

[4] K. Ong, E. Lau, J. Suggs, S. Kurtz, M. Manley, Risk of subsequent revision after primary and revision total joint arthroplasty, Clin. Orthop. Relat. Res. 468 (11) (2010) 3070–3076.

[5] D. Urquhart, F. Hanna, S. Brennan, A. Wluka, K. Leder, P. Cameron, S. Graves, F. Cicuttini, Incidence and risk factors for deep surgical site infection after primary total hip arthroplasty: a systematic review, J. Arthroplast. 25 (8) (2010) 1216–1222.e3.

[6] New York State Department of Health, Hospital Acquired Infections: New York State 2013, Retrieved November 16, 2015, from: https://www.health.ny.gov/statistics/facilities/hospital/hospital_acquired_infections/2013/docs/hospital_acquired_infection.pdf, 2014.

[7] J. Garrido-Gómez, M. Arrabal-Polo, M. Girón-Prieto, J. Cabello-Salas, J. Torres-Barroso, J. Parra-Ruiz, Descriptive analysis of the economic costs of periprosthetic joint infection of the knee for the public health system of andalusia, J. Arthroplast. 28 (7) (2013) 1057–1060.

[8] B. Kapadia, M. McElroy, K. Issa, A. Johnson, K. Bozic, M. Mont, The economic impact of periprosthetic infections following total knee arthroplasty at a specialized tertiary-care center, J. Arthroplast. 29 (5) (2014) 929–932.
[9] S. Kurtz, E. Lau, H. Watson, J.K. Schmier, J. Parvizi, Economic burden of periprosthetic joint infection in the united states, J. Arthroplast. 27 (8) (2012) 61–65.e1.
[10] D. Osmon, E. Berbari, A. Berendt, D. Lew, W. Zimmerli, J. Steckelberg, N. Rao, A. Hanssen, W. Wilson, Diagnosis and management of prosthetic joint infection: clinical practice guidelines by the infectious diseases society of America, Clin. Infect. Dis. 56 (1) (2012) e1–e25.
[11] J. Parvizi, T. Gehrke, A. Chen, Proceedings of the international consensus on periprosthetic joint infection, Bone Joint J. 95-B (11) (2013) 1450–1452.
[12] A.D. Hanssen, D.R. Osmon, C.L. Nelson, Prevention of deep periprosthetic joint infection, Instr. Course Lect. 46 (1997) 555–567.
[13] G. Maoz, M. Phillips, J. Bosco, J. Slover, A. Stachel, I. Inneh, R. Iorio, The Otto Aufranc Award: Modifiable versus Nonmodifiable risk factors for infection after hip arthroplasty, Clin. Orthop. Relat. Res. 473 (2) (2014) 453–459.
[14] E. Jämsen, H. Huhtala, T. Puolakka, T. Moilanen, Risk factors for infection after knee arthroplasty. A register-based analysis of 43,149 cases, J. Bone Joint Surg. Am. 91 (1) (2009) 38–47.
[15] G. Peersman, R. Laskin, J. Davis, M. Peterson, Infection in total knee replacement: a retrospective review of 6489 total knee replacements, Clin. Orthop. Relat. Res. 392 (2001) 15–23.
[16] C. Willis-Owen, A. Konyves, D. Martin, Factors affecting the incidence of infection in hip and knee replacement: an analysis of 5277 cases, J. Bone Joint Surg. Br. Vol. 92-B (8) (2010) 1128–1133.
[17] Y.H. Kim, S.H. Oh, J.S. Kim, Total hip arthroplasty in adult patients who had childhood infection of the hip, J. Bone Joint Surg. Am. 85-A (2) (2003) 198–204.
[18] J. Charnley, N. Eftekhar, Postoperative infection in total prosthetic replacement arthroplasty of the hip-joint. With special reference to the bacterial content of the air of the operating room, Br. J. Surg. 56 (9) (1969) 641–649.
[19] J.B. Jupiter, A.W. Karchmer, J.D. Lowell, W.H. Harris, Total hip arthroplasty in the treatment of adult hips with current or quiescent sepsis, J. Bone Joint Surg. Am. 63 (2) (1981) 194–200.
[20] J.C. Mulier, J. Vandepitte, J. Stuyck, A. Burssens, G. Vermaut, J. Van Ruymbeke, Long-term study of the perioperative infection rate in 1007 total hip replacements using prophylactic cefamandole and other precautions, Arch. Orthop Trauma Surg. 106 (3) (1987) 135–139.
[21] B. Konigsberg, C. Valle, N. Ting, F. Qiu, S. Sporer, Acute hematogenous infection following total hip and knee arthroplasty, J. Arthroplast. 29 (3) (2014) 469–472.
[22] D.M. LaPorte, B.J. Waldman, M.A. Mont, D.S. Hungerford, Infections associated with dental procedures in total hip arthroplasty, J. Bone Joint Surg. (Br.) 81 (1) (1999) 56–59.
[23] P.L. Graham, S.X. Lin, E.L. Larson, A U.S. population-based survey of Staphylococcus aureus colonization, Ann. Intern. Med. 144 (5) (2002) 318–325.
[24] J. Kluytmans, A. van Belkum, H. Verbrugh, Nasal carriage of Staphylococcus aureus: epidemiology, underlying mechanisms, and associated risks, Clin. Microbiol. Rev. 10 (3) (1997) 505–520.
[25] L. Bode, J. Kluytmans, H. Wertheim, D. Bogaers, C. Vandenbroucke-Grauls, R. Roosendaal, A. Troelstra, A. Box, A. Voss, I. van der Tweel, A. van Belkum, H. Verbrugh, M. Vos, Preventing surgical-site infections in nasal carriers of Staphylococcus aureus, N. Engl. J. Med. 362 (1) (2010) 9–17.

[26] M.D. Kalmeijer, H. Coertjens, P.M. van Nieuwland-Bollen, D. Bogaers-Hofman, G.A. de Baere, A. Stuurman, A. van Belkum, J.A. Kluytmans, Surgical site infections in orthopedic surgery: the effect of mupirocin nasal ointment in a double-blind, randomized, placebo-controlled study, Clin. Infect. Dis. 35 (4) (2002) 353–358.

[27] T.M. Perl, J.J. Cullen, R.P. Wenzel, M.B. Zimmerman, M.A. Pfaller, D. Sheppard, J. Twombley, P.P. French, L.A. Herwaldt, Mupirocin and the risk of staphylococcus aureus study team. Intranasal mupirocin to prevent postoperative Staphylococcus aureus infections, N. Engl. J. Med. 346 (24) (2002) 1871–1877.

[28] S. Huang, E. Septimus, K. Kleinman, J. Moody, J. Hickok, T. Avery, J. Lankiewicz, A. Gombosev, L. Terpstra, F. Hartford, M. Hayden, J. Jernigan, R. Weinstein, V. Fraser, K. Haffenreffer, E. Cui, R. Kaganov, K. Lolans, J. Perlin, R. Platt, Targeted versus Universal Decolonization to Prevent ICU Infection, N. Engl. J. Med. 368 (24) (2013) 2255–2265.

[29] S. Harbarth, C. Fankhauser, J. Schrenzel, J. Christenson, P. Gervaz, C. Bandiera-Clerc, G. Renzi, N. Vernaz, H. Sax, D. Pittet, (2008). Universal screening for methicillin-resistant Staphylococcus aureus at hospital admission and nosocomial infection in surgical patients, JAMA 299 (10) (2008) 1149–1157.

[30] J. McDanel, C. Murphy, D. Diekema, V. Quan, D. Kim, E. Peterson, K. Evans, G. Tan, M. Hayden, S. Huang, Chlorhexidine and mupirocin susceptibilities of methicillin-resistant staphylococcus aureus from colonized nursing home residents, Antimicrob. Agents Chemother. 57 (1) (2012) 552–558.

[31] S. Glick, D. Samson, E. Huang, V. Vats, S. Weber, N. Aronson, Screening for methicillin-resistant staphylococcus aureus (MRSA), AHRQ WebM&M (2013). Available at: http://www.ncbi.nlm.nih.gov/books/NBK153098/ (accessed 30.11.15).

[32] S.K. Fung, M. Louie, A.E. Simor, Combined topical and oral antimicrobial therapy for the eradication of methicillin-resistant Staphylococcus aureus (MRSA) colonization in hospitalized patients, Can. J. Infect. Dis. 13 (5) (2002) 287–292.

[33] D. Niven, G. Rubenfeld, A. Kramer, H. Stelfox, Effect of published scientific evidence on glycemic control in adult intensive care units, JAMA Intern. Med. 175 (5) (2015) 801.

[34] J. Chrastil, M. Anderson, V. Stevens, R. Anand, C. Peters, C. Pelt, Is hemoglobin a1c or perioperative hyperglycemia predictive of periprosthetic joint infection or death following primary total joint arthroplasty? J. Arthroplast. 30 (7) (2015) 1197–1202.

[35] B. Mraovic, D. Suh, C. Jacovides, J. Parvizi, Perioperative hyperglycemia and postoperative infection after lower limb arthroplasty, J. Diabetes Sci. Technol. 5 (2) (2011) 412–418.

[36] M. Marchant Jr., The impact of glycemic control and diabetes mellitus on perioperative outcomes after total joint arthroplasty, J. Bone Joint Surg. Am. 91 (7) (2009) 1621.

[37] J. Barrington, T. Barrington, What is the true incidence of dental pathology in the total joint arthroplasty population? J. Arthroplast. 26 (6) (2011) 88–91.

[38] E. Berbari, D. Osmon, B. Lahr, J. Eckel-Passow, G. Tsaras, A. Hanssen, T. Mabry, J. Steckelberg, R. Thompson, The mayo prosthetic joint infection risk score: implication for surgical site infection reporting and risk stratification, Infect. Control Hosp. Epidemiol. 33 (08) (2012) 774–781.

[39] T. Morrison, M. Figgie, A.O. Miller, S. Goodman, Periprosthetic joint infection in patients with inflammatory joint disease: a review of risk factors and current approaches to diagnosis and management, HSS Journal ® 9 (2) (2013) 183–194.

[40] J. Falakassa, A. Diaz, M. Schneiderbauer, Outcomes of total joint arthroplasty in HIV patients, Iowa Orthop. J. 34 (2014) 102–106.

[41] D.L. Powell, C.J. Whitener, C.E. Dye, J.O. Ballard, M.L. Shaffer, M.E. Eyster, Knee and hip arthroplasty infection rates in persons with haemophilia: a 27 year single center experience during the HIV epidemic, Haemophilia 11 (3) (2005) 233–239.

[42] C. Lin, A. Kuo, S. Takemoto, Comorbidities and perioperative complications in HIV-positive patients undergoing primary total hip and knee arthroplasty, J. Bone Joint Surg. Am. 95 (11) (2013) 1028.

[43] M. Dowsey, P. Choong, Obese diabetic patients are at substantial risk for deep infection after primary TKA, Clin. Orthop. Relat. Res. 467 (6) (2008) 1577–1581.

[44] R. Malinzak, M. Ritter, M. Berend, J. Meding, E. Olberding, K. Davis, Morbidly obese, diabetic, younger, and unilateral joint arthroplasty patients have elevated total joint arthroplasty infection rates, J. Arthroplast. 24 (6) (2009) 84–88.

[45] V. Groen, V. van de Graaf, V. Scholtes, S. Sprague, B. van Wagensveld, R. Poolman, Effects of bariatric surgery for knee complaints in (morbidly) obese adult patients: a systematic review, Obes. Rev. 16 (2) (2014) 161–170.

[46] M. Lui, C.A. Jones, M.D. Westby, Effect of non-surgical, non-pharmacological weight loss interventions in patients who are obese prior to hip and knee arthroplasty surgery: a rapid review, Syst. Rev. 27 (4) (2015) 121.

[47] M. D'Apuzzo, W. Novicoff, J. Browne, The John Insall Award: morbid obesity independently impacts complications, mortality, and resource use after TKA, Clin. Orthop. Relat. Res. 473 (1) (2014) 57–63.

[48] D. Ward, L. Metz, P. Horst, H. Kim, A. Kuo, Complications of morbid obesity in total joint arthroplasty: risk stratification based on BMI, J. Arthroplast. 30 (9) (2015) 42–46.

[49] M.B. Cross, P.H. Yi, C.F. Thomas, J. Garcia, C.J. Della Valle, Evaluation of malnutrition in orthopaedic surgery, J. Am. Acad. Orthop. Surg. 22 (3) (2014) 193–199.

[50] R. Huang, M. Greenky, G. Kerr, M. Austin, J. Parvizi, The effect of malnutrition on patients undergoing elective joint arthroplasty, J. Arthroplast. 28 (8) (2013) 21–24.

[51] C. Nelson, N. Elkassabany, A. Kamath, J. Liu, Low albumin levels, more than morbid obesity, are associated with complications after TKA, Clin. Orthop. Relat. Res. 473 (10) (2015) 3163–3172.

[52] M.D. Smith, J. McCall, L. Plank, G.P. Herbison, M. Soop, J. Nygren, Preoperative carbohydrate treatment for enhancing recovery after elective surgery, Cochrane Database Syst. Rev. 8 (2014) CD009161.

[53] J.A. Sanford, R.L. Gallo, Functions of the skin microbiota in health and disease, Semin. Immunol. 25 (5) (2013) 370–377.

[54] D.D. Balci, N. Duran, B. Ozer, R. Gunesacar, Y. Onlen, J.Z. Yenin, High prevalence of Staphylococcus aureus cultivation and superantigen production in patients with psoriasis, Eur. J. Dermatol. 19 (3) (2009) 238–242.

[55] J. de Souza, É. Vieira, T. Cortez, A. Mondelli, H. Miot, L. Abbade, Clinical and microbiologic evaluation of chronic leg ulcers, Adv. Skin Wound Care 27 (5) (2014) 222–227.

[56] A. Penington, Ulceration and antihypertensive use are risk factors for infection after skin lesion excision, ANZ J. Surg. 80 (9) (2010) 642–645.

[57] M. Drancourt, J.N. Argenson, H. Tissot Dupont, J.M. Aubaniac, D. Raoult, Psoriasis is a risk factor for hip-prosthesis infection, Eur. J. Epidemiol. 13 (2) (1997) 205–207.

[58] S.H. Stern, J.N. Insall, R.E. Windsor, A.E. Inglis, D.M. Dines, Total knee arthroplasty in patients with psoriasis, Clin. Orthop. Relat. Res. 1989 (248) (1989) 108–110.

[59] K. Duchman, Y. Gao, A. Pugely, C. Martin, N. Noiseux, J. Callaghan, The effect of smoking on short-term complications following total hip and knee arthroplasty, J. Bone Joint Surg. Am. 97 (13) (2015) 1049–1058.

[60] S. Teng, C. Yi, C. Krettek, M. Jagodzinski, Smoking and risk of prosthesis-related complications after total hip arthroplasty: a meta-analysis of cohort studies, PLoS ONE 10 (4) (2015) e0125294.

[61] O. Sadr Azodi, R. Bellocco, K. Eriksson, J. Adami, The impact of tobacco use and body mass index on the length of stay in hospital and the risk of post-operative complications among patients undergoing total hip replacement, J. Bone Joint Surg. (Br.) 88 (10) (2006) 1316–1320.

[62] D. Lindström, O. Azodi, A. Wladis, H. Tønnesen, S. Linder, H. Nåsell, S. Ponzer, J. Adami, Effects of a perioperative smoking cessation intervention on postoperative complications, Ann. Surg. 248 (5) (2008) 739–745.

[63] K. Bradley, A. Rubinsky, H. Sun, C. Bryson, M. Bishop, D. Blough, W. Henderson, C. Maynard, M. Hawn, H. Tønnesen, G. Hughes, L. Beste, A. Harris, E. Hawkins, T. Houston, D. Kivlahan, Alcohol screening and risk of postoperative complications in male va patients undergoing major non-cardiac surgery, J. Gen. Intern. Med. 26 (2) (2010) 162–169.

[64] M. Eliasen, M. Grønkjær, L. Skov-Ettrup, S. Mikkelsen, U. Becker, J. Tolstrup, T. Flensborg-Madsen, Preoperative alcohol consumption and postoperative complications, Ann. Surg. 258 (6) (2013) 930–942.

[65] B. Habermann, C. Eberhardt, A. Kurth, Total joint replacement in HIV positive patients, J. Infect. 57 (1) (2008) 41–46.

[66] C. Lehman, M. Ries, G. Paiement, A. Davidson, Infection after total joint arthroplasty in patients with human immunodeficiency virus or intravenous drug use, J. Arthroplast. 16 (3) (2001) 330–335.

[67] K. Wieser, P. Zingg, M. Betz, G. Neubauer, C. Dora, Total hip replacement in patients with history of illicit injecting drug use, Arch. Orthop. Trauma Surg. 132 (7) (2012) 1037–1044.

[68] N. Sullivan, V. Sutter, M. Mims, V. Marsh, S. Finegold, Clinical aspects of bacteremia after manipulation of the genitourinary tract, J. Infect. Dis. 127 (1) (1973) 49–55.

[69] R. Duncan, Editorial commentary: prosthetic joint replacement: should orthopedists check urine because it's there? Clin. Infect. Dis. 59 (1) (2014) 48–50.

[70] P. Koulouvaris, P. Sculco, E. Finerty, T. Sculco, N. Sharrock, Relationship between perioperative urinary tract infection and deep infection after joint arthroplasty, Clin. Orthop. Relat. Res. 467 (7) (2008) 1859–1867.

[71] R. Sousa, E. Munoz-Mahamud, J. Quayle, L. Dias da Costa, C. Casals, P. Scott, P. Leite, P. Vilanova, S. Garcia, M. Ramos, J. Dias, A. Soriano, A. Guyot, Is asymptomatic bacteriuria a risk factor for prosthetic joint infection? Clin. Infect. Dis. 59 (1) (2014) 41–47.

[72] E. Hansen, K. Belden, R. Silibovsky, M. Vogt, W. Arnold, G. Bicanic, S. Bini, F. Catani, J. Chen, M. Ghazavi, K.M. Godefroy, P. Holham, H. Hosseinzadeh, K.I. Kim, K. Kirketerp-Møller, L. Lidgren, J.H. Lin, J.H. Lonner, C.C. Moore, P. Papagelopoulos, L. Poultsides, R.L. Randall, B. Roslund, K. Saleh, J.V. Salmon, E. Schwarz, J. Stuyck, A.W. Dahl, K. Yamada, Perioperative antibiotics, J. Orthop. Res. 32 (Suppl. 1) (2014) S31–S59.

[73] C. Bouvet, A. Lubbeke, C. Bandi, L. Pagani, R. Stern, P. Hoffmeyer, I. Uckay, Is there any benefit in pre-operative urinary analysis before elective total joint replacement? Bone Joint J. 96-B (3) (2014) 390–394.

[74] J. Webster, S. Osborne, Preoperative bathing or showering with skin antiseptics to prevent surgical site infection, Cochrane Database Syst. Rev. (2) (2015) CD004985.

[75] A. Johnson, J. Daley, M. Zywiel, R. Delanois, M. Mont, Preoperative chlorhexidine preparation and the incidence of surgical site infections after hip arthroplasty, J. Arthroplast. 25 (6) (2010) 98–102.

[76] M. Zywiel, J. Daley, R. Delanois, Q. Naziri, A. Johnson, M. Mont, Advance preoperative chlorhexidine reduces the incidence of surgical site infections in knee arthroplasty, Int. Orthop. 35 (7) (2010) 1001–1006.
[77] D.J. Anderson, K.S. Kaye, D. Classen, K.M. Arias, K. Podgorny, H. Burstin, D.P. Calfee, S.E. Coffin, E.R. Dubberke, V. Fraser, D.N. Gerding, F.A. Griffin, P. Gross, M. Klompas, E. Lo, J. Marschall, L.A. Mermel, L. Nicolle, D.A. Pegues, T.M. Perl, S. Saint, C.D. Salgado, R.A. Weinstein, R. Wise, D.S. Yokoe, Strategies to prevent surgical site infections in acute care hospitals, Infect. Control Hosp. Epidemiol. 29 (S1) (2008) S51–S61.
[78] E. Berbari, A. Hanssen, M. Duffy, J. Steckelberg, D. Ilstrup, W. Harmsen, D. Osmon, Risk factors for prosthetic joint infection: case control study, Clin. Infect. Dis. 27 (5) (1998) 1247–1254.
[79] D.W. Bratzler, E. Dellinger, K. Olsen, T. Perl, P. Auwaerter, M. Bolon, D. Fish, L. Napolitano, R. Sawyer, D. Slain, J. Steinberg, R. Weinstein, Clinical practice guidelines for antimicrobial prophylaxis in surgery, Surg. Infect. 14 (1) (2013) 73–156.
[80] A. Mangram, T. Horan, M. Pearson, L. Silver, W. Jarvis, Guideline for prevention of surgical site infection, Am. J. Infect. Control 27 (2) (1999) 97–134.
[81] A. Tande, R. Patel, Prosthetic joint infection, Clin. Microbiol. Rev. 27 (2) (2014) 302–345.
[82] R. Elliott, H. Weatherly, N. Hawkins, G. Cranny, D. Chambers, L. Myers, A. Eastwood, M. Sculpher, An economic model for the prevention of MRSA infections after surgery: non-glycopeptide or glycopeptide antibiotic prophylaxis? Eur. J. Health Econ. 11 (1) (2009) 57–66.
[83] A. Sewick, A. Makani, C. Wu, J. O'Donnell, K. Baldwin, G. Lee, Does dual antibiotic prophylaxis better prevent surgical site infections in total joint arthroplasty? Clin. Orthop. Relat. Res. 470 (10) (2012) 2702–2707.
[84] American Academy of Orthopedic Surgeons, Recommendations for the use of intravenous antibiotic prophylaxis in primary total joint arthroplasty, Available at: http://www.aaos.org/about/papers/advistmt/1027.asp, 2015 (accessed 30.11.15).
[85] D.W. Bratzler, P.M. Houck, Antimicrobial prophylaxis for surgery: an advisory statement from the National Surgical Infection Prevention Project, Clin. Infect. Dis. 38 (12) (2004) 1706–1715.
[86] L. Prokuski, Prophylactic antibiotics in orthopaedic surgery, J. Am. Acad. Orthop. Surg. 16 (5) (2008) 283–293.
[87] D. Classen, R. Evans, S. Pestotnik, S. Horn, R. Menlove, J. Burke, The timing of prophylactic administration of antibiotics and the risk of surgical-wound infection, N. Engl. J. Med. 326 (5) (1992) 281–286.
[88] M. van Kasteren, J. Mannien, A. Ott, B. Kullberg, A. de Boer, I. Gyssens, Antibiotic prophylaxis and the risk of surgical site infections following total hip arthroplasty: timely administration is the most important factor, Clin. Infect. Dis. 44 (7) (2007) 921–927.
[89] R. Darouiche, M. Wall, K. Itani, Chlorhexidine-alcohol versus povidone-iodine for surgical-site antisepsis, J. Vasc. Surg. 51 (4) (2010) 1063.
[90] B. Swenson, T. Hedrick, R. Metzger, H. Bonatti, T. Pruett, R. Sawyer, Effects of preoperative skin preparation on postoperative wound infection rates: a prospective study of 3 skin preparation protocols, Infect. Control Hosp. Epidemiol. 30 (10) (2009) 964–971.
[91] A. Bonnet, M. Devienne, V. De Broucker, V. Duquennoy-Martinot, P. Guerreschi, Operating room fire: should we mistrust alcoholic antiseptics? Ann. Chir. Plast. Esthet. 60 (4) (2015) 255–261.
[92] E. Grice, J. Segre, The skin microbiome, Nat. Rev. Microbiol. 9 (4) (2011) 244–253.

[93] M. Lee, P. Pottinger, S. Butler-Wu, R. Bumgarner, S. Russ, F. Matsen, Propionibacterium persists in the skin despite standard surgical preparation, J. Bone Joint Surg. Am. 96 (17) (2014) 1447–1450.
[94] O. Schindler, R.F. Spencer, M.D. Smith, Should we use a separate knife for the skin? J. Bone Joint Surg. Br. Vol. 88-B (3) (2006) 382–385.
[95] M. McQueen, A. Littlejohn, S. Hughes, A comparison of systemic Cefuroxime and Cefuroxime loaded bone cement in the prevention of early infection after total joint replacement, Int. Orthop. 11 (3) (1987) 241–243.
[96] G. Josefsson, L. Kolmert, Prophylaxis with systematic antibiotics versus gentamicin bone cement in total hip arthroplasty, Clin. Orthop. Relat. Res. 292 (1993) 210–214.
[97] H. Buchholz, R. Elson, K. Heinert, Antibiotic-loaded acrylic cement, Clin. Orthop. Relat. Res. 190 (1984) 96–108.
[98] L. Engesæter, S. Lie, B. Espehaug, O. Furnes, S. Vollset, L. Havelin, Antibiotic prophylaxis in total hip arthroplasty Effects of antibiotic prophylaxis systemically and in bone cement on the revision rate of 22,170 primary hip replacements followed 0–14 years in the Norwegian Arthroplasty Register, Acta Orthop. 74 (6) (2003) 644–651.
[99] H. Malchau, P. Herberts, L. Ahnfelt, Prognosis of total hip replacement in Sweden: Follow-up of 92,675 operations performed 1978–1990, Acta Orthop. 64 (5) (1993) 497–506.
[100] W.A. Jiranek, A.D. Hanssen, A.S. Greenwald, Antibiotic-loaded bone cement for infection prophylaxis in total joint replacement, J. Bone Joint Surg. Am. 88 (11) (2006) 2487–2500.
[101] P.G. Hope, K.G. Kristinsson, P. Norman, R.A. Elson, Deep infection of cemented total hip arthroplasties caused by coagulase-negative staphylococci, J. Bone Joint Surg. (Br.) 71 (5) (1989) 851–855.
[102] L. Sanzén, A.S. Carlsson, G. Josefsson, L.T. Lindberg, Revision operations on infected total hip arthroplasties, Clin. Orthop. Relat. Res. 229 (1988) 165–172.
[103] F.Y. Chiu, C.F. Lin, Antibiotic-impregnated cement in revision total knee arthroplasty. A prospective cohort study of one hundred and eighty-three knees, J. Bone Joint Surg. Am. 91 (3) (2009) 628–633.
[104] J. Parvizi, K. Saleh, P. Ragland, A. Pour, M. Mont, Efficacy of antibiotic-impregnated cement in total hip replacement, Acta Orthop. 79 (3) (2008) 335–341.
[105] M. James, W.S. Khan, J.S. Nannaparaju, et al., Current evidence for the use of laminar flow in reducing infection rates in total joint arthroplasty, Open Orthop. J. 9 (S2:M7) (2015) 495–498.
[106] A.M. Spagnolo, G. Ottria, D. Amicizia, et al., Operating theater quality and prevention of surgical site infections, J. Prev. Med. Hyg. 54 (2013) 131–137.
[107] S.M. McHugh, A.D.K. Hill, H. Humphreys, Laminar airflow and the prevention of surgical site infection. More harm than good? Surgeon 13 (2015) 52–58.
[108] D. Hansen, C. Krabs, D. Benner, et al., Laminar air flow provides high air quality in the operating field even during real operating conditions, but personal protection seems to be necessary and operations with tissue combustion, Int. J. Hyg. Environ. Health 208 (6) (2005) 455–460.
[109] R.P. Evans, Current concepts for clean air and total joint arthroplasty: Laminar airflow and ultraviolet radiation, Clin. Orthop. Relat. Res. 496 (2011) 945–953.
[110] E.A. Salvati, R.P. Robinson, S.M. Zeno, et al., Infection rates after 3,175 total hip and total knee replacements performed with and without a horizontal unidirectional filtered air-flow system, J. Bone Joint Surg. Am. 64A (1982) 525–535.

[111] G.J. Hooper, A.G. Rothwell, C. Frampton, M.C. Wyatt, Does the use of laminar flow and space suits reduce early deep infection after total hip and knee replacement? The ten-year results of the New Zealand joint registry, J. Bone Joint Surg. (Br.) 93-B (2011) 85–90.

[112] D.C. Bosanquet, C.N. Jones, N. Gill, et al., Laminar flow reduces cases of surgical site infections in vascular patients, Ann. R. Coll. Surg. Engl. 95 (2013) 15–19.

[113] M. Rezapoor, J. Parvizi, Prevention of periprosthetic joint infection, J. Arthroplasty 30 (6) (2015) 902–907.

[114] A.O. Miller, B.D. Brause, Perioperative infection in the patient with rheumatic disease, Curr. Rheumatol. Rep. 15 (12) (2013) 379.

[115] C. Whall, Advisory Statement: Antibiotic prophylaxis for dental patient with total joint replacements: American Dental Association; American Association of Orthopaedic, J. Am. Dent. Assoc. 128 (7) (1997) 1004–1008.

[116] M.P. Rethman, W. Watters 3rd., E. Abt, P.A. Anderson, K.C. Carroll, R.P. Evans, H.C. Futrell, K. Garvin, S.O. Glenn, M.J. Goldberg, J. Hellstein, A. Hewlett, D. Kolessar, C. Moucha, R.J. O'Donnell, D.R. Osmon, J. O'Toole, A. Rinella, M.J. Steinberg, W.R. Martin 3rd., D.S. Cummins, S. Song, P. Sluka, K. Boyer, A. Woznica, Ristic, N.B. Hanson, The American Academy of Orthopaedic Surgeons and the American Dental Association clinical practice guideline on the prevention of orthopaedic implant infection in patients undergoing dental procedures, J. Bone Joint Surg. Am. 95 (8) (2013) 745–747.

[117] T. Sollecito, E. Abt, P. Lockhart, E. Truelove, T. Paumier, S. Tracy, M. Tampi, E. Beltrán-Aguilar, J. Frantsve-Hawley, The use of prophylactic antibiotics prior to dental procedures in patients with prosthetic joints, J. Am. Dent. Assoc. 146 (1) (2015) 11–16.e8.

[118] J. Wolf, C. Bennett, R. Dmochowski, B. Hollenbeck, M. Pearle, A. Schaeffer, Best practice policy statement on urologic surgery antimicrobial prophylaxis, J. Urol. 179 (4) (2008) 1379–1390.

Biofilm formation and the biological response

2

L. Drago[*,†], *M. Toscano*[*,†]
[*]IRCCS Galeazzi Institute, Milan, Italy, [†]University of Milan, Milan, Italy

2.1 Introduction

Nearly 99% of microorganisms live in microbial communities known as biofilms, which are particular biological structure formed by adhesion of cells to biotic and abiotic surfaces through an exopolymeric matrix [1]. Costerton et al. promulgated in the 1978 the general theory of biofilm, stating that the majority of bacteria grow in matrix enclosed in biofilms adherent to surfaces in all nutrients-sufficient aquatic ecosystems and that these sessile microorganisms differ deeply from their planktonic counterparts, regarding biological and biochemical pathways [2]. Biofilm formation is a developmental process that has been likened to differentiation in multicellular organisms, and it begins with the attachment of bacteria to a surface where they are able to grow and divide forming two layers under the control of specific biofilm genes [3]. Several studies have shown that biofilms may have greater than 100-fold increase in tolerance to antimicrobial agents when compared to the same bacteria in a planktonic state [4,5]. It is estimated that approximately 80% of the world's microbial mass resides in a biofilm state and that microbial biofilms cause more than 75% of all microbial infections found in humans [6]. Several forms of medically associated biofilm infections related to the abiotic surfaces involve intravascular, catheters, urinary catheters, and finally orthopedic prosthetic implants [7]. Biofilm organisms on these artificial surfaces are often derived from the skin flora of patients or medical personnel during device insertion or implantation. The predominant microorganisms include coagulase-negative staphylococci, *Staphylococcus aureus*, *Pseudomonas*, *Enterococcus*, *Stenotrophomonas,* and *Candida* on intravascular catheters; *Escherichia coli*, *Enterococcus*, *Pseudomonas*, *Klebsiella*, *Enterobacter*, *Proteus mirabilis,* and *Candida* on urinary catheters; and *Staphylococcus* spp., *Propionibacterium acnes* and some Gram negatives microorganisms on prosthetic hip or knee implants [7]. Because these devices often reside within patients for an extensive period of time, biofilm formation could result in regional and/or systemic infections [7]. Furthermore, oral biofilm represents an abundant reservoir of microorganisms that may spread via transient bacteremia, as demonstrated by the types of oral biofilms isolated from remote infections of oral cavity [8]. Shedding and subsequent hematogenous dissemination of oral biofilm bacteria have been associated with some forms of infective endocarditis, acute bacterial myocarditis, brain abscess, liver abscess, lung abscess, and cavernous sinus thrombosis [9–14]. The notion that oral biofilms may impact systemic health by inflammatory mechanisms is supported by cross-sectional studies reporting elevated systemic inflammatory markers

Management of Periprosthetic Joint Infections (PJIs). http://dx.doi.org/10.1016/B978-0-08-100205-6.00002-1

in patients with periodontitis [15,16]. Strong evidence supports the oral-systemic link between periodontitis and cardiovascular diseases, cerebrovascular diseases, and diabetes mellitus, all of which have an inflammatory etiology. Consistent associations between periodontitis and cardiovascular disease, irrespective of common underlying risk factors or confounders such as smoking, age, education, body mass index, and lifestyle factors, have been demonstrated [17]. In contrast, biofilm-like communities of the gastrointestinal and female urogenital tracts containing beneficial lactobacilli may have a protective role, as this structure facilitates long-term colonization and persistence in the organism [18]. A great challenge of the 21st century is the development of novel and effective strategies for the treatment of infections sustained by microbial biofilms, especially in the orthopedic field [19,20] (Fig. 2.1).

2.2 Biofilm formation

Biofilm production is a complex-multistep process that begins with the adhesion of planktonic bacteria to a surface where they initiate biofilm production in the presence of shear forces that dwarf of heart and exceed Reynolds numbers of 5000 [21]. The Reynolds number is a dimensionless number that describes the turbulent flow of a liquid; if this number is high, the turbulent flow exists; if the number is low, the laminar flow condition prevails. Scientists hypothesized that turbulent flow allows the bacterial adhesion to the surface, leading to the biofilm formation by impinging the planktonic cells on the surface. The ability of microorganisms to adhere to surfaces is fundamental for the biofilm production; indeed it has been observed that bacteria that lost their adhesion ability were not able to colonize biotic and abiotic surface, in particular very smooth one [22]. Bacterial adhesion includes [23]:

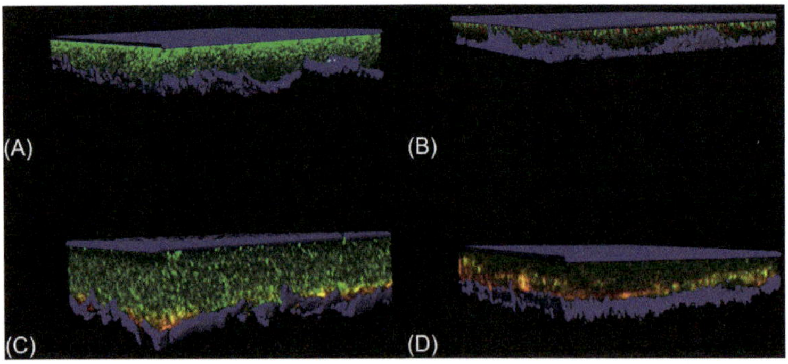

Fig. 2.1 Confocal laser scanning microscopy (CLSM) is a useful tool for the evaluation of the efficacy of antibiofilm treatments used in the orthopedic setting. The picture shows the 3D reconstruction of *S. aureus* (A, B) and *P. aeruginosa* (C, D) biofilms grown on prosthetic material *(blue)*, visualized through CLSM. Biofilms are stained with SYTO 9 *(green)* and propidium iodide *(red)*. On the left (A, C), untreated biofilms. On the right (B, D), biofilms treated with an antibiofilm substance, showing a visible biomass reduction. Area of each section: $2 \times 10^5 \mu m^2$.

(i) an initial attraction of cells toward a surface due to van der Waals attraction forces, Brownian motion, gravitational forces, electrostatic charges, and hydrophobic interactions;
(ii) molecular and cellular interactions by use of microbial surface structures such as pili, fimbriae, or extracellular polymeric substances [24,25].

The next phase of biofilm production is characterized by a microcolony arrangement on the attached surface obtained by the multiplication of free-floating cells. In this phase the cellular motility is reduced and exopolysaccharide production is activated to draw planktonic cells and nutrients, while several signal molecules are secreted in a cell density-dependent manner to coordinate the cellular responses by a process known as quorum sensing [26]. Also the production of a polysaccharide extracellular matrix is used by microorganisms embedded in biofilm to trap nutrients and planktonic cells [26]. The matrix is composed of polysaccharides, extracellular DNA (eDNA), and proteins; contributes to initial adhesion; triggers polysaccharide formation; and serves as a support linking molecules together in the biofilm matrix. The extracellular matrix differs greatly between the different microorganisms and is also influenced by growth conditions under which biofilm is formed [27]. In general, this matrix is composed of exopolysaccharides, proteins, and nucleic acids [27]. Proteinaceous components include cell surface adhesins, protein subunits of flagella and pili, secreted extracellular proteins, and proteins of outer membrane vesicles (OMVs) [27]. Pili and flagella contribute not only to the initial attachment of microorganisms to surfaces, but in some cases they are also involved in the migration along the surface, hereby improving the surface colonization [28,29]. Matrix proteins contribute to biofilm structure and stability, and their role has been highlighted by mutational studies, which showed that the absence or inactivation of these aforementioned proteins results in decreased biofilm formation, altering its architecture [30–32]. Matrix proteins show several and specific activities. Some proteins, indeed, exhibit enzymatic properties, such as the glycosyl hydrolase dispersin B that hydrolyzes polysaccharides, proteases, DNases that degrade extracellular nucleic acids, facilitating both the biofilm matrix degradation and biofilm matrix reorganization [33,34]. The last phase of biofilm formation is cell detachment, a process essential for the dissemination of single or clustered cells to other districts. Several mechanisms that promote cellular dispersion exist, such as sheer mechanical forces (blood flow), enzymes that digest the extracellular matrix, and nutrient limitation in the environment [26]. Dispersed cells exhibit increased adherence ability, increased pathogenicity, and filamentation, which allow them to establish new foci of infection [35]. Several studies highlighted the pivotal role of a particular class of molecules, called quorum sensing-controlled phenol-soluble modulins (PSMs) in the production of biofilm by staphylococci. These molecules participate both in the establishment of biofilm structure and in detachment process. PSMs are part of a novel toxin family with several roles in staphylococcal pathogenesis, as they promote the formation of biofilm channels and control the biofilm expansion, contributing to cellular dissemination [26]. During the dispersion of microorganisms from biofilm, PSMs act as surfactant-like peptides that inhibit cell-to-cell interactions at the biofilm surface, leading to the detachment of cells at the fluid-biofilm interphase and the subsequent systemic spread of biofilm-associated infections [36,37].

2.2.1 Matrix proteins

2.2.1.1 Biofilm-associated protein (Bap)

The Bap protein family represents one of the most studied groups of matrix proteins involved in the biofilm formation. Members of the Bap family are usually very large secreted proteins of up to several hundred kDA in molecular mass. The main unique characteristic of this protein family is the presence of multiple repeats of identical amino acid residues in the core region. Moreover, they contain an N-terminal secretion signal sequence, followed by a non-repetitive N-terminal Region B, which may be absent in some Bap orthologs [38]. The central region, instead, consists of multiple identical repeats that may contain amyloid-like peptide sequences [39]. Most of these Bap proteins can act both as virulence factors involved in pathogenesis and as matrix proteins mediating abiotic surface adhesion and subsequent biofilm production.

2.2.1.2 Outer membrane vesicles (OMVs)

OMVs are small spherical structures produced by Gram-negative bacteria through blabbing from the outer membrane. They generally contain periplasmic and cytoplasmic contents such as proteases, lipases, alkaline phosphatase, toxins, and several outer membrane proteins (OMPs). OMVs are involved in biofilm production, pathogenesis, horizontal gene transfer, and quorum sensing, and they are produced by a broad range of Gram-negative bacteria [40,41]. Interestingly, OMVs isolated from biofilm cultures are often different in quality and quantity if compared with those isolated from planktonic cultures, highlighting how OMVs play the different roles in different physiological conditions [42]. Moreover, OMVs are able to interact with secreted matrix proteins, through binding to OMPs located in OMVs, underlying the direct implication of these proteins in biofilm formation. A direct role of OMVs in biofilm production has been demonstrated in *Helicobacter pylori*, where OMVs production is linked to culture conditions and to the physiological state of bacterial cells. Indeed, the addition of purified OMV-fraction was able to induce biofilm formation in a dose-dependent manner [43]. The activity of OMVs in biofilm formation may be a multifactorial process, as these proteins can interact with several biofilm matrix components such as DNA, exopolysaccharides, and proteins, exerting in this way different roles in the biofilm formation. OMVs may also act as anchoring proteins able to influence the bacterial attachment on surfaces where OMVs are released.

2.2.1.3 Bacterial nucleoid-binding proteins

Nucleoid-binding proteins are involved not only in the maintenance of eDNA supercoiling and compaction but also in biofilm formation. DNABII family proteins can be divided into two subtypes: HU (histone-like protein), which are ubiquitous in Eubacteria, and IHF (integration host factor), which is only found in bacteria within the α- and γ-proteobacteria genera [44]. Extracellular-localized nucleoid-binding proteins have been found in association with eDNA in biofilm matrix from sputum samples of individuals suffering of cystic fibrosis. eDNA is directly involved in biofilm stability in several bacterial species, and it has been observed that the addition of anti-IHF serum to preformed biofilms is able to reduce the amount of produced biofilm [44].

2.3 Resistance to antimicrobial agents

The particular nature of biofilm architecture and the physiological attributes of microorganisms located in biofilm confer a strong resistance to antimicrobial agents, such as antibiotics, germicides, and disinfectants. The main mechanisms responsible for resistance are:

(i) Delayed penetration of the antimicrobial agent through the biofilm matrix;
(ii) Altered growth rate of microorganisms embedded in biofilm;

2.3.1 Delayed penetration of the antimicrobial agent

Extracellular matrix represents a diffusional barrier for antimicrobial molecules that must diffuse through the matrix in order to reach living microorganisms constituting biofilm. The polymeric substances of matrix are able to influence the rate of transport of molecules to the biofilm interior and the reaction of antimicrobial agents with the matrix material [22]. In *Pseudomonas aeruginosa*, for example, planktonic bacterial cells are 15 times more susceptible to tobramycin than sessile cells in biofilm [45].

2.3.2 Altered growth rate of biofilm organisms

It has been hypothesized that biofilm-associated cells grow significantly more slowly than planktonic cells, consequently adsorbing antimicrobial agents more slowly. *S. epidermidis* biofilm growth rates, indeed, deeply influenced the bacterial susceptibility to antimicrobial molecules. Faster is the rate of cell growth, more rapid is the rate of inactivation by ciprofloxacin [46]. Furthermore, old (10-day-old) chemostat-grown *P. aeruginosa* biofilms are significantly more resistant to tobramycin and piperacillin than young (2-day-old) biofilms [47].

2.4 Host response against biofilm

During infections, the human organism used innate immune defense to establish a rapid response mediated by effectors such as immune cells, receptors, and numerous humoral factors. The role of humoral factors, such as collectins, antibodies, and mannose binding lectins, is to seize invading microbes, promoting receptor-mediated recognition and phagocytosis by cells belonging to the innate immune system. Natural killer cells or professional phagocytes are able to recognize pathogen-associated molecular structure located on the surfaces of microbes and initiate a cascade of signaling pathways that induce phagocytosis, secretion of antimicrobial products and antiinflammatory cytokines, and other factors that are essential for the activation, maturation, and regulation of adaptive immune response [48]. Antimicrobial factors, such as lactoferrin and the human cationic host defense peptide LL-37 located on mucosal surfaces or in secondary granules of polymorphonuclear neutrophils (PMNs), are able to inhibit the formation of *P. aeruginosa* in vitro [26]. Lactoferrin, indeed, reduces the iron supply necessary for biofilm growth and promotes twitching motility [49], while LL-37 leads to a reduction of the initial attachment phase, increasing surface motility, and interfering with the quorum sensing system [50]. Moreover, LL-37 is involved in chemotaxis activation, angiogenesis, and chemokine secretion but is also able to inhibit biofilm formation by decreasing the adherence of microorganisms to surfaces, stimulating their motility.

When PMNs reached the microbial biofilm, a decrease in biofilm mass is due by phagocytosis, rapid degranulation of lactoferrin and elastase, and finally DNA release [51]. The adherence of PMNs to biofilms seems to be opsonization-independent, suggesting that biofilms could contain PMN signaling molecules [52]. Lactoferrin is also able to prevent biofilm formation by mean of its strong ion-chelating activity [53]. Decreased levels of lactoferrin might be predisposed to biofilm infections; indeed patients with cystic fibrosis have been observed to have reduced amount of this molecule. However, the presence of elastase can induce the production of cathepsin B, a protease that leads to the cleavage of lactoferrin and the lost of its bactericidal and antibiofilm activity [54,55].

2.4.1 How biofilm evades host' defenses

S. aureus is one of the main causes of community-associated infections, as it is frequent etiologic factor of biofilm-associated infections on medical devices as well as chronic infections such as endocarditis, rhinosinusitis, and osteomyelitis [56]. One of the main problems of *S. aureus*-associated infections is the emergence of drug-resistant strains, especially methicillin-resistant *S. aureus*, which lead to antibiotic therapy failure. Human leukocytes are able to penetrate into the biofilm using the nutrient channels in mature biofilm, but they exhibit a reduced ability to kill bacteria and an impaired phagocytosis [57]. Moreover, macrophage action is inhibited by the size and high complexity of biofilm structure, even if they are able to engulf disrupted biofilm material [58]. Furthermore, the biofilm microenvironment together with changes in pH due to bacteria and the release of toxins from the biofilm matrix contribute to thwart the antibacterial activity of macrophages. Moreover, staphylococcal biofilm can persist in the human body by skewing the immune response to favor antiinflammatory and profibrotic pathways. *P. aeruginosa* biofilm, instead, is able to influence negatively the migratory behavior of phagocytes, which become incapable of migrating and exerting their physiologic activity [59]. Interestingly, *P. aeruginosa* biofilm has been observed to be increased by 2- to 3.5-fold in the presence of PMNs, probably due to specific polymers produced by PMNs, such as DNA and actin [60]. Although, the main protective role in escaping the host response is exerted by the exopolysaccharide matrix, which is able to protect bacterial cells from leukocyte phagocytosis. In particular, the exopolysaccharide alginate is essential to protect biofilm from the immune system activity, as several studies highlighted that *P. aeruginosa* biofilm lacking the ability to produce biofilm was completely destroyed by human phagocytes [61].

2.5 Clinical significance of biofilm

Orthopedic infections represent a serious problem due to high number of total hip and knee arthroplasties that are performed worldwide. Even if artificial joints can significantly improve the quality of life of patients, failure of prosthetic can lead high morbidity and suffering [62]. Considering periprosthetic infections, there are two main routes and time scales associated with the penetration of bacteria. The first one is the perioperative period, where the source is the patient's endogenous bacteria or microorganisms belonging to the personnel or the environment of the operating room; the second one is from hematogenous spread occurring during the postoperative period [63]. The insertion of prosthesis in the human body increases the risk for infection that could finally mature into biofilm. The diagnosis of infection is still a difficult task for clinicians, as infections can rise in patients acutely or persist chronically for years after the orthopedic surgery [64]. Bacteria involved in biofilm-associated infections include coagulase-negative staphylococci, *S. aureus*, streptococci, enterococci, Gram-negative microorganisms, and anaerobes bacteria [65]. The aforementioned bacteria can produce biofilm exerting their pathogenic role in situ, but they can also leave the biofilm structure, spreading the infection within the host. Consequently, the bacterial dissemination plays a pivotal role in infections and can explain acute infectious episodes that appear as exacerbations of a chronic underlying process [66]. Bacterial cells residing in the external sides of biofilm are most exposed to the host's defenses and antibiotics, but these microorganisms developed several protective defenses. The matrix layers of cells within the biofilm create a physical barrier to slow the diffusion of antibiotics. Metabolic activity in the outer biofilm can create acidic or anorexic areas that might contribute to antibiotic degradation [67]. Furthermore, several secreted polymers located in the biofilm matrix can bind and deactivate antibiotics, forming an antibiotic "sink" [68]. In addition, the depletion of nutrients in the interior of biofilm can lead to starvation-induced dormancy and the development of resistant persister cells. Moreover, diffusion-limited transport, binding to matrix components and uptake by bacteria, can create a gradient of antibiotic concentration, which exposes a subpopulation of bacteria to a sublethal dose of antibiotics. As a consequence, this sublethal exposure can lead to an increase development of antibiotic tolerance by improving biofilm formation [69]. Optimal treatment for prosthetic joint infections requires antibiotic combinations, ideally including an agent acting on adhering stationary-phase isolates. Indeed, it has been observed that the combination of daptomycin or moxifloxacin with the macrolide antibiotic clarithromycin significantly increased the bacterial killing effect against biofilm produced by staphylococci [70]. It has also been hypothesized that some bacterial cells embedded in the biofilm adopt a distinct and protective biofilm phenotype not directly linked to nutrient limitation [71].

Biofilm poses also a notable problem in the microbiological diagnosis of infection because the matrix may hinder sampling and culturing of bacteria. In the 30% of cases, it may be difficult to detach biofilm-embedded bacteria from prosthetic surfaces. Recently new technology using chemical-debonding biofilm (dithiothreithol, DTT) has been developed [72,73].

2.6 Therapeutic strategies against biofilm

The onset of biofilm-associated infections after orthopedic implant surgery led to the need of innovative approach able to eradicate the clinical and economic impact of biofilm. One- and two-stage revision can serve as definitive treatment, and success rates have been documented in case-series to be 80% and 90%, respectively [71]. Furthermore, aggressive debridement and retention of the implant is an option in select cases of early-onset infection with no evidence of loosening or sinus tract formation [74]. This strategy in early-onset infection may be curative above all if associated with long-term antibiotics in up to 71% of cases. Finally, resection without reimplantation is a last therapeutic option. However, the cost and morbidity of periprosthetic joint infections and revision joint arthroplasty can be exorbitant. It is very difficult to completely kill microbial cells embedded in biofilm by a classical antibiotic treatment; for this reason in the last years new antibiofilm approaches have been developed by applying novel natural, synthetic, or bioengineered agents. One of the main advantages of these new compounds is that they are less susceptible to the emergence of resistance if compared with classical antibiotic therapies [75].

2.6.1 Biofilm matrix-degrading enzymes

Biofilm matrix-degrading enzymes are a class of promising biofilm-controlling strategy, especially DNase I, which degrades eDNA, and α-amylase and DspB, both involved in the degradation of extracellular polysaccharides. Several evidences highlighted the strong activity of DNase in degrading biofilms of several clinically important bacteria and fungi, as the cleavage of eDNA leads to an alteration of biofilm architecture and to a subsequent increased penetration of antibiotics into the biofilm [76,77]. α-Amylase, instead, is able not only to prevent the production of biofilm but also to be effective against the mature biofilm [78]. Regarding the activity of DspB, this enzyme has a broad-spectrum activity to dissolve matrix exopolysaccharide of bacterial biofilms. As a consequence, the combination of DspB with antibiotics or disinfectant has a synergistic effect on bacterial pathogens such as *S. aureus* and *S. epidermidis*, which leads to a reduced bacterial colonization of medical devices and a prolonged antibiofilm and antimicrobial activity [79]. Therapeutic approaches using biofilm matrix-degrading enzymes aim to inhibit or reduce the cell-to-cell and cell-to-surface associations during the development of biofilms, removing the biofilm colonies and increase microbial cells sensitive to antibiotics or to host immune defense. However, the current cost of pure industrial enzyme production is very high and makes the application of this class of molecules enough expensive. Anyhow, the combination of antibiotics and biofilm matrix-degrading enzymes is a promising effective approach able to remove or lead to a significant reduction clinically relevant biofilms.

2.6.2 Ultrasonic treatment

The combination of antibiotics and low-frequency ultrasound seems to be effective against the biofilm; therapeutic ultrasound can enhance both the transport of antimicrobial molecules across biofilms by the fragmentation of the biofilm matrix and the uptake of antibiotics due to a significant increase in bacterial membrane permeability [80,81]. Furthermore, ultrasound can increase the rate of oxygen and nutrient transport to bacterial cells within biofilm, reactivating bacterial metabolism and leading to a significant increase in the susceptibility to antibiotics. Recently, ultrasound-targeted microbubbles (UTMs) have been developed, and they seem to be able in improving antibiotic activity against biofilms, suggesting that the combination of UTMs with antimicrobial agents may provide an efficient novel tool to treat device-related biofilm infections [82].

2.6.3 Bacteriophages

Bacteriophages are natural parasites of bacteria that have often used for the treatment of bacterial infections [83]. They seem to be effective against the negative impact of biofilm by two main mechanisms:

1. the release of a great number of phages during the lytic cycle that lead to a radial cell lysis around the infection site [84];
2. the production of polysaccharide depolymerases that degrade the biofilm matrix [84].

Moreover, as bacteriophages have a great host specificity, they leave the physiological microbiota undisturbed not leading side effects to the host organism. Progeny phages are able to propagate through biofilm, and theoretically, they may be employed as an efficient therapy for the treatment of biofilm-associated infections. However, if phages can be frequently isolated from human gastrointestinal tract and oral cavity, several safety concerns of phage medical application still exist; above all for the existence of unknown risks of introducing in the human body, an organism able to replicate and be transmitted to other patients [85]. The association of antibiotics to phages has a synergistic effect as the modification of biofilm matrix structure induced by bacteriophages leads to an increase in bacteria susceptibility to antibiotics [86]. Interestingly, using synthetic biology technologies, phages can be engineered to express the biofilm matrix-degrading enzyme DspB, which enhances the efficacy of biofilm destruction by this enzymatic phage strategy. Finally, the use of the aforementioned approach is able not only to increase bacterial killing but also to reduce the incidence if antibiotic resistance among bacteria.

2.6.4 Quorum sensing inhibitors

These kind of molecules are able to inhibit the quench action of quorum sensing that is directly involved in the establishment and maintenance of biofilm, as bacteria use it to coordinate their behavior in a cell density-dependent manner. The combination of quorum sensing inhibitors (QSI) and antibiotics has no a synergistic effect against planktonic cells, but in contrast, this association is able to inhibit the formation of biofilm and the production of the exopolysaccharide matrix [87]. The use of QSI, indeed, leads to deep modifications in the matrix and biofilm structure, increasing the susceptibility of bacteria to antibiotics [88].

2.6.5 Silver nanoparticles

The antibacterial activity of silver has long been studied, and nanoparticles of this metal seem to be less toxic than silver ions [88]. The use of silver nanoparticles against bacterial biofilm leads to a significant reduction in its biomass in 24 h. Furthermore, these nanoparticles are able not only to kill bacteria embedded in mature biofilm but also to inhibit the formation of biofilm, acting on both in-formation and mature biofilm [88]. Silver nanoparticles can also enhance the antibacterial and antibiofilm activity of antibiotics, acting in synergy with several antimicrobial agents, such as ampicillin, kanamycin, streptomycin, and vancomycin against *P. aeruginosa* and *E. coli* [89].

2.7 Future trends

Biofilm has been clearly observed to be associated to a broad range of human infections, and it is often the direct cause of therapy failure. Moreover, the specific mechanism involved in the formation of biofilm and in its maintenance is still object of study, but to date we have detailed information about specific proteins and signaling molecules involved in the development of biofilm. As the production of the exopolysaccharide matrix is one of the main steps of biofilm formation, it has been hypothesized that use of biofilm matrix-degrading enzymes may be an efficient strategy to defeat biofilm-associated infections. Other new antibiofilm approaches comprise inhibitors of quorum sensing, as well as bacteriophages and ultrasonic treatment, which are able to induce significant changes in the biofilm structure, leading to a significant increase in the susceptibility of bacteria to antimicrobial agents. Better characterization of the biofilm matrix proteome, regulatory circuits involved into biofilm formation, specific molecules of quorum sensing, and proteins involved in the attachment of bacteria to biotic surface will allow the development of specific antibiofilm therapies. It is clear that we can no longer rely only on antibiotics and surgery to treat biofilm-associated infections due to increased patient costs and morbidity. The application of different technologies and disciplines can advance the field of biofilm disrupting technology. Innovative antibiofilm treatments are the forefront of medical research as scientists continue to look for new approaches able to definitively fight the complicated and serious problem of biofilm-associated infections.

References

[1] J.W. Costerton, P.S. Stewart, E.P. Greenberg, Bacterial biofilms: a common cause of persistent infections, Science 284 (5418) (1999) 1318–1322.

[2] J.W. Costerton, G.G. Geesey, K.J. Cheng, How bacteria stick, Sci. Am. 238 (1) (1978) 86–95.

[3] T. Beikler, T. Flemming, Oral biofilm-associated diseases: trends and implications for quality of life, systemic health and expenditures, Periodontol. 2000 55 (2000) 87–103.
[4] M.E. Olson, H. Ceri, D.W. Morck, A.G. Buret, R.R. Read, Biofilm bacteria, formation and comparative susceptibility to antibiotics, Can. J. Vet. Res. 66 (2) (2002) 86–92.
[5] H. Ceri, M.E. Olson, D.W. Morck, D. Storey, R. Read, A. Buret, et al., The MBEC assay system: multiple equivalent biofilms for antibiotic and biocide susceptibility testing, Methods Enzymol. 337 (2001) 377–384.
[6] D. Davies, Understanding biofilm resistance to antibacterial agents, Nat. Rev. Drug Discov. 2 (2) (2003) 114–122.
[7] C.S. Hung, J.P. Henderson, Emerging concepts of biofilms in infectious diseases, Mo. Med. 106 (4) (2009) 292–296.
[8] X. Li, K.M. Kolltveit, L. Tronstad, I. Olsen, Systemic diseases caused by oral infection, Clin. Microbiol. Rev. 13 (4) (2000) 547–558.
[9] C.A. Bartzokas, R. Johnson, M. Jane, M.V. Martin, P.K. Pearce, Y. Saw, Relation between mouth and haematogenous infection in total joint replacements, BMJ 309 (6953) (1994) 506–508.
[10] A.A. Mueller, B. Saldamli, S. Stubringer, C. Walter, U. Flückiger, A. Merlo, et al., Oral bacterial cultures in nontraumatic brain abscesses: results of a first-line study, Oral Surg. Oral Med. Oral Pathol. Oral Radiol. Endod. 107 (4) (2009) 469–476.
[11] A.I. Mylonas, F.H. Tzerbos, M. Mihalaki, D. Rologis, I. Boutsikakis, Cerebral abscess of odontogenic origin, J. Craniomaxillofac. Surg. 35 (1) (2007) 63–67.
[12] N.B. Parahitiyawa, L.J. Jin, W.K. Leung, W.C. Yam, L.P. Samaranayake, Microbiology of odontogenic bacteremia: beyond endocarditis, Clin. Microbiol. Rev. 22 (1) (2009) 46–64.
[13] S. Ulivieri, G. Olivieri, G. Filosomi, Brain abscess following dental procedures. Case report, Minerva Stomatol. 56 (5) (2007) 303–305.
[14] K.W. Wagner, R. Schon, M. Schumacher, R. Schmelzeisen, D. Schulze, Case report: brain and liver abscesses caused by oral infection with *Streptococcus intermedius*, Oral Surg. Oral Med. Oral Pathol. Oral Radiol. Endod. 102 (4) (2006) e21–e23.
[15] I.L. Chapple, The impact of oral disease upon systemic health-symposium overview, J. Dent. 37 (8) (2009) S568–S5571.
[16] R.G. Craig, J.K. Yip, M.K. So, R.J. Boylan, S.S. Socransky, A.D. Haffajee, Relationship of destructive periodontal disease to the acute-phase response, J. Periodontol. 74 (7) (2003) 1007–1016.
[17] G.R. Persson, R.E. Persson, Cardiovascular disease and periodontitis: an update on the associations and risk, J. Clin. Periodontol. 35 (8 Suppl.) (2008) 362–379.
[18] S.E. Jones, J. Versalovic, Probiotic *Lactobacillus reuteri* biofilms produce antimicrobial and anti-inflammatory factors, BMC Microbiol. 9 (2009) 35.
[19] C.L. Romanò, M. Toscano, D. Romanò, L. Drago, Antibiofilm agents and implant-related infections in orthopaedics: where are we? J. Chemother. 25 (2) (2013) 67–80.
[20] L. Drago, S. Agrappi, M. Bortolin, M. Toscano, C.L. Romanò, E. De Vecchi, How to study biofilms after microbial colonization of materials used in orthopaedic implants, Int. J. Mol. Sci. 17 (3) (2016) 293.
[21] W. Characklis, K.C. Marshall, Biofilms, John Wiley & Sons, New York, 1990.
[22] R.M. Donlan, J.W. Costerton, Biofilms: survival mechanisms of clinically relevant microorganisms, Clin. Microbiol. Rev. 15 (2) (2002) 167–193.
[23] B. Gottenbos, H.J. Busscher, H.C. Van Der Mei, P. Nieuwenhuis, Pathogenesis and prevention of biomaterial centered infections, J. Mater. Sci. Mater. Med. 13 (8) (2002) 717–722.
[24] L.A. Pratt, R. Kolter, Genetic analysis of *Escherichia coli* biofilm formation: roles of flagella, motility, chemotaxis and type I pili, Mol. Microbiol. 30 (2) (1998) 285–293.

[25] J. Wingender, T. Neu, H.-C. Flemming, What are bacterial extracellular polymer substances? in: J. Wingender, T. Neu, H.-C. Flemming (Eds.), Bacterial Extracellular Polymer Substances, Springer, Berlin, 1999.
[26] E. Roilides, M. Simitsopoulou, A. Katragkou, T.J. Walsh, How biofilms evade host defenses, Microbiol. Spectr. 3 (3) (2015).
[27] J.N.C. Fong, F.H. Yildiz, Biofilm matrix proteins, Microbiol. Spectr. 3 (2) (2015) 10.
[28] J.C. Fong, K. Karplus, G.K. Schoolnik, F.H. Yildiz, Identification and characterization of RbmA, a novel protein required for the development of rugose colony morphology and biofilm structure in *Vibrio cholerae*, J. Bacteriol. 188 (3) (2006) 1049–1059.
[29] J.C. Fong, F.H. Yildiz, The rbmBCDEF gene cluster modulates development of rugose colony morphology and biofilm formation in *Vibrio cholerae*, J. Bacteriol. 189 (6) (2007) 2319–2330.
[30] A. Ostrowski, A. Mehert, A. Prescott, T.B. Kiley, N.R. Stanley-Wall, YuaB functions synergistically with the exopolysaccharide and TasA amyloid fibers to allow biofilm formation by *Bacillus subtilis*, J. Bacteriol. 193 (18) (2011) 4821–4831.
[31] K. Kobayashi, M. Iwano, BslA(YuaB) forms a hydrophobic layer on the surface of *Bacillus subtilis* biofilms, Mol. Microbiol. 85 (1) (2012) 51–66.
[32] B.R. Borlee, A.D. Goldman, K. Murakami, R. Samudrala, D.J. Wozniak, M.R. Parsek, Pseudomonas aeruginosa uses a cyclic-di-GMP-regulated adhesin to reinforce the biofilm extracellular matrix, Mol. Microbiol. 75 (4) (2010) 827–842.
[33] J.B. Kaplan, C. Ragunath, N. Ramasubbu, D.H. Fine, Detachment of *Actinobacillus actinomycetemcomitans* biofilm cells by an endogenous beta-hexosaminidase activity, J. Bacteriol. 185 (16) (2003) 4693–4698.
[34] M. Marti, M.P. Trotonda, M.A. Tormo-Mas, M. Vergara-Irigaray, A.L. Cheung, I. Lasa, J.R. Penadés, Extracellular proteases inhibit protein-dependent biofilm formation in *Staphylococcus aureus*, Microbes Infect. 12 (1) (2010) 55–64.
[35] P. Uppuluri, A.K. Chaturvedi, A. Srinivasan, M. Banerjee, A.K. Ramasubramaniam, J.R. Kohler, D. Kadosh, J.L. Lopez-Ribot, Dispersion as an important step in the *Candida albicans* biofilm developmental cycle, PLoS Pathog. 6 (3) (2010) e1000828.
[36] R. Wang, B.A. Khan, G.Y. Cheung, T.H. Bach, M. Jameson-Lee, K.F. Kong, S.Y. Queck, M. Otto, *Staphylococcus epidermidis* surfactant peptides promote biofilm maturation and dissemination of biofilm-associated infection in mice, J. Clin. Invest. 121 (1) (2011) 238–248.
[37] S. Periasamy, H.S. Joo, A.C. Duong, T.H. Bach, V.Y. Tan, S.S. Chatterjee, G.Y. Cheung, M. Otto, How *Staphylococcus aureus* biofilms develop their characteristic structure, Proc. Natl. Acad. Sci. U. S. A. 109 (4) (2012) 1281–1286.
[38] C. Latasa, A. Roux, A. Toledo-Arana, J.M. Ghigo, C. Gamazo, J.R. Penades, I. Lasa, BapA, a large secreted protein required for biofilm formation and host colonization of *Salmonella enterica* serovar Enteritidis, Mol. Microbiol. 58 (5) (2005) 1322–1339.
[39] P. Lembre, C. Vendrely, P.D. Martino, Identification of an amyloidogenic peptide from the Bap protein of *Staphylococcus epidermidis*, Protein Pept. Lett. 21 (1) (2014) 75–79.
[40] T.J. Beveridge, Structures of Gram-negative cell walls and their derived membrane vesicles, J. Bacteriol. 181 (16) (1999) 4725–4733.
[41] A. Kulp, M.J. Kuehn, Biological functions and biogenesis of secreted bacterial outer membrane vesicles, Annu. Rev. Microbiol. 64 (2010) 163–184.
[42] M. Toyofuku, B. Roschitzki, K. Riedel, L. Eberl, Identification of proteins associated with the *Pseudomonas aeruginosa* biofilm extracellular matrix, J. Proteome Res. 11 (10) (2012) 4906–4915.

[43] H. Yonezawa, T. Osaki, S. Kurata, M. Fukuda, H. Kawakami, K. Ochiai, T. Hanawa, S. Kamiya, Outer membrane vesicles of *Helicobacter pylori* TK1402 are involved in biofilm formation, BMC Microbiol. 9 (2009) 197.

[44] S.D. Goodman, K.P. Obergfell, J.A. Jurcisek, L.A. Novotny, J.S. Downey, E.A. Ayala, et al., Biofilms can be dispersed by focusing the immune system on a common family of bacterial nucleoid-associated proteins, Mucosal Immunol. 4 (6) (2011) 625–637.

[45] B.D. Hoyle, K.W. Wong, J.W. Costerton, Disparate efficacy of tobramycin on Ca^{+2}-, Mg^{+2}-, and HEPES-treated *Pseudomonas aeruginosa* biofilms, Can. J. Microbiol. 38 (11) (1992) 1214–1218.

[46] I.G. Duguid, E. Evans, M.R. Brown, P. Gilbert, Growth rate-dependent killing by ciprofloxacin of biofilm-derived *Staphylococcus epidermidis*; evidence for cell-cycle dependency, J. Antimicrob. Chemother. 30 (6) (1992) 791–802.

[47] H. Anwar, J.L. Strap, K. Chen, J.W. Costerton, Dynamic interactions of biofilms of mucoid *Pseudomonas aeruginosa* with tobramycin and piperacillin, Antimicrob. Agents Chemother. 36 (6) (1992) 1208–1214.

[48] J. Banchereau, F. Briere, C. Caux, J. Davoust, S. Lebecque, Y.J. Liu, et al., Immunobiology of dendritic cells, Annu. Rev. Immunol. 18 (2000) 767–811.

[49] P.K. Singh, M.R. Parsek, E.P. Greenberg, M.J. Welsh, A component of innate immunity prevents bacterial biofilm development, Nature 417 (6888) (2002) 552–555.

[50] J. Overhage, A. Campisano, M. Bains, E.C. Torfs, B.H. Rehm, R.E. Hancock, Human host defense peptide LL-37 prevents bacterial biofilm formation, Infect. Immun. 76 (9) (2008) 4176–4182.

[51] F. Günther, G.H. Wabnitz, P. Stroh, B. Prior, U. Obst, Y. Samstag, Host defense against *Staphylococcus aureus* biofilms infection: phagocytosis of biofilms by polymorphonuclear neutrophils (PMN), Mol. Immunol. 46 (8–9) (2009) 1805–1813.

[52] E. Meyle, G. Brenner-Weiss, U. Obst, B. Prior, G.M. Hänsch, Immune defense against *S. epidermidis* biofilms: components of the extracellular polymeric substance activate distinct bactericidal mechanisms of phagocytic cells, Int. J. Artif. Organs 35 (10) (2012) 700–712.

[53] C.Y. O'May, K. Sanderson, L.F. Roddam, S.M. Kirov, D.W. Reid, Iron-binding compounds impair *Pseudomonas aeruginosa* biofilm formation, especially under anaerobic conditions, J. Med. Microbiol. 58 (Pt 6) (2009) 765–773.

[54] B.E. Britigan, M.B. Hayek, B.N. Doebbeling, R.B. Fick Jr., Transferrin and lactoferrin undergo proteolytic cleavage in the *Pseudomonas aeruginosa*-infected lungs of patients with cystic fibrosis, Infect. Immun. 61 (12) (1993) 5049–5055.

[55] M.P. Rogan, C.C. Taggart, C.M. Greene, P.G. Murphy, S.J. O'Neill, N.G. McElvaney, Loss of microbial activity and increased formation of biofilm due to decreased lactoferrin activity in patients with cystic fibrosis, J. Infect. Dis. 190 (7) (2004) 1245–1253.

[56] N.K. Archer, M.J. Mazaitis, J.W. Costerton, J.G. Leid, M.E. Powers, M.E. Shirtliff, *Staphylococcus aureus* biofilms: properties, regulation, and roles in human disease, Virulence 2 (5) (2011) 445–459.

[57] J.G. Leid, M.E. Shirtliff, J.W. Costerton, P. Stoodley, Human leukocytes adhere to, penetrate, and respond to *Staphylococcus aureus* biofilms, Infect. Immun. 70 (11) (2002) 6339–6345.

[58] L.R. Thurlow, M.L. Hanke, T. Fritz, A. Angle, A. Aldrich, S.H. Williams, et al., *Staphylococcus aureus* biofilms prevent macrophage phagocytosis and attenuate inflammation in vivo, J. Immunol. 186 (11) (2011) 6585–6596.

[59] A.J. Jesaitis, M.J. Franklin, D. Berglund, M. Sasaki, C.I. Lord, J.B. Bleazard, et al., Compromised host defense on *Pseudomonas aeruginosa* biofilms: characterization of neutrophil and biofilm interactions, J. Immunol. 171 (8) (2003) 4329–4339.

[60] T.S. Walker, K.L. Tomlin, G.S. Worthen, K.R. Poch, J.G. Lieber, M.T. Saavedra, et al., Enhanced *Pseudomonas aeruginosa* biofilm development mediated by human neutrophils, Infect. Immun. 73 (6) (2005) 3693–3701.

[61] J.G. Leid, C.J. Willson, M.E. Shirtliff, D.J. Hassett, M.R. Parsek, A.K. Jeffers, The exopolysaccharide alginate protects *Pseudomonas aeruginosa* biofilm bacteria from IFN-gamma-mediated macrophage killing, J. Immunol. 175 (11) (2005) 7512–7518.

[62] S. Kurtz, K. Ong, E. Lau, F. Mowat, M. Halpern, Projections of primary and revision hip and knee arthroplasty in the United States from 2005 to 2030, J. Bone Joint Surg. Am. 89 (4) (2007) 780–785.

[63] L.R. Greene, Guide to the elimination of orthopedic surgery surgical site infections: an executive summary of the Association for Professionals in Infection Control and Epidemiology elimination guide, Am. J. Infect. Control 40 (4) (2012) 384–386.

[64] J. Parvizi, T. Gehrke, A.F. Chen, Proceedings of the international consensus on periprosthetic joint infection, Bone Joint J. 95-B (11) (2013) 1450–1452.

[65] R. Pandey, A.R. Berendt, N.A. Athanasou, Histological and microbiological findings in non-infected and infected revision arthroplasty tissues. The OSIRIS Collaborative Study Group. Oxford Skeletal Infection Research and Intervention Service, Arch. Orthop. Trauma Surg. 120 (10) (2000) 570–574.

[66] B.R. Boles, A.R. Horswill, Agr-mediated dispersal of *Staphylococcus aureus* biofilms, PLoS Pathog. 4 (4) (2008) e1000052.

[67] C.T. Huang, F.P. Yu, G.A. Mcfeters, P.S. Stewart, Nonuniform spatial patterns of respiratory activity within biofilms during disinfection, Appl. Environ. Microbiol. 61 (6) (1995) 2252–2256.

[68] N. Høiby, Recent advances in the treatment of *Pseudomonas aeruginosa* infections in cystic fibrosis, BMC Med. 9 (2011) 32.

[69] K.K. Jefferson, D.A. Goldmann, G.B. Pier, Use of confocal microscopy to analyze the rate of vancomycin penetration through *Staphylococcus aureus* biofilms, Antimicrob. Agents Chemother. 49 (6) (2005) 2467–2473.

[70] C. Jacqueline, J. Caillon, Impact of bacterial biofilm on the treatment of prosthetic joint infections, J. Antimicrob. Chemother. 69 (Suppl. 1) (2014) 37–40.

[71] G.S. Biring, T. Kostamo, D.S. Garbuz, B.A. Masri, C.P. Duncan, Two-stage revision arthroplasty of the hip for infection using an interim articulated Prostalac hip spacer: a 10- to 15-year follow-up study, J. Bone Joint Surg. (Br.) 91 (11) (2009) 1431–1437.

[72] L. Drago, C.L. Romanò, R. Mattina, V. Signori, E. De Vecchi, Does dithiothreitol improve bacterial detection from infected prostheses? A pilot study, Clin. Orthop. Relat. Res. 470 (10) (2012) 2915–2925.

[73] L. Drago, V. Signori, E. De Vecchi, C. Vassena, E. Palazzi, L. Cappelletti, D. Romanò, C.L. Romanò, Use of dithiothreitol to improve the diagnosis of prosthetic joint infections, J. Orthop. Res. 31 (11) (2013) 1694–1699.

[74] D.R. Osmon, E.F. Berbari, A.R. Berendt, D. Lew, W. Zimmerli, J.M. Steckelberg, N. Rao, et al., Infectious Diseases Society of America, Diagnosis and management of prosthetic joint infection: clinical practice guidelines by the Infectious Diseases Society of America, Clin. Infect. Dis. 56 (1) (2013) e1–e25.

[75] F. Sun, F. Qu, Y. Ling, P. Mao, P. Xia, H. Chen, et al., Biofilm-associated infections: antibiotic resistance and novel therapeutic strategies, Future Microbiol. 8 (7) (2013) 877–886.

[76] J.B. Kaplan, K. Lovetri, S.T. Cardona, S. Madhyastha, I. Sadovskaya, S. Jabbouri, et al., Recombinant human DNase I decreases biofilm and increases antimicrobial susceptibility in staphylococci, J. Antibiot. 65 (2) (2012) 73–77.

[77] M. Martins, M. Henriques, J.L. Lopez-Ribot, R. Oliveira, Addition of DNase improves the in vitro activity of antifungal drugs against *Candida albicans* biofilms, Mycoses 55 (1) (2012) 80–85.
[78] G.V. Tetz, N.K. Artemenko, V.V. Tetz, Effect of DNase and antibiotics on biofilm characteristics, Antimicrob. Agents Chemother. 53 (3) (2009) 1204–1209.
[79] R.O. Darouiche, M.D. Mansouri, P.V. Gawande, S. Madhyastha, Antimicrobial and antibiofilm efficacy of triclosan and DispersinB combination, J. Antimicrob. Chemother. 64 (1) (2009) 88–93.
[80] J. Bartley, D. Young, Ultrasound as a treatment for chronic rhinosinusitis, Med. Hypotheses 73 (1) (2009) 15–17.
[81] Y. Dong, S. Chen, Z. Wang, N. Peng, J. Yu, Synergy of ultrasound microbubbles and vancomycin against *Staphylococcus epidermidis* biofilm, J. Antimicrob. Chemother. 68 (4) (2013) 816–826.
[82] N. He, J. Hu, H. Liu, T. Zhu, B. Huang, X. Wang, et al., Enhancement of vancomycin activity against biofilms by using ultrasound-targeted microbubble destruction, Antimicrob. Agents Chemother. 55 (11) (2011) 5331–5337.
[83] B. Burrowes, D.R. Harper, J. Anderson, M. Mcconville, M.C. Enright, Bacteriophage therapy: potential uses in the control of antibiotic-resistant pathogens, Expert Rev. Anti Infect. Ther. 9 (9) (2011) 775–785.
[84] G.W. Hanlon, S.P. Denyer, C.J. Olliff, L.J. Ibrahim, Reduction in exopolysaccharide viscosity as an aid to bacteriophage penetration through *Pseudomonas aeruginosa* biofilms, Appl. Environ. Microbiol. 67 (6) (2001) 2746–2753.
[85] R.M. Donlan, Preventing biofilms of clinically relevant organisms using bacteriophage, Trends Microbiol. 17 (2) (2009) 66–72.
[86] W. Fu, T. Forster, O. Mayer, J.J. Curtin, S.M. Lehman, R.M. Donlan, Bacteriophage cocktail for the prevention of biofilm formation by *Pseudomonas aeruginosa* on catheters in an in vitro model system, Antimicrob. Agents Chemother. 54 (1) (2010) 397–404.
[87] G. Brackman, P. Cos, L. Maes, H.J. Nelis, T. Coenye, Quorum sensing inhibitors increase the susceptibility of bacterial biofilms to antibiotics in vitro and in vivo, Antimicrob. Agents Chemother. 55 (6) (2011) 2655–2661.
[88] K. Markowska, A.M. Grudniak, K.I. Wolska, Silver nanoparticles as an alternative strategy against bacterial biofilms, Acta Biochim. Pol. 60 (4) (2013) 523–530.
[89] K.I. Wolska, K. Grześ, A. Kurek, Synergy between novel antimicrobials and conventional antibiotics or bacteriocins, Pol. J. Microbiol. 61 (2) (2012) 95–104.

Biomaterials in treatment of orthopedic infections

3

T.A. van Vugt*, J.A.P. Geurts*, J.J. Arts*, N.C. Lindfors†
*Maastricht University Medical Centre, Maastricht, The Netherlands
†University of Helsinki, Helsinki, Finland

3.1 Orthopedic implant-related infections

In the past decades a major increase in the usage of orthopedic implants is seen. Most of these implants are used for bone fixation in (orthopedic) trauma surgery or for total joint arthroplasty in orthopedic surgery [1,2]. In trauma surgery, internal and external fixation devices are used for temporary fixation to treat fractures, and these devices can, if needed, be removed after consolidation of the fractured parts. In orthopedic surgery, these implants are generally used for total joint arthroplasty in patients with irreversibly damaged joints to relieve them from pain and increase their functionality, for instance, due to osteoarthritis.

Implant-related infections are unfortunately a common seen complication. The infection rates of these procedures vary from less than 1% in primary total arthroplasty procedures to 5% in revision surgery and even up to 16% in trauma surgery after severe open fractures [3–5]. Implant-related infections are associated with high morbidity, high mortality, and high costs because these kind of infections are difficult to treat [6]. Major difficulties in treatment of these infections are the protective biofilms formed on the surface of the orthopedic implants by microorganisms resulting in natural defence against antibiotics and the natural host response. Thereby, local compromised host defence mechanisms and the existence of local compromised tissue leading to poor vasculature and ischemia also cause treatment difficulties [7–9].

Improper or delayed treatment of these implant-related infections could lead to chronic infection of the bone, chronic osteomyelitis [10]. This chronic osteomyelitis can be localized in a small part of the bone, but it can also affect the bone marrow, periosteum, and local soft tissues. Due to the presence of the inflammatory processes bone necrosis, destruction of trabeculae and destruction of the bone matrix occur and vascularity is obliterated which leads to ischemia [7–9]. The necrosis of these different tissues leads to avascular dead bone, called sequestra, where microorganisms can harbor despite different forms of antibiotic therapies [11]. Just like the implant-associated infections, chronic osteomyelitis is often associated with microorganisms protected by a biofilm layer. These biofilm layers frequently form on the surface of dead bone and the sequestra, which creates an intrinsically resistance for conventional antibiotic therapies up to 800-fold compared to planktonic microorganisms [12,13].

Antimicrobial therapy of implant-related infections and chronic osteomyelitis should have antibacterial effects against surface-adhering, slow-growing, and

Management of Periprosthetic Joint Infections (PJIs). http://dx.doi.org/10.1016/B978-0-08-100205-6.00003-3

biofilm-producing microorganisms [14]. Thereby, it is important that the microorganism is sensitive for the type of antibiotic and that the antibiotic has a good bioavailability, good bone penetration, good penetration in the surrounding tissues, and the systemic side effects should be acceptable [15]. In order to eliminate the biofilm, in most cases, rifampicin (or other rifamycins) is administered because of its biofilm-penetrating properties on *staphylococci* biofilms [16,17]. Sometimes fluoroquinolones are administered because of their effect on biofilms of Gramm-negative bacteria [18,19]. To fully eradicate the infection and to conquer the rising resistance of bacteria, the biofilm-penetrating antibiotics are administered in combination with another antibiotic, for example, penicillin or clindamycin depending on whether the microorganism is susceptible [9,11].

Comparing implant-related infection and chronic osteomyelitis to other infections like pneumonia and urinary tract infections, these infections never spontaneously heal and even oral antibiotic therapy will fail in most cases [9,20]. These infections should be treated with a combination of surgery and antibiotics. Multiple treatment algorithms are developed to treat implant-related infection, for instance, debridement with implant retention, one-stage revision, two-stage revision with short interval, and two-stage revision with long interval [10]. Treatment of these implant-related infections will be further discussed in upcoming chapters. In treatment of chronic osteomyelitis, it is important to combine debridement surgery with antibiotic administration. During the debridement surgery, the surgeon has to perform an extensive removal of all sequestra and all the dead, infected, and scarred bone and soft tissue [21]. Due to the bone loss after these debridements, dead space will remain. This is a major disadvantage because these bone defects may result in bone instability and hematoma formation. These bone defects require appropriate management or reconstruction to prevent complications like fractures. A few methods described to treat the bone defects are: defect filling with autologous bone grafts, nonfilling with closure using muscle and myocutaneous flaps, or temporary filling with polymethylmethacrylate (PMMA) beads impregnated with antibiotics before reconstruction with bone grafts [11,22].

To obtain a good eradication of chronic osteomyelitis, appropriate antibiotic treatment is crucial and, in most cases, systemic and local administration is necessary. Systemic antibiotics can be administered oral and parenteral depending on type of antibiotic, for instance, clindamycin could be administered oral where vancomycin should be administered parenteral [23]. In case of local antibiotic therapy, different biomaterials with several antimicrobial properties are used. The gold standard in local treatment are the PMMA beads impregnated with antibiotics, but in the past years, other biomaterials such as bioactive glass, calcium sulfates, and calcium phosphates are also used [24].

3.2 Biomaterials

In treatment of implant-related infections and chronic osteomyelitis, the role of biomaterials is considerable and increasing. These biomaterials are used for reconstruction/filling of bone defects, or for local antimicrobial therapy, or both.

The combination of the wide application and the strong desire for an improvement of infection treatment caused a numerous amount of biomaterials available for treatment of implant-related infections and osteomyelitis. Biomaterials differ in characteristics and properties as: composition, (micro)structure, phase state, surface, and shape. The different properties make these biomaterials applicable for several causes, but it is important to recognize that not all materials are suitable for all applications. Major principles for local antimicrobial therapy combined with bone defect filling are the choice of the delivering biomaterial combined with the choice of the antimicrobial mechanism/agent [25].

3.2.1 Local antibiotic delivery

In case of implant-related infections and chronic osteomyelitis, the infection site is often surrounded by avascular bone and necrotic or sclerotic tissues. This results in a poor reachability for systemic antibiotics when treating these infections with solely oral or intravenous antibiotics. This makes local antibiotic treatment necessary [10,20]. For local administration of antibiotics or other local antimicrobial therapies, different biomaterials with different properties are available. Important considerations in local antibiotic therapy concern the combination of biomaterial characteristics and antibiotic properties [24].

Two types of local delivery devices of antibiotics can be distinguished based on their biomaterial properties, namely, nondegradable and degradable biomaterials. The nondegradable biomaterials are applied for years in treatment of orthopedic infections, where the application of degradable antibiotics delivery devices is a relatively novel technique [26].

The most common applied nondegradable biomaterial is the polymer PMMA. PMMA is widely applied because it is a host-friendly nontoxic biomaterial with limited adverse reactions, hypersensitivity, or allergic reactions. Thereby, PMMA is easy to handle and applicable with a broad spectrum of antibiotics [26]. The disadvantages of PMMA beads are the necessity of a two-stage surgery to remove them, and furthermore, PMMA can become a substratum for colonization of bacteria when antibiotic concentrations drop and the material remains in vivo [27]. Therefore, the use of biodegradable biomaterials for local antibiotic delivery is an attractive alternative, because ideally these biomaterials release their antibiotics during degradation and they obliterate the bone defects resulting from the debridement surgery without remaining a substratum for new bacterial colonization. Biomaterials suitable as biodegradable local antibiotic delivery device can be categorized in different groups based on their composition: bone grafts, polymers, and bone graft substitutes [26].

When using biomaterials for local antibiotic administration, it is important to consider the biomaterial-specific release profiles [24]. These biomaterial-specific release profiles are dependent on the mutual effects of the diffusion-related antibiotic release and the degradation of the biomaterial. Degradation can be the dominant factor in antibiotic release which occurs in two ways, by bulk-eroding degradation and by surface-eroding degradation. Surface-eroding materials erode layer by layer, where bulk-eroding materials erode throughout. Thereby is known that surface-eroding biomaterials degrade faster but are less permeable. This results in lower hydration compared to bulk-eroding biomaterials [28]. These properties result in two different antibiotic release profiles: surface-eroding materials release their antibiotics from their surface resulting in sustained and relatively constant levels of release, whereas bulk-eroding materials have a delayed increase of release due to the diffusion after increasing the porosity due to the erosion [26]. In summary, there are three main mechanisms for antibiotics release in degradable biomaterials, namely, diffusion-based, bulk-erosion-based, and surface-erosion-based release (Fig. 3.1). It is also important to remember that degradation of biomaterials may be associated with a change of the surrounding tissue, for instance, a decreasing pH, which again may affect degradation speed and the antibiotics release [29].

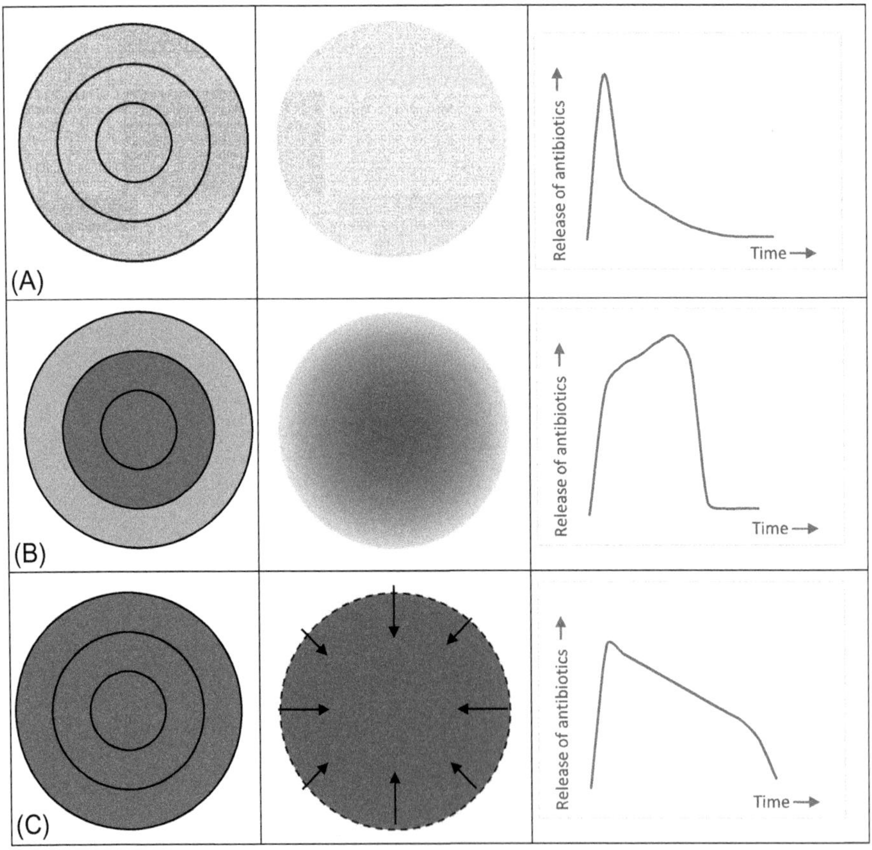

Fig. 3.1 Biomaterial degradation mechanisms. (A) Diffusion-based erosion. (B) Bulk erosion. (C) Surface erosion.

When using antibiotic therapy in treatment of osteomyelitis or implant-related infection, high local concentrations are necessary to exceed the minimal inhibitory (or bactericidal) concentrations (MIC) of the causative microorganisms. Major advantage of the application of local antibiotic therapy is the ability to achieve these high concentrations of antibiotics with the relatively low systemic side effects as nephrotoxicity or ototoxicity in osteomyelitis treatment with aminoglycosides [30,31]. In treatment of these infections, it is important to consider which type of antibiotic is necessary. In proper osteomyelitis treatment, the antibiotic selection is based on microbial resistance patterns, microorganism killing patterns, and penetration depth in surrounding tissue. The penetration depth must ideally reach approximately 1 cm and attain a concentration as high as possible before harming the surrounding tissues [31]. The antimicrobial activity patterns of antibiotics are based on two different mechanisms, time-dependent killing and concentration-dependent killing [32] (Fig. 3.2). Time-dependent killing is based on the duration of exposure, where the concentration of the antibiotics exceeds the MIC (T>MIC). This time-dependent killing is characteristic for macrolides, β-lactam antibiotics, and clindamycin. Concentration-dependent killing is based on attaining the highest possible local antibiotic concentration (AUC_{24}/MIC or C_{max}/MIC) and is characteristic for aminoglycosides, quinolones, and vancomycin. Ideally the concentration-dependent killing concentration is achieved for the first 2 days of treatment, where time-dependent killing concentrations should be maintained for the next weeks [24]. Last important issue about using antibiotics in treatment of osteomyelitis or implant-related infections is that these local antibiotic concentrations should not exceed the damaging concentration; otherwise they can affect the levels of osteoblast activity and bone-healing capacities [33,34].

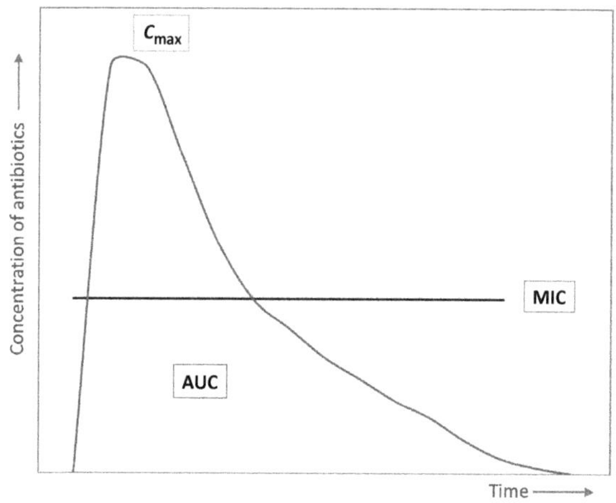

Fig. 3.2 The time-related pharmacodynamics and pharmacokinetics of antibiotics. C_{max}, maximum concentration; *MIC*, minimal inhibitory concentration; *AUC*, area under the curve.

3.2.2 Bone defect filling

Biomaterials are expected to become even more important in filling bone defects after debridement surgery in cases of infection. In order to reconstruct the bony defects, bone regeneration is usually performed using autologous or allogeneic bone, because these are most effective in promoting bone formation by osteoinduction and osteoconduction [35,36]. Despite the effective bone regeneration of autografts and allografts, these methods are associated with some disadvantages, respectively, limited availability and donor morbidity in autograft procedures or disease transmission or bacterial infection in allograft procedures [37,38]. Therefore, the use and applicability of biomaterials, that is, bone graft substitutes, are becoming increasingly important.

Concerning bone regeneration and biomaterials, biocompatibility, biodegradability, and mechanical properties are important and frequently used terms. Biocompatibility is about the ability of a biomaterial to react on the host responses in specific applications [39]. Within biocompatibility, osteoinduction and osteoconduction are important concepts for the regeneration of bone. Osteoinduction is the stimulation of primitive, undifferentiated, and pluripotent cells to differentiate to osteoprogenitor cells to form bone tissue in any environment, where osteoconduction means that the biomaterial has properties to support tissue ingrowth and stimulation of growth of osteoprogenitor cells to form new bone [39].

Biodegradability or biodegradation is a commonly used term, which has no clear definition but is mainly used for describing the chemical breakdown of biomaterials that occur over days or years and causes changes in physical properties of the environmental tissues. Degradation of biomaterials can occur in different ways as the materials can solubilize, crumble, become rubbery, or just become rigid. These pathways of degradation differ per type of material, where degradation is commonly seen in metals, polymers, ceramics, and composites. The degradation of a biomaterial depends mainly on its chemical composition and microstructure. For example, in ceramics, tricalcium phosphates ($Ca_3(PO_4)_2$), hydroxyapatite (HA), and calcium sulfates ($CaSO_4$) are degraded fairly rapid, while bioactive glasses degrade relatively slow. On the other hand, porosity of these materials will influence the degradation rates, where fully dense materials degrade much slower compared to highly porous (microporous) materials. Besides the biomaterial-related properties of degradation, the effects of in vivo factors cannot be misunderstood. Variations in degradation rates have been observed on different implantation sites, possibly cause by different cellular activity, phagocytosis, or release of free radicals. Thereby it is important to realize that degradation of some biomaterials can cause hypersensitivity, toxicity, and other adverse reactions.

Regarding the mechanical properties of different biomaterials, the material strength, deformation, and toughness are distinguishing characteristics which determine the specific applicability. Material strength is determined by the principles of stress and strain combined with the amount and type of deformity measured. The ratio between stress and strain determines the elasticity of a biomaterial that can be measured with Young's modulus of elasticity, where the higher the modulus, the stiffer the biomaterial. Deformity of biomaterials occurs due to a force resulting on the biomaterial. This deformity can be reversible (elastic deformity) or irreversible (plastic deformity). Another important property of biomaterials is the toughness, which determines the amount of energy per volume the biomaterial can absorb before failure/fracture. For example, ceramics as bioactive glasses have a very high Young's modulus because they have a low ability to deform before material failure. These specific properties make ceramics very brittle biomaterials. Metals like titanium have a lower Young's modulus because they allow much more strain compared to ceramics. This is because metals have a higher ability of plastic deformity before biomaterial failure, what makes metals like titanium ductile biomaterials. In case of filling load-bearing defects, bone graft substitutes must have good mechanical load capacities to retain strength and prevent fracture. But the biomaterials must have a Young's modulus compared to bone, because when this is much higher the bone can fracture because it cannot handle the stress of the biomaterial.

3.2.3 Materials for local antimicrobial treatment

3.2.3.1 Polymers

In the development of local antibiotic carrier systems, polymers have been studied widely, and they can be divided into the natural (protein-based) polymers and the synthetic polymers [26]. Synthetic polymers or polyesters have been investigated worldwide because their biocompatibility and their biodegradable properties in treatment of bone defects for decades, but in the past years many synthetic polymers, have been studied for their applicability as a local antibiotic delivery system. Polymers based on polylactide acid and/or polyglycolide acid are used in different chemical forms and combinations to attain maximum strength and good antibiotic release profiles [40]. Despite the fact that these materials are studied in ex vivo and in in vivo animal models, these materials are not manufactured for commercial use and clinical testing in in vivo human models [41]. Natural polymers are protein based and derived from biologic tissues obtained from autograft, allograft, or xenograft materials. Collagen (type 1), chitosan, gelatine, thrombin, and even autologous blood clots are clinically applied as natural polymers. These polymers function as protein or scaffold to bind antibiotics and provide local antibiotic release. Elution rates of these natural polymers are high, and they tend to release their antibiotics within hours or several days [41]. Collagen is the most applied natural polymer in clinical practice, because it has most supportive data, has low costs, has good biocompatibility, and has a high availability [26].

Collagen

Collagen-based polymers are often produced from equine, porcine, or bovine skin or tendons, of which the type 1 microfibrillar collagen structures are harvested to form a matrix for the collagen meshes or fleeces [41,42]. These collagen fleeces are able to bind and release drugs, and these abilities can be modified by changing properties as porosity and chemical preparation. Antibiotic release of collagen fleeces occurs in three stages, namely, rapid release due to the porosity of the fleeces, intermediate release due to opening of the closed pores in the fleeces, and prolonged release after the release out of the collagen structures during fleece resorption [42,43]. Data from ex vivo elution studies have shown that collagen fleeces release up to 95% of their antibiotics within the first 2h and that these fleeces are only useful for a maximum of 48h [44,45]. Despite these data, in vivo animal studies showed that collagen fleeces reach effective local antibiotic therapy up to 28 days [46]. These fleeces are fully degraded approximately 8 weeks after implantation, which means that their release time is much shorter compared to the degradation time [47,48]. The differences in the antibiotic release speed and the fleece degradation rates combined with the fact that the fleece may not fully resorb can cause a risk of infection sustainment. Besides their antibiotic-releasing capacities, collagen fleeces are associated with bone healing by the proliferation of osteoblasts, production of callus, and stimulatory effects on bone mineralisation [49,50].

3.2.3.2 Ceramics, composites, and bioactive glasses

Synthetic bone graft substitutes are ceramic or composite biomaterials that are used in orthopedic surgery for years and are commonly used as bone filler. With the increasing call for local antibiotic therapy, the search for antibiotic-loaded biodegradable ceramics or composites is rising. These ceramics and composites, also called (synthetic) bone graft substitutes, have been investigated in many in vitro and in vivo studies over the past years. Most bone graft substitutes applied in local antibiotic therapy are based on calcium sulfates, calcium phosphates, and bioactive glass [42].

Calcium sulfates

Calcium sulfate-based bone graft substitutes are used in a broad spectrum of clinical problems in orthopedic and trauma surgery, for instance, bone defect filling, treatment of nonunion fractures, bone graft expanders, and prevention and treatment of infections [51]. Calcium sulfate is a natural ceramic that is commonly used in its alpha-hemihydrate from ($CaSO4{\cdot}0.5H_2O$), because this form is stronger and harder, which makes it more useful for bone defect filling [26]. These calcium sulfates behave as bone defect fillers by restoring the morphological bone contours and forming an osteoconductive matrix in the cavity, which enables the ingrowth of blood vessels and bone and disables the ingrowth of soft tissue [52]. Several studies have shown that the degradation of the calcium sulfate pellets implanted in a bony defect is a relatively fast process occurring in a matter of weeks to months, not years. In vivo studies showed a degradation rate of 100% after a maximum of approximately 3–6 months in both animal and human studies [53–55]. The antibiotic load of these bone graft substitutes is based on their chemical composition, because calcium sulfates can be loaded with water-soluble antibiotics such as aminoglycoside, vancomycin, and teicoplanin. Antibiotic release proceeds relatively fast. Different in vitro and in vivo (animal) studies showed that the release of antibiotics by calcium sulfates varies from 17% up to 80% of the total antibiotic content [56–58]. However, some of these studies also showed that the local antibiotics concentrations remain effective and above the MIC for minimal 10–42 days [56,58,59]. The disadvantages of the calcium sulfate hemihydrates are the fast degradation and mild cytotoxic reactions reported in some cases in human studies [60].

Calcium phosphates

Calcium phosphate-based bone graft substitutes (mostly tricalcium phosphate or hydroxyapatite) are bioceramics that show the greatest similarity to the minerals found in bone. This is what gives calcium phosphates excellent biocompatibility, biodegradability, and osteoconductivity [61,62]. Osteoconductivity of calcium phosphates is based on their capacity to form molecular interaction with the surrounding tissues which results in the formation of an apatite layer on its surface [63]. These calcium phosphate-based bone graft substitutes degrade relatively slow compared to calcium sulfates. Thereby, different calcium phosphate compositions show different degradation rates, for instance, β-tricalcium phosphate degrade faster compared to hydroxyapatite (HA) [64]. After implantation they stimulate osteoid formation onto its surface, which is the first step in the bone remodeling mechanism of calcium phosphate [65]. Resorption of the calcium phosphate biomaterials takes place after the activation of osteoclastic and multinuclear cells [61]. Calcium phosphates suitable for local antibiotic delivery are generally produced as beads, which are able to bind antibiotics only on their surface in contrast with calcium phosphate cements or calcium sulfates that are able to bind antibiotics trough their entire structure [45,66].

Bioactive glasses

Bioactive glasses are bioceramics that are solid, nonporous, and hard materials, which consist of the main component silicon dioxide (or silicate) and three other basic components: sodium dioxide, calcium oxide, and phosphorous. By varying all of these components, different forms of bioactive glasses can be made [67]. Bioactive glasses have been investigated for decades and have shown good results in bone regeneration. In all these years several types of bioactive glasses are developed: the conventional silicate glass (45S5 bioactive glass or Bioglass®), glass ceramics (S53P4 bioactive glass or BonAlive®), and borate-based glasses (19-93B3 bioactive glass) [68]. The biocompatibility of bioactive glasses generally depends on the silicate part of the glass and reaches an optimum graft-bone bonding between when the glass contains 45–52% silicate [69]. The process of graft-bone bonding of bioactive glasses starts with the release of soluble ions after which a silica gel layer is formed on the bioactive glass surface. After the formation of the silica gel layer, amorphous calcium phosphates precipitate on this layer where they form a natural hydroxyapatite layer due to crystallization. The hydroxyapatite layer activates the osteoblastic cells to start the formation of new bone [64,70,71]. This graft-bone bonding is one of the two mechanisms involving degradation of the bioactive glasses, where the second mechanism is based on the dissolution of different ions and thereby activating osteogenesis [70,72,73]. The degradation speed of bioactive glasses is depending on the composition of the glasses and can vary from several hours to months. For instance, of borate bioactive glass is known that it degrades much faster than silicate bioactive glass [64,74].

In the past decade the popularity of bioactive glasses in the treatment of infections increased. Their versatility in composition and physiological properties enables us to use bioactive glasses in treatment of infection based on different antibacterial mechanisms. These antibacterial properties can be divided into three principles [64]. The first principle is based on the change of local physiological conditions due to dissolution of ions that results in a bactericidal environment. S53P4 bioactive glass (commercially available as BonAlive® bioactive glass) is a bioactive glass with the ability to inhibit bacterial growth. This antibacterial effect is caused by the dissolution of surface (alkali) ions that cause an increase of pH and a higher osmotic pressure [36,75]. This mechanism is proven bactericidal on different bacterial strains, such as *Staphylococcus aureus*, without causing any signs of cytotoxic effects or adverse reactions on their environmental tissues [64,76,77]. Second principle is based on adding element as silver to the bioactive glass during the manufacturing process because of their bactericidal effects. As well silicate as borate bioactive glasses are tested as silver-doped antibiotic glasses [78,79]. Silver-doped silicate bioactive glasses showed slow degradation rates and uncontrollable silver release profiles, where silver-doped borate bioactive glasses showed that due to the manufacturing process there is a controlled release of silver, which is favorable in infection treatment [64,79]. Despite the good in vitro results for eradication of infection, there is a little in vivo evidence for silver-doped bioactive glasses; this might be explained due to the toxicity and adverse reactions when using high concentrations of silver that is favorable in infection treatment. The third and last principle uses antibiotic-loaded bioactive glass that elutes antibiotics during degradation compared to the previously mentioned bone graft substitutes. The bioactive glasses used as an antibiotic delivery device have a different composition compared to conventional bioactive glasses because the conventional glasses are nonporous and too dense [64]. The bioactive glasses applied for antibiotic delivery are mainly composed as sol–gel-derived carriers or glass-composite carriers. The sol–gel-derived glasses have low release rates of antibiotics (only 20–25% after 3 months) and have a low strength what makes them less applicable in treatment of bone infections [80]. The composite glasses showed better release rates compared to gel–sol-derived glasses. An in vitro study showed release rates up to 80–85% after 21 days for a chitosan borate glass composite, where an in vivo study showed release rates up to 90% after approximately 14 days for phosphate-cement borate glass composites [81,82]. In treatment of bone infections with bioactive glasses, only the silicate bioactive glass S53P4 has shown antimicrobial effect by itself and has been extensively tested compared to the other antibacterial bioactive glasses. However, borate bioactive glass may be an interesting carrier for antibiotic delivery, but the evidence is not circumstantial.

3.3 Latest clinical evidence treatment osteomyelitis

The results of all surgical methods used in the past years provide moderate results, and appropriate bone defect management combined with eradication of the chronic osteomyelitis remains a challenge. Therefore, three biodegradable antimicrobial biomaterials have been tested in a clinical situation, that is, antibiotic-loaded collagen fleeces, antibiotic-loaded bone graft substitutes, and S53P4 bioactive glass. The results of all these studies will be discussed in the sections below.

3.3.1 *Collagen fleeces*

Collagen is one of the most widely established biodegradable antibiotic carrier materials applied in the local treatment of different kinds of surgical infections [92]. There are several prefabricated commercially available antibiotic-loaded collagen fleeces or sponges. This collagen is harvested from achilles tendons or skin from equine, bovine, or porcine origin [26,93]. Some of these commercial available collagen fleeces are Septocoll®, Refocoll® (or Jason G®), Collatamp (or Innocoll®), and Gentafleece®.

Regarding the treatment of chronic osteomyelitis, there are nine studies which tested antibiotic-loaded collagen fleeces on its clinical applicability (Table 3.1). In these nine studies, one study tested Septocoll® [83], seven studies tested Collatamp® [84–90], and one study did not specify which product they used [91]. Of all nine studies only two randomized controlled trials were performed where debridement surgery-combined local antibiotic treatment with gentamicin-loaded collagen fleeces was compared to either debridement and antibiotic-loaded PMMA beads or debridement without local antibiotic treatment [83,89]. The other seven studies had a descriptive design and were either retrospective outcome studies or case series reports where treatment was not compared to any kind of control group [84–88,90,91].

The results of all studies, concerning treatment of chronic osteomyelitis with antibiotic-loaded collagen fleeces, show eradication rates varying from 63% to 100% (see Table 3.2). Buehler et al. reported an eradication rate 95.5% in treatment with debridement surgery combined with implantation of a mixture of Septocoll® and autologous bone graft versus 87.8% in treatment with debridement surgery and only autologous bone graft implantation ($p=0.093$) [83]. Letsch et al. reported eradication in 80% of patients treated with debridement surgery and Collatamp® implantation versus 90% in patients treated with debridement surgery and antibiotic-loaded PMMA implantation. This was without statistical differences due to the small study groups (10 vs 10). The pharmacokinetics and elution characteristics are studied in most studies as a secondary outcome. In all studies was seen that serum concentrations showed systemic levels of gentamicin high enough to achieve the MIC, without any signs of gentamicin-related adverse events or toxicity. Thereby, wound exudate concentrations were many times higher compared to systemic concentrations, so local treatment exceeded the MIC easily.

Failure of infection treatment was seen in a recurrence rate of 41 patients out of 308 patients (13.3%), in all studies, treated with debridement surgery and antibiotic-loaded

Table 3.1 Study characteristics of studies concerning antibiotic-loaded collagen fleeces in treatment of chronic osteomyelitis

Study	Clinical indication	Study design	Primary outcome	Secondary outcome	Outcome measures
Buehler et al. [83]	Chronic osteomyelitis	Randomized controlled trial	Eradication of infection	Complications/adverse events	Clinical examination, hematological analysis, radiographic imaging
Ascherl et al. [84]	Orthopedic infections (i.e., (chronic) osteomyelitis, prosthetic joint infections	Case series report	Eradication of infection	Pharmacokinetics	Hematological analysis, wound exudate analysis, radiographic imaging, clinical examination
Dieckman et al. [85]	Patients (children) with acute or chronic osteomyelitis	Retrospective outcome study	Eradication of infection	None	Clinical examination, hematological analysis, radiographic imaging
Feil et al. [86]	Orthopedic infections (i.e., (chronic), osteomyelitis, prosthetic joint infections	Case series report	Pharmacokinetics	Eradication of infection, complications	Hematological analysis, wound exudate analysis, radiographic imaging, clinical examination
Ipsen et al. [87]	Chronic osteomyelitis	Case series report	Eradication of infection	Complications/adverse events	Clinical examination, hematological analysis, radiographic imaging
Kwasny et al. [88]	Chronic osteomyelitis, osteitis, soft-tissue infections	Case series report	Eradication of infection	Pharmacokinetics	Hematological analysis, wound exudate analysis, radiographic imaging
Letsch et al. [89]	Chronic osteomyelitis	Randomized controlled trial	Eradication of infection	Biocompatibility, safety/toxicity	Clinical examination, hematological analysis, radiographic imaging
Wernet et al. [90]	Chronic osteomyelitis	Case series report	Eradication of infection	Pharmacokinetics/elution characteristics	Not mentioned
Walz [91]	Chronic osteomyelitis	Case series report	Eradication of infection	None	Not mentioned

Table 3.2 Results and outcomes of studies concerning antibiotic-loaded collagen fleeces

Study	Infection eradication (%)	Other outcomes	Complications
Buehler et al. [83]	95.5	No secondary outcomes described	2× Recurrent infection, 2× superficial wound infection, 2× new fistulae, 1 collagen graft removal
Ascherl et al. [84]	63	Harmless levels of gentamicin concentration in serum and wound exudate. No signs of toxicity	10× Recurrent infection, 8× new fistulae, wound-healing problems, prolonged wound leakage
Dieckmann et al. [85]	83.3	No secondary outcomes described	6× Recurrent infection
Feil et al. [86]	93.3	Harmless levels of gentamicin concentration in wound exudate. No signs of toxicity	5× Recurrent infection, 11× wound-healing problem, 1× coronary infarction
Ipsen et al. [87]	100	Harmless levels of gentamicin concentration in urine, serum and wound exudate. No signs of toxicity	None
Kwasny et al. [88]	95	Harmless levels of gentamicin concentration in serum and wound exudate. No signs of toxicity	4× Recurrent infection, 1 Fracture
Letsch et al. [89]	80	Harmless levels of gentamicin concentration in urine, serum, and wound exudate. No signs of toxicity	2× Recurrent infection, 1× amputation
Wernet et al. [90]	77	Harmless levels of gentamicin concentration in serum and wound exudate. No signs of toxicity	11× Recurrent infection, 1× prolonged wound leakage
Walz [91]	89	None	1× Recurrent infection

collagen fleece implantation. Other problems described in the different studies were new formed fistulae, superficial wound infections, prolonged wound leakage, and in one case fracture of the treated bone region.

Eradication of infection is described in all studies, but neither one of the studies discussed bone defect healing or in vivo fleece degradation. The collagen fleeces were mixed with autologous bone grafts in multiple studies, which might suggest that treatment with fleeces alone is not sufficient for bone regeneration of the resulting bone defect. Although, there was only one fracture reported in all studies [88].

The different studies showed quite satisfying results in treating chronic osteomyelitis, low levels of complications, and no gentamicin-related adverse events or toxicity.

Despite these results, the quality and validity of the different studies and their results should be taken into account. Most studies lack in a good treatment protocol, especially regarding systemic antibiotics administration. Thereby their study designs induce moderate to high risks of different types of bias, for example, due to the absence of control groups or due to selective outcome reporting. This makes it difficult to assess if antibiotic-loaded gentamicin fleeces are suitable for clinical application in treatment of patients with chronic osteomyelitis.

3.3.2 Antibiotic-loaded bone graft substitutes (ceramics)

Antibiotic-loaded bone graft substitutes are deemed to become the future of local antibiotic treatment of chronic osteomyelitis. Therefore, there are numerous bone graft substitutes tested for clinical application of local antibiotic delivery. Several of these bone graft substitutes are commercially available, that is, Cerament-G® (Cerament-V®), Herafill-G®, OsteoSet-T®, PerOssal®, and Stimulan-G®. All of these biomaterials are composites based on calcium sulfates, but they differ in terms of: composition with other biomaterials (e.g., hydroxyapatite, calcium carbonate), application forms, and in the possibility of being combined with different types of antibiotics (see Table 3.3).

All five mentioned products are available for clinical application in local treatment of chronic osteomyelitis, but only three of these products (Herafill-G®, OsteoSet-T®, and PerOssal®) have been described in publications of clinical studies [106]. A total of twelve studies is performed to assess the applicability of commercially available bone graft substitutes in chronic osteomyelitis treatment. Only one randomized controlled trial was performed by McKee et al. in 2010, where they compared debridement surgery and OsteoSet-T® implantation with debridement and antibiotic-loaded PMMA implantation [100]. Of the remaining eleven studies, two were retrospective cohort studies comparing antibiotic-loaded BGS with a control group, where all other studies were case series or descriptive outcome studies (see Table 3.4). Chang at al. performed a retrospective cohort study where they compared debridement surgery and OsteoSet-T® implantation with only a debridement surgery without local antibiotic treatment [96], where Romano et al. compared three different biomaterials, that is, S53P4 bioactive glass (BonAlive®), antibiotic-loaded hydroxyapatite calcium sulfates (PerOssal®), and tricalcium phosphates (Calcibon®) combined with antibiotic-loaded bone matrix (Targobone®) [104].

All studies report eradication of infection in all or a majority of all patients treated with an antibiotic-loaded BGS. These eradication rates are comparable for the different BGS and vary from 86% to 100% for OsteoSet-T®, from 80% to 100% for PerOssal® and from 80% to 100% for Herafill-G®. Of the studies that compared antibiotic-loaded BGS with a control group, only Chang et al. showed significant better results. They compared debridement and OsteoSet-T versus debridement without local treatment (eradication of 94% vs. 59%; $p=0.024$) in a subgroup of patients with the same osteomyelitis classification (Cierny-Mader 1A). McKee et al. did found an eradication rate of 86% in both groups (OsteoSet-T and antibiotic-loaded PMMA), and Romano et al. found a eradication rate of 88.9% in the PerOssal group versus 92.6% and 86.3% in the control groups, which did not show any statistical significance. A comparison of

Table 3.3 Commercially available bone graft substitutes in treatment of chronic osteomyelitis

Product name	Composition	Antimicrobial mechanism	Antibiotic type	Studies
Cerament-G® (Cerament-V®)	Hydroxyapatite (40%) with alpha-hemihydrate calcium sulfate (60%)	Antibiotic impregnated bone graft substitute	Gentamicin (Vancomycin)	None
Herafill-G®	Calcium sulfate, calcium carbonate, hydrogenated triglyceride	Antibiotic impregnated bone graft substitute	Gentamicin	[94,95]
OsteoSet-T®	Alpha-hemihydrate calcium sulfate	Antibiotic impregnated bone graft substitute	Tobramycin	[96–102]
PerOssal®	Nanocrystalline hydroxyapatite (51.5%), calcium sulfate (48.5%)	Antibiotic impregnated bone graft substitute	Different types of antibiotics (surgeon's choice)	[103–105]
Stimulan®	Hemihydrate form of calcium sulfate	Antibiotic impregnated bone graft substitute	Gentamicin, vancomycin, tobramycin	None

the other studies showed that studies with a higher number of participants show little lower eradication rates. For example, Ferguson et al. showed an eradication rate of 90.8% in 195 cases treated with OsteoSet-T [97], where some smaller studies report eradication rates of 100%. All studies with bone healing as an (secondary) outcome parameter showed bone regeneration, measured by the presence of bone formation on X-ray images, after implantation of the different BGSs. Degradation of BGS was a secondary outcome in most studies, and they all reported full or partial degradation of the different types of BGSs at the end of follow-up. Complications were reported in most studies. Major complications (i.e., death and venous thrombosis) were seen but were not related to the infection or BGS implantation. Common seen complications related to this treatment were recurrence of infection (8.4%), prolonged wound leakage (8.9%), and fracture (3.1%) in all patients treated with antibiotic-loaded BGSs. Less frequently seen complications were wound-healing problems as a result of superficial wound infections or other causes.

Wound leakage is described in several studies as a problem directly related to calcium-based BGSs and is not directly related to a recurrence of infection. Ferguson et al. showed that recurrence of infection in 18 patients was not significantly related to a prolonged wound leakage ($p=0.109$) [97,104]. Bone growth following the

Table 3.4 Study characteristics of studies concerning antibiotic-loaded bone graft substitutes in treatment of chronic osteomyelitis

Study	Indication	Study type	Primary outcomes	Secondary outcomes	Outcome measurements
Chang et al. [96]	Chronic osteomyelitis	Retrospective cohort study	Eradication of infection	BGS degradation	Clinical examination, haematological analysis, radiographic imaging
Ferguson et al. [97]	Chronic osteomyelitis	Retrospective outcome study	Eradication of infection	Bone defect filling, complication rates	Clinical examination, haematological analysis, radiographic imaging
Gitelis and Brebach [98]	Chronic osteomyelitis	Prospective outcome study	Eradication of infection	BGS degradation, bone growth	Hematological and microbiological analysis, radiographic imaging
Humm et al. [99]	Chronic osteomyelitis	Retrospective outcome study	Eradication of infection	Complications	Clinical examination, hematological and microbiological analysis, radiographic imaging
McKee et al. [100]	Chronic osteomyelitis	Randomized controlled trial	Eradication of infection	BGS degradation, bone growth, complications	Clinical examination, hematological and microbiological analysis, radiographic imaging
Tsai et al. [101]	Chronic osteomyelitis/infected nonunions	Case series report	Eradication of infection	BGS degradation, bone growth	Clinical examination, hematological analysis, radiographic imaging

von Stechow et al. [102]	Chronic osteomyelitis/ spondylitis	Retrospective outcome study	Eradication of infection	BGS degradation, bone growth, complications	Hematological, radiographic imaging
Berner et al. [103]	Chronic osteomyelitis	Case report	Eradication of infection	BGS degradation, bone growth	Clinical examination and radiographic imaging
Romano et al. [104]	Chronic osteomyelitis	Retrospective cohort study	Eradication of infection	Complications	Clinical examination, hematological and microbiological analysis, radiographic imaging
von Stechow and Rauschmann [105]	Chronic osteomyelitis/ spondylitis	Prospective outcome study	Eradication of infection	BGS degradation, bone growth, complications	Haematological analysis, radiographic imaging
Fleiter et al. [94]	(Chronic) osteomyelitis/osteitis	Prospective outcome study	Pharmacokinetics of BGS	Eradication of infection, BGS degradation, complications	Haematological analysis, radiographic imaging
Franceschini et al. [95]	Chronic osteomyelitis	Case series report	Eradication of infection	BGS degradation, bone growth, complications	Clinical examination, hematological analysis, radiographic imaging

degradation of the BGSs is varying among the different studies. All studies concerning OsteoSet-T showed a full degradation of BGSs within several weeks to months, which can be the reason for varying bone growth rates [96–102]. In the studies concerning PerOssal, lower degradation rates and higher bone growth rates were seen [104,105]. These differences may be due to the fact that OsteoSet-T is made of calcium sulfates where PerOssal is a composite of calcium sulfates combined with hydroxyapatite (Table 3.5).

Although the relatively good outcome results reported in all studies, the poor methodological quality of most of these studies should be taken into account. Due to the large differences between the studies, the large amount of poor study designs, the lack of study protocols in most studies, the inconclusive outcome reporting, the lack of statistical analysis, the risks of bias within these studies are worth mentioning. This makes it difficult to interpret and compare all outcomes for good evidence-based decision-making in the local treatment of chronic osteomyelitis with antibiotic-loaded bone graft substitutes.

3.3.3 Bioactive glass

In treatment of chronic osteomyelitis, bioactive glass is used as a local antimicrobial biomaterial and as a bone defect filling biomaterial as well [71]. Regarding to the different types of bioactive glass used in chronic osteomyelitis treatment, there is only one type of bioactive glass which is tested in clinical trials, the S53P4 bioactive glass. As mentioned before, this S53P4 bioactive glass does not contain any kind of antibiotics, but the antimicrobial properties of this glass are based on the increasing pH and higher osmotic pressure caused by, respectively, the exchange of alkali ions and release of salt ions [36,70].

S53P4 bioactive glass has been investigated in three different clinical studies so far. Romano et al. did a retrospective cohort study where they compared treatment of chronic osteomyelitis with debridement combined with systemic and local antimicrobial therapy in three different groups [104,107]. In this study they compared a treatment group treated with S53P4 bioactive glass, with two control groups. A second study was performed by Lindfors et al., and they performed a multicentre case study of eleven patients treated with debridement surgery and S53P4 bioactive glass implantation combined with systemic antibiotic therapy [108]. The third study was performed by McAndrew et al., and they treated three chronic osteomyelitis patients with comparable debridement surgery and S53P4 implantation combined with systemic antibiotics [109] (Table 3.6).

The success rates of these different studies show that treatment of chronic osteomyelitis succeeded in 93% (25 of 27 patients) [104], 82% (10 of 11 patients) [108], and 100% (3 of 3 patients) [109], which results in a total of 93% (38 of 41) absence of chronic osteomyelitis at the end of follow-up that varied between an average of 17.3 months to 24 months (see Table 3.7). Romano et al. compared their treatment group ($n=27$) with two control groups, one with antibiotic-loaded hydroxyapatite with tricalcium phosphate ($n=27$) and one group with antibiotic-loaded tricalcium phosphates combined with teicoplanin-loaded demineralised bone matrix ($n=22$). At the end of follow-up, no

Table 3.5 Results and outcomes of studies concerning antibiotic-loaded collagen fleeces in treatment of chronic osteomyelitis

Study	Infection eradication in % of patients	Bone growth in % of patients	BGS degradation in % of patients	Complications
Chang et al. [96]	80%	40%	100%	5× Recurrent infection
Ferguson et al. [97]	90.8%	63.4%	100%	18× Recurrent infection, 9× fractures, 17× prolonged wound leakage, 2× muscle flap complications
Gitelis and Brebach [98]	100%	100%	100%	No complication
Humm et al. [99]	95.2%	Not mentioned	Not mentioned	1× Recurrent infection, 7× prolonged wound leakage
McKee et al. [100]	86%	100%	100%	1× Recurrent infection, 2× fracture, 1× superficial wound infection, 1× neuropraxy
von Stechow et al. [102]	87.5%	87.50%	100%	2× Patients died due thromboembolism, 2× prolonged wound leakage, 3× other complications
Tsai et al. [101]	100%	100%	100%	No complications
Berner et al. [103]	100%	100%	100%	No complications
Romano et al. [104]	88.9%	Not mentioned	100%	3× Recurrent infection, 8× prolonged wound leakage, 1× wound-healing problems, 1× venous thrombosis
von Stechow and Rauschmann [105]	100%	100%	100%	1× Venous thrombosis, 3× screw loosening due to osteoporosis
Fleiter et al. [94]	80%	Not mentioned	100%	4× Recurrent infection
Franceschini [95]	100%	100%	100%	No complications

Table 3.6 Study characteristics of studies concerning S53P4 bioactive glass in treatment of chronic osteomyelitis

Study	Indication	Study type	Primary outcomes	Secondary outcomes	Outcome measurement
Romano et al. [104,107]	Chronic osteomyelitis	Retrospective cohort study	Eradication of infection	Bone healing, S53P4 degradation	Clinical examination, haematological analysis, radiographic imaging
Lindfors et al. [108]	Chronic osteomyelitis	Retrospective outcome study	Eradication of infection	Bone healing, S53P4 degradation	Haematological analysis, radiographic imaging
McAndrew et al. [109]	Chronic osteomyelitis	Retrospective case study	Eradication of infection	Bone healing, S53P4 degradation	Clinical examination, haematological analysis, radiographic imaging

Table 3.7 **Results and outcomes of studies concerning S53P4 bioactive glass in treatment of chronic osteomyelitis**

Study	Number of treated patients	Number of successful treatments	Bone healing/ incorporation of S53P4	Complications	Mean follow-up months (range)
Romano et al. [104,107]	27	25 (93%)	Partial incorporation of S53P4	2 Recurrence of infection, 1 muscle flap complication, 1 deep venous thrombosis, 1 death (pneumonia)	21.8 (12–36)
Lindfors et al. [108]	11	10 (82%)	Good incorporation of S53P4, callus formation, proper bone healing	1 Recurrence of infection, 1 muscle flap complication, 1 fracture, 1 lung embolism	24.0 (10–38)
McAndrew et al. [109]	3	3 (100%)	Good incorporation of S53P4	None	17.3 (14–21)

statistical significant differences regarding eradication/recurrence of infection, hospital stay, and additional antibiotic therapy was seen. Although, Romano et al. showed a significant decrease in prolonged wound leakage (>2 weeks) and a significant better wound healing.

Lindfors et al. and McAndrew et al. also reported good eradication of infection and clinical outcomes. Thereby, all three studies described good incorporation of the S53P4 bioactive glass in the bone cavity resulting in proper bone healing on plain radiographic imaging.

Recurrence of infection was reported in a total of three cases, and muscle flap or soft tissue complications were described in two cases [104,108]. In two cases of recurrence of infections, they were described to improper filling of the bone defect by both studies. The other recurrence of infection was due to a combination of a soft tissue infection of compromised soft tissue and the presence of a mixed bacterial flora (containing methicillin resistant *Staphylococcus aureus* and *Pseudomonas aeruginosa* and an *Enterococcus* spp.). Romano et al. reported one spontaneous fracture at the infection site. Complications not related to S53P4 described in the different studies were deep venous thrombosis, lung embolism, and death due to pneumonia in a patient with severe comorbidities [104,108].

Regarding the latest clinical evidence for treatment of chronic osteomyelitis by using S53P4 bioactive glass, it indicates that this form of treatment is at least as successful compared to treatment with gentamicin-loaded PMMA beads or other antibiotic-loaded bone graft substitutes [71,104,110]. The advantage of S53P4 bioactive is its different antimicrobial pathway, so it does not influence resistance to antibiotics of bacteria, and no resistance to this treatment has been shown so far.

Despite these high eradication rates, good clinical outcomes, and relatively low complication rates, the level of evidence and methodological quality of these studies should be taken into consideration. None of the studies was a prospective randomized controlled trial, no blinding was applied, and only one study had control groups. Thereby, based on the methodological quality, there is a considerably risk of different types of bias [106]. For good evidence-based decision-making, it is preferable that randomized controlled trials with larger study populations are performed, where treatment with S53P4 bioactive glass is compared to, for instance, treatment with antibiotic-loaded PMMA beads.

3.4 Summary

In the treatment of either implant-related infections or chronic osteomyelitis, several major difficulties occur. These difficulties are: the biofilm layer, the inflammatory-induced bone necrosis, destruction of trabeculae, destruction of the bone matrix occur and vascularity is obliterated which leads to ischemia. Due to these factors, local tissue is compromised and systemic antibiotic treatment is insufficient to eradicate infections, so local antibacterial treatment is necessary. Nondegradable and degradable biomaterials are used as (carrier) materials for local antibacterial therapy. The last years, biodegradable biomaterials have been studied extensively to evaluate their applicability in orthopedic infection treatment. Important considerations in local antibiotic therapy concern the combination of biomaterial characteristics and antibiotic properties. It is important, when treating with local and systemic antibiotics, that the concentrations exceed the MIC of the causative microorganisms. In order to treat the bacterial infection, the bacteria can be killed using two different mechanisms that are antibiotic dependent, that is, time-dependent killing and concentration-dependent killing. Biomaterials used for local antibiotic therapy should have good biocompatibility, biodegradability, and mechanical properties in order to treat the infection and heal the bone defect persisting after debridement surgery. In the development of these local antibacterial products, different biomaterials are used. Not only polymer-based biomaterials are used to develop, for example, antibiotic-loaded collagen fleeces, but also ceramic or composite-based biomaterials are developed. These ceramics and/or composites are usually calcium based, called bone graft substitutes as: calcium sulfates, calcium phosphates, and hydroxyapatite. Thereby, the interests in bioactive glasses are increasing because they can have a complete different mechanism of action in antibacterial therapies.

The clinical results of treatment of chronic osteomyelitis with antibiotic-loaded collagen fleeces show eradication rates varying from 63% to 100%, which are quite promising, but bone regeneration is not described in neither of the studies using product. Patients treated with an antibiotic-loaded BGS had comparable eradication rates varying from 80% to 100% for all commercially available and clinically tested products. When using S53P4 bioactive glass in local treatment of chronic osteomyelitis, the eradication rates were similar to the other discussed products, ranged from 82% to 100%. One of the differences of treatment with the different product was prolonged wound leakage, which is more extensive in biomaterials that are calcium based. The poor methodological quality of these studies makes it difficult to interpret and compare all outcomes and different products.

Taken together, the S53P4 bioactive glass and the antibiotic-loaded fleeces and BGS enable the possibility of a one-stage surgery and can cause lower patient burden, shorter hospital stay, and lower costs. Thereby, the bone healing capacities of the different types of treatment may enable remodeling to normal bone over time, which is important because many patients require additional (orthopedic) surgery, for example, joint replacement surgery. For good evidence-based decision-making, it is preferable that randomized controlled trials with larger study populations are performed, where treatment with one of the tested local antibacterial biomaterials is compared to, for instance, treatment with antibiotic-loaded PMMA beads.

References

[1] W.H. Harris, C.B. Sledge, Total hip and total knee replacement (2), N. Engl. J. Med. 323 (12) (1990) 801–807.

[2] M.E. Muller, Internal fixation for fresh fractures and for non-union, Proc. R. Soc. Med. 56 (1963) 455–460.

[3] R.B. Gustilo, R.L. Merkow, D. Templeman, The management of open fractures, J. Bone Joint Surg. Am. 72 (2) (1990) 299–304.

[4] S.M. Mortazavi, et al., Revision total knee arthroplasty infection: incidence and predictors, Clin. Orthop. Relat. Res. 468 (8) (2010) 2052–2059.

[5] A. Trampuz, W. Zimmerli, Diagnosis and treatment of infections associated with fracture-fixation devices, Injury 37 (Suppl. 2) (2006) S59–S66.

[6] M. Hellmann, et al., The estimated magnitude and direct hospital costs of prosthetic joint infections in the United States, 1997 to 2004, J. Arthroplasty 25 (5) (2010). 766–71.e1.

[7] A.G. Gristina, Biomaterial-centered infection: microbial adhesion versus tissue integration, Science 237 (4822) (1987) 1588–1595.

[8] W. Zimmerli, P. Sendi, Pathogenesis of implant-associated infection: the role of the host, Semin. Immunopathol. 33 (3) (2011) 295–306.

[9] W. Zimmerli, A. Trampuz, P.E. Ochsner, Prosthetic-joint infections, N. Engl. J. Med. 351 (16) (2004) 1645–1654.

[10] W. Zimmerli, Clinical presentation and treatment of orthopaedic implant-associated infection, J. Intern. Med. 276 (2) (2014) 111–119.

[11] D.P. Lew, F.A. Waldvogel, Osteomyelitis, Lancet 364 (9431) (2004) 369–379.

[12] J.W. Costerton, et al., Microbial biofilms, Annu. Rev. Microbiol. 49 (1995) 711–745.

[13] P.S. Stewart, Mechanisms of antibiotic resistance in bacterial biofilms, Int. J. Med. Microbiol. 292 (2) (2002) 107–113.

[14] J.W. Costerton, P.S. Stewart, E.P. Greenberg, Bacterial biofilms: a common cause of persistent infections, Science 284 (5418) (1999) 1318–1322.

[15] B. Spellberg, B.A. Lipsky, Systemic antibiotic therapy for chronic osteomyelitis in adults, Clin. Infect. Dis. 54 (3) (2012) 393–407.

[16] O. El-Helou, et al., Functional assessment of Toll-like receptor 2 and its relevance in patients with *Staphylococcus aureus* infection of joint prosthesis, Hum. Immunol. 72 (1) (2011) 47–53.

[17] W. Zimmerli, et al., Role of rifampin for treatment of orthopedic implant-related staphylococcal infections: a randomized controlled trial. Foreign-Body Infection (FBI) Study Group, JAMA 279 (19) (1998) 1537–1541.

[18] P.H. Hsieh, et al., Gram-negative prosthetic joint infections: risk factors and outcome of treatment, Clin. Infect. Dis. 49 (7) (2009) 1036–1043.

[19] A.F. Widmer, et al., Killing of nongrowing and adherent Escherichia coli determines drug efficacy in device-related infections, Antimicrob. Agents Chemother. 35 (4) (1991) 741–746.

[20] C.M. Brandt, et al., *Staphylococcus aureus* prosthetic joint infection treated with debridement and prosthesis retention, Clin. Infect. Dis. 24 (5) (1997) 914–919.

[21] J.J. Eckardt, P.Z. Wirganowicz, T. Mar, An aggressive surgical approach to the management of chronic osteomyelitis, Clin. Orthop. Relat. Res. 298 (1994) 229–239.

[22] B. Parsons, E. Strauss, Surgical management of chronic osteomyelitis, Am. J. Surg. 188 (Suppl. 1A) (2004) 57–66.

[23] L.O. Conterno, M.D. Turchi, Antibiotics for treating chronic osteomyelitis in adults, Cochrane Database Syst. Rev. 9 (2013). CD004439.

[24] J. Geurts, J.J. Chris Arts, G.H. Walenkamp, Bone graft substitutes in active or suspected infection. Contra-indicated or not? Injury 42 (Suppl. 2) (2011) S82–S86.

[25] C.G. Zalavras, M.J. Patzakis, P. Holtom, Local antibiotic therapy in the treatment of open fractures and osteomyelitis, Clin. Orthop. Relat. Res. 427 (2004) 86–93.

[26] O.S. Kluin, et al., Biodegradable vs non-biodegradable antibiotic delivery devices in the treatment of osteomyelitis, Expert Opin. Drug Delivery 10 (3) (2013) 341–351.

[27] D. Neut, et al., Residual gentamicin-release from antibiotic-loaded polymethylmethacrylate beads after 5 years of implantation, Biomaterials 24 (10) (2003) 1829–1831.

[28] J. Siepmann, A. Gopferich, Mathematical modeling of bioerodible, polymeric drug delivery systems, Adv. Drug Deliv. Rev. 48 (2–3) (2001) 229–247.
[29] K.W. Leong, B.C. Brott, R. Langer, Bioerodible polyanhydrides as drug-carrier matrices. I: characterization, degradation, and release characteristics, J. Biomed. Mater. Res. 19 (8) (1985) 941–955.
[30] H.W. Buchholz, H. Engelbrecht, Depot effects of various antibiotics mixed with Palacos resins, Chirurg 41 (11) (1970) 511–515.
[31] G.H. Walenkamp, T.B. Vree, T.J. van Rens, Gentamicin-PMMA beads. Pharmacokinetic and nephrotoxicological study, Clin. Orthop. Relat. Res. 205 (1986) 171–183.
[32] M.R. Jacobs, Optimisation of antimicrobial therapy using pharmacokinetic and pharmacodynamic parameters, Clin. Microbiol. Infect. 7 (11) (2001) 589–596.
[33] P.D. Holtom, et al., Inhibitory effects of the quinolone antibiotics trovafloxacin, ciprofloxacin, and levofloxacin on osteoblastic cells in vitro, J. Orthop. Res. 18 (5) (2000) 721–727.
[34] P.M. Huddleston, et al., Ciprofloxacin inhibition of experimental fracture healing, J. Bone Joint Surg. Am. 82 (2) (2000) 161–173.
[35] W.R. Moore, S.E. Graves, G.I. Bain, Synthetic bone graft substitutes, ANZ J. Surg. 71 (6) (2001) 354–361.
[36] D. Zhang, et al., Antibacterial effects and dissolution behavior of six bioactive glasses, J. Biomed. Mater. Res., Part A 93 (2) (2010) 475–483.
[37] C. Delloye, et al., Bone allografts: what they can offer and what they cannot, J. Bone Joint Surg. (Br.) 89 (5) (2007) 574–579.
[38] J.A. Goulet, et al., Autogenous iliac crest bone graft. Complications and functional assessment, Clin. Orthop. Relat. Res. 339 (1997) 76–81.
[39] T. Albrektsson, C. Johansson, Osteoinduction, osteoconduction and osseointegration, Eur. Spine J. 10 (Suppl. 2) (2001) S96–S101.
[40] K. Garvin, C. Feschuk, Polylactide-polyglycolide antibiotic implants, Clin. Orthop. Relat. Res. 437 (2005) 105–110.
[41] A.C. McLaren, Alternative materials to acrylic bone cement for delivery of depot antibiotics in orthopaedic infections, Clin. Orthop. Relat. Res. 427 (2004) 101–106.
[42] M. El-Husseiny, et al., Biodegradable antibiotic delivery systems, J. Bone Joint Surg. (Br.) 93 (2) (2011) 151–157.
[43] H. Uludag, et al., rhBMP-collagen sponges as osteoinductive devices: effects of in vitro sponge characteristics and protein pI on in vivo rhBMP pharmacokinetics, Ann. N. Y. Acad. Sci. 875 (1999) 369–378.
[44] T.S. Sorensen, A.I. Sorensen, S. Merser, Rapid release of gentamicin from collagen sponge. In vitro comparison with plastic beads, Acta Orthop. Scand. 61 (4) (1990) 353–356.
[45] H. van de Belt, et al., Surface roughness, porosity and wettability of gentamicin-loaded bone cements and their antibiotic release, Biomaterials 21 (19) (2000) 1981–1987.
[46] J.S. Humphrey, et al., Pharmacokinetics of a degradable drug delivery system in bone, Clin. Orthop. Relat. Res. 349 (1998) 218–224.
[47] H.J. Rutten, P.H. Nijhuis, Prevention of wound infection in elective colorectal surgery by local application of a gentamicin-containing collagen sponge, Eur. J. Surg. Suppl. 578 (1997) 31–35.
[48] A. Stemberger, et al., Local treatment of bone and soft tissue infections with the collagen-gentamicin sponge, Eur. J. Surg. Suppl. 578 (1997) 17–26.
[49] K.P. Rao, Recent developments of collagen-based materials for medical applications and drug delivery systems, J. Biomater. Sci. Polym. Ed. 7 (7) (1995) 623–645.

[50] A.H. Reddi, Implant-stimulated interface reactions during collagenous bone matrix-induced bone formation, J. Biomed. Mater. Res. 19 (3) (1985) 233–239.

[51] M.J. Beuerlein, M.D. McKee, Calcium sulfates: what is the evidence? J. Orthop. Trauma 24 (Suppl. 1) (2010) S46–S51.

[52] M. Sidqui, et al., Osteoblast adherence and resorption activity of isolated osteoclasts on calcium sulphate hemihydrate, Biomaterials 16 (17) (1995) 1327–1332.

[53] C.M. Kelly, et al., The use of a surgical grade calcium sulfate as a bone graft substitute: results of a multicenter trial, Clin. Orthop. Relat. Res. 382 (2001) 42–50.

[54] T.M. Turner, et al., Radiographic and histologic assessment of calcium sulfate in experimental animal models and clinical use as a resorbable bone-graft substitute, a bone-graft expander, and a method for local antibiotic delivery. One institution's experience, J. Bone Joint Surg. Am. A 83 (Suppl. 2, Pt. 1) (2001) 8–18.

[55] W.R. Walsh, et al., Response of a calcium sulfate bone graft substitute in a confined cancellous defect, Clin. Orthop. Relat. Res. 406 (2003) 228–236.

[56] S.S. Aiken, et al., Local release of antibiotics for surgical site infection management using high-purity calcium sulfate: an in vitro elution study, Surg Infect (Larchmt) 16 (1) (2015) 54–61.

[57] A.C. McLaren, et al., The effect of sampling method on the elution of tobramycin from calcium sulfate, Clin. Orthop. Relat. Res. 403 (2002) 54–57.

[58] T.A. Wichelhaus, et al., Elution characteristics of vancomycin, teicoplanin, gentamicin and clindamycin from calcium sulphate beads, J. Antimicrob. Chemother. 48 (1) (2001) 117–119.

[59] R. Roberts, S.J. McConoughey, J.H. Calhoun, Size and composition of synthetic calcium sulfate beads influence dissolution and elution rates in vitro, J. Biomed. Mater. Res. B Appl. Biomater. 102 (4) (2014) 667–673.

[60] M.A. Rauschmann, et al., Nanocrystalline hydroxyapatite and calcium sulphate as biodegradable composite carrier material for local delivery of antibiotics in bone infections, Biomaterials 26 (15) (2005) 2677–2684.

[61] T.J. Blokhuis, et al., Properties of calcium phosphate ceramics in relation to their in vivo behavior, J. Trauma 48 (1) (2000) 179–186.

[62] R.Z. LeGeros, Biodegradation and bioresorption of calcium phosphate ceramics, Clin. Mater. 14 (1) (1993) 65–88.

[63] P. Ducheyne, Q. Qiu, Bioactive ceramics: the effect of surface reactivity on bone formation and bone cell function, Biomaterials 20 (23–24) (1999) 2287–2303.

[64] M.N. Rahaman, B.S. Bal, W. Huang, Review: emerging developments in the use of bioactive glasses for treating infected prosthetic joints, Mater. Sci. Eng., C 41 (2014) 224–231.

[65] S. Larsson, Calcium phosphates: what is the evidence? J. Orthop. Trauma 24 (Suppl. 1) (2010) S41–S45.

[66] M. Bohner, et al., Gentamicin-loaded hydraulic calcium phosphate bone cement as antibiotic delivery system, J. Pharm. Sci. 86 (5) (1997) 565–572.

[67] P.V. Giannoudis, H. Dinopoulos, E. Tsiridis, Bone substitutes: an update, Injury 36 (Suppl. 3) (2005) S20–S27.

[68] E. Garcia-Gareta, M.J. Coathup, G.W. Blunn, Osteoinduction of bone grafting materials for bone repair and regeneration, Bone 81 (2015) 112–121.

[69] V.V. Valimaki, H.T. Aro, Molecular basis for action of bioactive glasses as bone graft substitute, Scand. J. Surg. 95 (2) (2006) 95–102.

[70] J.R. Jones, Review of bioactive glass: from Hench to hybrids, Acta Biomater. 9 (1) (2013) 4457–4486.

[71] N.A. van Gestel, et al., Clinical applications of S53P4 bioactive glass in bone healing and osteomyelitic treatment: a literature review, Biomed. Res. Int. 2015 (2015) 684826.
[72] L.L. Hench, J.M. Polak, Third-generation biomedical materials, Science 295 (5557) (2002) 1014–1017.
[73] L.L. Hench, I. Thompson, Twenty-first century challenges for biomaterials, J. R. Soc. Interface 7 (Suppl. 4) (2010) S379–S391.
[74] Q. Fu, et al., Silicate, borosilicate, and borate bioactive glass scaffolds with controllable degradation rate for bone tissue engineering applications. II. In vitro and in vivo biological evaluation, J. Biomed. Mater. Res., Part A 95 (1) (2010) 172–179.
[75] O. Lepparanta, et al., Antibacterial effect of bioactive glasses on clinically important anaerobic bacteria in vitro, J. Mater. Sci. Mater. Med. 19 (2) (2008) 547–551.
[76] D.C. Coraca-Huber, et al., Efficacy of antibacterial bioactive glass S53P4 against S. aureus biofilms grown on titanium discs in vitro, J. Orthop. Res. 32 (1) (2014) 175–177.
[77] E. Munukka, et al., Bactericidal effects of bioactive glasses on clinically important aerobic bacteria, J. Mater. Sci. Mater. Med. 19 (1) (2008) 27–32.
[78] D.C. Clupper, L.L. Hench, Bioactive response of Ag-doped tape cast Bioglass 45S5 following heat treatment, J. Mater. Sci. Mater. Med. 12 (10–12) (2001) 917–921.
[79] S.H. Luo, et al., In vitro evaluation of cytotoxicity of silver-containing borate bioactive glass, J. Biomed. Mater. Res. B Appl. Biomater. 95 (2) (2010) 441–448.
[80] Z.R. Domingues, et al., Bioactive glass as a drug delivery system of tetracycline and tetracycline associated with beta-cyclodextrin, Biomaterials 25 (2) (2004) 327–333.
[81] W.T. Jia, et al., Elution characteristics of teicoplanin-loaded biodegradable borate glass/chitosan composite, Int. J. Pharm. 387 (1–2) (2010) 184–186.
[82] Z. Xie, et al., Treatment of osteomyelitis and repair of bone defect by degradable bioactive borate glass releasing vancomycin, J. Control. Release 139 (2) (2009) 118–126.
[83] G.K. Schmidt Hergo, M. Buehler, V. Vecsei, U. Gruessner, Controlled, single-blind, international, multicentre study on the efficacy and tolerability of Gentamicin-Collagen-Fleece Septocoll® in subjects with autologous cancellous bone grafts following chronic osteomyelitis and/or infected non-union, 2002.
[84] R. Ascherl, et al., Local treatment of infection with collagen gentamicin, Aktuelle Probl. Chir. Orthop. 34 (1990) 85–93.
[85] R. Dieckmann, et al., Treatment of acute and chronic osteomyelitis in children, Z. Orthop. Unfall. 146 (3) (2008) 375–380.
[86] J. Feil, et al., Bioresorbable collagen-gentamicin compound as local antibiotic therapy, Aktuelle Probl. Chir. Orthop. 34 (1990) 94–103.
[87] T. Ipsen, et al., Gentamicin-collagen sponge for local applications. 10 cases of chronic osteomyelitis followed for 1 year, Acta Orthop. Scand. 62 (6) (1991) 592–594.
[88] O. Kwasny, G. Bockhorn, V. Vecsei, The use of gentamicin collagen floss in the treatment of infections in trauma surgery, Orthopedics 17 (5) (1994) 421–425.
[89] R. Letsch, E. Rosenthal, T. Joka, Local antibiotic administration in osteomyelitis treatment—a comparative study with two different carrier substances, Aktuelle Traumatol. 23 (7) (1993) 324–329.
[90] E. Wernet, et al., Antibiotic-containing collagen sponge in therapy of osteitis, Unfallchirurg 95 (5) (1992) 259–264.
[91] M. Walz, Use of a collagen-gentamicin compound in open spongiosaplasty in chronic post-traumatic osteomyelitis, Aktuelle Probl. Chir. Orthop. 34 (1990) 104–106.
[92] W.K. Chang, et al., Gentamicin-collagen implants to reduce surgical site infection: systematic review and meta-analysis of randomized trials, Ann. Surg. 258 (1) (2013) 59–65.

[93] Z. Ruszczak, W. Friess, Collagen as a carrier for on-site delivery of antibacterial drugs, Adv. Drug Deliv. Rev. 55 (12) (2003) 1679–1698.
[94] N. Fleiter, et al., Clinical use and safety of a novel gentamicin-releasing resorbable bone graft substitute in the treatment of osteomyelitis/osteitis, Bone Joint Res. 3 (7) (2014) 223–229.
[95] M. Franceschini, et al., Treatment of a chronic recurrent fistulized tibial osteomyelitis: administration of a novel antibiotic-loaded bone substitute combined with a pedicular muscle flap sealing, Eur. J. Orthop. Surg. Traumatol. 22 (Suppl. 1) (2012) 245–249.
[96] W. Chang, et al., Adult osteomyelitis: debridement versus debridement plus Osteoset T pellets, Acta Orthop. Belg. 73 (2) (2007) 238–243.
[97] J.Y. Ferguson, et al., The use of a biodegradable antibiotic-loaded calcium sulphate carrier containing tobramycin for the treatment of chronic osteomyelitis: a series of 195 cases, Bone Joint J. 96-B (6) (2014) 829–836.
[98] S. Gitelis, G.T. Brebach, The treatment of chronic osteomyelitis with a biodegradable antibiotic-impregnated implant, J. Orthop. Surg. (Hong Kong) 10 (1) (2002) 53–60.
[99] G. Humm, et al., Adjuvant treatment of chronic osteomyelitis of the tibia following exogenous trauma using OSTEOSET((R))-T: a review of 21 patients in a regional trauma centre, Strategies Trauma Limb Reconstr. 9 (3) (2014) 157–161.
[100] M.D. McKee, et al., A prospective, randomized clinical trial comparing an antibiotic-impregnated bioabsorbable bone substitute with standard antibiotic-impregnated cement beads in the treatment of chronic osteomyelitis and infected nonunion, J. Orthop. Trauma 24 (8) (2010) 483–490.
[101] Y.H. Tsai, et al., Treatment of infected tibial nonunion with tobramycin-impregnated calcium sulfate: report of two cases, Chang Gung Med. J. 27 (7) (2004) 542–547.
[102] D. von Stechow, D. Scale, M.A. Rauschmann, Minimizing the surgical approach in patients with spondylitis, Clin. Orthop. Relat. Res. 439 (2005) 61–67.
[103] A. Berner, et al., Treatment of lower limb osteomyelitis by a local bone substitute supplemented with antibiotics, Z. Orthop. Unfall. 146 (3) (2008) 371–374.
[104] C.L. Romano, et al., A comparative study of the use of bioactive glass S53P4 and antibiotic-loaded calcium-based bone substitutes in the treatment of chronic osteomyelitis: a retrospective comparative study, Bone Joint J. 96-B (6) (2014) 845–850.
[105] D. von Stechow, M.A. Rauschmann, Effectiveness of combination use of antibiotic-loaded PerOssal with spinal surgery in patients with spondylodiscitis, Eur. Surg. Res. 43 (3) (2009) 298–305.
[106] T.A. van Vugt, J. Geurts, J.J. Arts, Clinical application of antimicrobial bone graft substitute in osteomyelitis treatment: a systematic review of different bone graft substitutes available in clinical treatment of osteomyelitis, Biomed. Res. Int. 2016 (2016) 6984656.
[107] L. Drago, et al., Bioactive glass BAG-S53P4 for the adjunctive treatment of chronic osteomyelitis of the long bones: an in vitro and prospective clinical study, BMC Infect. Dis. 13 (2013) 584.
[108] N.C. Lindfors, et al., Bioactive glass S53P4 as bone graft substitute in treatment of osteomyelitis, Bone 47 (2) (2010) 212–218.
[109] J. McAndrew, et al., Through the looking glass; bioactive glass S53P4 (BonAlive(R)) in the treatment of chronic osteomyelitis, Ir. J. Med. Sci. 182 (3) (2013) 509–511.
[110] G.H. Walenkamp, L.L. Kleijn, M. de Leeuw, Osteomyelitis treated with gentamicin-PMMA beads: 100 patients followed for 1–12 years, Acta Orthop. Scand. 69 (5) (1998) 518–522.

S53P4 bioactive glass

4

D.J. Hulsen, N.A. van Gestel*†, J.A.P. Geurts*, J.J. Arts*,†*
*Maastricht University Medical Centre, Maastricht, The Netherlands, †Eindhoven University of Technology, Eindhoven, The Netherlands

4.1 Introduction

Bioactive glasses (BAGs) are synthetic bone graft substitutes that have been investigated extensively in the past decades. BAGs are bone-bonding, osteoconductive materials, and they are used for various clinical applications. S53P4 is a specific composition of BAG that is the main topic of this chapter, in which its development, working mechanism, effect on angiogenesis, antibacterial effect, and current clinical applications are discussed in detail. Although BAGs can also be manufactured as microspheres, fibers, and porous implants, the focus of this chapter is on BAG in common granular form.

In the early 1970s, Hench investigated phase diagrams for ceramics in order to design the first compositions of bone-bonding glass-ceramics [1]. A selected composition (45S5) was proven to develop a hydroxyapatite (HA) layer in in vitro test solutions (Fig. 4.1). This material was subsequently investigated in vivo and shown to bond strongly to bone, thereby proving the hypothesis. In the following years, the nature of the bone–glass bond and the responses of other tissues to the material were studied extensively. BAGs were proven nontoxic to tissue, and they formed a bond to the host bone that was reported equal to or greater than the strength of the bond between HA and bone [2–6].

Directly after the first findings of Professor Hench, other research centers started with the development of similar materials. In the 1980s, various possible BAGs with a wide range of compositions were designed at Åbo Akademi and the University of Turku, Finland. These different compositions were investigated by Andersson et al. on their in vivo and in vitro behavior, their surface reactions, and their acceptance in bone [7–9].

Andersson developed a mathematical approach to describe bioactivity of glasses in terms of glass composition [10]. Based on the results of these studies, S53P4 was selected as the glass that was both most suitable and most interesting for clinical use. S53P4 BAG has a nominal oxide composition of (wt%/mol%) 23.0/22.7 Na_2O, 20.0/21.8 CaO, 4.0/1.7 P_2O_5, and 53.0/53.9 SiO_2. Clinically used granules of this glass are manufactured by melting the oxides and dissolving silica into the melt. The obtained bulk glass is then crushed, and the particles are sieved to obtain granules of the desired size. S53P4 granules gained European approval for orthopedic use as a bone graft substitute in 2006 and are today manufactured as BonAlive® (BonAlive Biomaterials Ltd., Turku, Finland).

Management of Periprosthetic Joint Infections (PJIs). http://dx.doi.org/10.1016/B978-0-08-100205-6.00004-5

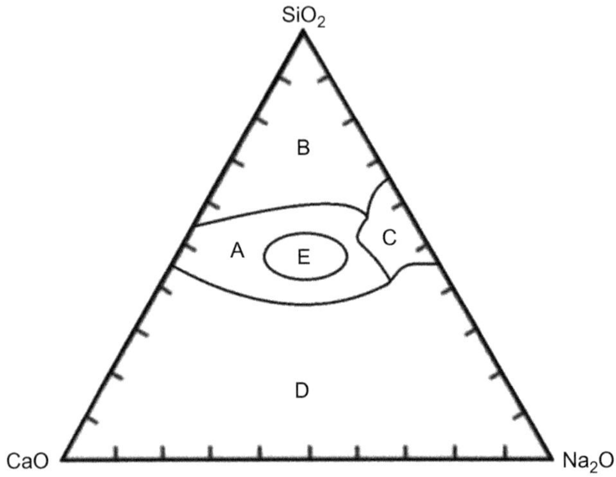

Fig. 4.1 Compositional diagram for bone-bonding glasses. The compositions of both S53P4 and 45S5 are in the E region, in which glasses bond to both bone and soft tissue and are gene activating.
Adapted from L.L. Hench, The story of bioglass, J. Mater. Sci. Mater. Med. 17 (11) (2006) 967–978.

4.2 Working mechanism

S53P4 BAG has some important characteristics that make it an appropriate bone graft substitute. First, the term bioactive means that the material has characteristics that allow it to form a bond with living tissue. This bond is initialized by processes on the surface of BAG when it is exposed to extracellular fluids. Moreover, BAG provides a surface for cells to attach, proliferate, and deposit matrix on and even shows stimulating effects on the recruitment and differentiation of osteoblasts. Bone forms and proliferates in BAG-filled defects, and finally the glass is degradated on the long term.

4.2.1 Surface reactions

As was hypothesized by Hench, the formation of a HA layer has been proven to be essential for BAG to bond to bone. This layer is formed on BAG by surface reactions directly after it is exposed to a biological environment. The reactions that lead to the formation of HA can be subdivided in five stages (Table 4.1) [11–13].

The first reaction that takes place after BAG exposure to a biologically active environment is the exchange of Na^+ ions from the surface of the glass with H^+ or H_3O^+ ions from the solution. This results in an alkaline environment and the formation of silanol (SiOH) near the glass surface, according to the following reaction [3]:

$$SiONa^+ + H^+ \leftrightarrow SiOH^+ + Na^+ \tag{4.1}$$

Table 4.1 Working mechanism of BAG divided in stages, including a timeline and the paragraph in which the stage is discussed

Time	Stage		Discussed in paragraph
Minutes	1	Exchange of ions	4.2.1. Surface reactions
	2	Network dissolution and formation of silanol	
	3	Silica-gel polymerization	
	4	Chemisorption of amorphous Ca, PO_2, CO_2	
	5	Crystallization of HA layer	
Hours	6	Biochemical adsorption of biologic structures	4.2.2. Bond to bone
	7	Inflammation and macrophage action	
	8	Attachment of stem cells	
Days	9	Proliferation and differentiation of stem cells	4.2.3. Osteostimulation
	10	Generation of bone matrix	4.2.2. Bond to bone
	11	Crystallization of bone matrix	
Weeks	12	Proliferation and growth of bone	4.2.4. Bone proliferation

Shortly following Na^+, other cations (silicate, phosphate, calcium) also start to leach, further increasing the solution's alkalinity. This leads to further network breakdown of the glass and additional silanol formation in the solution:

$$SiOSi + H^+ + OH^- \leftrightarrow 2\,SiOH \quad (4.2)$$

During the third stage, the dissolved silanol starts to repolymerize on the glass surface, and a silica-rich gel layer is formed:

$$SiOH + SiOH \leftrightarrow SiOSi + H_2O \quad (4.3)$$

Subsequently, chemisorption of amorphous calcium, phosphate, and carbonate occurs in the growing silica gel layer. Thereafter, the absorbed calcium, phosphate, and carbonate nucleate and crystallize to eventually form a HA layer on the glass surface within several hours. The resulting glass surface is now covered with a 100–120 μm thick SiO_2-rich layer and a 30 μm thick HA layer (Fig. 4.2).

4.2.2 Bond to bone

Following the formation of a HA layer on BAG surface, a BAG-bone bond is established by biological processes that can be subdivided in six stages (stage 6–11 in Table 4.1) [1, 11, 12, 14].

Stage 6, the adsorption of biologic structures on the BAG surface, is a reaction stage that occurs concurrently with the crystallization and growth of the HA layer.

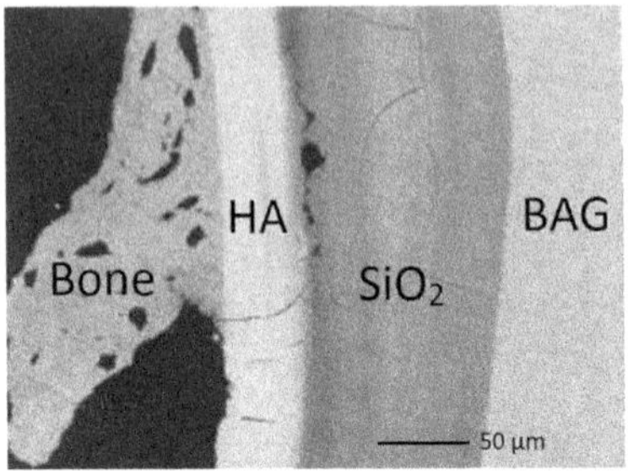

Fig. 4.2 Scanning electron microscopy profile of a BAG-bone interface, 28 days after implantation in a rat model.
Adapted from P. Virolainen, et al., Histomorphometric and molecular biologic comparison of bioactive glass granules and autogenous bone grafts in augmentation of bone defect healing, J. Biomed. Mater. Res. 35 (1) (1997) 9–17.

The structures that become embedded within the growing HA layer include growth factors, collagen, and other proteins. Collagen facilitates further apatite deposition between the adsorbed fibers, leading to an integrated structure. Furthermore, S53P4 BAG has been shown to selectively adsorb proteins, which appears the initial step in establishing biocompatibility and bonding of tissues to the implant surface [15].

The extracellular proteins that are adsorbed during stage 6 attract macrophages, mesenchymal stem cells, and osteoprogenitor cells. First, the attracted macrophages prepare the implant site for tissue repair by phagocytosis of cellular debris, small ceramic particles, and possibly pathogens (stage 7). Subsequently, both surface-mediated and solution-mediated mechanisms stimulate the attachment of stem cells (stage 8). The attachment is shortly followed by differentiation of stem cells (stage 9), and within weeks the generation and crystallization of bone matrix take place (stage 10 and 11). The resulting bond between S53P5 BAG and bone has been verified in vivo and in vitro using scanning electron microscopy, energy dispersive X-ray analysis, and push-out tests [7, 8, 16, 17].

4.2.3 Osteostimulation

When osteogenic cells approach the BAG granules, they attach to the bone-like surface, completely with embedded organic components. Following attachment, the cells should be stimulated to proliferate and differentiate in order to form new bone around the BAG granules (stage 9–11 in Table 4.1). The proosteogenic effect of BAG has been studied intensively, and the results have shown that BAG indeed has stimulating effects on both proliferation and differentiation of osteoblast progenitor cells and bone matrix formation. These effects are believed to separate BAG from mere osteoconductive materials and have led to the claim that the material is osteostimulative (also termed osteoproductive or osteopromotive). This novel classification denotes that BAG enhances bone regeneration in a fashion superior to osteoconductive materials, but that it cannot be considered osteoinductive. The big difference with osteoinductivity is the fact that osteostimulative materials need a bony environment to stimulate bone formation. However, it should be noted that a clear distinction and wide acceptation of the term has not yet been well established in literature.

In definition, a material is considered osteostimulative when it possesses the following three properties:

1. It stimulates the recruitment and differentiation of osteoblasts
   Several in vivo studies have confirmed that BAG indeed increases osteogenic activity of rat osteoblasts [18–20]. Based on histological, biochemical, and morphological parameters, Vrouwenvelder et al. also concluded that BAG leads to a superior osteoblastic response compared to HA, titanium, and stainless steel [21]. Furthermore, it was shown by Waselau et al. that S53P4 BAG stimulates osteogenic differentiation of human adipose stem cells [22].
2. It activates osteoblasts to produce new bone
   In vitro research has shown that, in addition to differentiation, BAG can also support human primary osteoblasts to form mineralized nodules [23–25]. Mineralized nodule formation is an initial process of intramembranous ossification, by which bone grows in a BAG-filled defect [14].
3. Its stimulation is a consequence of ionic dissolution from the material
   It has been found that specific concentrations of soluble ionic constituents released from BAG cause up regulation of certain genes in osteoblasts [26]. This is believed to be the key phenomenon in the effect of BAG on cell proliferation and differentiation of the osteogenic cell lineage [27, 28]. Ojansivu et al. found that S53P4 BAG ionic extracts could stimulate osteogenic differentiation in a 5% human serum medium only if the medium extract was supplemented with osteogenic factors [29].

4.2.4 Bone proliferation in a BAG graft layer

After the formation of a HA layer, the establishment of a bond to bone, and osteostimulation, bone proliferation takes place to fill up a defect and ultimately restore bone anatomy. Rapid bone formation in BAG-filled defects has been described by various authors in animal studies [30–32]. The regenerated defects were described as a large network of growing bone between BAG granules.

Bone formation in a BAG-filled defect differs from bone formation around autograft because of differences in osteoinductivity and osteogenicity. Bone growth in autograft layers is homogeneous, but it grows in a centripetal pattern in BAG graft layers [14, 33]. From the defect surface, bone grows toward the center from granule to granule [17, 33]. This fashion of bone proliferation is similar to what is observed in defects filled with other synthetic graft materials.

Although there are studies that have tried to compare the filler effect of BAG to that of allograft and autograft, their results remain inconclusive. In a rat model by Virolainen et al., it was confirmed that BAG is osteostimulative, but that bone growth proceeds faster in autograft-filled defects [34]. Turunen et al. observed that in human maxillary sinus floor augmentation, mixture of S53P4 granules and autograft showed more rapid bone repair with thicker trabeculae compared to autograft alone [35]. In a later study, Conejero et al. found that a mixture of 60% BAG and 40% autograft led to superior bone formation compared to an 80% BAG and 20% autograft mixture in a rabbit cranial defect model [36]. More recently, Gunn et al. did not find a difference in filler effect between BAG and allograft in rabbits [37]. The filler effect of BAG has also been compared with that of other synthetic graft materials, such as HA and calcium carbonate. These animal studies have concluded that BAG has a stronger or similar filling effect as HA, and that BAG outperforms calcium carbonate, which is resorbed too quickly [16, 37–40]. In summary, literature shows that the filler effect of BAG is at least similar to that of other synthetic alternatives, although clinical studies are needed to confirm this.

4.2.5 BAG degradation

Although the mineral phase of healthy bone can be degraded by osteoclastic activity to allow for bone remodeling, it has been shown that this is not the case for BAG, and that the surface of BAG is resistant to cell-mediated degradation [41]. Furthermore, physicochemical degradation is limited to the outer layer of a BAG granule, where dissolution and reprecipitation take place. Therefore, mechanical degradation is the initial phase in the degradation of solid granules. Stresses on the brittle material result in cracking, exposing more BAG surface to physiological fluids and subsequently leading to more physicochemical degradation (Fig. 4.3). These processes are relatively slow, and therefore degradation of BAG granules in vivo can take years. An ideal bone graft substitute would be degraded as fast as new bone can be formed in its place, leading to creeping substitution. However, the degradation rate of BAG in bone is low, and clinical results show that BAG granules are still present in the body years after grafting [42]. Although this is considered a drawback for a graft material, it has been shown that nondegraded cores of granules are well incorporated in the bone. These particles seem to be stable, do not migrate, and do not inhibit bone growth [33, 43, 44]. Moreover, in richly vascularized tissues, BAG granules have been shown to turn completely to silica gel, which is slowly dissolved [45]. Therefore, BAG granules that migrate out of the bone and into soft tissues shortly after implantation are no major

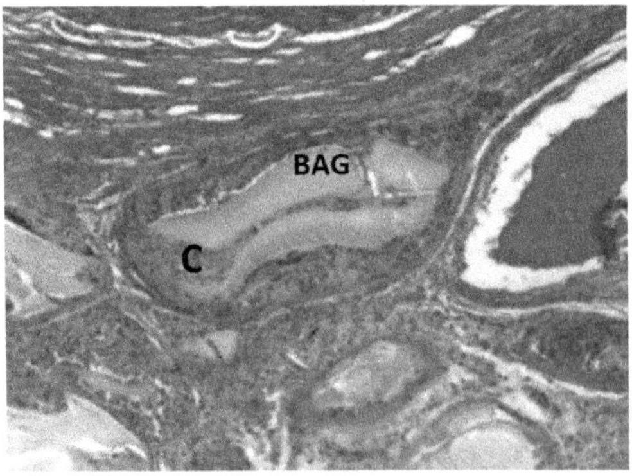

Fig. 4.3 Histological image of cells (C) inside a crack in a 45S5 BAG granule. Cells were observed to migrate in the crack.
Adapted from J.A. Conejero, J.A. Lee, J.A. Ascherman, Cranial defect reconstruction in an experimental model using different mixtures of bioglass and autologous bone, J. Craniofac. Surg. 18 (6) (2007) 1290–1295.

threat to the soft tissue's health. Most important, BAG granules are clearly not degraded too fast, and hence they do not leave empty space in its wake.

4.3 Antibacterial effects

Bacterial infections have always been a major problem in orthopedics and trauma surgery. The incidence of infections after total joint revision, for example, ranges from 5% to 20% [46]. Traumatologic interventions, which often involve open fracture treatment, show even higher numbers [47, 48]. In spite of advances in both antibiotic therapies and operative techniques, ever-increasing number of interventions and an increase in antibiotic resistant bacteria make that absolute numbers of infections are still rising [49]. The current gold standard in local treatment of bone infections is placement of PMMA beads loaded with antibiotics. This approach has the drawback that it requires several distinct operations to replace and/or remove the PMMA beads. Furthermore, because infection is a common cause for graft layer failure, infection is a contraindication for bone grafting. A unique property of BAG, and arguably the most important advantage over other graft materials, is its ability to fight bacterial infections in vivo. This means that the material can concurrently serve as both local antibiotic-like treatment and a permanent graft layer.

Extensive in vitro research has been carried out on the antibacterial properties of BAG. It was shown that powdered BAGs have a strong antibacterial effect on a large panel of clinically important bacteria [50–53]. S53P4 was shown to possess the fastest bacterial growth inhibitory effect of various studied BAGs [50, 51]. The effects of S53P4 BAG on microorganisms have also been investigated in detail by Stoor et al., who focused on craniomaxillofacial and otolaryngologic infection-associated microorganisms [54–56]. This group observed favorable in vitro results, and no BAG associated infections in patients that were treated with BAG in otolaryngologic surgery.

The antimicrobial effects of BAG depend primarily on the pH increase in the solution that is caused by its surface reactions [57]. It is believed that the combination of a high pH and an increased osmotic pressure is to what bacteria are most sensible. Scanning electron microscopy analysis of bacteria showed cell shrinkage and membrane damage after exposure to BAG [58]. A recent study has also shown the antibacterial (antibiofilm) effects of S53P4 BAG on antibiotic-resistant bacteria [59].

The amount of BAG that dissolves depends on granule size (surface/volume ratio), which explains why powdered BAG has such a strong bactericidal effect [60]. However, powdered glass is of little use as bone graft material as the material dissolves completely [61].

4.4 Effect on angiogenesis

Angiogenesis—the formation and proliferation of new blood vessels—plays a crucial role in bone regeneration. Newly formed bone can only survive when it is properly vascularized, as cells that are located too far away from blood vessels are destined to die. Because of this, angiogenic potential is an important property of a synthetic bone graft material. The ability of BAG to stimulate angiogenesis has therefore become an important research topic, and there is growing evidence that BAG actively improves neovascularization [62]. Although the complete mechanism of this process remains unknown, it has been shown in vitro that BAG stimulates angiogenic growth factor (VEGF and bFGF) secretion, increases endothelial cell proliferation, and stimulates endothelial tubule formation and anastomosis [62, 63]. These findings are supported by in vivo experiments by various researchers, although Gorustovich et al. emphasize that further dedicated research is needed [63]. A more recent study performed at the university of Erlangen showed that VEGF expression was marginally inhibited by granules of a size over 2 mm, which is a clinically relevant size [64].

4.5 Current clinical applications

The main advantages of BAG over allograft and autograft are its unlimited availability, not necessitating a second incision, and standardized mechanical properties and biological regenerative potential. Compared to other synthetic materials, it has the advantage of being osteostimulative, stimulating angiogenesis, and protecting itself and surrounding tissue to bacterial infections. Because of these advantageous properties, BAG has been used in clinical practice for decades, albeit to a lesser extent than ceramic materials such as tricalciumphosphate or HA.

In current clinical practice, S53P4 BAG is being used for various clinical purposes [65]. A large amount of clinical publications about S53P4 BAG are in the field of craniomaxillofacial and otolaryngologic surgery [35, 66–71]. It has been used successfully in grafting procedures in the skull, and in treatment of infections in those regions. In a retrospective study on 843 cases, of which 133 were treated with S53P4 BAG, Schimanski concluded that the approach using BAG led to the least abnormal postoperative findings, as compared to other materials [72].

Recently, long-term results of S53P4 BAG granules in spine surgery have been published [73, 74]. The authors concluded that S53P4 BAG could safely be used in spondylodesis although a mixture of BAG and allograft could have been even more useful in their opinion.

Another clinical application of BAG granules that is still subject to intensive research is osteomyelitis treatment. Lindfors et al. showed in 2010 that S53P4 BAG could be considered as a usable bone graft substitute in one-stage osteomyelitis treatment. Preliminary data from McAndrew et al. showed that the material gave excellent results in patients that had underwent multiple previous debridements and antimicrobial regimens to no avail [44, 75]. A retrospective study by Romano and a Prospective studies by Drago have shown equivalent results that reinforce previous findings on the efficacy of S53P4 in the treatment of chronic osteomyelitis [76, 77].

Lindfors et al. have reported successful usage of S53P4 BAG granules in the treatment of benign bone tumors and aneurysmal bone cysts [33, 42, 43]. Last, S53P4 BAG granules have been reported to lead to similar results as autograft bone in the operative treatment of depressed tibial plateau fracture [78, 79].

References

[1] L.L. Hench, The story of bioglass, J. Mater. Sci. Mater. Med. 17 (11) (2006) 967–978.
[2] T. Fujiu, M. Ogino, Difference of bond bonding behavior among surface active glasses and sintered apatite, J. Biomed. Mater. Res. 18 (7) (1984) 845–859.
[3] L.L. Hench, Bioactive ceramics, Ann. N. Y. Acad. Sci. 523 (1988) 54–71.
[4] J. Wilson, et al., Toxicology and biocompatibility of bioglasses, J. Biomed. Mater. Res. 15 (6) (1981) 805–817.
[5] L.L. Hench, et al., Analysis of bioglass fixation of hip prostheses, J. Biomed. Mater. Res. 11 (2) (1977) 267–282.
[6] G. Piotrowski, et al., Mechanical studies of the bone bioglass interfacial bond, J. Biomed. Mater. Res. 9 (4) (1975) 47–61.
[7] Ö.H. Andersson, Calcium-phosphate formation at the surface of bioactive glass in vivo, J. Non-Cryst. Solids 119 (3) (1990) 290–296.
[8] Ö.H. Andersson, I. Kangasniemi, Calcium phosphate formation at the surface of bioactive glass in vitro, J. Biomed. Mater. Res. 25 (8) (1991) 1019–1030.
[9] Ö.H. Andersson, et al., Evaluation of the acceptance of glass in bone, J. Mater. Sci. Mater. Med. 3 (2) (1992) 145–150.
[10] Ö.H. Andersson, et al., In vivo behaviour of glasses in the SiO2-Na2O-CaO-P2O5-Al2O3-B2O3 system, J. Mater. Sci. Mater. Med. 1 (4) (1990) 219–227.
[11] V.V. Valimaki, H.T. Aro, Molecular basis for action of bioactive glasses as bone graft substitute, Scand. J. Surg. 95 (2) (2006) 95–102.
[12] J. Will, L.C. Gerhardt, A.R. Boccaccini, Bioactive glass-based scaffolds for bone tissue engineering, Adv. Biochem. Eng. Biotechnol. 126 (2012) 195–226.
[13] L. Hupa, Melt-derived bioactive glasses, in: H.O. Ylanen (Ed.), Bioactive Glasses, Woodhead Publishing Limited, Cambridge, GBR, 2011, pp. 3–28.
[14] J.T. Heikkila, Use of bioactive glasses as bone substitutes in orthopaedics and traumatology, in: H.O. Ylanen (Ed.), Bioactive Glasses, Woodhead Publishing Limited, Cambridge, GBR, 2011, pp. 189–208.
[15] E. Soderling, et al., Protein adsorption to a bioactive glass with special reference to precorrosion, J. Biomed. Mater. Res. 31 (4) (1996) 525–531.
[16] J.T. Heikkila, et al., Bioactive glass versus hydroxylapatite in reconstruction of osteochondral defects in the rabbit, Acta Orthop. Scand. 64 (6) (1993) 678–682.
[17] J.T. Heikkila, et al., Bone formation in rabbit cancellous bone defects filled with bioactive glass granules, Acta Orthop. Scand. 66 (5) (1995) 463–467.
[18] C. Loty, et al., Bioactive glass stimulates in vitro osteoblast differentiation and creates a favorable template for bone tissue formation, J. Bone Miner. Res. 16 (2) (2001) 231–239.
[19] T. Matsuda, J.E. Davies, The in vitro response of osteoblasts to bioactive glass, Biomaterials 8 (4) (1987) 275–284.
[20] W.C. Vrouwenvelder, C.G. Groot, K. de Groot, Behaviour of fetal rat osteoblasts cultured in vitro on bioactive glass and nonreactive glasses, Biomaterials 13 (6) (1992) 382–392.

[21] W.C. Vrouwenvelder, C.G. Groot, K. de Groot, Histological and biochemical evaluation of osteoblasts cultured on bioactive glass, hydroxylapatite, titanium alloy, and stainless steel, J. Biomed. Mater. Res. 27 (4) (1993) 465–475.
[22] M. Waselau, et al., Effects of bioactive glass S53P4 or beta-tricalcium phosphate and bone morphogenetic protein-2 and bone morphogenetic protein-7 on osteogenic differentiation of human adipose stem cells, J. Tissue Eng. 3 (1) (2012). 2041731412467789.
[23] J.E. Gough, J.R. Jones, L.L. Hench, Nodule formation and mineralisation of human primary osteoblasts cultured on a porous bioactive glass scaffold, Biomaterials 25 (11) (2004) 2039–2046.
[24] J.E. Gough, I. Notingher, L.L. Hench, Osteoblast attachment and mineralized nodule formation on rough and smooth 45S5 bioactive glass monoliths, J. Biomed. Mater. Res. A 68 (4) (2004) 640–650.
[25] Z.H. Zhou, et al., Mineralization and osteoblast response to bioactive glass in vitro, J. Med. Eng. Technol. 34 (4) (2010) 285–290.
[26] I.D. Xynos, et al., Gene-expression profiling of human osteoblasts following treatment with the ionic products of Bioglass 45S5 dissolution, J. Biomed. Mater. Res. 55 (2) (2001) 151–157.
[27] I.D. Xynos, et al., Ionic products of bioactive glass dissolution increase proliferation of human osteoblasts and induce insulin-like growth factor II mRNA expression and protein synthesis, Biochem. Biophys. Res. Commun. 276 (2) (2000) 461–465.
[28] I.D. Xynos, et al., Bioglass 45S5 stimulates osteoblast turnover and enhances bone formation in vitro: implications and applications for bone tissue engineering, Calcif. Tissue Int. 67 (4) (2000) 321–329.
[29] M. Ojansivu, et al., Bioactive glass ions as strong enhancers of osteogenic differentiation in human adipose stem cells, Acta Biomater. 21 (2015) 190–203.
[30] N.C. Lindfors, A.J. Aho, Tissue response to bioactive glass and autogenous bone in the rabbit spine, Eur. Spine J. 9 (1) (2000) 30–35.
[31] N.C. Lindfors, A.J. Aho, Granule size and composition of bioactive glasses affect osteoconduction in rabbit, J. Mater. Sci. Mater. Med. 14 (4) (2003) 365–372.
[32] T. Turunen, et al., Effect of bioactive glass granules and polytetrafluoroethylene membrane on repair of cortical bone defect, J. Mater. Sci. Mater. Med. 6 (11) (1995) 639–641.
[33] N.C. Lindfors, et al., Bioactive glass and autogenous bone as bone graft substitutes in benign bone tumors, J. Biomed. Mater. Res. B Appl. Biomater. 90 (1) (2009) 131–136.
[34] P. Virolainen, et al., Histomorphometric and molecular biologic comparison of bioactive glass granules and autogenous bone grafts in augmentation of bone defect healing, J. Biomed. Mater. Res. 35 (1) (1997) 9–17.
[35] T. Turunen, et al., Bioactive glass granules as a bone adjunctive material in maxillary sinus floor augmentation, Clin. Oral Implants Res. 15 (2) (2004) 135–141.
[36] J.A. Conejero, J.A. Lee, J.A. Ascherman, Cranial defect reconstruction in an experimental model using different mixtures of bioglass and autologous bone, J. Craniofac Surg. 18 (6) (2007) 1290–1295.
[37] J.M. Gunn, et al., Comparison of the osteoconductive properties of three particulate bone fillers in a rabbit model: allograft, calcium carbonate (Biocoral(R)) and S53P4 bioactive glass, Acta Odontol. Scand. 71 (5) (2013) 1238–1242.
[38] M.J. Peltola, et al., In vivo model for frontal sinus and calvarial bone defect obliteration with bioactive glass S53P4 and hydroxyapatite, J. Biomed. Mater. Res. 58 (3) (2001) 261–269.
[39] E. Schepers, et al., Bioactive glass particulate material as a filler for bone lesions, J. Oral Rehabil. 18 (5) (1991) 439–452.

[40] T. Turunen, et al., Bioactive glass and calcium carbonate granules as filler material around titanium and bioactive glass implants in the medullar space of the rabbit tibia, Clin. Oral Implants Res. 8 (2) (1997) 96–102.
[41] T. Wilson, et al., Intact surface of bioactive glass S53P4 is resistant to osteoclastic activity, J. Biomed. Mater. Res. A 77 (1) (2006) 67–74.
[42] N.C. Lindfors, et al., A prospective randomized 14-year follow-up study of bioactive glass and autogenous bone as bone graft substitutes in benign bone tumors, J. Biomed. Mater. Res. B Appl. Biomater. 94 (1) (2010) 157–164.
[43] N.C. Lindfors, Treatment of a recurrent aneurysmal bone cyst with bioactive glass in a child allows for good bone remodelling and growth, Bone 45 (2) (2009) 398–400.
[44] N.C. Lindfors, et al., Bioactive glass S53P4 as bone graft substitute in treatment of osteomyelitis, Bone 47 (2) (2010) 212–218.
[45] R. Penttinen, Cell interaction with bioactive glasses and ceramics, in: H.O. Ylanen (Ed.), Bioactive Glasses, Woodhead Publishing Limited, Cambridge, GBR, 2011, pp. 53–84.
[46] S.M. Mortazavi, et al., Revision total knee arthroplasty infection: incidence and predictors, Clin. Orthop. Relat. Res. 468 (8) (2010) 2052–2059.
[47] G.I. Im, S.K. Tae, Distal metaphyseal fractures of tibia: a prospective randomized trial of closed reduction and intramedullary nail versus open reduction and plate and screws fixation, J. Trauma 59 (5) (2005) 1219–1223. discussion 1223.
[48] M. Ronga, U.G. Longo, N. Maffulli, Minimally invasive locked plating of distal tibia fractures is safe and effective, Clin. Orthop. Relat. Res. 468 (4) (2010) 975–982.
[49] J. Geurts, J.J. Chris Arts, G.H. Walenkamp, Bone graft substitutes in active or suspected infection. Contra-indicated or not? Injury 42 (Suppl. 2) (2011) S82–S86.
[50] O. Lepparanta, et al., Antibacterial effect of bioactive glasses on clinically important anaerobic bacteria in vitro, J. Mater. Sci. Mater. Med. 19 (2) (2008) 547–551.
[51] E. Munukka, et al., Bactericidal effects of bioactive glasses on clinically important aerobic bacteria, J. Mater. Sci. Mater. Med. 19 (1) (2008) 27–32.
[52] D. Zhang, et al., Antibacterial effects and dissolution behavior of six bioactive glasses, J. Biomed. Mater. Res. A 93 (2) (2010) 475–483.
[53] D.C. Coraca-Huber, et al., Efficacy of antibacterial bioactive glass S53P4 against *S. aureus* biofilms grown on titanium discs in vitro, J. Orthop. Res. 32 (1) (2014) 175–177.
[54] P. Stoor, E. Soderling, J.I. Salonen, Antibacterial effects of a bioactive glass paste on oral microorganisms, Acta Odontol. Scand. 56 (3) (1998) 161–165.
[55] P. Stoor, E. Soderling, R. Grenman, Bioactive glass S53P4 in repair of septal perforations and its interactions with the respiratory infection-associated microorganisms *Haemophilus influenzae* and *Streptococcus pneumoniae*, J. Biomed. Mater. Res. 58 (1) (2001) 113–120.
[56] P. Stoor, E. Soderling, R. Grenman, Interactions between the bioactive glass S53P4 and the atrophic rhinitis-associated microorganism klebsiella ozaenae, J. Biomed. Mater. Res. 48 (6) (1999) 869–874.
[57] D. Zhang, et al., Factors controlling antibacterial properties of bioactive glasses, Key Eng. Mater. 330–332 (2007) 173–176.
[58] L. Drago, et al., Antimicrobial activity and resistance selection of different bioglass S53P4 formulations against multidrug resistant strains, Future Microbiol 10 (2015) 1293–1299.
[59] L. Drago, et al., In vitro antibiofilm activity of bioactive glass S53P4, Future Microbiol. 9 (5) (2014) 593–601.
[60] I. Gergely, et al., The in vitro antibacterial effect of S53P4 bioactive glass and gentamicin impregnated polymethylmethacrylate beads, Acta Microbiol. Immunol. Hung. 61 (2) (2014) 145–160.

[61] D. Zhang, M. Hupa, L. Hupa, In situ pH within particle beds of bioactive glasses, Acta Biomater. 4 (5) (2008) 1498–1505.

[62] A.A. Gorustovich, J.A. Roether, A.R. Boccaccini, Effect of bioactive glasses on angiogenesis: a review of in vitro and in vivo evidences, Tissue Eng. B 16 (2) (2010) 199–207.

[63] R.M. Day, Bioactive glass stimulates the secretion of angiogenic growth factors and angiogenesis in vitro, Tissue Eng. 11 (5–6) (2005) 768–777.

[64] R. Detsch, et al., Increase in VEGF secretion from human fibroblast cells by bioactive glass S53P4 to stimulate angiogenesis in bone, J. Biomed. Mater. Res. A 102 (11) (2014) 4055–4061.

[65] N.A. van Gestel, et al., Clinical applications of S53P4 bioactive glass in bone healing and osteomyelitic treatment: a literature review, Biomed. Res. Int. 2015 (2015) 684826.

[66] K. Aitasalo, et al., Repair of orbital floor fractures with bioactive glass implants, J. Oral Maxillofac. Surg. 59 (12) (2001) 1390–1395. discussion 1395–1396.

[67] M. Peltola, et al., Bioactive glass S53P4 in frontal sinus obliteration: a long-term clinical experience, Head Neck 28 (9) (2006) 834–841.

[68] M.J. Peltola, et al., Long-term microscopic and tissue analytical findings for 2 frontal sinus obliteration materials, J. Oral Maxillofac. Surg. 66 (8) (2008) 1699–1707.

[69] J. Sarin, et al., Bioactive glass S53P4 in mastoid obliteration surgery for chronic otitis media and cerebrospinal fluid leakage, Ann. Otol. Rhinol. Laryngol. 121 (9) (2012) 563–569.

[70] J.T. Silvola, Mastoidectomy cavity obliteration with bioactive glass: a pilot study, Otolaryngol. Head Neck Surg. 147 (1) (2012) 119–126.

[71] P. Stoor, J. Pulkkinen, R. Grenman, Bioactive glass S53P4 in the filling of cavities in the mastoid cell area in surgery for chronic otitis media, Ann. Otol. Rhinol. Laryngol. 119 (6) (2010) 377–382.

[72] G. Schimanski, E. Schimanski, Obliteration of mastoid cavities: 30 years of experience with recommendations for surgical strategy, HNO 63 (8) (2015) 538–545.

[73] J. Frantzen, et al., Instrumented spondylodesis in degenerative spondylolisthesis with bioactive glass and autologous bone: a prospective 11-year follow-up, J. Spinal Disord. Tech. 24 (7) (2011) 455–461.

[74] J. Rantakokko, et al., Posterolateral spondylodesis using bioactive glass S53P4 and autogenous bone in instrumented unstable lumbar spine burst fractures. A prospective 10-year follow-up study, Scand. J. Surg. 101 (1) (2012) 66–71.

[75] J. McAndrew, et al., Through the looking glass; bioactive glass S53P4 (BonAlive(R)) in the treatment of chronic osteomyelitis, Ir. J. Med. Sci. 182 (3) (2013) 509–511.

[76] L. Drago, et al., Bioactive glass BAG-S53P4 for the adjunctive treatment of chronic osteomyelitis of the long bones: an in vitro and prospective clinical study, BMC Infect. Dis. 13 (2013) 584.

[77] C.L. Romano, et al., A comparative study of the use of bioactive glass S53P4 and antibiotic-loaded calcium-based bone substitutes in the treatment of chronic osteomyelitis: a retrospective comparative study, Bone Joint J. 96-B (6) (2014) 845–850.

[78] K. Pernaa, et al., Bioactive glass S53P4 and autograft bone in treatment of depressed tibial plateau fractures—a prospective randomized 11-year follow-up, J. Long-Term Eff. Med. Implants 21 (2) (2011) 139–148.

[79] J.T. Heikkila, et al., Bioactive glass granules: a suitable bone substitute material in the operative treatment of depressed lateral tibial plateau fractures: a prospective, randomized 1 year follow-up study, J. Mater. Sci. Mater. Med. 22 (4) (2011) 1073–1080.

Experimental models in orthopedic infection research

5

*J.C.E. Odekerken**,†, *T.J. Welting**, *G.H.I.M. Walenkamp**, *J.J. Arts**
*Maastricht University Medical Centre, Maastricht, The Netherlands,
†Zuyd University of Applied Sciences, Heerlen, The Netherlands

5.1 Osteomyelitis

Osteomyelitis, the infection of bone tissue, is probably the most severe complications in orthopedic and trauma surgery. The disease itself is probably as old as the existence of vertebrates, especially since its occurrence has been demonstrated in dinosaurs and Neanderthals [1–3]. Also its symptoms, disease progression, and eventual treatment have been described in ancient- (Egyptian, Greek, Roman, and Mayan) and late- renaissance literature (Paracelsus, Paré, and Scultetus) [2,4,5].

However, Longmore made one of the first reports on osteomyelitis in "modern era" literature, where he described it as a complication after gunshot wounds in the American civil war, with the initial focus at the morphological changes of the diseased bone [6]. The possible bacterial cause of osteomyelitis (a micrococcus, later described as *Staphylococcus aureus*) was identified in 1883 by Becker (group of Robert Koch) [7]. They were also the first to culture *S. aureus* and to use it in an experimental osteomyelitis in which they inoculated animals with bone fractures with the cultured bacterium, resulting in pus formation and animal death within 14 days after inoculation [7]. These early studies already indicated the relation between foreign objects in the soft and bony tissue and the development of an osteomyelitis.

5.2 Prostheses, osteosynthesis, and infection

Current medical practice concerning orthopedic and trauma surgery uses prostheses and osteosynthesis on a daily basis to restore joint function or reconstruct bone defects, respectively. There is a broad collection of implantable devices, which all are relatively large foreign objects, being implanted in the patient for either the stabilization and fixation of bones or to maintain the movement of the body. Currently there are prostheses available for the hip, knee, ankle, shoulder, elbow, wrist, finger, and toe. Fixation devices like screws, plates, and wires are mainly used in trauma surgery. Intervertebral body spacers and intramedullary nails are examples of implants for stabilization.

Orthopedic prostheses and fixation devices (including intramedullary nails) are mostly made of titanium, cobalt chrome, or stainless steel alloys, while ultra-high-molecular-weight polyethylene (UHMWPE) or ceramics are used for the articulating surface in order to minimize wear and friction.

Management of Periprosthetic Joint Infections (PJIs). http://dx.doi.org/10.1016/B978-0-08-100205-6.00005-7

The major issue with orthopedic implants (as with many other medical implants like pacemakers, cranial defect plates, stents, dental implants) is the foreign body reaction of the host against the implant (and wear debris in case of prostheses) and the risk for bacterial infection with the accompanying biofilm formation [8–11].

In general, hip and knee and shoulder arthroplasties account for the majority of the prostheses, while trauma surgery uses the majority of the fixation devices [12,13]. Currently about 0.5–1% of the total hip and 1–4% of the total knee arthroplasties eventually result in an implant infection, while the deep infection rate of osteosynthesis can range up to 30% depending on the indication [12–14].

With an aging population and more patients being operated at younger age, an increase in the number of orthopedic implants being implanted per year is to be expected [15–17]. So even if the incidence of implant infections remains unchanged, the prevalence of implant infections will increase [15–17].

Prosthesis infections are mainly initiated by perioperative contamination with bacteria or via the hematogenous route during postoperative bacteraemia, resulting in prosthesis colonization and subsequently infection [13,18]. However, an osteomyelitis can also develop due to trauma or operation without the use of an implant; the treatment regimen however remains the same [13,19,20].

Implant infections can be classified by onset; acute or early infection [most frequently initiated by strains of *S. aureus* (Table 5.1)] if the symptoms develop in the first 3 postoperative months. Delayed infections (mainly initiated by *Staphylococcus epidermidis* strains) manifest between 3 months and 2 years after surgery, while low-grade or late infections manifest after 2 years [13,18]. Another infection classification system is the classification established by the Centers of Disease Control and Prevention (CDC, USA), which is based on the condition of the tissue rather than on the time of infection onset [21]. If the early infections are not treated successfully, they can result in a chronic infection which may result in loss of function of the affected joint or limb, with disability (e.g., by amputation) or in some cases even death as a consequence [22–25].

Table 5.1 Distribution of microorganism causing prostheses infections [13]

Microorganism	Frequency (%)
Staphylococcus aureus	33–43
Coagulase-negative *Staphylococci* (e.g., *Staphylococcus epidermidis*)	17–21
Streptococci (e.g., *Streptococcus pneumoniae*)	11–12
Gram-negative bacilli (e.g., *Pseudomonas aeruginosa*)	5–14
Enterococci (e.g., *Enterococcus faecalis* and *E. faecium*)	3–7
Anaerobes (e.g., *Propionibacterium acnes*)	2–5
Polymicrobial	5–13
Unknown	5–6

5.3 Treatment

The local use of antibiotic carriers in orthopedic surgery to treat and prevent orthopedic infections is in many clinics considered to be common practice, originating from the initial ideas of Buchholz (loading of bone cement with antibiotics, for implant fixation) and Klemm (the use of gentamicin-loaded PMMA-beads to treat infections locally) in the 1970s [26–30]; however, general consensus is lacking. Still it is a proven fact that antibiotic-loaded bone cement functions as a reliable prophylactic measure for orthopedic infections [26,31,32].

In general there are two surgical approaches for implant infection treatment, a 1-stage (1 surgical intervention; implant retraction, debridement, and new prosthesis implantation) or a 2-stage revision (2 surgical interventions; implant retraction, debridement, wound closure, and after a couple of weeks the implantation of a new cemented prosthesis). Both revision methods can be combined with antibiotics as a treatment/prophylaxis.

This results in mainly four types of actual infection treatment procedures:

- 1-Stage revision; direct revision of implant with(out) conventional bone cement.
  (Clearing of infection: 58–59%) [33,34]
- 1-Stage revision; direct revision of implant with antibiotic-loaded bone cement.
  (Clearing of infection: 82–86%) [33,34]
- 2-Stage revision; implant removal with debridement and eventual implant placement with (out) conventional bone cement.
  (Clearing of infection: 82–86%) [33,34]
- 2-Stage revision; implant removal with debridement and local antibiotics (spacers/beads) and eventual implant placement with antibiotic-loaded bone cement.
  (Clearing of infection: 91–93%) [33,34]

The success-rates (clearing of infection) of these treatments indicate the importance of the use of local antibiotics in implant infection treatment [30,33–35]. Still the above methods mainly describe antibiotic treatment and prophylaxis in cemented prostheses, while prophylaxis and treatment with uncemented prostheses remains a challenging topic. A possible solution is the application of an antibiotic- or antiseptic-releasing coating on the implant surface. Up till now a few studies have been published, reporting on the in vitro [36–38] and in vivo [39–43] efficacy of these antibiotic- (gentamicin, vancomycin, tobramycin, and rifampin) releasing coatings, without clinical reports.

Although antibiotic treatment with surgical debridement remains the most effective treatment in the majority of the infection cases, unfortunately the number of antibiotic-resistant strains as methicillin-resistant *Staphylococcus aureus* (MRSA), methicillin-resistant *Staphylococcus epidermidis* (MRSE), and vancomycin-resistant *Staphylococcus aureus* (VRSA) is increasing and is frequently the cause for unsuccessful infection treatment [13,20,44]. Therefore the use of an antiseptic compound instead of an antibiotic could be preferable.

5.4 Infection development and prevention

The tolerance of the body toward an implanted foreign material (prostheses or osteosynthesis) is expressed by the growth of host tissue around or on the surface of such a device. The principle of timely adherence of host cells or bacteria on the implant surface is known as "*the race for the surface*" (Fig. 5.1) [10]. This hypothesis postulates that when the surface of an implanted inert biomaterial is occupied by host tissue cells (before bacterial adhesion to that same surface), the implant surface would be less susceptible for bacterial colonization [8–11].

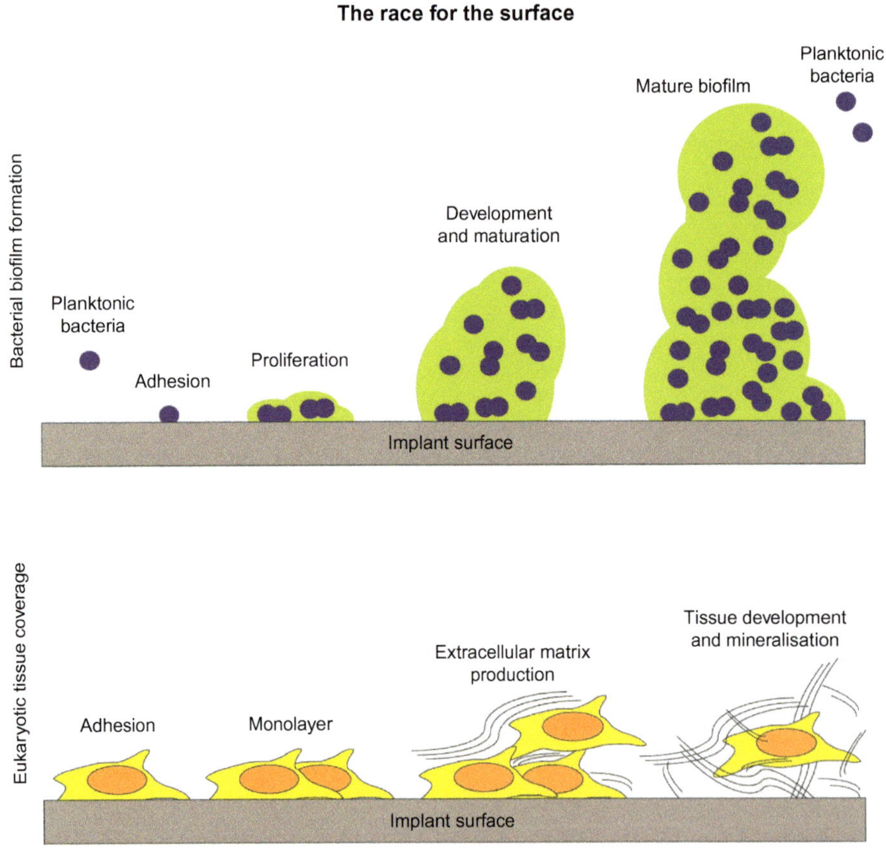

Fig. 5.1 The general principle of "The race for the surface." *Bacteria*: Bacterial colonization of the surface followed by encapsulation with a protecting matrix, enabling maturation toward a mature biofilm which releases bacteria (including toxins and virulence factors) from the biofilm for secondary colonization sites. *Eukaryotic cells*: Cell adhesion on the surface followed by proliferation, differentiation with matrix-production leading to tissue development surrounding the surface [45].

In healthy aseptic situations, the host cells (in the case of orthopedic implants, osteoblasts) adhere to the surface of the implant and start to proliferate and differentiate with the production of a collagenous matrix. The calcification of this matrix (carried out by osteoblasts) will eventually result in bone apposition on the implant surface (Fig. 5.1) [46].

However, in the case of unfortunate septic conditions, bacteria will settle on the implant surface, encapsulating themselves in a protective layer called a biofilm. In such a biofilm bacteria are protected from the host's immune system and many antibiotics [8–11,47]. Bacteria are able to proliferate within the biofilm, colonizing the implant surface. When the biofilm becomes mature, it starts to excrete toxins, and other virulence factors and planktonic bacteria are being released from the biofilm to initiate new biofilm formation elsewhere in the body (Fig. 5.1) [8,9,11]. This indicates the importance of the prevention of biofilm formation or to disrupt the already existing biofilm on the implant surface.

The established method for infection prevention is the systemic and local use of prophylactic antibiotics to hamper bacterial growth and colonization [13]. Also disruption of the already present biofilm, by antibiotics (rifampin), can be an effective treatment option [18,48]. Another important possible method to achieve biofilm prevention could be the modification of the implant surface to promote adherence of, for example, osteoblasts to the implant surface and subsequent bone apposition and so prevent adhesion by bacteria, and win the race for that surface. Taken together current regimens do not offer full protection against implant infection, especially when an already infected prosthesis is being removed and should be replaced by a new prosthesis.

In such cases the risk for infection remains high [19,47]. An ideal solution would be an implant coating which allows bone apposition as well as the local release of an antimicrobial compound in a high dosage to prevent bacterial colonization at the implant location. Currently such coatings are not yet commercially available for clinical use [12].

5.5 Experimental models for orthopedic infections

As described before, osteomyelitis is an old disease which still affects many patients worldwide. The first experimental animal model to study osteomyelitis was developed by Becker et al. in 1883 [7]. They identified the bacterial cause of osteomyelitis (a micrococcus, later described as *S. aureus*). They cultured the bacterium and used it in an experimental osteomyelitis in which they inoculated animals with bone fractures with the cultured bacterium, which resulted in pus formation and the subsequent death of the animals within 14 days after inoculation [7].

The first actually published report dedicated on an experimental hematogenic osteomyelitis was conducted by Rodet, who in 1884 reported that the intravenous injection of a micrococcus (presumably *S. aureus*) in rabbits resulted in development of osteomyelitis. He described periosteal lifting, loss of cortical integrity, and the presence of sequesters in the affected bone tissue [49]. Dedicated models to study osteomyelitis were established in the early 1970s by Norden [50,51], Bowers [52], and Andriole [53,54]. The model of Norden focused on the development of osteomyelitis by the intramedullary administration of *S. aureus* in rabbit tibiae and was followed-up by evaluating blood and bone cultures, the differences in erythrocyte sedimentation rate and whole white blood cell count, X-ray, and histology [50,51]. The model of Bowers focused on antibiotic prophylaxis for osteomyelitis in a dog model; in this model the osteomyelitis was initiated by an intramedullary peroperative contamination with *S. aureus* of the femur and assessed by bacterial cultures of the affected tissue and blood [52]. The model described by Andriole was the first to include an intramedullary nail to study implant-related infections. The main procedure was the perioperative contamination of the intramedullary cavity of the tibia of a rabbit in the presence or absence of an intramedullary nail, with or without a fracture (the implantation of a foreign object initiates the host's immune response). The infection was assessed by bacterial culture, X-rays, and histology [53,54].

To facilitate the initiation of osteomyelitic development many studies included the use of a sclerosing agent like sodium morrhuate to locally initiate vascular thrombosis (less angiogenesis), resulting in local aseptic necrosis of the bone and a decreased transport of nutrients and host immune cells, making the bone more susceptible to an infection [50,53,55–57]. However the necessity of a sclerosing agent is debatable since many studies did not require a sclerosing agent to obtain a reproducible osteomyelitic lesion [41,52,58–66]. Still it appears that the use of a sclerosing agent speeds up the development of an osteomyelitic lesion by the forced absence of a local host immune response [50,53,55–57].

Postmortem bacterial culture is currently the golden standard for infection diagnosis in these experimental models. Since bacterial culture allows false negative cultures, the diagnostic pallet is often supplemented with the accompanying radiological and histological findings to increase the accuracy of these models [18]. Furthermore, the assessment of sequential time points during follow-up, instead of only endpoint measurements, could monitor disease progression during the experimental follow-up [67–71]. Besides, such a sequential follow-up would also allow detailed monitoring of infection prophylaxis and treatment efficacy [67,68].

An overview of the different models studying osteomyelitis and its treatment modalities is presented in Table 5.2. Overall these models describe the induction of osteomyelitis by perioperative bacterial contamination of the surgical site or by bacteraemia caused by i.v. injection. The *radius*, *ulna*, *humerus*, *tibia,* and *femur* have been used as an area of interest in mice, rats, guinea pigs, chinchilla's, rabbits, chickens, dogs, sheep, and goats.

It is to be expected that sequential and precise measurements of a broad spectrum of parameters can provide detailed information about the progression of an infection and the efficacy of its treatment and even could deliver unexpected findings. The use of sequential measurements of such multiple parameter follow-up may reduce the amount of animals necessary to achieve statistically relevant data, since the sacrifice of animals at multiple time points is avoided. For example, the use of several calcium-binding fluorophores enable the histological detection of bone apposition on multiple time points during the experimental follow-up in each animal. This could also be achieved by the sequential use of radiological (X-ray) or nuclear imaging techniques (^{18}F-FDG microPET), X-ray radiographs indicate bone morphological changes, and ^{18}F-FDG microPET would detect metabolic activity in the implant area, monitoring disease progression during the experimental follow-up. So this approach enhances the power of the study and reduces the amount of animals needed to achieve statistically relevant data.

5.6 Scoring of orthopedic infections in experimental models—(semi) systemic parameters

Clinical infection diagnostics are multifactorial (Fig. 5.2), consisting of a wide pallet of, for example, clinical, hematological, bacteriological parameters in combination with imaging modalities [18]. The most general symptoms describing the general condition of the patient and the local aspect of the wound are in case of orthopedic implant infections: redness of the periimplant tissue, soft tissue swelling, and a locally elevated tissue temperature. Laboratory parameters like leucocyte differentiation, erythrocyte sedimentation rate (ESR), and C-reactive protein (CRP) levels are systemic factors which are indicative for an infection [18].

The leucocyte differentiation describes the percentage of lymphocytes, neutrophilic granulocytes, monocytes, basophilic granulocytes, and eosinophilic granulocytes in the blood pool. In case of a bacterial infection, an increase in neutrophils (shift to the left) is typically seen. For example, lymphocytes are more prone to react to viral infections, while neutrophilic granulocytes are more prone to react to bacterial presence, and thus the percentage of neutrophillic granulocytes is expected to increase at the expense of the percentage of lymphocytes. Eosinophilic and basophilic granulocytes are mainly seen in case of allergic responses, and monocytes support the neutrophilic granulocytes in a reaction against bacteria. This suggests that the differentiation of leucocytes is indicative for the diagnosis (bacterial or viral infection or allergy) [18,167,168].

The ESR is a nonspecific parameter for infection, which is based on the sedimentation of erythrocytes as function of the fibrinogen content in the blood. In case of an infection the fibrinogen content in the blood increases causing agglutination of the erythrocytes which leads to an increased sedimentation rate [167,168].

C-reactive protein is an acute phase protein which is produced in the liver as a result of the release of, for example, interleukin-6 after the infiltration of bacteria in the host-organism. CRP is also a marker for inflammation [167,168]. These parameters are relatively straightforward to interpret, for example, in a graph in combination with statistical analysis. Scoring of imaging data and histological sections is more laborious and depending on the interpretation of the presented data by specialists.

Table 5.2 Overview of experimental osteomyelitis models and their experimental parameters

Author, year	Experimental animal	Sclerosing agent	Implant material	Antimicrobial compound	Bacterial contaminant	Inoculation route
Becker, 1883	Rabbit	No	–	–	*S. aureus*	i.v.
Rodet, 1884	Rabbit	No	–	–	*S. aureus*	i.v.
Lexer, 1894	Rabbit, dog	No	–	–	*S. aureus* *S. pyogenes*	i.v.
Lexer, 1896	Rabbit	No	–	–	*S. aureus* *S. pyogenes*	i.v.
Thompson, 1938	Rabbit	No	–	–	*S. aureus*	i.v.
Scheman, 1941	Rabbit	Yes	–	–	*S. aureus*	i.v. local
Norden, 1970	Rabbit	Yes	–	–	*S. aureus*	Local
Andriole, 1973	Rabbit	No	Stainless steel	–	*S. aureus*	Local
Bowers, 1973	Dog	No	–	Cephaloridine	*S. aureus*	Local
Andriole, 1974	Rabbit	No	Stainless steel	–	*S. aureus*	Local
Elson, 1977	Rat	No	PMMA	Fucidin, Gentamicin	*S. aureus* *P. aeruginosa* *P. mirabilis* *S. canis*	Local
Norden, 1980	Rabbit	Yes	–	Sisomicin, Carbenicillin	*P. aeruginosa*	Local
Blomgren, 1980	Rabbit	No	PMMA	–	*S. aureus*	i.v.
Blomgren, 1980	Rabbit	No	TKA, PMMA	–	*S. aureus*	i.v.
Blomgren, 1981	Rabbit	No	TKA, PMMA	Gentamicin	*S. aureus*	i.v.
Blomgren, 1981	Rabbit	No	TKA	–	*P. acnes*	i.v.
Zimmerli, 1982	Guinea pig	No	PMMA, PTFE	–	*S. aureus*	Local
Fitzgerald, 1983	Dog	No	PMMA	Gentamicin	*S. aureus*	Local
Norden, 1983	Rabbit	Yes	–	Vancomycin, Rifampin	*S. aureus*	Local
Emslie, 1983	Chicken	No	–	–	*S. aureus*	i.v.
Rodeheaver, 1983	Rabbit	No	PMMA	Erythromycin, Colistin	*S. aureus*, *E. coli*	Local
Norden, 1983	Rabbit	Yes	–	Rifampin, Sisomicin, Cephalothin, Trimethoprim	*S. aureus*	Local
Norden, 1985	Rabbit	Yes	–	Ciprofloxacin	*P. aeruginosa*	Local
Petty, 1985	Dog	No	Stainless steel, Cobalt chrome, HD-PE, PMMA	–	*S. aureus*, *S. epidermidis*, *E. coli*	Local
Southwood, 1985	Rabbit	No	pHA (fem. comp.), PMMA	–	*S. aureus*	i.v.
Petty, 1988	Dog	No	PMMA	Gentamicin	*S. aureus*, *S. epidermidis*, *E. coli*	Local
Worlock, 1988	Rabbit	No	Stainless steel	–	*S. aureus*	Local
Nelson, 1990	Rat	No	Bone wax	–	*S. aureus*	Local
Power, 1990	Rat	Yes	–	–	*S. aureus*	Local
Jacob, 1991	Rabbit	Yes	PL:PG	Ampicillin	*S. aureus*	Local
Gerhart, 1993	Rat	No	TCP, PMMA	–	*S. aureus*	Local

Clinical parmeters (weight, temp.)	Hematology (ESR, CRP, Leucoc.)	Radiology (X-ray (x), CT (c), MRI (m))	Nuclear imaging (Scint. (s), PET (p))	Bioluminescence	*Postmortem* culture	Calcium-binding fluorophores	Histology	Reference
					x			[7,72]
					x			[49,73]
					x		x	[74]
		x			x		x	[75]
x	x	x			x			[76]
					x		x	[77]
x	x	x			x			[50]
		x			x		x	[53]
x					x			[52]
		x			x			[54]
					x			[78]
x		x			x			[56]
					x			[58]
		x			x		x	[59]
					x			[79]
		x	s		x		x	[60]
	x				x			[80]
		x			x		x	[81]
		x			x			[82]
x					x		x	[83]
		x			x		x	[84]
					x			[85]
		x			x			[86]
					x		x	[63]
x					x			[65]
					x			[87]
		x			x		x	[88]
					x		x	[89]
					x		x	[57]
		x			x			[90]
		x			x			[91]

Continued

Table 5.2 Continued

Author, year	Experimental animal	Sclerosing agent	Implant material	Antimicrobial compound	Bacterial contaminant	Inoculation route
Laurencin, 1993	Rat		PolyA, PMMA	Gentamicin	*S. aureus*	Local
Melcher,1994	Rabbit & chinchilla	No	Stainless steel	–	*S. aureus*	Local
Garvin, 1994	Dog	No	PMMA, PL:PG	Gentamicin	*S. aureus*	Local
Sanzén, 1995	Rabbit	No	Titanium, PMMA	–	*S. aureus*, *S. epidermidis*	Local
Fialkov, 1996	Rabbit	No	Bone graft	–	*S. aureus*	Local
Nelson, 1997	Rabbit	No	PolyA	Gentamicin	*S. aureus*	Local
An, 1997	Rabbit	No	Titanium-BSA	–	*S. epidermidis*	Local
Calhoun, 1997	Rabbit	No	PLA-PL:CG	Vancomycin	*S. aureus*	Local
Smeltzer, 1997	Rabbit	No	Bone graft	–	*S. aureus*	Local
Kaarsemakers, 1997	Sheep	Yes	Gelatin sponge	–	*S. aureus*	Local
Awasthi, 1998	Rabbit	Yes	–	–	*S. aureus*	Local
Nie, 1998	Rabbit	No	PL:PG	Ofloxacin	*P. aeruginosa*	Local
Vogely, 2000	Rabbit	No	Titanium-HA	–	*S. aureus*	Local
Nijhof, 2000	Rabbit	No	PMMA	Tobramycin, Cefazolin	*S. aureus*	Local
Gürsel, 2001	Rabbit	No	PHA	Sulbactam, Cefoperazone, Ampicillin	*S. aureus*	Local
Lucke, 2002	Rat	No	Stainless steel	–	*S. aureus*	Local
Mader, 2002	Rabbit	No	PMMA, Fibrin	Tobramycin	*S. aureus*	Local
Shirtliff, 2002	Rabbit	Yes	PMMA, HA	Vancomycin	MRSA	Local
Rutledge, 2003	Rabbit	Yes	PCL	Tobramycin	S. aureus	Local
Huneault, 2004	Dog	No	CLHAS	Ciprofloxacin	*S. aureus*	Local
Ambrose, 2004	Rabbit	No	PLGA, PMMA	Tobramycin, Cefazolin	*S. aureus*	Local
Koort, 2004	Rabbit	No	PMMA	–	*S. aureus*	Local
Koort, 2005	Rabbit	No	PLA	Ciprofloxacin	*S. aureus*	Local
Craig, 2005	Rabbit	No	Stainless steel, PMMA, UHMWPE	–	MRSA	Local
Mendel, 2005	Rat	Yes	Collagen sponge, PMMA	Cefazolin, Gentamicin	*S. aureus*	Local
Fukushima, 2005	Rat	No	–	–	*S. aureus*	Local
Jones-Jackson, 2005	Rabbit	No	–	–	*S. aureus*	Local
Mäkinen, 2005	Rat	Yes	–	–	*S. aureus*	Local
Mäkinen, 2005	Rabbit	No	PLGA,	Ciprofloxacin	*S. aureus*	Local
Beardmore, 2005	Goat	No	PMMA, Calcium sulfate, Demineral. bone	Tobramycin	*S. aureus*	Local
Joosten, 2005	Rabbit	Yes	HAC,	Vancomycin	*S. aureus*, *MRSA*	Local
Salgado, 2005	Goat	Yes	–	–	*S. aureus*	Local
Alt, 2006	Rabbit	No	Stainless steel	–	*S. aureus*	Local

Clinical parmeters (weight, temp.)	Hematology (ESR, CRP, Leucoc.)	Radiology (X-ray (x), CT (c), MRI (m))	Nuclear imaging (Scint. (s), PET (p))	Biolumi-nescence	*Postmortem* culture	Calcium-binding fluoro-phores	Histology	Reference
					x		x	[92]
		x			x			[93]
					x		x	[94]
x		x			x		x	[95]
					x	x	x	[96]
					x		x	[97]
		x			x		x	[98]
		x			x			[61]
		x			x		x	[64]
x	x	x			x		x	[55]
		x	s		x		x	[99]
		x			x		x	[100]
		x			x		x	[66]
x	x	x			x		x	[62]
		x			x			[101]
x	x	x			x		x	[102]
x		x			x		x	[103]
		x			x			[104]
					x		x	[105]
		x			x		x	[106]
		x			x		x	[107]
		x, c	p		x		x	[108]
		c	p		x		x	[109]
					x			[110]
		x			x		x	[111]
x		x			x		x	[112]
		x	p		x		x	[113]
		x, c	p		x		x	[114]
		x	p		x		x	[115]
					x			[116]
x	x				x		x	[117]
x		x			x		x	[118]
					x		x	[40]

Continued

Table 5.2 Continued

Author, year	Experimental animal	Sclerosing agent	Implant material	Antimicrobial compound	Bacterial contaminant	Inoculation route
Kälicke, 2006	Rabbit	No	Titanium-PLLA	Rifampin, Fusidic acid, Octenidin, Irgasan	*S. aureus*	Local
Darouiche, 2007	Rabbit	No	Titanium-Ab	Minocycline, Rifampin	*S. aureus*	Local
Poultsides, 2008	Rabbit	No	Tantalum	–	MRSA	i.v.
Williams, 2008	Rabbit	No	Titanium, Tantalum	–	*S. aureus*	Local
Laure, 2008	Sheep	No	Stainless steel-HA, Stainless steel, PMMA	–	*S. epidermidis*	Local
Kanellakopoulou, 2008	Rabbit	No	PLA,	Ciprofloxacin	*P. aeruginosa*	Local
Hamel, 2008	Rabbit	No	Stainless steel	–	MRSA	Local
Moojen, 2009	Rabbit	No	Titanium-PA,	Tobramycin	*S. aureus*	Local
Moojen, 2009	Rabbit	No	PMMA, Silver	–	*S. aureus*	Local
Bernthal, 2010	Mouse	No	Stainless steel	–	*S. aureus*	Local
Moskowitz, 2010	Rabbit	No	PMMA, Titanium-Ab	Gentamicin	*S. aureus*	Local
Jacqueline, 2010	Rabbit	No	–	Ceftaroline	MRSA	Local
Stewart, 2010	Rat	No	Stainless steel, PPF	–	*S. aureus* *E. coli*	Local
Robinson, 2011	Rat	No	Stainless steel	–	*S. aureus*	Local
Vergidis, 2011	Rat	No	Titanium	Linezolid, Vancomycin, Rifampin	MRSA	Local
Gaudin, 2011	Rabbit	No	–	–	MRSA	Local
Alt, 2011	Rat	No	Stainless steel	–	S. aureus	Local
Poeppl, 2011	Rat	No	–	Fosfomycin	MRSA	Local
Lankinen, 2012	Rabbit	Yes	PMMA	–	*S. aureus*, *S. epidermidis*	Local
Pribaz, 2012	Mouse	No	Stainless steel	–	*S. aureus*	Local
Niska, 2012	Mouse	No	Stainless steel, Titanium	Daptomycin, Tigecycline, Vancomycin	*S. aureus*, MRSA	Local
Horst, 2012	Mouse	No	–		*S. aureus*	i.v.
Li, 2012	Rabbit	Yes	–	L-PRP	MRSA	Local
Johansen, 2013	Pig	No	–	–	*S. aureus*	i.v.
Yin, 2013	Rabbit	Yes	–	–	MRSA, *K. pneumoniae*, *P. aeruginosa*, *A. baumannii*	Local
Xing, 2013	Rabbit	No	Alginate beads	Vancomycin	*S. aureus*	Local
Xie, 2013	Rabbit	Yes	Bioactive glass	Gentamicin	*E. coli*	Local
Søe, 2013	Rat	No	Stainless steel, HD-PE	–	*S. aureus*	Local
Sinclair, 2013	Sheep	No	Titanium	–	MRSA	Local
Tran, 2013	Goat	No	Stainless steel	Silver	*S. aureus*	Local
Odekerken, 2013–2014	Rabbit	No	Titanium	–	*S. aureus*	Local
Helbig, 2014	Rabbit	No	Titanium	Vancomycin	MRSA	Local

Clinical parmeters (weight, temp.)	Hematology (ESR, CRP, Leucoc.)	Radiology (X-ray (x), CT (c), MRI (m))	Nuclear imaging (Scint. (s), PET (p))	Biolumi-nescence	*Postmortem* culture	Calcium-binding fluoro-phores	Histology	Reference
					x		x	[43]
					x			[119]
x	x	x			x		x	[120]
					x			[121]
		x			x		x	[122]
					x		x	[123]
		x			x			[124]
x	x				x		x	[41]
x	x				x		x	[125]
				x	x		x	[126]
					x			[42]
					x		x	[127]
		x			x		x	[128]
		x			x		x	[129]
					x			[130]
					x		x	[131]
		x, c			x		x	[132]
		x			x		x	[133]
		x, c	p		x		x	[68]
				x	x			[134]
				x	x			[135]
x	x	x, m			x		x	[136]
		c			x		x	[137]
		x			x			[138]
		x			x			[139]
		x, c			x		x	[140]
		x			x		x	[141]
x	x	x			x		x	[142]
		x, c			x		x	[143]
	x	x, c			x		x	[144]
x	x	x, c	p		x	x	x	[69,70]
x	x				x		x	[145]

Continued

Table 5.2 Continued

Author, year	Experimental animal	Sclerosing agent	Implant material	Antimicrobial compound	Bacterial contaminant	Inoculation route
Crémieux, 2014	Rabbit	Yes	–	–	MRSA	Local
Varrone, 2014	Mouse	No	Stainless steel	Gmd-1C11	MRSA	Local
McLaren, 2014	Sheep	No	PLGA-PEG	Gentamicin, Clindamycin	*S. aureus*	Local
Inanmaz, 2014	Rat	No	Stainless steel	Teicoplanin	*S. aureus*	Local
Gahukamble, 2014	Rabbit	No	Stainless steel	–	*P. acnes* *S. lugdunensis*	Local
Zhu, 2014	Mouse	No	–	Anisomycin	MRSA	Local
Freiberg, 2014	Rabbit	Yes	Dextran		*S. pyogenes*	Local
Chen, 2014	Dog	No	Gelatin, Titanium	Cefazolin, Gentamicin	*S. aureus*	Local
Ding, 2014	Rabbit	Yes	Bioactive glass, Calcium sulfate	Vancomycin	MRSA	Local
Cui, 2014	Rabbit	Yes	Bioactive glass, Calcium sulfate	Vancomycin	MRSA	Local
Odekerken, 2014	Rabbit	No	–	–	*S. aureus*	Local
Shishatskaya, 2014	Chinchilla, rabbit	No	P3HB, P3HB/HA, Xenograft	–	*S. aureus*	Local
Poeppl, 2014	Rat	No	Titanium	Fosfomycin, Vancomycin	MRSA	Local
Gatin, 2014	Rabbit	No	Silicon	Ceftaroline-Fosamil, Vancomycin, Rifampin	MRSA	Local
Ahtinen, 2014	Rat	Yes	PTFE	–	*S. aureus* *S. epidermidis*	Local
Alt, 2014	Rabbit	No	Stainless steel	Rifampin Fosfomycin	*S. aureus* MRSA	Local
Chatziioannou, 2015	Rabbit	No	Stainless steel	Daptomycin	*S. aureus*	Local
Neut, 2015	Rabbit, dog	No	Titanium, PLGA, HA	Gentamicin	*S. aureus*	Local
Shiels, 2015	Rat	No	Titanium	–	*S. aureus*	Local
Nielsen, 2015	Pig	No	–	–	*S. aureus*	i.v.
Yan, 2015	Rabbit	Yes	HA	Rifapentine	*S. aureus*	Local
Arens, 2015	Rabbit	No	Stainless steel	–	*S. aureus*	Local
Odekerken, 2015	Rabbit	No	Titanium	Chlorhexidine	*S. aureus*	Local

General abbreviations: CRP, C-reactive protein; CT, computed tomography; ESR, erythrocyte sedimentation rate; i.v., intravenous; leucoc., leucocyte count/differentiation; PET, positron emission tomography; scint., scintigraphy; temp., temperature.
Abbreviations "Implant material" and "Antimicrobial compound": Ab, coated with antibiotic; BSA, coated with bovine serum albumin; HA, coated with hydroxyapatite; PA, coated with periapatite; PLLA, coated with poly-L-lactide; CLHAS, crosslinked high amylose starch; Gmd-1C11, antiglucosamidase monoclonal antibody; HA(C), hydroxyapatite (cement); HD-PE, high-density polyethylene; L-PRP, leucocyte- and platelet-rich plasma; P3HB, poly-3-hydroxybutyrate; P3HB/HA, poly-3-hydroxybutyrate/hydroxyapatite; PCL, polycarprolactone; pHA, partial hip arthroplasty; PHA, polyhydroxyalkanoate; PLA, polylactic acid; PLA-PL:CG, polylactic acid-polylactide/polyglycolide; PL:PG, polylactide/polyglycolide; PLGA, polylactic-co-glycolic acid; PLGA-PEG, polylactic-co-glycolic acid-polyethylene glycol; PMMA, polymethylmethacrylate; PolyA, polyanhydride; PPF, polypropylene fumarate; PTFE, polytetrafluoroethylene; TCP, tricalciumphosphate; TKA, Total knee arthroplasty; UHMWPE, ultra-high-molecular-weight polyethylene.
Abbreviations "Bacterial contaminant": *E. coli*, *Escherichia coli*; *MRSA*, Methicillin-resistant *Staphylococcus aureus*; *P. acnes*, *Propionibacterium acnes*; *P. aeruginosa*, *Pseudomonas aeriginosa*; *P. mirabilis*, *Proteus mirabilis*; *S. aureus*, *Staphylococcus aureus*; *S. canis*, *Streptococcus canis*; *S. epidermidis*, *Staphylococcus epidermidis*.

Clinical parmeters (weight, temp.)	Hematology (ESR, CRP, Leucoc.)	Radiology (X-ray (x), CT (c), MRI (m))	Nuclear imaging (Scint. (s), PET (p))	Biolumi-nescence	*Postmortem* culture	Calcium-binding fluoro-phores	Histology	Reference
	x				x		x	[146]
		c		x	x		x	[147]
		c			x		x	[148]
		x			x			[149]
	x	x			x		x	[150]
x	x				x		x	[151]
					x		x	[152]
		m			x			[153]
		x			x		x	[154]
		x			x		x	[155]
x	x	x, c	p		x	x	x	[71]
		x			x		x	[156]
					x		x	[157]
					x			[158]
		c	p		x		x	[159]
		x			x		x	[160]
		c	p		x		x	[161]
	x				x		x	[162]
		x, c			x		x	[163]
	x	c	s, p		x		x	[164]
	x	x			x		x	[165]
		x			x		x	[166]
x	x	x, c	p		x	x	x	[45]

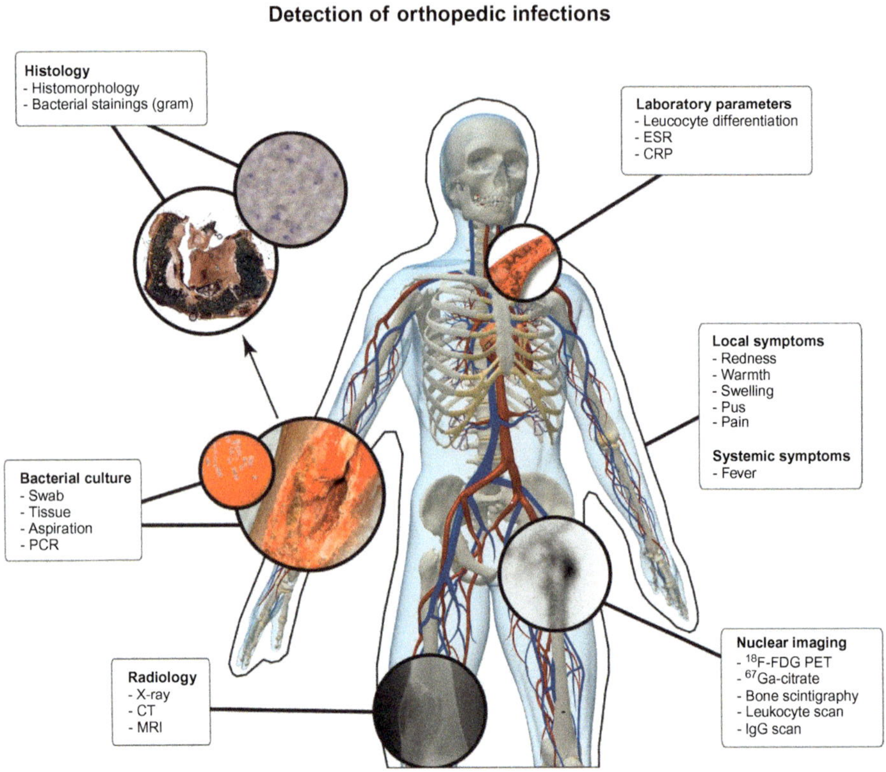

Fig. 5.2 Clinical modalities to detect orthopedic infections.

5.7 Scoring of orthopedic infections in experimental models—imaging and histology

Several imaging modalities can be applied to gather insight in the local morphological aspect concerning orthopedic infection. This regards two aspects: is an infection present and where is it located. X-ray radiography and its volumetric equivalent computed tomography (CT) are frequently used in infection diagnostics and are based on absorption of radiation as a result of the density of the tissue (bone is more dense than soft tissue and is therefore capable of absorbing the radiation when passing through bone tissue). As a result X-ray and CT only describe bone morphological changes as a result of an infection; they do not provide direct insight on the activity or the progression of an infection. These morphological changes can be used as a scoring system to monitor the impact of the infection on the bone tissue and mineralization (Fig. 5.3A–D).

With respect to infection activity or progression nuclear imaging, however, is a potent screening modality. Nuclear imaging is being applied since the early 1970s for the detection of infections (Table 5.3) [169]. The first (nonspecific) tracer compound for this application was ^{67}Ga-citrate which binds to transferrin as an iron analogue and uses the transferrin receptors (CD71) to access the cell [169–171].

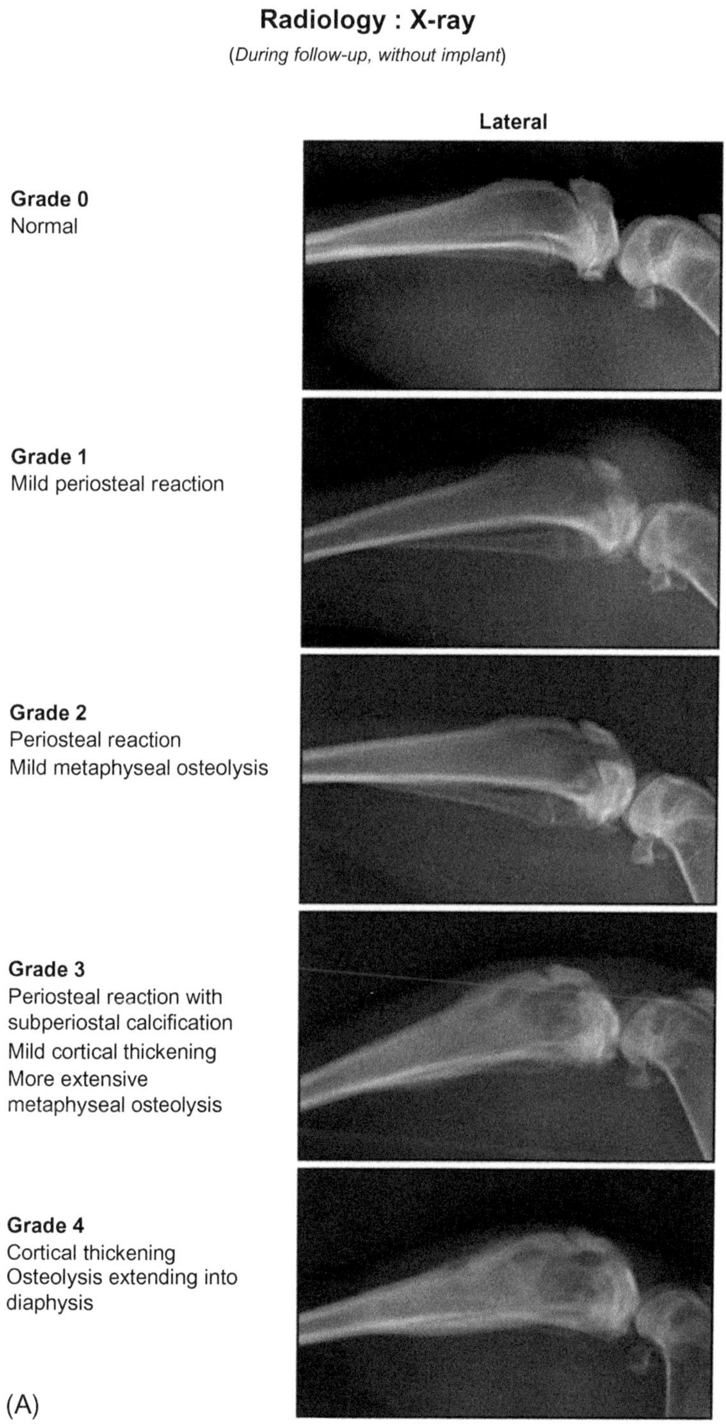

Fig. 5.3 Radiological scoring system for (A) osteomyelitis by X-ray,

(Continued)

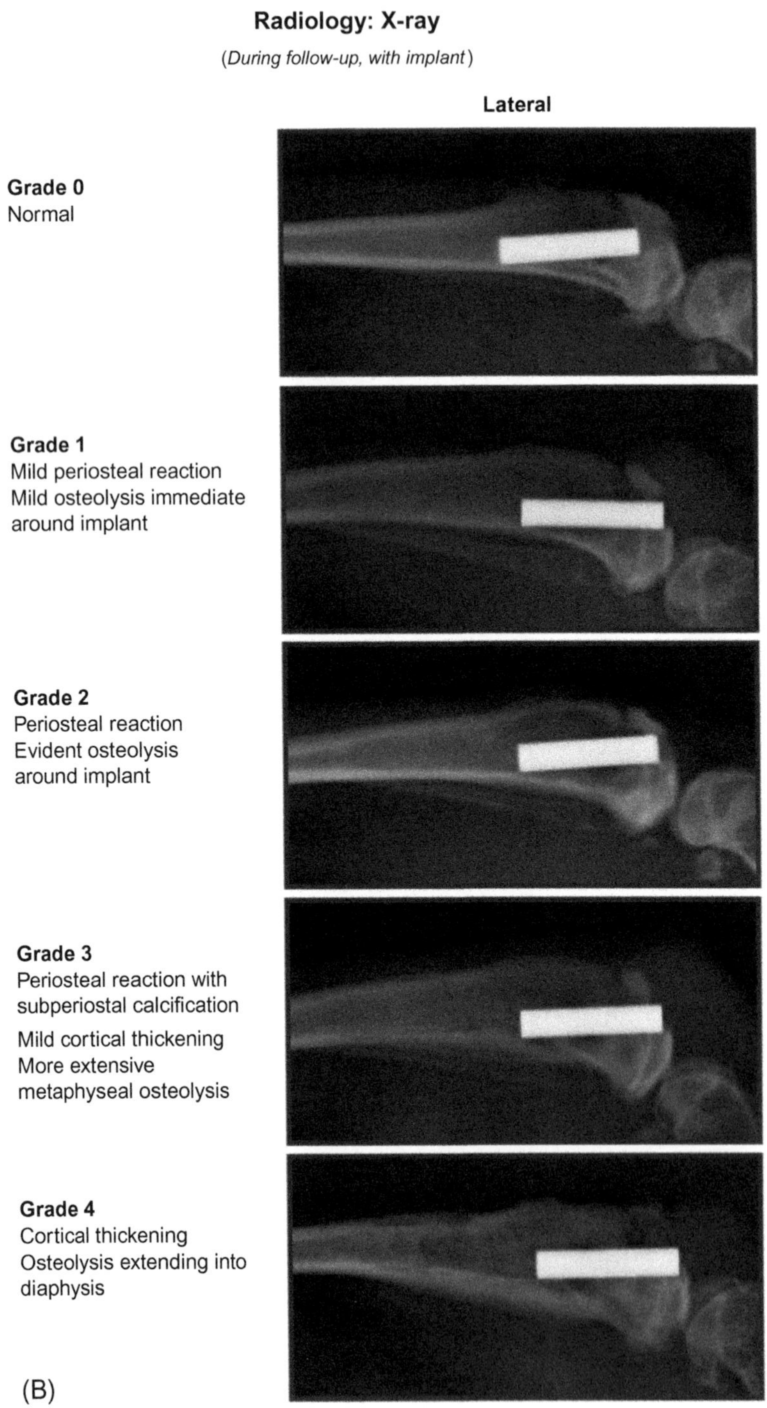

Fig. 5.3 Continued. (B) implant-related osteomyelitis by X-ray,

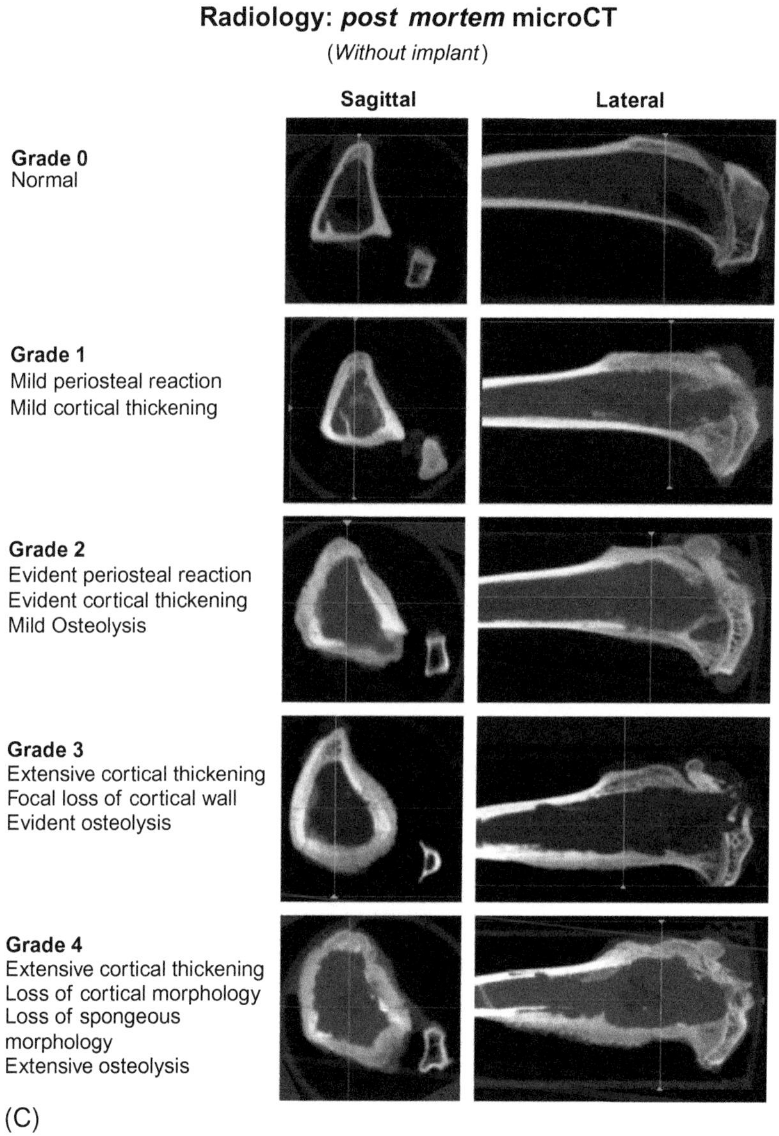

Fig. 5.3 Continued. (C) osteomyelitis by microCT, and

(Continued)

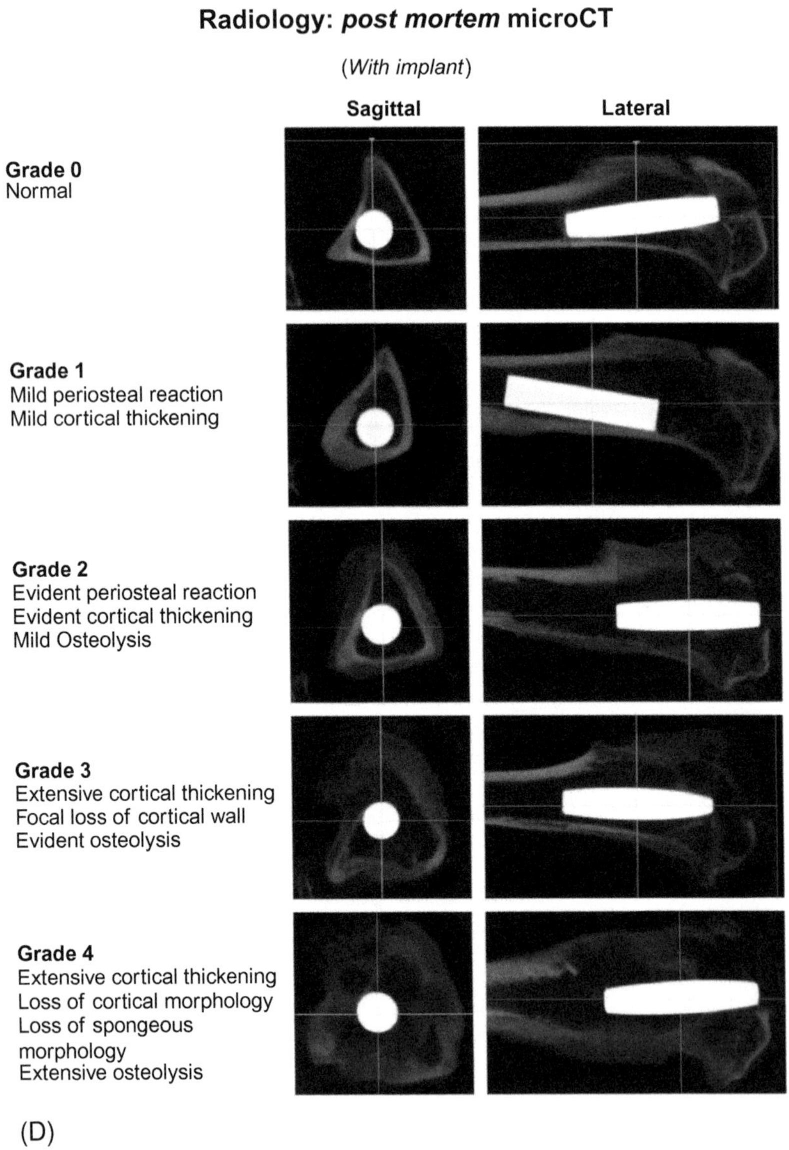

Fig. 5.3 Continued. (D) implant-related osteomyelitis by microCT.

^{67}Ga also binds to lactoferrin which is highly present in infected tissue [169–171]. The binding to transferrin results in a low specificity and a high background signal; furthermore, ^{67}Ga emits high-energy γ-radiation and has a relatively long half-life of 3.26 days [169–171].

The labeling of diphosphonates with technetium results in a tracer which allows imaging in three phases: the perfusion phase (local hyper perfusion), the circulation

Table 5.3 **Nuclear imaging approaches for infection detection [114,169–172]**

Imaging modality	Technique	Measures	Sensitivity	Specificity
^{67}Ga-citrate	Scintigraphy/ SPECT	Blood flow Transferrin/lactoferrin	+	±
3-Phase bone scan (^{99m}Tc)	Scintigraphy/ SPECT	Focal hyper perfusion Focal hyperaemia Focal bone uptake	++	−
Labeled IgG (^{99m}Tc & ^{111}In)	Scintigraphy/ SPECT	Leucocyte localization	+	±
Labeled leucocytes (^{99m}Tc & ^{111}In)	Scintigraphy/ SPECT	Leucocyte localization	++	±
^{68}Ga	PET	Blood flow Transferrin/lactoferrin	+	±
^{18}F-FDG	PET	Metabolic glucose uptake	++	±

phase (local hyperemia), and the bone imaging phase (local bone remodeling) [169–171]. Although the sensitivity is high, the specificity of the 3-phase bone scan is mainly depending on bone remodeling, limited to not only osteomyelitis-mediated remodeling but also bone apposition on an implant surface or physiological bone remodeling [169–171].

Another approach is the visualization of leucocytes (directly or indirect), since they are involved in the host immune response toward an infection and are expected to be present in elevated levels at the infection site, resulting in elevated tracer levels. In general these methods are sensitive and relatively specific. However the direct labeling of leucocytes requires isolation of the patient's leucocytes which are subsequently labeled in the laboratory and then injected back in the patient before the initial imaging can be performed, making it a laborious and expensive approach [169–171]. Instead of using patient material, it is possible to label leukocyte antibodies. Unfortunately the antibodies diffuse slowly into the infected area (making it less sensitive than the directly labeled leucocytes), and there is a slow systemic clearance of the labeled antibodies [169–171].

All the above mentioned tracers are used in scintigraphy or SPECT (single photon emission-computed tomography) imaging, which is based on the detection of single emitted photons, making exact localization of the source of emission more difficult compared to positron emission tomography (PET), which detects two photons emitted in opposite directions [169–171].

PET is a nuclear imaging modality originally applied in tumor detection and which is still considered as a possible tool for the detection of orthopedic infections. PET is an upcoming imaging modality in the clinic since its initial development in the mid 1970s [173,174]. The working mechanism is based on the annihilation of a proton (originating from the tracer) with an electron (originating from the organism) resulting in the formation of two photons (511 keV) traveling in the opposite direction of each other. These two photons are then detected in coincidence by two scintillation detectors (instead of one detector with SPECT) which allows accurate calculation of the origin of the annihilation [173–175] and thus the infection. There are various tracers available for clinical applications concerning PET-imaging. One of those tracers is ^{68}Ga, of which the physiological principle is comparable to that of ^{67}Ga [114]. The most common PET tracer however is ^{18}fluor-fluorodeoxyglucose (^{18}F-FDG), a radioactively labeled glucose-derivative [175–177]. This tracer originates from the field of oncology mainly due to the high glucose uptake of tumor tissue, which allows highly specific imaging of malignant cells [175–177]. Since acute implant infections are considered highly metabolically active, ^{18}F-FDG is proposed as an infection tracer. Currently clinical application of this tracer for the detection of orthopedic implant infection is limited [18,178], in experimental procedures; however they have proven their efficacy [68,70,71].

Data management and quantification is also an important aspect for the analysis of nuclear imaging data. There are two main approaches to analyze these imaging data, depending on the method of acquisition and the preferred algorithm.

A widely used method for the quantification of the imaging data is the calculation of the standardized uptake value (SUV), which is calculated from the formula below [175].

$$\mathrm{SUV} = \frac{\text{Detected activity in the tissue by the PET scanner}\left(\mathrm{Bq}/\mathrm{g}\right)}{\left(\dfrac{\text{Injected systemic activity}\left(\mathrm{Bq}\right)}{\text{Body weight}\left(\mathrm{g}\right)}\right)}$$

The SUV provides in a quantifiable value for local PET tracer uptake in static detection conditions. In the case of an experimental model, nuclear imaging by means of ^{18}F-FDG PET can distinguish uninfected tissue from infected tissue within 2–3 weeks after inoculation [70,71].

The other approach is the use of dynamic acquisition in combination with the calculation of the rate constant K. This constant describes the distribution of the tracer in the tissue, based on the injected activity in the artery, the unbound tracer fraction in the tissue, and the bound tracer fraction in the tissue.

While imaging modalities in general provide insight in the in vivo situation during infection, histological analysis mainly describes the impact of the infection ex vivo. This analysis is often described by the use of single staining methods, each focusing on specific tissues like bone or soft tissue or to check for bacterial presence. These stainings mainly describe endpoint measurements; however some chemical components, like calcein labels, can be used during follow-up to "pinpoint" the morphological situation at the time of injection (Fig. 5.4). The use of these calcium-binding fluorophores allows the assessment

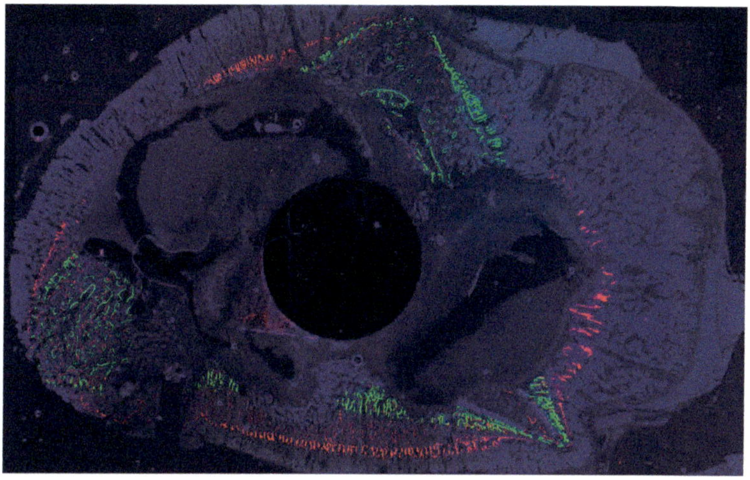

Fig. 5.4 The use of calcein labels during follow-up. Calcein Green at 2 weeks after bacterial contamination, Xylenol Orange at 4 weeks after contamination, and Calcein Blue at 6 weeks after contamination. The migration of the labels outward, from the implant in the middle, indicates the presence of periosteal lifting and in this situation the presence of osteomyelitis. The loss of staining in some regions indicates the presence of osteolysis.

of periosteal lifting and osteolysis due to osteomyelitis on sequential time points during follow-up, by ex vivo visualization.

The combination of multiple staining methods on sequential histological sections does allow quantification of infection-status by the scoring system described in Table 5.4 and Fig. 5.5.

Table 5.4 Histological scoring for (implant-related) osteomyelitis based on sequential sections with different stainings

Morphological abnormality	Histological staining	Score (per abnormality)
Cortical thickening Presence of microabcesses Enlarged Haversian canals	Masson-Goldner	0: Absent 1: Mild to moderate 2: Moderate to severe
Periosteal elevation	Calcium-binding fluorophores	
Gram stain	Modified Gram stain	0: Negative 2: Positive
Total histological score		***Osteomyelitis grade***
0–3		Not infected
>3–5		Mild
>5–7		Moderate
>7–10		Severe

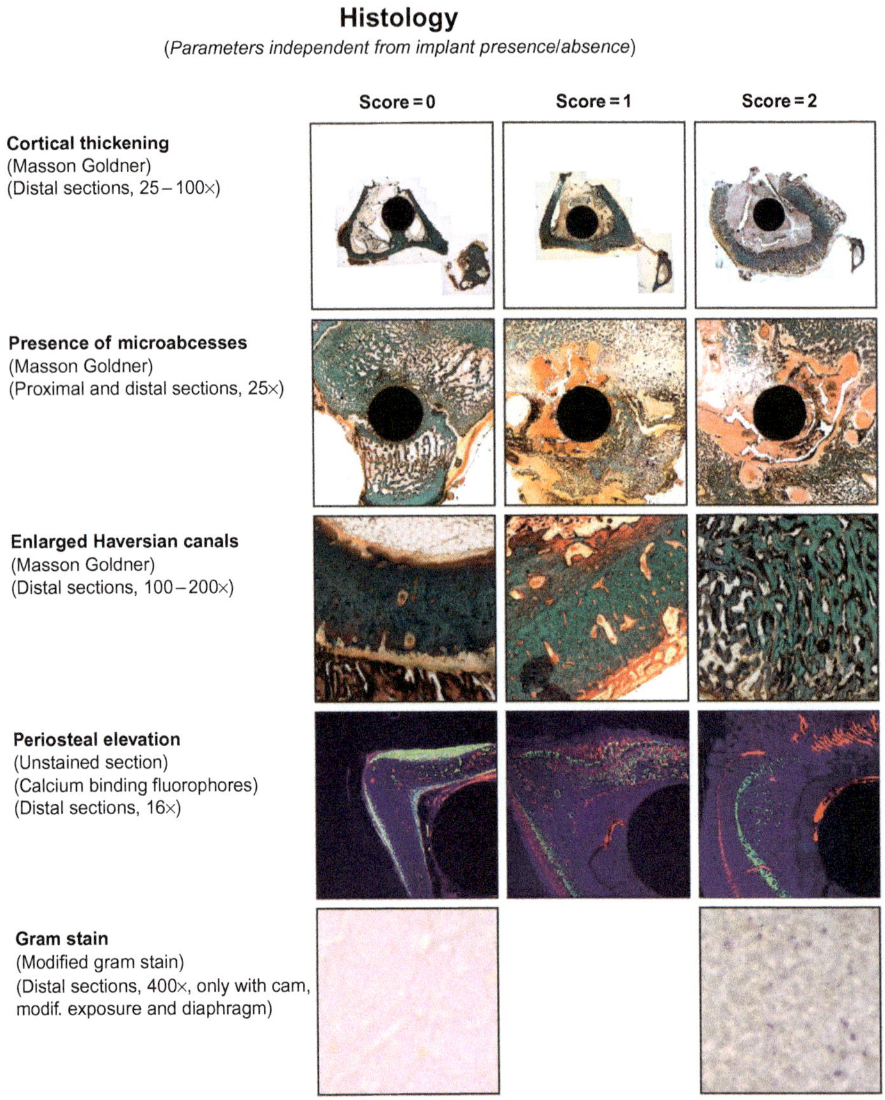

Fig. 5.5 Histological scoring for (implant-related) osteomyelitis based on sequential sectioning and the use of different stainings.

5.8 Concluding remarks

Osteomyelitis and implant infection remain to be devastating diseases for the patient. In this chapter an overview was presented about the available experimental models for implant-related infection, trauma-related infection, as well as osteomyelitis research. It is clear that there are numerous methods of investigating the efficacy of a new coating or a therapeutic agent for prophylaxis or treatment of orthopedic infections. Still the choice for which model to use is depending on various factors like:

- the intended clinical indication,
- the outcome parameters of the model of choice, and
- the variation within the experimental study groups.

As such, the choice of the preferred animal species depends on the size of the implant and/or the procedure to be used, still rabbits are the most commonly used animals for these experimental models. Whereas implant-associated infection models usually focus on early postoperative infections, prophylaxis, and possible treatment. Trauma-related infection models, however, are considerably more difficult to use due to the multifactorial difficulties like standardization of the established trauma, containment of the infection and impaired soft tissue, impaired bone stability, and impaired angiogenesis. These difficulties can considerably interfere with the analysis of experimental data like:

- blood values (elevated marker levels due to spreading of the infection),
- imaging modalities (due to impaired angiogenesis and soft tissue swelling), and
- containment of infection (local infection can become systemic or result in open lesions).

Still, both models can be very challenging for the animals and may result in unnecessary animal loss. Therefore, there is an important factor to consider; the use of multiple analytical methods in an experimental model in combination with the assessment of these parameters on multiple time points during follow-up may significantly strengthen the power of the collected dataset and could possibly reduce the amount of animals required for the experiment. It may furthermore enhance insight in coating functionality and efficacy during follow-up by including the infection onset and progression into the equation, without focusing on endpoint parameters only. The herein described scoring systems may contribute to the quantification of sequentially gathered experimental data.

References

[1] E. Rega, R.B. Holmes, A. Tirabaso, Habitual locomotor behaviour inferred from manual pathology in two Late Cretaceous chasmosaurine ceratopsid dinosaurs, Chamsosaurus irvinensis (CMN41357) and Chasmosaurus belli (ROM 843), in: M.J. Ryan (Ed.), New Perspectives on Horned Dinosaurs: The Royal Tyrrell Museum Ceratopsian Symposium, Life of the PastIndiana University Press, Bloomington, IN, USA, 2010.
[2] C. Burri, Posttraumatische Osteitis, Hans Huber, Bern, 1979.
[3] A.S. Schulp, G.H.I.M. Walenkamp, P.A.M. Hofman, Y. Stuip, B.M. Rothschild, Chronic bone infection in the jaw of Mosasaurus hoffmanni (Squamata), Oryctos 6 (2006) 41–52.
[4] J.H. Breasted, The Edwin Smith Surgical Papyrus: Published in Fascimile and Hieroglyphic Transliteration With Translation and Commentary in 2 Volumes, University of Chicago, 1930.
[5] J. Scultetus, Armamentarium Chirurgicum, 1655, Ulmae Suevorum imp: B. Kühnemann.

[6] T. Longmore, Remarks upon osteo-myelitis consequent on gunshot wounds of the upper and lower extremities, and especially upon the treatment of stumps affected with osteomyelitis after amputation necessitated by such injuries, Med. Chir. Trans. 48 (1865) 43–64.
[7] W.W. Vanarsdale II, On the present state of knowledge in bacterial science in its surgical relations (continued): osteomyelitis, Ann. Surg. 3 (3) (1886) 221–226.
[8] C.R. Arciola, D. Campoccia, P. Speziale, L. Montanaro, J.W. Costerton, Biofilm formation in Staphylococcus implant infections. A review of molecular mechanisms and implications for biofilm-resistant materials, Biomaterials 33 (26) (2012) 5967–5982.
[9] H.J. Busscher, H.C. van der Mei, G. Subbiahdoss, P.C. Jutte, J.J. van den Dungen, S.A. Zaat, et al., Biomaterial-associated infection: locating the finish line in the race for the surface, Sci. Transl. Med. 4 (153) (2012) 153rv10.
[10] A.G. Gristina, Biomaterial-centered infection: microbial adhesion versus tissue integration, Science 237 (4822) (1987) 1588–1595.
[11] N. Hoiby, O. Ciofu, H.K. Johansen, Z.J. Song, C. Moser, P.O. Jensen, et al., The clinical impact of bacterial biofilms, Int. J. Oral Sci. 3 (2) (2011) 55–65.
[12] L. Montanaro, P. Speziale, D. Campoccia, S. Ravaioli, I. Cangini, G. Pietrocola, et al., Scenery of Staphylococcus implant infections in orthopedics, Future Microbiol. 6 (11) (2011) 1329–1349.
[13] A. Trampuz, W. Zimmerli, Antimicrobial agents in orthopaedic surgery: prophylaxis and treatment, Drugs 66 (8) (2006) 1089–1105.
[14] B. Adeli, J. Parvizi, Strategies for the prevention of periprosthetic joint infection, J. Bone Joint Surg. (Br.) 94 (11 Suppl. A) (2012) 42–46.
[15] Y.P. Acklin, A.F. Widmer, R.M. Renner, R. Frei, T. Gross, Unexpectedly increased rate of surgical site infections following implant surgery for hip fractures: problem solution with the bundle approach, Injury 42 (2) (2011) 209–216.
[16] H. Dale, G. Hallan, B. Espehaug, L.I. Havelin, L.B. Engesaeter, Increasing risk of revision due to deep infection after hip arthroplasty, Acta Orthop. 80 (6) (2009) 639–645.
[17] S.M. Kurtz, E. Lau, K.L. Ong, L. Carreon, H. Watson, T. Albert, et al., Infection risk for primary and revision instrumented lumbar spine fusion in the Medicare population, J. Neurosurg. Spine 17 (4) (2012) 342–347.
[18] W. Zimmerli, Infection and musculoskeletal conditions: prosthetic-joint-associated infections, Best Pract. Res. Clin. Rheumatol. 20 (6) (2006) 1045–1063.
[19] J. Geurts, J.J. Chris Arts, G.H. Walenkamp, Bone graft substitutes in active or suspected infection. Contra-indicated or not? Injury 42 (Suppl. 2) (2011) S82–S86.
[20] A. Trampuz, W. Zimmerli, Diagnosis and treatment of infections associated with fracture-fixation devices, Injury 37 (Suppl. 2) (2006) S59–S66.
[21] T.C. Horan, M. Andrus, M.A. Dudeck, CDC/NHSN surveillance definition of health care-associated infection and criteria for specific types of infections in the acute care setting, Am. J. Infect. Control 36 (5) (2008) 309–332.
[22] J. Sanders, C. Mauffrey, Long bone osteomyelitis in adults: fundamental concepts and current techniques, Orthopedics 36 (5) (2013) 368–375.
[23] R.O. Darouiche, Treatment of infections associated with surgical implants, N. Engl. J. Med. 350 (14) (2004) 1422–1429.
[24] F.A. Waldvogel, G. Medoff, M.N. Swartz, Osteomyelitis: a review of clinical features, therapeutic considerations and unusual aspects, N. Engl. J. Med. 282 (4) (1970) 198–206.
[25] G.A. Noskin, R.J. Rubin, J.J. Schentag, J. Kluytmans, E.C. Hedblom, C. Jacobson, et al., National trends in *Staphylococcus aureus* infection rates: impact on economic

burden and mortality over a 6-year period (1998–2003), Clin. Infect. Dis. 45 (9) (2007) 1132–1140.
[26] H.W. Buchholz, H. Engelbrecht, Depot effects of various antibiotics mixed with Palacos resins, Chirurg 41 (11) (1970) 511–515.
[27] K. Klemm, Gentamicin-PMMA-beads in treating bone and soft tissue infections (author's transl), Zentralbl. Chir. 104 (14) (1979) 934–942.
[28] G.H. Walenkamp, T.J. van Rens, Chains of gentamicin-PMMA beads: a new method for the local treatment of osteomyelitis, Ned. Tijdschr. Geneeskd. 126 (47) (1982) 2136–2142.
[29] G.H. Walenkamp, T.B. Vree, T.J. van Rens, Gentamicin-PMMA beads. Pharmacokinetic and nephrotoxicological study, Clin. Orthop. Relat. Res. 205 (1986) 171–183.
[30] G.H. Walenkamp, L.L. Kleijn, M. de Leeuw, Osteomyelitis treated with gentamicin-PMMA beads: 100 patients followed for 1–12 years, Acta Orthop. Scand. 69 (5) (1998) 518–522.
[31] J. Parvizi, K.J. Saleh, P.S. Ragland, A.E. Pour, M.A. Mont, Efficacy of antibiotic-impregnated cement in total hip replacement, Acta Orthop. 79 (3) (2008) 335–341.
[32] L.B. Engesaeter, S.A. Lie, B. Espehaug, O. Furnes, S.E. Vollset, L.I. Havelin, Antibiotic prophylaxis in total hip arthroplasty: effects of antibiotic prophylaxis systemically and in bone cement on the revision rate of 22,170 primary hip replacements followed 0–14 years in the Norwegian Arthroplasty Register, Acta Orthop. Scand. 74 (6) (2003) 644–651.
[33] K.L. Garvin, B.G. Evans, E.A. Salvati, B.D. Brause, Palacos gentamicin for the treatment of deep periprosthetic hip infections, Clin. Orthop. Relat. Res. 298 (1994) 97–105.
[34] F. Langlais, Can we improve the results of revision arthroplasty for infected total hip replacement? J. Bone Joint Surg. (Br.) 85 (5) (2003) 637–640.
[35] G.H. Walenkamp, Gentamicin PMMA beads and other local antibiotic carriers in two-stage revision of total knee infection: a review, J. Chemother. 1 (1) (2001) 66–72.
[36] C. Castelli, P. Marone, V. Monzillo, K. Segú, Antistaphylococcal activity of antibiotic impregnated bone cement, Knee 2 (2) (1995) 219–222.
[37] G. Lewis, S. Janna, The in vitro elution of gentamicin sulfate from a commercially available gentamicin-loaded acrylic bone cement, VersaBond AB, J. Biomed. Mater. Res. B Appl. Biomater. 71 (1) (2004) 77–83.
[38] D. Neut, O.S. Kluin, B.J. Crielaard, H.C. van der Mei, H.J. Busscher, D.W. Grijpma, A biodegradable antibiotic delivery system based on poly-(trimethylene carbonate) for the treatment of osteomyelitis, Acta Orthop. 80 (5) (2009) 514–519.
[39] V. Alt, A. Bitschnau, F. Bohner, K.E. Heerich, E. Magesin, A. Sewing, et al., Effects of gentamicin and gentamicin-RGD coatings on bone ingrowth and biocompatibility of cementless joint prostheses: an experimental study in rabbits, Acta Biomater. 7 (3) (2011) 1274–1280.
[40] V. Alt, A. Bitschnau, J. Osterling, A. Sewing, C. Meyer, R. Kraus, et al., The effects of combined gentamicin-hydroxyapatite coating for cementless joint prostheses on the reduction of infection rates in a rabbit infection prophylaxis model, Biomaterials 27 (26) (2006) 4627–4634.
[41] D.J. Moojen, H.C. Vogely, A. Fleer, P.G. Nikkels, P.A. Higham, A.J. Verbout, et al., Prophylaxis of infection and effects on osseointegration using a tobramycin-periapatite coating on titanium implants—an experimental study in the rabbit, J. Orthop. Res. 27 (6) (2009) 710–716.
[42] J.S. Moskowitz, M.R. Blaisse, R.E. Samuel, H.P. Hsu, M.B. Harris, S.D. Martin, et al., The effectiveness of the controlled release of gentamicin from polyelectrolyte

multilayers in the treatment of *Staphylococcus aureus* infection in a rabbit bone model, Biomaterials 31 (23) (2010) 6019–6030.

[43] T. Kalicke, J. Schierholz, U. Schlegel, T.M. Frangen, M. Koller, G. Printzen, et al., Effect on infection resistance of a local antiseptic and antibiotic coating on osteosynthesis implants: an in vitro and in vivo study, J. Orthop. Res. 24 (8) (2006) 1622–1640.

[44] A. Patel, R.P. Calfee, M. Plante, S.A. Fischer, N. Arcand, C. Born, Methicillin-resistant *Staphylococcus aureus* in orthopaedic surgery, J. Bone Joint Surg. (Br.) 90 (11) (2008) 1401–1406.

[45] J. Odekerken, Orthopaedic Infections: Pre-clinical Models, Diagnostic Tools and Novel Coatings, Jim C.E. Odekerken, Klimmen, The Netherlands, 2015, 267 pp.

[46] T. Albrektsson, C. Johansson, Osteoinduction, osteoconduction and osseointegration, Eur. Spine J. 10 (Suppl. 2) (2001) S96–S101.

[47] R.A. Brady, J.G. Leid, J.H. Calhoun, J.W. Costerton, M.E. Shirtliff, Osteomyelitis and the role of biofilms in chronic infection, FEMS Immunol. Med. Microbiol. 52 (1) (2008) 13–22.

[48] W. Zimmerli, C. Moser, Pathogenesis and treatment concepts of orthopaedic biofilm infections, FEMS Immunol. Med. Microbiol. 65 (2) (2012) 158–168.

[49] A. Rodet, The classic. An experimental study on infectious osteomyelitis, Clin. Orthop. Relat. Res. 99 (96) (1973) 3–4.

[50] C.W. Norden, Experimental osteomyelitis. I. A description of the model, J. Infect. Dis. 122 (5) (1970) 410–418.

[51] C.W. Norden, Experimental osteomyelitis. II. Therapeutic trials and measurement of antibiotic levels in bone, J. Infect. Dis. 124 (6) (1971) 565–571.

[52] W.H. Bowers, F.C. Wilson, W.B. Greene, Antibiotic prophylaxis in experimental bone infections, J. Bone Joint Surg. Am. 55 (4) (1973) 795–807.

[53] V.T. Andriole, D.A. Nagel, W.O. Southwick, A paradigm for human chronic osteomyelitis, J. Bone Joint Surg. Am. 55 (7) (1973) 1511–1515.

[54] V.T. Andriole, D.A. Nagel, W.O. Southwick, Chronic staphylococcal osteomyelitis: an experimental model, Yale J. Biol. Med. 47 (1) (1974) 33–39.

[55] S. Kaarsemaker, G.H. Walenkamp, A.E. vd Bogaard, New model for chronic osteomyelitis with *Staphylococcus aureus* in sheep, Clin. Orthop. Relat. Res. 339 (1997) 246–252.

[56] C.W. Norden, E. Keleti, Experimental osteomyelitis caused by *Pseudomonas aeruginosa*, J. Infect. Dis. 141 (1) (1980) 71–75.

[57] M.E. Power, M.E. Olson, P.A. Domingue, J.W. Costerton, A rat model of *Staphylococcus aureus* chronic osteomyelitis that provides a suitable system for studying the human infection, J. Med. Microbiol. 33 (3) (1990) 189–198.

[58] G. Blomgren, U. Lindgren, Postoperative infections resulting from bacteremia. An experimental study in rabbits, Acta Orthop. Scand. 51 (5) (1980) 761–765.

[59] G. Blomgren, U. Lindgren, The susceptibility of total joint replacement to hematogenous infection in the early postoperative period: an experimental study in the rabbit, Clin. Orthop. Relat. Res. 151 (1980) 308–312.

[60] G. Blomgren, H. Lundquist, C.E. Nord, U. Lindgren, Late anaerobic haematogenous infection of experimental total joint replacement. A study in the rabbit using *Propionibacterium acnes*, J. Bone Joint Surg. (Br.) 63B (4) (1981) 614–618.

[61] J.H. Calhoun, J.T. Mader, Treatment of osteomyelitis with a biodegradable antibiotic implant, Clin. Orthop. Relat. Res. 341 (1997) 206–214.

[62] M.W. Nijhof, H.P. Stallmann, H.C. Vogely, A. Fleer, L.M. Schouls, W.J. Dhert, et al., Prevention of infection with tobramycin-containing bone cement or systemic cefazolin in an animal model, J. Biomed. Mater. Res. 52 (4) (2000) 709–715.

[63] W. Petty, S. Spanier, J.J. Shuster, C. Silverthorne, The influence of skeletal implants on incidence of infection. Experiments in a canine model, J. Bone Joint Surg. Am. 67 (8) (1985) 1236–1244.
[64] M.S. Smeltzer, J.R. Thomas, S.G. Hickmon, R.A. Skinner, C.L. Nelson, D. Griffith, et al., Characterization of a rabbit model of staphylococcal osteomyelitis, J. Orthop. Res. 15 (3) (1997) 414–421.
[65] R.T. Southwood, J.L. Rice, P.J. McDonald, P.H. Hakendorf, M.A. Rozenbilds, Infection in experimental hip arthroplasties, J. Bone Joint Surg. (Br.) 67 (2) (1985) 229–231.
[66] H.C. Vogely, C.J. Oosterbos, E.W. Puts, M.W. Nijhof, P.G. Nikkels, A. Fleer, et al., Effects of hydrosyapatite coating on Ti-6A1-4V implant-site infection in a rabbit tibial model, J. Orthop. Res. 18 (3) (2000) 485–493.
[67] M. Patel, Y. Rojavin, A.A. Jamali, S.J. Wasielewski, C.J. Salgado, Animal models for the study of osteomyelitis, Semin. Plast. Surg. 23 (2) (2009) 148–154.
[68] P. Lankinen, K. Lehtimaki, A.J. Hakanen, A. Roivainen, H.T. Aro, A comparative ^{18}F-FDG PET/CT imaging of experimental *Staphylococcus aureus* osteomyelitis and *Staphylococcus epidermidis* foreign-body-associated infection in the rabbit tibia, EJNMMI Res. 2 (1) (2012) 41.
[69] J.C. Odekerken, J.J. Arts, D.A. Surtel, G.H. Walenkamp, T.J. Welting, A rabbit osteomyelitis model for the longitudinal assessment of early post-operative implant infections, J. Orthop. Surg. Res. 8 (1) (2013) 38.
[70] J.C. Odekerken, B.T. Brans, T.J. Welting, G.H. Walenkamp, ^{18}F-FDG microPET imaging differentiates between septic and aseptic wound healing after orthopedic implant placement. A longitudinal study of an implant osteomyelitis in the rabbit tibia, Acta Orthop. 85 (3) (2014) 9.
[71] J.C.E. Odekerken, G.H.I.M. Walenkamp, B.T. Brans, T.J.M. Welting, J.J.C. Arts, The longitudinal assessment of osteomyelitis development by molecular imaging in a rabbit model, Biomed. Res. Int. 2014 (2014) 424652.
[72] Becker, Dtsch. Med. Wochenschr. 9 (46) (1883).
[73] A. Rodet, Etude expérimentale sur l'osteomyelite infectieuse, C. R. Hebd. Seances Acad. Sci. 99 (1884) 569–571.
[74] E. Lexer, Zur experimentellen erzeugung osteomyelitischer herde, Archiv für Klinische Chirurgie 48 (1894) 181–200.
[75] E. Lexer, Experimente uber osteomyelitis, Archiv für Klinische Chirurgie 53 (1896) 266–277.
[76] R.H. Thompson, R.J. Dubos, Production of experimental osteomyelitis in Rabbits by intravenous injection of *Staphylococcus aureus*, J. Exp. Med. 68 (2) (1938) 191–206.
[77] L. Scheman, M. Janota, P. Lewin, The production of experimental osteomyelitis: preliminary report, JAMA 117 (18) (1941) 1525–1529.
[78] R.A. Elson, A.E. Jephcott, D.B. McGechie, D. Verettas, Bacterial infection and acrylic cement in the rat, J. Bone Joint Surg. (Br.) 59-B (4) (1977) 452–457.
[79] G. Blomgren, U. Lindgren, Late hematogenous infection in total joint replacement: studies of gentamicin and bone cement in the rabbit, Clin. Orthop. Relat. Res. 155 (1981) 244–248.
[80] W. Zimmerli, P.D. Lew, F.A. Waldvogel, Pathogenesis of foreign body infection. Evidence for a local granulocyte defect, J. Clin. Investig. 73 (4) (1984) 1191–1200.
[81] R.H. Fitzgerald Jr., Experimental osteomyelitis: description of a canine model and the role of depot administration of antibiotics in the prevention and treatment of sepsis, J. Bone Joint Surg. Am. 65 (3) (1983) 371–380.

[82] C.W. Norden, M. Shaffer, Treatment of experimental chronic osteomyelitis due to *Staphylococcus aureus* with vancomycin and rifampin, J. Infect. Dis. 147 (2) (1983) 352–357.

[83] K.R. Emslie, S. Nade, Acute hematogenous staphylococcal osteomyelitis. A description of the natural history in an avian model, Am. J. Pathol. 110 (3) (1983) 333–345.

[84] G.T. Rodeheaver, D. Rukstalis, M. Bono, W. Bellamy, A new model of bone infection used to evaluate the efficacy of antibiotic-impregnated polymethylmethacrylate cement, Clin. Orthop. Relat. Res. 178 (1983) 303–311.

[85] C.W. Norden, Experimental chronic staphylococcal osteomyelitis in rabbits: treatment with rifampin alone and in combination with other antimicrobial agents, Rev. Infect. Dis. 5 (Suppl. 3) (1983) S491–S494.

[86] C.W. Norden, E. Shinners, Ciprofloxacin as therapy for experimental osteomyelitis caused by *Pseudomonas aeruginosa*, J. Infect. Dis. 151 (2) (1985) 291–294.

[87] W. Petty, S. Spanier, J.J. Shuster, Prevention of infection after total joint replacement. Experiments with a canine model, J. Bone Joint Surg. Am. 70 (4) (1988) 536–539.

[88] P. Worlock, R. Slack, L. Harvey, R. Mawhinney, An experimental model of post-traumatic osteomyelitis in rabbits, Br. J. Exp. Pathol. 69 (2) (1988) 235–244.

[89] D.R. Nelson, T.B. Buxton, Q.N. Luu, J.P. Rissing, The promotional effect of bone wax on experimental *Staphylococcus aureus* osteomyelitis, J. Thorac. Cardiovasc. Surg. 99 (6) (1990) 977–980.

[90] E. Jacob, J.A. Setterstrom, D.E. Bach, J.R. Heath III, L.M. McNiesh, G. Cierny III, Evaluation of biodegradable ampicillin anhydrate microcapsules for local treatment of experimental staphylococcal osteomyelitis, Clin. Orthop. Relat. Res. 1991 (267) (1991) 237–244.

[91] T.N. Gerhart, R.D. Roux, P.A. Hanff, G.L. Horowitz, A.A. Renshaw, W.C. Hayes, Antibiotic-loaded biodegradable bone cement for prophylaxis and treatment of experimental osteomyelitis in rats, J. Orthop. Res. 11 (2) (1993) 250–255.

[92] C.T. Laurencin, T. Gerhart, P. Witschger, R. Satcher, A. Domb, A.E. Rosenberg, et al., Bioerodible polyanhydrides for antibiotic drug delivery: in vivo osteomyelitis treatment in a rat model system, J. Orthop. Res. 11 (2) (1993) 256–262.

[93] G.A. Melcher, B. Claudi, U. Schlegel, S.M. Perren, G. Printzen, J. Munzinger, Influence of type of medullary nail on the development of local infection. An experimental study of solid and slotted nails in rabbits, J. Bone Joint Surg. (Br.) 76 (6) (1994) 955–959.

[94] K.L. Garvin, J.A. Miyano, D. Robinson, D. Giger, J. Novak, S. Radio, Polylactide/polyglycolide antibiotic implants in the treatment of osteomyelitis. A canine model, J. Bone Joint Surg. Am. 76 (10) (1994) 1500–1506.

[95] L. Sanzen, L. Linder, Infection adjacent to titanium and bone cement implants: an experimental study in rabbits, Biomaterials 16 (16) (1995) 1273–1277.

[96] J.A. Fialkov, J.H. Phillips, S.L. Walmsley, I. Morava-Protzner, The effect of infection and lag screw fixation on revascularization and new bone deposition in membranous bone grafts in a rabbit model, Plast. Reconstr. Surg. 98 (2) (1996) 338–345.

[97] C.L. Nelson, S.G. Hickmon, R.A. Skinner, Treatment of experimental osteomyelitis by surgical debridement and the implantation of bioerodable, polyanhydride-gentamicin beads, J. Orthop. Res. 15 (2) (1997) 249–255.

[98] Y.H. An, J. Bradley, D.L. Powers, R.J. Friedman, The prevention of prosthetic infection using a cross-linked albumin coating in a rabbit model, J. Bone Joint Surg. (Br.) 79 (5) (1997) 816–819.

[99] V. Awasthi, B. Goins, R. Klipper, R. Loredo, D. Korvick, W.T. Phillips, Imaging experimental osteomyelitis using radiolabeled liposomes, J. Nucl. Med. 39 (6) (1998) 1089–1094.

[100] L. Nie, D.P. Nicolau, P.R. Tessier, H.P. Kourea, B.D. Browner, C.H. Nightingale, Use of a bioabsorbable polymer for the delivery of ofloxacin during experimental osteomyelitis treatment, J. Orthop. Res. 16 (1) (1998) 76–79.
[101] I. Gursel, F. Korkusuz, F. Turesin, N.G. Alaeddinoglu, V. Hasirci, In vivo application of biodegradable controlled antibiotic release systems for the treatment of implant-related osteomyelitis, Biomaterials 22 (1) (2001) 73–80.
[102] M. Lucke, G. Schmidmaier, S. Sadoni, B. Wildemann, R. Schiller, A. Stemberger, et al., A new model of implant-related osteomyelitis in rats, J. Biomed. Mater. Res. B Appl. Biomater. 67 (1) (2003) 593–602.
[103] J.T. Mader, C.M. Stevens, J.H. Stevens, R. Ruble, J.T. Lathrop, J.H. Calhoun, Treatment of experimental osteomyelitis with a fibrin sealant antibiotic implant, Clin. Orthop. Relat. Res. 403 (2002) 58–72.
[104] M.E. Shirtliff, J.H. Calhoun, J.T. Mader, Experimental osteomyelitis treatment with antibiotic-impregnated hydroxyapatite, Clin. Orthop. Relat. Res. 401 (2002) 239–247.
[105] B. Rutledge, D. Huyette, D. Day, J. Anglen, Treatment of osteomyelitis with local antibiotics delivered via bioabsorbable polymer, Clin. Orthop. Relat. Res. 411 (2003) 280–287.
[106] L.M. Huneault, B. Lussier, P. Dubreuil, L. Chouinard, C. Desevaux, Prevention and treatment of experimental osteomyelitis in dogs with ciprofloxacin-loaded crosslinked high amylose starch implants, J. Orthop. Res. 22 (6) (2004) 1351–1357.
[107] C.G. Ambrose, T.A. Clyburn, K. Louden, J. Joseph, J. Wright, P. Gulati, et al., Effective treatment of osteomyelitis with biodegradable microspheres in a rabbit model, Clin. Orthop. Relat. Res. 421 (2004) 293–299.
[108] J.K. Koort, T.J. Makinen, J. Knuuti, J. Jalava, H.T. Aro, Comparative ^{18}F-FDG PET of experimental *Staphylococcus aureus* osteomyelitis and normal bone healing, J. Nucl. Med. 45 (8) (2004) 1406–1411.
[109] J.K. Koort, T.J. Makinen, E. Suokas, M. Veiranto, J. Jalava, J. Knuuti, et al., Efficacy of ciprofloxacin-releasing bioabsorbable osteoconductive bone defect filler for treatment of experimental osteomyelitis due to *Staphylococcus aureus*, Antimicrob. Agents Chemother. 49 (4) (2005) 1502–1508.
[110] M.R. Craig, K.A. Poelstra, J.C. Sherrell, M.S. Kwon, E.L. Belzile, T.E. Brown, A novel total knee arthroplasty infection model in rabbits, J. Orthop. Res. 23 (5) (2005) 1100–1104.
[111] V. Mendel, H.J. Simanowski, H.C. Scholz, H. Heymann, Therapy with gentamicin-PMMA beads, gentamicin-collagen sponge, and cefazolin for experimental osteomyelitis due to *Staphylococcus aureus* in rats, Arch. Orthop. Trauma Surg. 125 (6) (2005) 363–368.
[112] N. Fukushima, K. Yokoyama, T. Sasahara, Y. Dobashi, M. Itoman, Establishment of rat model of acute staphylococcal osteomyelitis: relationship between inoculation dose and development of osteomyelitis, Arch. Orthop. Trauma Surg. 125 (3) (2005) 169–176.
[113] L. Jones-Jackson, R. Walker, G. Purnell, S.G. McLaren, R.A. Skinner, J.R. Thomas, et al., Early detection of bone infection and differentiation from post-surgical inflammation using 2-deoxy-2-[^{18}F]-fluoro-D-glucose positron emission tomography (FDG-PET) in an animal model, J. Orthop. Res. 23 (6) (2005) 1484–1489.
[114] T.J. Makinen, P. Lankinen, T. Poyhonen, J. Jalava, H.T. Aro, A. Roivainen, Comparison of ^{18}F-FDG and ^{68}Ga PET imaging in the assessment of experimental osteomyelitis due to *Staphylococcus aureus*, Eur. J. Nucl. Med. Mol. Imaging 32 (11) (2005) 1259–1268.

[115] T.J. Makinen, M. Veiranto, J. Knuuti, J. Jalava, P. Tormala, H.T. Aro, Efficacy of bioabsorbable antibiotic containing bone screw in the prevention of biomaterial-related infection due to *Staphylococcus aureus*, Bone 36 (2) (2005) 292–299.
[116] A.A. Beardmore, D.E. Brooks, J.C. Wenke, D.B. Thomas, Effectiveness of local antibiotic delivery with an osteoinductive and osteoconductive bone-graft substitute, J. Bone Joint Surg. Am. 87 (1) (2005) 107–112.
[117] U. Joosten, A. Joist, G. Gosheger, U. Liljenqvist, B. Brandt, C. von Eiff, Effectiveness of hydroxyapatite-vancomycin bone cement in the treatment of *Staphylococcus aureus* induced chronic osteomyelitis, Biomaterials 26 (25) (2005) 5251–5258.
[118] C.J. Salgado, A.A. Jamali, S. Mardini, K. Buchanan, B. Veit, A model for chronic osteomyelitis using *Staphylococcus aureus* in goats, Clin. Orthop. Relat. Res. 436 (2005) 246–250.
[119] R.O. Darouiche, M.D. Mansouri, D. Zakarevicz, A. Alsharif, G.C. Landon, In vivo efficacy of antimicrobial-coated devices, J. Bone Joint Surg. Am. 89 (4) (2007) 792–797.
[120] L.A. Poultsides, L.K. Papatheodorou, T.S. Karachalios, L. Khaldi, A. Maniatis, E. Petinaki, et al., Novel model for studying hematogenous infection in an experimental setting of implant-related infection by a community-acquired methicillin-resistant *S. aureus* strain, J. Orthop. Res. 26 (10) (2008) 1355–1362.
[121] D. Williams, R. Bloebaum, C.A. Petti, Characterization of *Staphylococcus aureus* strains in a rabbit model of osseointegrated pin infections, J. Biomed. Mater. Res. A 85 (2) (2008) 366–370.
[122] B. Laure, J.M. Besnier, A.M. Bergemer-Fouquet, N. Marquet-Van Der Mee, F. Damie, R. Quentin, et al., Effect of hydroxyapatite coating and polymethylmethacrylate on stainless steel implant-site infection with *Staphylococcus epidermidis* in a sheep model, J. Biomed. Mater. Res. A 84 (1) (2008) 92–98.
[123] K. Kanellakopoulou, G.C. Thivaios, M. Kolia, I. Dontas, L. Nakopoulou, E. Dounis, et al., Local treatment of experimental *Pseudomonas aeruginosa* osteomyelitis with a biodegradable dilactide polymer releasing ciprofloxacin, Antimicrob. Agents Chemother. 52 (7) (2008) 2335–2339.
[124] A. Hamel, J. Caillon, C. Jacqueline, J.M. Rogez, G. Potel, Internal device decreases antibiotic's efficacy on experimental osteomyelitis, J. Child. Orthop. 2 (3) (2008) 239–243.
[125] D.J. Moojen, H.C. Vogely, A. Fleer, A.J. Verbout, R.M. Castelein, W.J. Dhert, No efficacy of silver bone cement in the prevention of methicillin-sensitive Staphylococcal infections in a rabbit contaminated implant bed model, J. Orthop. Res. 27 (8) (2009) 1002–1007.
[126] N.M. Bernthal, A.I. Stavrakis, F. Billi, J.S. Cho, T.J. Kremen, S.I. Simon, et al., A mouse model of post-arthroplasty *Staphylococcus aureus* joint infection to evaluate in vivo the efficacy of antimicrobial implant coatings, PLoS ONE 5 (9) (2010). e12580.
[127] C. Jacqueline, G. Amador, J. Caillon, V. Le Mabecque, E. Batard, A.F. Miegeville, et al., Efficacy of the new cephalosporin ceftaroline in the treatment of experimental methicillin-resistant *Staphylococcus aureus* acute osteomyelitis, J. Antimicrob. Chemother. 65 (8) (2010) 1749–1752.
[128] R.L. Stewart, J.T. Cox, D. Volgas, J. Stannard, L. Duffy, K.B. Waites, et al., The use of a biodegradable, load-bearing scaffold as a carrier for antibiotics in an infected open fracture model, J. Orthop. Trauma 24 (9) (2010) 587–591.
[129] D.A. Robinson, J.E. Bechtold, C.S. Carlson, R.B. Evans, M.G. Conzemius, Development of a fracture osteomyelitis model in the rat femur, J. Orthop. Res. 29 (1) (2011) 131–137.

[130] P. Vergidis, M.S. Rouse, G. Euba, M.J. Karau, S.M. Schmidt, J.N. Mandrekar, et al., Treatment with linezolid or vancomycin in combination with rifampin is effective in an animal model of methicillin-resistant *Staphylococcus aureus* foreign body osteomyelitis, Antimicrob. Agents Chemother. 55 (3) (2011) 1182–1186.
[131] A. Gaudin, G. Amador Del Valle, A. Hamel, V. Le Mabecque, A.F. Miegeville, G. Potel, et al., A new experimental model of acute osteomyelitis due to methicillin-resistant *Staphylococcus aureus* in rabbit, Lett. Appl. Microbiol. 52 (3) (2011) 253–257.
[132] V. Alt, K.S. Lips, C. Henkenbehrens, D. Muhrer, M.C. Oliveira Cavalcanti, U. Sommer, et al., A new animal model for implant-related infected non-unions after intramedullary fixation of the tibia in rats with fluorescent in situ hybridization of bacteria in bone infection, Bone 48 (5) (2011) 1146–1153.
[133] W. Poeppl, S. Tobudic, T. Lingscheid, R. Plasenzotti, N. Kozakowski, A. Georgopoulos, et al., Efficacy of fosfomycin in experimental osteomyelitis due to methicillin-resistant *Staphylococcus aureus*, Antimicrob. Agents Chemother. 55 (2) (2011) 931–933.
[134] J.R. Pribaz, N.M. Bernthal, F. Billi, J.S. Cho, R.I. Ramos, Y. Guo, et al., Mouse model of chronic post-arthroplasty infection: noninvasive in vivo bioluminescence imaging to monitor bacterial burden for long-term study, J. Orthop. Res. 30 (3) (2012) 335–340.
[135] J.A. Niska, J.H. Shahbazian, R.I. Ramos, J.R. Pribaz, F. Billi, K.P. Francis, et al., Daptomycin and tigecycline have broader effective dose ranges than vancomycin as prophylaxis against a *Staphylococcus aureus* surgical implant infection in mice, Antimicrob. Agents Chemother. 56 (5) (2012) 2590–2597.
[136] S.A. Horst, V. Hoerr, A. Beineke, C. Kreis, L. Tuchscherr, J. Kalinka, et al., A novel mouse model of *Staphylococcus aureus* chronic osteomyelitis that closely mimics the human infection: an integrated view of disease pathogenesis, Am. J. Pathol. 181 (4) (2012) 1206–1214.
[137] G.Y. Li, J.M. Yin, H. Ding, W.T. Jia, C.Q. Zhang, Efficacy of leukocyte- and platelet-rich plasma gel (L-PRP gel) in treating osteomyelitis in a rabbit model, J. Orthop. Res. 31 (6) (2013) 949–956.
[138] L.K. Johansen, E.L. Svalastoga, D. Frees, B. Aalbaek, J. Koch, T.M. Iburg, et al., A new technique for modeling of hematogenous osteomyelitis in pigs: inoculation into femoral artery, J. Investig. Surg. 26 (3) (2013) 149–153.
[139] L.Y. Yin, M.M. Manring, J.H. Calhoun, A rabbit osteomyelitis model to simulate multibacterial war wound infections, Mil. Med. 178 (6) (2013) 696–700.
[140] J. Xing, T. Hou, B. Luobu, F. Luo, Q. Chen, Z. Li, et al., Anti-infection tissue engineering construct treating osteomyelitis in rabbit tibia, Tissue Eng. A 19 (1–2) (2013) 255–263.
[141] Z. Xie, X. Cui, C. Zhao, W. Huang, J. Wang, C. Zhang, Gentamicin-loaded borate bioactive glass eradicates osteomyelitis due to Escherichia coli in a rabbit model, Antimicrob. Agents Chemother. 57 (7) (2013) 3293–3298.
[142] N.H. Soe, N.V. Jensen, B.M. Nurnberg, A.L. Jensen, J. Koch, S.S. Poulsen, et al., A novel knee prosthesis model of implant-related osteomyelitis in rats, Acta Orthop. 84 (1) (2013) 92–97.
[143] K.D. Sinclair, T.X. Pham, D.L. Williams, R.W. Farnsworth, C.M. Loc-Carrillo, R.D. Bloebaum, Model development for determining the efficacy of a combination coating for the prevention of perioperative device related infections: a pilot study, J. Biomed. Mater. Res. B Appl. Biomater. 101 (7) (2013) 1143–1153.
[144] N. Tran, P.A. Tran, J.D. Jarrell, J.B. Engiles, N.P. Thomas, M.D. Young, et al., In vivo caprine model for osteomyelitis and evaluation of biofilm-resistant intramedullary nails, Biomed. Res. Int. 2013 (2013) 674378.

[145] L. Helbig, H.G. Simank, H. Lorenz, C. Putz, C. Wolfl, A.J. Suda, et al., Establishment of a new methicillin resistant *Staphylococcus aureus* animal model of osteomyelitis, Int. Orthop. 38 (4) (2014) 891–897.
[146] A.C. Cremieux, A. Saleh-Mghir, C. Danel, F. Couzon, O. Dumitrescu, T. Lilin, et al., alpha-Hemolysin, not Panton-Valentine leukocidin, impacts rabbit mortality from severe sepsis with methicillin-resistant *Staphylococcus aureus* osteomyelitis, J. Infect. Dis. 209 (11) (2014) 1773–1780.
[147] J.J. Varrone, K.L. de Mesy Bentley, S.N. Bello-Irizarry, K. Nishitani, S. Mack, J.G. Hunter, et al., Passive immunization with anti-glucosaminidase monoclonal antibodies protects mice from implant-associated osteomyelitis by mediating opsonophagocytosis of *Staphylococcus aureus* megaclusters, J. Orthop. Res. 32 (10) (2014) 1389–1396.
[148] J.S. McLaren, L.J. White, H.C. Cox, W. Ashraf, C.V. Rahman, G.W. Blunn, et al., A biodegradable antibiotic-impregnated scaffold to prevent osteomyelitis in a contaminated in vivo bone defect model, Eur. Cell. Mater. 27 (2014) 332–349.
[149] M.E. Inanmaz, M. Uslu, C. Isik, E. Kaya, T. Tas, R. Bayram, Extracorporeal shockwave increases the effectiveness of systemic antibiotic treatment in implant-related chronic osteomyelitis: experimental study in a rat model, J. Orthop. Res. 32 (6) (2014) 752–756.
[150] A.D. Gahukamble, A. McDowell, V. Post, J. Salavarrieta Varela, E.T. Rochford, R.G. Richards, et al., *Propionibacterium acnes* and *Staphylococcus lugdunensis* cause pyogenic osteomyelitis in an intramedullary nail model in rabbits, J. Clin. Microbiol. 52 (5) (2014) 1595–1606.
[151] C. Zhu, J. Wang, T. Cheng, Q. Li, H. Shen, H. Qin, et al., The potential role of increasing the release of mouse beta-defensin-14 in the treatment of osteomyelitis in mice: a primary study, PLoS ONE 9 (1) (2014) e86874.
[152] J.A. Freiberg, K.S. McIver, M.E. Shirtliff, In vivo expression of streptococcus pyogenes immunogenic proteins during tibial foreign body infection, Infect. Immun. 82 (9) (2014) 3891–3899.
[153] W.H. Chen, Y.J. Kang, L.Y. Dai, B. Wang, C. Lu, J. Li, et al., Bacteria detected after instrumentation surgery for pyogenic vertebral osteomyelitis in a canine model, Eur. Spine J. 23 (4) (2014) 838–845.
[154] H. Ding, C.J. Zhao, X. Cui, Y.F. Gu, W.T. Jia, M.N. Rahaman, et al., A novel injectable borate bioactive glass cement as an antibiotic delivery vehicle for treating osteomyelitis, PLoS ONE 9 (1) (2014) e85472.
[155] X. Cui, C. Zhao, Y. Gu, L. Li, H. Wang, W. Huang, et al., A novel injectable borate bioactive glass cement for local delivery of vancomycin to cure osteomyelitis and regenerate bone, J. Mater. Sci. Mater. Med. 25 (3) (2014) 733–745.
[156] E.I. Shishatskaya, I.V. Kamendov, S.I. Starosvetsky, Y.S. Vinnik, N.N. Markelova, A.A. Shageev, et al., An in vivo study of osteoplastic properties of resorbable poly-3-hydroxybutyrate in models of segmental osteotomy and chronic osteomyelitis, Artif. Cells Nanomed. Biotechnol. 42 (5) (2014) 344–355.
[157] W. Poeppl, T. Lingscheid, D. Bernitzky, U.Y. Schwarze, O. Donath, T. Perkmann, et al., Efficacy of fosfomycin compared to vancomycin in treatment of implant-associated chronic methicillin-resistant *Staphylococcus aureus* osteomyelitis in Rats, Antimicrob. Agents Chemother. 58 (9) (2014) 5111–5116.
[158] L. Gatin, A. Saleh-Mghir, J. Tasse, I. Ghout, F. Laurent, A.C. Cremieux, Ceftaroline-fosamil efficacy against methicillin-resistant *Staphylococcus aureus* rabbit prosthetic joint infection, Antimicrob. Agents Chemother. (2014).

[159] H. Ahtinen, J. Kulkova, L. Lindholm, E. Eerola, A.J. Hakanen, N. Moritz, et al., (68)Ga-DOTA-Siglec-9 PET/CT imaging of peri-implant tissue responses and staphylococcal infections, EJNMMI Res. 4 (2014) 45.

[160] V. Alt, K. Kirchhof, F. Seim, I. Hrubesch, K.S. Lips, H. Mannel, et al., Rifampicin-fosfomycin coating for cementless endoprostheses: antimicrobial effects against methicillin-sensitive *Staphylococcus aureus* (MSSA) and methicillin-resistant *Staphylococcus aureus* (MRSA), Acta Biomater. 10 (10) (2014) 4518–4524.

[161] S. Chatziioannou, O. Papamichos, M.N. Gamaletsou, A. Georgakopoulos, N.G. Kostomitsopoulos, S. Tseleni-Balafouta, et al., 18-Fluoro-2-deoxy-D-glucose positron emission tomography/computed tomography scan for monitoring the therapeutic response in experimental *Staphylococcus aureus* foreign-body osteomyelitis, J. Orthop. Surg. Res. 10 (2015) 132.

[162] D. Neut, R.J. Dijkstra, J.I. Thompson, C. Kavanagh, H.C. van der Mei, H.J. Busscher, A biodegradable gentamicin-hydroxyapatite-coating for infection prophylaxis in cementless hip prostheses, Eur. Cell. Mater. 29 (2015) 42–55. discussion 55-6.

[163] S.M. Shiels, K.M. Bedigrew, J.C. Wenke, Development of a hematogenous implant-related infection in a rat model, BMC Musculoskelet. Disord. 16 (2015) 255.

[164] O.L. Nielsen, P. Afzelius, D. Bender, H.C. Schonheyder, P.S. Leifsson, K.M. Nielsen, et al., Comparison of autologous (111)In-leukocytes, (18)F-FDG, (11)C-methionine, (11)C-PK11195 and (68)Ga-citrate for diagnostic nuclear imaging in a juvenile porcine haematogenous *Staphylococcus aureus* osteomyelitis model, Am. J. Nucl. Med. Mol. Imaging 5 (2) (2015) 169–182.

[165] L. Yan, D.M. Jiang, Z.D. Cao, J. Wu, X. Wang, Z.L. Wang, et al., Treatment of *Staphylococcus aureus*-induced chronic osteomyelitis with bone-like hydroxyapatite/poly amino acid loaded with rifapentine microspheres, Drug Des. Devel. Ther. 9 (2015) 3665–3676.

[166] D. Arens, M. Wilke, L. Calabro, S. Hackl, S. Zeiter, I. Zderic, et al., A rabbit humerus model of plating and nailing osteosynthesis with and without *Staphylococcus aureus* osteomyelitis, Eur. Cell. Mater. 30 (2015) 148–161. discussion 61–2.

[167] M. Paakkonen, M.J. Kallio, P.E. Kallio, H. Peltola, C-reactive protein versus erythrocyte sedimentation rate, white blood cell count and alkaline phosphatase in diagnosing bacteraemia in bone and joint infections, J. Paediatr. Child Health 49 (3) (2013) E189–E192.

[168] L.Y. Shih, J.J. Wu, D.J. Yang, Erythrocyte sedimentation rate and C-reactive protein values in patients with total hip arthroplasty, Clin. Orthop. Relat. Res. 225 (1987) 238–246.

[169] A.W. Glaudemans, F. Galli, M. Pacilio, A. Signore, Leukocyte and bacteria imaging in prosthetic joint infection, Eur. Cell. Mater. 25 (2013) 61–77.

[170] C.J. Palestro, C. Love, K.K. Bhargava, Labeled leukocyte imaging: current status and future directions, Q. J. Nucl. Med. Mol. Imaging 53 (1) (2009) 105–123.

[171] C.J. Palestro, C. Love, T.T. Miller, Infection and musculoskeletal conditions: imaging of musculoskeletal infections, Best Pract. Res. Clin. Rheumatol. 20 (6) (2006) 1197–1218.

[172] A. Al-Zahrani, K. El-Saban, H. Al-Sakhri, Diagnosis of bone infection by complementary role of technetium-99m MDP and technetium-99m hexamethylpropylene-amineoxime-leukocytes, Indian J. Nucl. Med. 27 (3) (2012) 164–171.

[173] E.J. Hoffmann, M.E. Phelps, N.A. Mullani, C.S. Higgins, M.M. Ter-Pogossian, Design and performance characteristics of a whole-body positron transaxial tomograph, J. Nucl. Med. 17 (6) (1976) 493–502.

[174] M.E. Phelps, E.J. Hoffman, N.A. Mullani, M.M. Ter-Pogossian, Application of annihilation coincidence detection to transaxial reconstruction tomography, J. Nucl. Med. 16 (3) (1975) 210–224.

[175] E.E. Kim, Clinical PET: Principles and Applications, Springer, New York, 2004. xviii, 394 pp.
[176] D.J. Schlyer, PET tracers and radiochemistry, Ann. Acad. Med. Singap. 33 (2) (2004) 146–154.
[177] C.P. Bleeker-Rovers, F.J. Vos, F.H. Corstens, W.J. Oyen, Imaging of infectious diseases using [18°F] fluorodeoxyglucose PET, Q. J. Nucl. Med. Mol. Imaging 52 (1) (2008) 17–29.
[178] A. Schmitz, H.J. Risse, T. Kalicke, F. Grunwald, O. Schmitt, FDG-PET for diagnosis and follow-up of inflammatory processes: initial results from the orthopedic viewpoint, Z. Orthop. Ihre Grenzgeb. 138 (5) (2000) 407–412.

Part Two

Types of periprosthetic joint infections

Periprosthetic infection in the hip joint

6

S. Alazzawi*, O. Khan†, F.S. Haddad‡,§
*The Royal London Hospital, London, United Kingdom, †Guy's and St. Thomas' NHS Foundation Trust, London, United Kingdom, ‡Institute of Sport, Exercise and Health, London, United Kingdom, §University College Hospital, London, United Kingdom

6.1 Introduction

Total hip replacement (THR) is one of the most successful operations performed today. Approximately 95% of patients report an improvement in their hip-related pain symptoms following their surgery. Periprosthetic joint infection (PJI) is one of the leading causes of failure following hip replacement surgery and places an immense physical and emotional burden on the patient and enormous financial pressures on society. Eradication of PJI remains an on-going challenge and one, which today is receiving even more intense efforts to overcome.

6.2 Definition

PJI of the hip joint or arthroplasty-related infection can be defined as the growth of microorganisms in the hip joint in the presence of a hip replacement prosthesis. It represents deep joint infection and does not include isolated superficial wound infection. The affected hip prosthesis may include components of primary or revision THR, hip resurfacing, or hip hemiarthroplasty. The incidence of PJI is 0.5–2% after primary total hip arthroplasty (THA) and 5–7% after revision THA.

The Musculoskeletal Infection Society proposed specific criteria for the definition of PJI during its 21st annual meeting in 2011 [1]. These criteria include:

- Presence of a sinus tract communicating with the prosthesis; or
- A pathogen is isolated by culture from at least two separate tissue or fluid samples obtained from the affected prosthetic joint; or
- Four of the following six criteria exist:
  - Elevated serum erythrocyte sedimentation rate (ESR) and serum C-reactive protein (CRP) concentration
  - Elevated synovial leukocyte count
  - Elevated synovial neutrophil percentage (PMN%)
  - Presence of purulence in the affected joint
  - Isolation of a microorganism in one culture of periprosthetic tissue or fluid
  - Greater than five neutrophils per high-power field in five high-power fields observed from histologic analysis of periprosthetic tissue at 9400 magnification.

Management of Periprosthetic Joint Infections (PJIs). http://dx.doi.org/10.1016/B978-0-08-100205-6.00006-9

Despite these criteria, making the correct diagnosis of PJI remains a difficult challenge due to the heterogeneity of the clinical presentations. Haddad et al. [2] identified three potential problems with using the criteria listed above for defining PPI. First, on many occasions there is no sinus tract communicating between the deep joint space and the extracorporeal environment. Second, studies have shown good specificity is achieved with three cultures positive for the same pathogen from six tissue or fluid samples [3]. The third problem is having specific cut-off limits for certain tests such as an ESR > 30 mm/h and a CRP > 10 mg/L. These low thresholds can improve the sensitivity at the expense of the specificity particularly if there is an underlying inflammatory arthropathy. They propose that the MSIS definition may need to be revised in the future to detect more cases of PPI that currently would not satisfy the above criteria. They have suggested that the definition be based on the certainty of making the diagnosis, divided into four grades:

- Definite infection
- Probable infection
- Possible infection, and
- Infection unlikely

6.3 Classification

PJI is commonly classified based on the time it takes for the infection to manifest itself from the index operation. This method of classification has been popularized as it helps determine the management. In 1975 Coventry [4] described one of the earliest methods of PJI classification based on the chronological order from the time of the initial procedure. It includes three stages. Stage 1 (the immediate postoperative period) occurs within the first 30 days and arises as a result of an infected haematoma or superficial wound infection that progresses deeper. Stage 2 or delayed infection occurs 6–24 months after surgery and is due to intraoperative contamination perhaps due to a smaller inoculum or organism of low virulence. The third stage occurs late (more than 2 years after surgery) and is usually due to haematogenous spread; also, changes in host immunity may trigger the growth of bacteria that were implanted at the original surgery but which had remained dormant. This classification has been modified over the years by Toms et al. [5] and Tsukayama et al. [6] as shown in Tables 6.1–6.3, respectively.

Table 6.1 Classification of periprosthetic infection in total joint arthroplasty [5]

Classification of periprosthetic infection in total joint arthroplasty (as described by Toms et al. [5])	
Type I	Infections that occur within 6 weeks post surgery (acute infections)
Type II	Stage II—Delayed presentation with signs of chronic infection irrespective of time from surgery, for example, sinus
Type III	Infection that occurs suddenly in an otherwise well-functioning hip—usually secondary to haematogenous spread from a distant source

Table 6.2 Classification of infection after total hip arthroplasty [6]

Classification of infection after total hip arthroplasty (as described by Tsukayama et al. [6])	
Type I	Early postoperative infection that developed less than 4 weeks after the operation
Type II	Late chronic infection was one that developed 4 weeks or more after the index operation and that had an insidious clinical course
Type III	Acute hematogenous infection associated with a antecedent bacteremia and characterized by an acute onset of joint symptoms in a previously well-functioning joint replacement
Type IV	Positive intraoperative cultures that had been obtained during revision arthroplasty operation for aseptic loosening, and infection was not obvious clinically at the time of the revision

Table 6.3 Classification of infection after total hip arthroplasty based on the mechanism of infection

Classification of infection after total hip arthroplasty based on the mechanism of infection (as described by Hernigou et al. [18])	
Type I	Intraoperative contamination
Type II	Delayed contamination
Type III	Contamination by direct spread
Type IV	Infections related to reactivation of an indolent infection

6.4 Causative organisms

Bacteria are the usual cause of PJI's and *Staphylococcus aureus* is the most commonly associated organism (Table 6.4). The implanted prosthetic joint has no immune response, and its surface may act as a binding site for bacteria. The bacteria can arise from direct contamination during the original surgery or from a haematogenous spread [8]. More recent studies have shown a changing trend from a single pathogen growth to a more polymicrobial one, indicating a change in the complexity of PPI. Table 6.5 summarizes the causative organisms from 125 patients who were treated for infected THR (results from University College Hospital, London).

In immuno compromised individuals fungal infection should also be considered. In many cases the causative organism remains unknown, and these patients are more likely to have a poor outcome following revision surgery because appropriately targeted antimicrobial therapy is difficult to accomplish.

6.4.1 Biofilm formation

When the implant in a joint becomes colonised by micro-organisms, the latter will produce an extra-cellular matrix or glycocalix, comprised of multiple polysaccharides, also known as the "biofilm". The term "biofilm" was first used by Mack et al. in 1975

Table 6.4 Types of organisms obtained in patients with infected total hip arthroplasty

Type of organism	UK (%)	Australia (%)	Germany (%)	USA (%)
Methicillin-sensitive *Staphylococcus aureus* (MSSA)	44	30	13	31
Coagulase-negative *Staphylococcus aureus* (CNS)	47	45	39.3	20.2
Methicillin-resistant *Staphylococcus aureus* (MRSA)	8	45	48.1	12.1
Polymicrobial infection	47	36	–	–
Anaerobic organisms	7	1.7	9	0.9
Culture negative infections	–	–	16.1	15.8

Results reported by Sandiford et al. [7].

Table 6.5 Causative organisms from 125 patients who were treated for infected THR

Organism	Before 2004	After 2004	*P* value
Staphylococcus aureus	20	9	
Methicillin-resistant *Staphylococcus aureus*	6	14	
Coagulase-negative *Staphylococcus*	6	15	0.04
Methicillin-resistant staph epidermidis	4	10	
Polymicrobial	4	12	0.038
Streptococcus	6	5	
No growth	8	5	
Gram-negative bacteria	5	14	0.031
Anaerobic	3	5	
Fungal	1	1	
Mycobacterial	0	0	

Results from University College Hospital, London [9].

(Microbial film development in a trickling filter, Microbial Ecology) [10]. However, the process was described much earlier by Anthony van Leewenhoek in 1684 from his work on the accumulation of microorganisms in dental plaque.

The Biofilm formation can start within minutes of implanting the prosthesis and was described by Myrvik and his colleagues [19] as "the race for the surface." The biofilm resists antibiotic penetration and provides a degree of protection for the bacteria from antibiotics and macrophages. It also helps the bacteria spread into surrounding tissues.

Antibiotics are very effective in eliminating bacteria in their free floating (planktonic) states. Bacteria existing within a biofilm are protected by the exopolysaccharide matrix, which increases bacterial resistance to antibiotics up to a 1000-fold. Other factors within the biofilm also help maintain bacterial resistance. The bacteria within the biofilm are exposed to subminimal inhibitory concentrations of antibiotics, inducing resistance; the process of horizontal gene transfer is facilitated within the biofilm. General stress responses of the bacteria within the biofilm to their local environment (e.g., local heat, pH, and osmolality) may induce more tolerant phenotypes.

The biofilm helps in the progression of chronic infection by allowing individual cells to disperse from the original biofilm and seed new sites of infection. The bacteria that are in these free (planktonic) states are more susceptible to antibiotics and are currently being analyzed to allow for the development of dispersal-mediated treatment options.

It is the difficulty to recreate the complex biofilm environment in the laboratory, which makes the search for effective eradication therapies much more challenging.

6.5 Diagnosis

6.5.1 History and clinical examination

There are a few risk factors in a patient's history that should alert the surgeon to the possibility of PJI. These include a history of recent infection, revision surgery, multiple operations to the same joint, history of previous PJI, diabetes, renal failure, or immune-comprised patients.

The clinical presentation is determined mainly by the type and the severity of the infection. The patient usually presents with a painful joint and difficulty in weight bearing. The history of fever, night sweat, and general malaise should raise the suspicion of joint infection.

On examination, local induration, warm, and erythematous skin may be present. A discharging surgical wound is usually associated with acute early infection, while the presence of a sinus is a sign of chronic infection.

6.5.2 Laboratory blood tests

Basic blood tests are used as a first line for investigating the possibility of an infected joint. These include:

- White blood cell (WBC) count and differential
- ESR > 30 mm/h has sensitivity of 82%, a specificity of 85%, a positive predictive value of 58%, and a negative predictive value of 95% [11]. It rises following uncomplicated joint replacement and takes up to 90 days to return to normal levels. ESR levels are not useful for workup of acute PJI (<6 weeks).
- CRP > 10 mg/L has 96% sensitivity, 92% specificity, 74% positive predictive value, and a 99% negative predictive value [5]. It returns to a normal level within 3 weeks following surgery.
- Combination of both ESR and CRP increased the sensitivity to 95–96%. Normal ESR combined with normal CRP is reliably specific for excluding infection.

Other newer serum biomarkers include procalcitonin, tumor necrosis factor-alpha, and interleukin-6 (IL-6), which are increased in serum during bacterial infections. Hoell et al. [12] reported their results from monitoring IL-6 in 55 patients with PPI. Their results showed that IL-6 ≥ 13 pg/mL could be considered as the cut-off value to confirm the presence of PPI with a positive-predictive value of 90.9%. A serum IL-6 ≤ 8 pg/mL can exclude the presence of PPI with a negative predictive value of 92.1%. Despite ongoing research, the "perfect" biomarker is yet to be found; and every so often, a "new" biomarker is unveiled with an even "better" predictive value.

6.5.3 Joint aspiration (synovial fluid tests)

In order to achieve maximum probability of a positive result, the patient should be off antibiotics for 2 weeks prior to diagnostic aspiration or obtaining tissue specimens.

- Gram stain microscopy has low sensitivity, but its value comes from its high specificity of almost 100%, which makes it an excellent tool for the rapid confirmation of PJI when positive.
- Culture and antibiotics sensitivity. The sensitivity of the cultures of the aspirate ranges from 11% to 100% but on average it is 70%.
- Synovial WBC count and polymorphonuclear neutrophil (PMN) percentage: acute PJI has raised synovial WBC of >10,000 cells/μL and >90% synovial PMN. In chronic infection the synovial WBC is >3000 cells/μL with >80% synovial PMN.
- Synovial Leukocyte esterase is another useful marker that can be used in establishing the diagnosis of PJI. It has a sensitivity of 81–93% and specificity of 87–100%.
- Alpha-defensin (α-defensin) in synovial fluid is a new biomarker that can help diagnose PPI. The cut-off positive value for this test is 5.2 mg/L [13]. It has the advantage of not being affected by antibiotic treatment and is reported to have high sensitivity and specificity for the diagnosis of PPI. The results from the Rothman Institute are based on using laboratory immunoassay to detect the α-defensin biomarker. However, it is important to note that the sensitivity drops to 69% while maintaining a specificity of 100% when leukocyte esterase reagent test strip was used. The same group analyzed the synovial fluid of 149 patients and showed that a combination of synovial fluid a-defensin and synovial CRP tests increased the sensitivity to 97% and the specificity to 100% for the diagnosis of PPI.
- Polymerase chain reaction is a laboratory technique that studies the 16s rRNA gene of bacteria. It is usually helpful when direct identification of microorganisms is not possible. For instance, when patients were already treated with courses of antibiotics. The sample is heated to a point that causes the DNA strands to unfold and separate. A short chain of predetermined nucleic acid sequence (called a primer), specific for the organism of interest, is then added and binds only to the strands that match its sequence, allowing bacterial detection. However, this technique does not differentiate between live and dead bacteria. Hence, there may be a high number of false-positive results [14].

6.5.4 Radiological investigations

- This includes an anterioposterior image of the pelvis and lateral view of the affected hip joint. Usually comparing serial images over a long period of time when feasible, they can help identify periprosthetic bone resorption (endosteal scalloping) over a short period of time, which should always raise the alarm of possible PJI. However, normal radiographs cannot exclude infection.
- Ultrasound is a useful tool to identify any collection around or within the hip joint. It is noninvasive, safe, and a cost-effective method of investigation. It can aid with the joint aspiration if it is required, but aspiration of a prosthetic joint must only be performed in an operating theater.
- Magnetic resonance imaging is useful especially in joints with metal on metal adverse tissue reaction associated with possible joint infection. The use of metal artifact reduction

sequence software allows the identification of adverse soft tissue reaction in the form of a mass, fluid collection, or muscle destruction. However, it can be unreliable in differentiating aseptic from septic loosening.

- CT scan is usually helpful to identify the available bone stock and can reveal areas of significant bone loss especially on the acetabular side.
- Bone scan has become less popular as it has low sensitivity and specificity. The rates can be improved when a dual tracer technique such as an indium In 111-labeled leukocyte scan performed simultaneously with a technetium Tc 99 m diphosphonate scan.
- Single photon emission computed tomography (SPECT) is the current popular modality. Several radio-materials have been used in combination with the SPECT scan including 99mTc-labeled diphosphonates, radio-labeled WBCs with 111In, 99mTc-HMPAO, or 67Ga citrate. The SPECT scan has good sensitivity and the specificity rates reach 93%. It has the added ability to identify the focus of the infection and whether it involves only the soft tissue, or both the soft tissue and the bone [15].
- 18F-Fluorodeoxyglucose (FDG) positron emission tomography is based on detecting the presence of inflammatory cells with an increased glucose uptake in areas of infection. Patients have to fast for a few hours prior to the test. Results from a few studies have reported a sensitivity of 80–100% and specificity of nearly 90–100% [15].
- Fistulography can be used in the presence of chronic infection with a well-formed fistula that may be communicating with the hip joint. Contrast is injected into the opening of the discharging sinus and subsequent radiographs are performed.

6.5.5 Intraoperative tissue specimens

- Intraoperative frozen sections of tissue samples can be used during single-stage revision for patients in whom the diagnosis of PJI has not been established or excluded. The finding of acute inflammation, defined as 5 or more WBCs (polymorphonuclear leukocytes) per high-power field. The main benefits are a relatively low cost and a quick time to obtain the results. However, it has the disadvantages of low sensitivity and being operator dependant. As a result, most centers no longer use frozen section as part of their routine practice.
- During the revision surgery, the administration of any antibiotics should be withheld until multiple tissue specimens (minimum of 5–6 as per Atkins et al. report, 1998) are taken during the operation. The samples should be taken from different layers and areas. These specimens are sent for histological and microbiological analyses. The latter are usually cultured using broth or solid medium.

6.6 Other new diagnostic tools

6.6.1 Microcalorimetry

This method aims to detect the bacterial activity by measuring bacterial heat production, which is generated by metabolic activity and cell division, that is, replicating microorganisms in culture.

6.6.2 Mass spectrometry

This technique involves the use of matrix-assisted laser desorption ionization time-of-flight mass spectrometry devices. It helps identify bacteria directly from clinical samples rather than from culture.

6.6.3 Fluorescence in situ hybridization

Fluorescence in situ hybridization has been developed to obtain rapid detection of pathogenic bacteria implicated in urine and blood. The time from culture positivity to identification is <1 h.

6.6.4 Sonication of explanted prosthesis

This method relies on sonication of explanted prosthesis followed by culture of the fluid sample obtained. Sonication dislodges bacteria, which are present in their protective biofilm from the surface of the prosthesis. It is useful when dealing with low-grade infection or in situations where patients have received previous antimicrobial therapy.

After extracting the prosthesis during the operation, the removed implant is aseptically placed in a sterile container in the operating room and transported to the microbiological laboratory. A normal saline or Ringer's solution is added to cover at least 75% of the implant surface. It is then mixed for 30 seconds, sonicated for 1 minute and mixed again for 30 seconds before plating the sonication fluid.

6.7 Management

6.7.1 Multidisciplinary team working

PJI is a potentially catastrophic problem that has significant health and financial burdens for the patient and the health care provider, respectively. The two most important factors to ensure successful outcomes are a multidisciplinary team approach and to have the treatment in an institution, which has the required expertise to deal with PJI. The necessary team members and resources are:

- Nursing staff that are familiar with revision hip surgery
- Physiotherapist
- Occupational therapist
- Microbiologists (infectious disease specialists) with lab facilities that enable them to perform the diagnostic investigations, which have been discussed earlier
- Clinical pharmacists
- Musculoskeletal radiologists with access to various imaging modalities
- Revision hip surgeon with experience in dealing with PJI
- Plastic surgeon
- Experienced anesthetists that can deal with high-risk patients and an access to high dependency and intensive care units
- Pain care team.

6.7.2 Preoperative planning and decision making

McPherson and his colleagues [16] from the University of Southern California described a staging System for the management of periprosthetic infections in 2002. It is based on the type of infection, the medical health of the patient (subdivided into host A–C), and the condition of the wound (Table 6.5). This staging system has been shown to improve the outcome of treatment, which can be tailored for a particular class of patient. It is now commonly used in clinical practice to help decide on the appropriate management of PJI.

Staging system for periprosthetic infections according to McPherson et al. [16]

Infection type	Systemic host grade	Local extremity grade
I: Early postoperative infection (<4 postoperative weeks)	A: Uncompromised	1: Uncompromised
II: Hematogenous infection (<4 weeks duration)	B: Compromised (1–2 compromising factors)	2: Compromised (1–2 compromising factors)
III: Late chronic infection (>4 weeks duration)	C: Significant compromise (>2 compromising factors) or one of – Absolute neutrophil count <1000 – CD4 T-cell count <100 – Intravenous drug abuse – Chronic active infection at another site – Dysplasia or neoplasm of the immune system	3: Significant compromise (>2 compromising factors)
	Compromising factors:	Local extremity grade (wound) compromising factors:
	– Age > 80	– Active infection present >3–4 months
	– Immunosuppressive drugs	– Multiple incision with skin bridges
	– Alcoholism	– Soft tissue loss from prior trauma
	– Malignancy	– Subcutaneous abscess > 8 cm^2
	– Chronic active dermatitis or cellulites	– Synovial cutaneous fistula
	– Pulmonary insufficiency	– Prior periarticular fracture or trauma about a joint
	– Chronic indwelling catheter	– Prior local irradiation
	– Renal failure requiring dialysis	– Vascular insufficiency to extremity
	– Chronic malnutrition	
	– Systemic inflammatory disease	
	– Current nicotin use	
	– Systemic immune compromise	
	– Diabetes	
	– Hepatic insufficiency	

Source: Table from periprosthetic total hip infection: outcomes using a staging system. McPherson et al. [16].

6.7.3 Debridement and implant retention (with or without exchange of mobile parts)

This option is preferable for early postoperative (less than 3–4 weeks) or late acute haematogenous infection. It necessitates a stable implant, good soft tissue cover; patients that are not immune comprised, ability to identify the causative organism(s) and the availability of sensitive antibiotics.

The procedure involves an open approach with aggressive debridement of all infected tissue and extensive synovectomy. A through and copious washout of minimum of 9L of warm normal saline using low pressure pulsed lavage. Multiple tissue specimens should be taken throughout the procedure, and the antibiotics should be withheld until all specimens were taken. The tissue samples should be sent for both microbiological and histological analyses. The surgical team may wish to rescrub and apply new drapes prior to the start of the closure.

The exchange of mobile parts with new ones is preferable as it may help reduce the chances of failure. Repeat the washout may be required to achieve good results.

The patient will then commence on intravenous (IV) broad-spectrum antibiotics, which has to be modified according to the microbiology results. A minimum period of 6 weeks of IV antibiotics is recommended.

The benefits of this treatment approach are fewer operations, less costs, and preservation of bone stock. However, the success rate is widely variable between 26% and 71% [5]. These variable results could be related to patient selection, surgical techniques, causative organism(s), different postoperative treatment regimes, and difficulty in identifying the actual onset of infection especially in patients with late, acute haematogenous infection.

6.7.4 Single-stage revision (direct exchange)

This option involves the removal of the entire prosthesis and implanting a new one during the same operation. It has the advantage of one single operation with a success rate reported between 87% and 100%. However, the key to its success depends on identifying the causative organism(s), availability of antibiotic sensitivity, patients who are host A according to McPherson et al., presence of good soft tissue cover, and a surgical team and hospital setting that are familiar with this strategy.

The procedure involves an open approach with removal of all implants, aggressive debridement of all infected tissue, extensive synovectomy, thorough and copious washout with a minimum of 9L of warm normal saline using low pressure pulsed lavage. Multiple tissue specimens should be taken throughout the procedure, and antibiotics should be withheld until all specimens are taken. For cemented prosthesis, all cement fragments should be removed.

The surgical team will rescrub, apply new drapes, and use new instruments. Broad-spectrum antibiotics will be administered at this stage as per local protocol, and the new prosthesis will be implanted.

If the patient has had an earlier aspiration and organisms were identified, then the antibiotics to which the organism is sensitive can be initiated. Otherwise, the patient will be commenced on broad-spectrum, IV antibiotics, which can be modified according to the impending microbiology results. A minimum period of 6 weeks of IV antibiotics is recommended. The patient usually requires a central venous catheter to facilitate IV access.

As the success rate of this approach is higher than debridement and implant retention, it is more preferable in acute early infection for uncemented prostheses. This allows for a relatively easier direct exchange of the implants, prior to complete osteointegration, which would make revision surgery more challenging. However, it is important to emphasize that the success of this approach is heavily reliant on the available surgical expertise and the multidisciplinary approach to it. Therefore, single stage revision should be reserved for centers dealing with high volumes of PJI and have experience with this procedure.

6.7.5 Two-stage revision and implantation of a cement spacer

This strategy has remained the gold standard for treating PJI, particularly for chronic infection; following the failure of other treatment modalities, where the causative organism remains unknown, host B or C patients and where there is poor soft tissue coverage. The reported success varies from 88% to 95% [7].

The first operation involves removal of all implants, aggressive debridement of all infected tissue, excision of any existing sinuses, extensive synovectomy, reaming of the medullary canal, and thorough, copious washout with a minimum of 9L of warm normal saline using low pressure pulsed lavage. Multiple tissue specimens should be taken throughout the procedure, and antibiotics should be withheld until all specimens are taken.

The next step is to implant a cement spacer. The cement used in these spacers is impregnated with antibiotics, which help deliver antibiotics locally within the infected area. The antibiotics can be added while mixing the cement during the operation. Most of the readily antibiotic-impregnated cements have prophylaxis concentrations of antibiotics and do not contain an adequate dose of antibiotic for the management of existing infection.

Several choices of antibiotic are available for combining with cement. These include amikacin, ceftazidime, ciprofloxacin, clindamycin, daptomycin, erythromycin, gentamicin, lincomycin, methicillin, ticarcillin, tobramycin, and vancomycin. They should be bactericidal, released gradually over a period of time, water-soluble, and thermodynamically stable (to withstand the heat of cement polymerization).

The use of the spacer helps prevent retraction and contracture at the joint site. It maintains the joint space with the formation of less scar tissue. There are two types of spacers based on the mode of articulation. It can be either nonarticulating which has the disadvantage of limited joint movement and hence muscle atrophy or a dynamic spacer which maintains some degree of joint movement and is more convenient to the patient than the fixed ones. However, the latter is associated with a greater risk of dislocation or mechanical failure.

The spacer is usually made during the surgery by using antibiotic mixed with cement, with the aid of molds or prefabricated cement spacers. However, there are ready-made options available which are made of metal and plastic coated with antibiotic-impregnated cement.

The patient will then commence intravenous (IV) broad-spectrum antibiotics, which have to be modified according to the microbiology results. The patient usually receives a long line to facilitate intravenous access. Serial readings of inflammatory markers are obtained. It is also important to monitor the kidney and liver function of the patients.

It remains difficult to identify the ideal time for proceeding to the second stage. A combination of factors may aid the decision. They include: normal value of inflammatory markers, a dry and fully healed wound, the absence of local joint symptoms such as pain or warmth, and no evidence of systemic symptoms such as fever or malaise. On average it may take 6–12 weeks. If these factors are not achievable, then second stage should be postponed and a repeat of stage one may be required.

After completion of the planned course of antibiotics, the patient will have a period of 2 weeks without antibiotics. Repeat clinical assessment and inflammatory markers are performed to confirm resolution prior to the second stage. More recently, many surgeons have started to proceed directly to the second-stage implantation without the antibiotics-free interval.

During the second stage, the same surgical approach is used. The spacer is removed and further tissue specimens should be taken for microbiology analysis. Frozen section analysis can be performed during the operation to rule out persistent infection. If there is any obvious evidence of active infection, then the second stage should be abandoned and a repeat of stage one should be performed.

If a cemented prosthesis is to be implanted, then a maximum of 2 g of antibiotic powder per 40 g of cement can be used to avoid significant weakening of the bone cement.

If after the second stage the cultures are negative, the intravenous antibiotics are converted to oral antibiotics as per guidance of the microbiologist. Otherwise, IV antibiotics may need to be continued if the specimens show additional growth of organisms.

6.7.6 Long-term suppressive antibiotics

This option is usually preserved for those who are medically unfit to have further surgical intervention or who refused surgery. It is important to be aware that patients may develop resistance to the antibiotics over a period of time. If they are asymptomatic, then antibiotics may be discontinued. However, close monitoring should be undertaken to preempt any rapid deterioration and the development of sepsis.

6.7.7 Salvage procedures (amputation/arthrodesis/excisional arthroplasty-girdlestones procedures)

These options are less likely to be used nowadays. They can be considered in life-saving situations where the patient is systematically unwell and cannot tolerate multiple complex surgeries.

6.8 Management of periprosthetic fracture in the presence of chronic PJI

Patients with chronic PJI have progressive bone resorption. If the joint is left untreated, it may increase the risk of periprosthetic fracture. The management here is complex and has an associated high morbidity and mortality rate. These patients have usually undergone multiple previous surgeries with failure to clear the infection. They also tend to be medically unfit for further major surgery, and hence the infected joint remains untreated. The other challenges include limited bone stock and poor soft tissue coverage.

Managing these patients should be undertaken in a unit with a well-established service for PJI. The treatment goal will be to optimize the general condition of the patient and then to perform a two-stage revision. The chronicity and bone resorption may indicate that most of the proximal femur will be resected. If that is the case, then custom-made spacers will be required for the first stage. The second stage may involve proximal femoral replacement.

If the patient is unfit to have multiple operations, then one of the salvage operations mentioned earlier may be used.

6.9 The eradication of biofilm: Goals for the future

The prevention of biofilm formation and the eradication of preformed biofilm require further research. Tactics may include special, bactericidal coating of implant surfaces. Development of combination antibiotic therapy, with or without photodynamic treatment (which helps detach the biofilm), may help remove the biofilm bacteria. Certain antibiotics have already been found to be active against biofilm, agents like rifampicin, ciprofloxacin, and fosfomycin [17]. Research into dispersal-mediated treatments may improve the effectiveness of antibiotic penetration of the biofilm and subsequent clearance.

6.10 Conclusion

The management of PJI remains a great challenge. With the development of classification systems, the availability of new techniques to help make the diagnosis, the recognition of the important role of a multidisciplinary team approach, and the use of specialized centers that deal with PJI; the future, successful management of this potentially catastrophic complication is promising.

References

[1] J. Parvizi, B. Zmistowski, E.F. Berbari, T.W. Bauer, B.D. Springer, C.J. Della Valle, et al., New definition for periprosthetic joint infection: from the Workgroup of the Musculoskeletal Infection Society, Clin. Orthop. Relat. Res. 469 (11) (2011) 2992–2994.

[2] S. Oussedik, K. Gould, I. Stockley, F.S. Haddad, Defining peri-prosthetic infection: do we have a workable gold standard? J. Bone Joint Surg. (Br.) 94 (11) (2012) 1455–1456.

[3] B.L. Atkins, N. Athanasou, J.J. Deeks, D.W. Crook, H. Simpson, T.E. Peto, et al., Prospective evaluation of criteria for microbiological diagnosis of prosthetic-joint infection at revision arthroplasty. The OSIRIS Collaborative Study Group, J. Clin. Microbiol. 36 (10) (1998) 2932–2939.

[4] M.B. Coventry, Treatment of infections occurring in total hip surgery, Orthop. Clin. North Am. 6 (4) (1975) 991–1003.

[5] A.D. Toms, D. Davidson, B.A. Masri, C.P. Duncan, The management of peri-prosthetic infection in total joint arthroplasty, J. Bone Joint Surg. (Br.) 88 (2) (2006) 149–155.

[6] D.T. Tsukayama, R. Estrada, R.B. Gustilo, Infection after total hip arthroplasty. A study of the treatment of one hundred and six infections, J. Bone Joint Surg. Am. 78 (4) (1996) 512–523.

[7] N.A. Sandiford, C.P. Duncan, D.S. Garbuz, B.A. Masri, Two-stage management of the infected total hip arthroplasty, Hip Int. 25 (4) (2015) 308–315.

[8] S. Esposito, S. Leone, Prosthetic joint infections: microbiology, diagnosis, management and prevention, Int. J. Antimicrob. Agents 32 (4) (2008) 287–293.

[9] M.S. Ibrahim, S. Raja, M.A. Khan, F.S. Haddad, A multidisciplinary team approach to two-stage revision for the infected hip replacement: a minimum five-year follow-up study, Bone Joint J. 96-B (10) (2014) 1312–1318.

[10] W.N. Mack, J.P. Mack, A.O. Ackerson, Microbial film development in a trickling filter, Microb. Ecol. 2 (3) (1975) 215–226.

[11] M.J. Spangehl, B.A. Masri, J.X. O'Connell, C.P. Duncan, Prospective analysis of preoperative and intraoperative investigations for the diagnosis of infection at the sites of two hundred and two revision total hip arthroplasties, J. Bone Joint Surg. Am. 81 (5) (1999) 672–683.

[12] S. Hoell, L. Borgers, G. Gosheger, R. Dieckmann, D. Schulz, J. Gerss, et al., Interleukin-6 in two-stage revision arthroplasty: what is the threshold value to exclude persistent infection before re-implanatation? Bone Joint J. 97-B (1) (2015) 71–75.

[13] C. Deirmengian, K. Kardos, P. Kilmartin, A. Cameron, K. Schiller, R.E. Booth Jr., et al., The alpha-defensin test for periprosthetic joint infection outperforms the leukocyte esterase test strip, Clin. Orthop. Relat. Res. 473 (1) (2015) 198–203.

[14] R. Trebse, The diagnostic protocol for evaluation of periprosthetic joint infection, Hip Int. 22 (Suppl. 8) (2012) S25–S35.

[15] B. Yue, T. Tang, The use of nuclear imaging for the diagnosis of periprosthetic infection after knee and hip arthroplasties, Nucl. Med. Commun. 36 (4) (2015) 305–311.

[16] E.J. McPherson, C. Woodson, P. Holtom, N. Roidis, C. Shufelt, M. Patzakis, Periprosthetic total hip infection: outcomes using a staging system, Clin. Orthop. Relat. Res. 403 (2002) 8–15.

[17] R.P. Baker, U. Furustrand Tafin, O. Borens, Patient-adapted treatment for prosthetic hip joint infection, Hip Int. 25 (4) (2015) 316–322.

[18] P. Hernigou, C.H. Flouzat-Lachianette, R. Jalil, S. Uirassu Batista, I. Guissou, A. Poignard, Treatment of infected hip arthroplasty, Open Orthop. J. 4 (2010) 126–131.

[19] A.G. Gristina, P. Naylor, Q. Myrvik, Infections from biomaterials and implants: a race for the surface, Med. Prog. Technol. 14 (3–4) (1988–1989) 205–224.

Infection in total knee arthroplasty 7

K.N. Malizos, S.E. Varitimidis
University of Thessaly, Medical School, Biopolis, Larissa, Greece

7.1 Total knee arthroplasty today

Arthritis of the knee is one of the most common clinical entities in the musculoskeletal system, affecting the health-related quality of life of millions of people with pain, disability, and suffering. General risk factors include hereditary predisposition, older age, female gender, ethnicity, high body mass index (BMI), inflammatory diseases, and trauma [1]. The variety of risk factors suggests that different pathophysiologic etiologies may cause primary arthritis, and their identification may lead toward different preventive approaches. Local factors such as chronic repeated overloading, ligament instability, neuromuscular impairment, and joint deformity may accelerate the degenerative process. The progress of the disease varies but at the end stage leads to joint degeneration debilitating the patient with pain, restricted range of motion, and deformity. People with strenuous physical work and with a high BMI are at particularly high risk for severely disabling OA of the knee and should be targeted with effective preventive measures. Encouraged by statistics based upon revision rates alone, total knee arthroplasty (TKA) has been proven one of the most successful operations of all medicine and a durable surgical treatment with very good and long-lasting functional outcome in older patients. The number of TKAs performed worldwide has increased dramatically, particularly in patients under the age of 60 years.

7.2 Infection risks and prevention

Prosthetic joint infection (PJI) after TKA is a very serious and challenging complication that can drastically affect patients' lives, as it may lead to persistent pain and disability, multiple operations with attendant morbidity, and prolonged convalescent periods. Infection may occur in the wound or deep around the prosthesis. It may happen while in the hospital or after various periods of time, even years later. The reported incidence for TKA varies from 1% to 2% at 2 years postoperatively but becomes higher with longer f-up nearing to 7% after revision surgery. Although the rate appears small, to the increasing numbers of TKAs will lead to a great number of infected patients. An infected TKA also adds to the economic costs and psychological burden of the patient [2–6]. A variety of etiology-associated factors may lead to infection of the knee arthroplasty. The presence of systemic or local active infection in an arthritic knee can lead to significantly higher rates of infection, either hematogenous or through direct seeding of the implant following TKA; therefore, elective arthroplasty should be delayed until infections are confirmed to be eradicated [7–9].

Management of Periprosthetic Joint Infections (PJIs). http://dx.doi.org/10.1016/B978-0-08-100205-6.00007-0

Patients with previous operative procedures in the same knee present a compromised local wound environment which may contribute to the development of an infection after the knee arthroplasty [10]. Diabetes mellitus and an uncontrolled hyperglycemia during the perioperative period are associated with increased postoperative complications and higher number of infections [11–15]. Malnutrition has an adverse effect on the outcome of a ioint arthroplasty through the poor wound healing with persistent drainage and increased susceptibility to infections. Parameters to evaluate nutritional status include serum albumin (normal 3.5–5.0 g/dL), serum transferrin (normal 204–360 mg/dL), serum prealbumin (normal 15–35 mg/dL), and total serum lymphocyte count (800–2000/mm^3); should be checked prior to an elective knee replacement; and corrected appropriately with administration of high protein supplements, vitamin and mineral supplementation, increased consumption of calories, mobilization, and physiotherapy [16–20].

Reported patient factors include previous fractures about the knee [21], obesity [22–24], male sex [25], and rheumatoid arthritis [26]. The decision to perform elective knee arthroplasty in morbidly obese patients with BMI $\geq$ 40.0 kg/m^2 should be weighed only after careful consideration of the benefits versus the risk of complications including infection [27,28]. In a meta-analysis of data pooled from 15 observational studies demonstrated that smoking cessation led to fewer wound-healing complications (RR = 0.73, CI = 0.61–0.87). Singh et al. found that current smokers undergoing joint replacement were more likely to present with wound infections, whereas prior smokers were not associated with such high risk. Smoking intervention programs, instituted 6 weeks prior to elective surgery, may diminish the risk of infectious and wound-healing complications [29–33]. The incidence of SSI is reported significantly higher in patients with excessive alcohol consumption, and it is reasonable to delay elective arthroplasty until they reduce consumption, or till dependance is controlled [34,35].

Comorbidities and immunosuppression, although it is ill defined, are a significant risk factor for PJIs. Examples of immunosuppressive agents include glucocorticoids, cytostatics including cyclophosphamide and methotrexate, drugs that act on immunophilins such as tacrolimus, and others agents such as interferons and tumor necrosis factor (TNF)-α inhibiting agents [36]. Patients with end-stage renal disease on hemodialysis are at high risk for severe complications and death. An elective knee arthroplasty should be rather postponed until a successful renal transplantation is carried out [36–38]. An active liver disease may also predispose to a higher infection rate after TKA [39,40]. Patients with history of intravenous drug abuse and with painful knee arthritis present a difficult treatment dilemma, as they are at a very high risk for postoperative infections and other complications [41]. The “International PJIs consensus” is of the opinion that active IV drug abusers should not be offered elective knee arthroplasty. In the case of patients with HIV, orthopedic surgeons should work closely with infectious disease specialists in monitoring CD4 counts and viral loads and that decisions to undertake TKA be made on an individual basis. Patients with CD4 counts greater than 400 cells/mL and undetectable viral loads may be appropriate candidates for elective TKA, as the risk of subsequent SSIs may be decreased. Habermann et al. reported no difference in functional outcome following TJA between patients with or without HIV [37].

Apart from the operative time, other surgeon-related factors associated with PJI after TKA include surgeon case volume [42] and use of antibiotic-impregnated cement [43]. Especially in the elderly population, PJI may result in a higher incidence of mortality as well [42]. Hence, it is imperative to continue research strategies that minimize or prevent PJI. Airborne particulate bacteria are a major source of contamination in the operating room environment and that bacteria shed by personnel are the predominant source of these particles. The operating room air is proportionally contaminated with an increased circulation of personnel, and the recommendation that operating room traffic should always be kept to a minimum reaches a unanimous consensus [44].

7.2.1 Antibiotic prophylaxis in TKA

The scientific rationale for antibiotic prophylaxis is to inhibit or eliminate contaminating microorganisms that gain access to the surgical site during the procedure, which reduces the probability of infection. The optimal prophylactic antibiotic should be bactericidal (penicillin, cephalosporin, vancomycin, or aminoglycosides), not simply bacteriostatic. The agent should also have a half-life that covers the decisive interval (the first 2 h after incision or contamination) with therapeutic concentrations from time of incision to wound closure. Thus, the goal of administering preoperative antibiotics is to allow for adequate tissue (blood, soft tissue, and bone) concentrations by the time of implantation of the prosthesis. A first- or second-generation cephalosporin should be administered for routine perioperative surgical prophylaxis. These antibiotics cover Gram-positive organisms and clinically important aerobic Gram-negative bacilli and anaerobic gram-positive organisms [44]. Failure to maintain tissue concentrations above the MIC increases the risk of wound infection [45]. Surveillance measures are critical in ensuring clinician compliance. Patients such as those residing in nursing homes, the ones dependent on dialysis, patients who have been in the intensive care unit, or in regions with a high prevalence of MRSA, and in the healthcare workers, the risk for infection after a knee arthroplasty is higher and should be considered for preoperative screening. When routine antibiotic prophylaxis cannot be administered, teicoplanin, clindamycin, or vancomycin are reasonable alternatives. Vancomycin should be considered for patients who are current MRSA carriers or have anaphylactic allergy to penicillins, but routine use of vancomycin for preoperative prophylaxis is not recommended [46]. Antibiotic prophylaxis should not be administered for greater than 24 h after surgery because of the possibility of added antimicrobial toxicity, selection of resistant organisms, and unnecessary expense [47]. The adherence to the implementation of a World Health Organization checklist in the OR increased the appropriate preoperative antibiotic administration from 5% to 83% and decreased the incidence of SSI significantly from 6.2% to 3.4% ($p<0.001$) [48]. In another study from the Netherlands, in hospitals with a high standard of care, performing the surgical patient safety system check list's pre-, intra-, and postoperative elements also reduced the incidence of SSI (from 3.8% to 2.7%, $p=0.006$) as well as other major postoperative complications [49]. In a prospective study, it was observed that many evidence-based measures for SSI reduction (prophylactic antibiotic timing, maintaining normothermia during surgery, appropriate urinary tract catheterisztion, and hand hygiene) were not applied adequately for arthroplasty procedures and the situation was even worse for fracture surgeries [50]. There is evidence that the addition of antibiotics to PMMA cement leads to a reduction in the incidence of PJI and all-time failure of the prostheses after elective knee arthroplasty [51]. Most common bacteria are the staphylococci, with the prevalence of CNS to be increasing, while that of *Staphylococcus aureus* and other organisms is decreasing. Vancomycin and teicoplanin were the most effective antibiotics, with overall sensitivity rates of 100% and 96%, respectively [52]. In hematogenous infections, *S. aureus* is the dominating pathogen [53].

7.3 Diagnosis

Given the rising numbers of joint replacements performed annually and the ever increasing number of patients who have previously undergone a TKA, an understanding of the clinical presentation and outcomes of this infection is important. TKA is defined as infected when two cultures of periprosthetic tissue turn positive with phenotypically identical organisms, or in the presence of sinus tract communicating with the joint (Fig. 7.1), or having three of the following minor criteria: (1) an elevated serum C-reactive protein (CRP) and erythrocyte sedimentation rate (ESR), (2) an elevated synovial fluid white blood cell (WBC) count, (3) ++change on the leukocyte esterase (LE) test strip, or a positive alpha-defensin test in the joint aspirate, (4) elevated synovial fluid polymorphonuclear neutrophil percentage (PMN%), (5) a positive histological analysis of periprosthetic tissue, and (6) a single positive culture. Clinically, an infected TKA may be present without meeting these criteria, specifically in the case of less virulent organisms (e.g., *Proprioniobacter acnes*) [54].

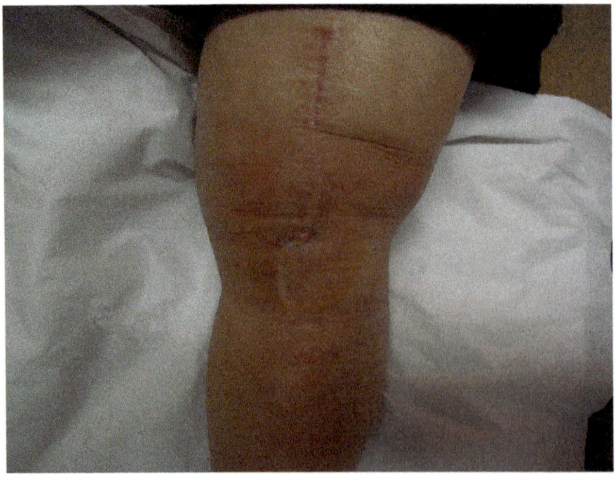

Fig. 7.1 Acute neglected infection of TKA in a female patient 6 months after the primary procedure with a draining sinus at the lower part of the incision.

7.3.1 Acute hematogenous infections of a TKA

There is a paucity of information presently available on the management and outcomes of patients treated for an acute hematogenous infection, with most reports describing small numbers of patients or discussed hematogenous infections as a subset of a larger cohort of patients who were treated for an infected TKA [55–66].

ESR and CRP are both commonly elevated in the acute postoperative time period regardless of infection status. The existing literature used 6 weeks as definition of the acute postoperative time period. However, ESR and CRP are likely still elevated up to 90 days following surgery. These inflammatory markers are highly elevated in the majority of patients with an acute hematogenous infection. The ESR is elevated in the majority and CRP > 100 mg/L in all patients, with a Synovial WBC count > 10,000 cells/μL, of which PMN% > 90% (Fig. 7.2A and B). These markers are helpful in

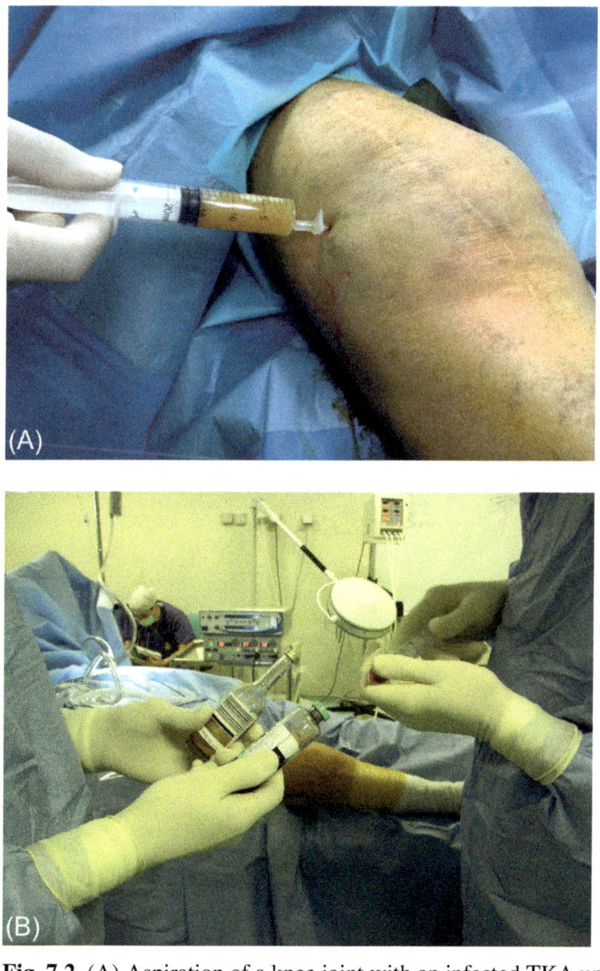

Fig. 7.2 (A) Aspiration of a knee joint with an infected TKA under sterile conditions in the operating room and a syringe full of pus. (B) The aspirate is injected into blood culture flasks and sent for cultures without delay.

diagnosing this type of infection in several series. Blood culture should be performed in patients suspected of an acute hematogenous infection, as they can confirm the diagnosis and identification of bacteria and may warrant more aggressive treatment and monitoring for patients with additional prosthetic joints. An acute hematogenous infection may be a marker of poor general health predisposing to this complication and has been associated to some major medical comorbidities that rendered them immunocompromised with high rate of early mortality. Patients who develop an acute hematogenous infection have variable outcomes following operative debridement with prosthetic retention. For patients who are infected with an organism other than *Staphylococcus* species, prosthetic retention was seen in a high percentage of cases with a single debridement. However, acute infection of a knee arthroplasty from *Staphylococcus* is highly predictive of failure, likely due to the rapid establishment of biofilm and resistance to antibiotics. Unfortunately, antibiotic therapy is normally chosen according to conventional susceptibility testing which may be inadequate in the setting of biofilm. If the organism is identified as *Staphylococcus* prior to surgery, it may be wiser to choose to have a two-stage resection and reimplantation rather than irrigation, debridement, and polyethylene exchange. For patients who develop recurrent infection following an attempt at prosthetic retention in the setting of an acute hematogenous infection, the risk of complications and reinfection is high with a failure rate greater than 50% [61,63,66–69].

7.4 Chronic TKA infection

Patients with a well functioning knee arthroplasty, now presenting with a history of recently established persistent pain or stiffness, have a higher probability and should raise suspicion for a chronic infection. Recent bacteraemia, multiple surgeries on the same joint, history of joint infection, comorbidities predisposing to an immunocompromised state, for example, diabetes mellitus, inflammatory arthropathy, or malnutrition, intravenous drug use, poor wound conditions, psoriasis, chronic venous stasis, or skin ulceration, are common risk factors that increase pathogen exposure to the joint or impair the body's ability to eradicate pathogens [70–72]. Physical examination findings such as joint warmth, redness, tenderness, or swelling, gradual limitation in the previously achieved range of motion, are not specific, but they are suggestive of a peri-PJI and further increase the suspicion.

In a patient with a painful knee arthroplasty after the 6th postoperative week, we should proceed, in the preoperative setting, with Erythrocyte Sedimentation Rate—ESR and C-Reactive Protein—CRP screening, followed by joint aspiration. This approach taken as a first step, and examined by a separate multicriteria decision analysis, has been proven as the most cost-effective method for diagnosing PJI [73]. The cut-off values for the ESR > 30 mm/h, the serum CRP > 10 mg/L, the Synovial fluid WBC count > 3000 cells\μL, of which PolyMorphoNucleate cells > 80%. Limited evidence suggests that no difference exists in the thresholds of ESR, CRP, or synovial fluid WBC count and differential to diagnose PJI in patients with and without inflammatory arthropathies [74]. Both the ESR and CRP when negative are better tests for ruling out an infection of the Joint. More recent reports have demonstrated the diagnostic value of newer Synovial fluid markers such as the leucocide esterase and the α-Defensin as predictors of periprosthetic infection. Alpha-defensin test has been demonstrated to be highly accurate in the diagnosis of PJI and was found as more effective than current diagnostic testing in predicting positive cultures. The combined measurement of synovial fluid α-defensin and CRP levels correctly diagnosed 99% of the cases as aseptic or infected. This is achieved despite the inclusion of patients with systemic inflammatory disease and those receiving treatment with antibiotics. Alpha-defensin may be an effective adjunct in the workup of shoulder, hip, and knee PJIs. The alpha-defensin test provides consistent results regardless of the organism type, Gram type, species, or virulence of the organism and is seriously considered a standard diagnostic tool in the evaluation for PJI.

The performance of this test in specific clinical scenarios such as the immediate postoperative period, in the setting of severe immunocompromise, and in the setting of a native joint has not yet been established. However, the synovial fluid α-defensin immunoassay outperformed the LE colorimetric test strip and provided reliable results even when the LE test strip failed as a result of blood interference [75–78].

Samples from the joint aspirate should be injected into blood culture flasks and sent for cultures, the sensitivity of which is significantly increased when culture time in attempts to diagnose PJI is extended to 2 weeks, while not increasing the risk of contaminants. While there is no evidence to determine the cost-effectiveness of 2-week versus 1-week cultures in presumed aseptic cases, the incidence of clinically significant positive results is not insignificant. The majority of common infecting organisms can be isolated within a few days of conventional cultures. There is no reason to extend the duration of culture in patients in whom the infecting organism has been isolated preoperatively. For patients with suspected PJI, culture negative cases, and patients who may be infected with low virulence organisms, the culture should be maintained for a prolonged period.

Preoperative biopsy of the knee although an invasive diagnostic tool with a theoretical risk of contaminating a previously aseptic joint has also an established accuracy in diagnosing PJI and should be limited to those cases with a high probability of PJI but inconclusive aspirate results [79–82]. Clinical judgment should not be outweighed by use of the diagnostic algorithm proposed or any of the individual tests. A sinus tract communicating with the joint is considered a pathognomonic physical examination finding for PJI.

Imaging is an additional tool in the preoperative diagnostic approach but with a low specificity and sensitivity. It is important to note that plain radiographs are generally normal in the setting of PJI. Radiographic signs suggestive of TKA infection include signs of loosening of previously well-fixed components, osteolysis, or bone resorption around the prosthetic components, subperiosteal elevation, or transcortical sinus tracts, particularly if seen at less than 5 years postoperatively and should not be considered to be related to wear of the bearing surface. It is not currently recommended to utilize CT to evaluate for PJI when other imaging and noninvasive tests have proven efficacy. The role of nuclear imaging in the work-up for an infected TKA is debatable and should be limited [83]. Planning to return the patient to the operating room will allow for joint visualization, periprosthetic tissue culture, and possible explant sonication. With the multitude of more cost-effective measures, the utility for nuclear imaging is rare.

Investigation for the PJI of the knee is also continued intraoperatively with the collection of tissue or fluid samples from representative sites, preferably from the interface. Each sample is taken with an unused instrument. We strongly recommend against swab cultures from wound or periarticular tissues. In the only known quantitative analysis, it was found that sensitivity and specificity are maximized with 5 or 6 periprosthetic samples being collected [81]. Perioperative prophylactic antibiotics should be withheld only in cases with a high suspicion for PJI in which an infecting organism has not been isolated, but others recommend the antibiotics to be withheld for 2 days to 2 weeks prior to obtaining cultures. Use of sonication of the explants increases the likelihood of isolating pathogens without increasing the rate of contaminants but should be limited to cases of suspected or proven PJI (based upon presentation and other testing) in which preoperative aspiration does not yield positive culture and in cases where antibiotics have been administered within the previous 2 weeks [84].

In cases with high clinical suspicion of infection but negative cultures or other diagnostic tests, molecular techniques (PCR) with or without sonication may help identify the unknown pathogens or antibiotic sensitivity for targeting antimicrobial therapies. While molecular techniques have shown some promise in identifying genes associated with antibiotic resistance [82], they do not yet match the clinical applicability of testing the antibiotic susceptibility of organisms grown in culture. The cost and availability of this technology limit its broad application and therefore is not currently considered a standard tool in the work-up of PJI. Frozen sections, however, may help distinguish infection from aseptic failure with less potential morbidity than preoperative biopsy [85].

7.5 Management of the infected TKA

Management of the infected TKA remains difficult for both the surgeon and the patient. The possibility of successful eradication of the pathogen (one or multiple bacteria) depends on various factors including the general condition of the patient, his/her immune response, and the susceptibility of the pathogen to administration of antibiotics [85–87]. Immediately after diagnosis of the infection, the surgeon and the patient have to decide the mode of treatment. The timing of diagnosis of invasive surgical site infection (SSI) following joint replacement surgery is an important criterion used to determine subsequent medical and surgical management of PJIs, as treatment strategies depend on this parameter. In a recent report with 661 infected hip and knee arthroplasties, the diagnosis of invasive SSI following TKA was delayed (median) 42 days (21–114) following knee arthroplasty, compared with 25 days (interquartile range 17–48) following hip arthroplasty. We hypothesize that differences in symptom manifestation and disparities in access to care may contribute to the observed differential timing of diagnosis [86].

If the infection occurs within 3 weeks after the insertion of the implants, the arthroplasty can be saved by an immediate opening of the joint, meticulous debridement and curettage, and changing of the polyethylene insert [84,85,88]. If the infection occurs after a considerable time after the index procedure, the appropriate method of treatment will be removal of all implants and insertion of a new arthroplasty in one or two-stage procedure [89–91].

7.6 One-stage revision

There are certain indications and contraindications for one-stage revision [91]. Most recent reports indicate one-stage revision in cases with (1) preoperative diagnosis of infection with only one microorganism of low virulence, with documented susceptibility provided by an antibiogram; (2) patients in good general health status, without immunocompromising systemic conditions such as collagen disease, diabetes mellitus, or neoplasms; (3) the absence of any septic focus; (4) minimal bone loss in the femoral condyles or in the tibial plateau; and (5) the absence of any soft tissue defect or a condition that impedes direct closure of the wound at the end of the revision. Contraindications include: (1) soft tissue defects that preclude easy closure of the wound, (2) bone defects that make cement reconstruction impossible, (3) peripheral vascular disease, and (4) polymicrobial infection or infection with multiresistant bacteria like MRSA and MRSE (methicillin-resistant *Staphylococcus epidermidis*) [92,93].

The revision procedure starts with careful opening of the joint. Previous scars in the skin incision should be removed. If a new incision toward the lateral aspect of the knee is necessary, a safe distance should be kept from the previous skin incision to avoid skin necrosis. Also, acute angles with previous skin incisions should also be avoided. All implants, cement, and intramedullary restrictors should be removed carefully and then tissue debridement follows. Debridement should be meticulous and aggressive as in tumor surgery. Biofilm which may have spread and covered large parts of the joint must be removed aggressively (Fig. 7.3). No compromise is allowed. Any suspicious bone and soft tissue infected or dead should be removed. Particular attention should be given during debridement to the posterior part of the joint. A complete synovectomy must be performed, and all diseased tissues (bone and soft tissues) should also be removed. If the posterior cruciate and the collateral ligaments appear disseminated by the infection, they should be removed as well. The surgical debridement is the main factor for reduction of the bacterial load. Administration of antibiotics is important as an adjuvant factor, but can never cure the infection alone. Multiple tissue samples (6 to 10) are taken from different sites of the joint before administration of antibiotics and are sent for culture and histopathology. At the end of debridement, the joint is irrigated with hydrogen peroxide and then with pulsatile lavage with normal saline. At this point, new draping is placed and all instruments are replaced with new sterile ones. New implants are inserted with antibiotic-loaded bone cement according to antibiogram. Usually a constrained or rotating hinge prosthesis is implanted because the ligaments of the knee may have been excised in the

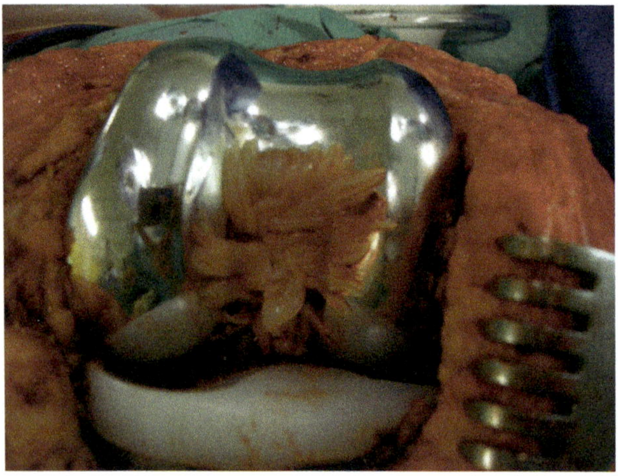

Fig. 7.3 Biofilm covers the femoral notch in a chronic infection of TKA which was negative in joint aspiration.

aggressive debridement. Vancomycin and gentamicin are commonly used in the preparation of the acrylic cement. The addition of antibiotic powder in the cement should be between 5% and 10%. For example, 2–4 g of Vancomycin should be mixed with 40 mg of cement powder. Intravenous and oral antibiotics are administered for 6 weeks postoperatively. Oral administration of antibiotics is allowed if the infecting microorganism(s) is sensitive to oral form of antibiotics. Infection parameters are evaluated weekly to document return to normal values. Mobilization of the knee is the same as in primary TKA [94–99].

Main advantages of this approach include: (1) shorter hospitalization of the patient because only one operative procedure will be required; (2) lower financial cost for the patient, the hospital, and the insurance carrier; (3) better functional outcome as the knee will be immediately free for mobilization; (4) better psychology and emotional status of the patient because his/her problem will be solved in considerably less time; and (5) shorter absence from home, work, and other activities [92,93].

7.7 Two-stage revision

Two-stage exchange still remains the preferred method of treatment for the infected TKA. Numerous studies report rates of curing the infection, with two-stage revision close to 90% [100–104]. Indications for two-stage revision include: (1) polymicrobial infection or infection with multiresistant bacteria like MRSA and MRSE, (2) soft tissue defects that preclude easy closure of the wound, (3) bone defects that make cement reconstruction impossible, and (4) reinfection [92,102].

First stage: The operative procedure starts with the removal of the implants and debridement of the knee joint (Fig. 7.4). Removal of the implants should be very careful to prevent unnecessary loss of bone stock. Oscillating saws and thin osteotomes are

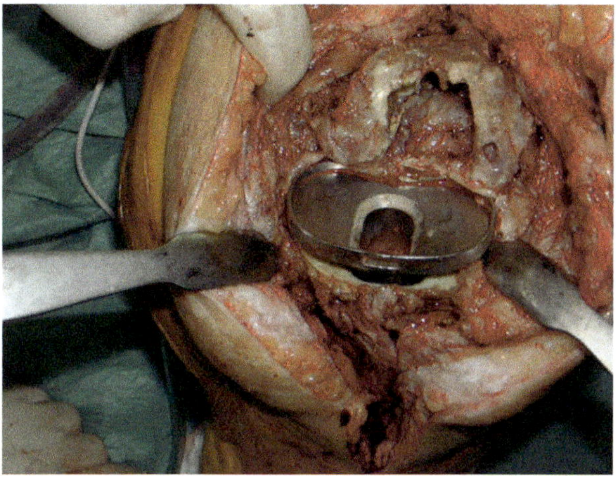

Fig. 7.4 After implants are removed, biofilm is noted covering the entire joint with extensive erosions of the condyles.

very useful at this stage. Debridement of the joint should be as meticulous and aggressive as in one-stage arthroplasty. All diseased tissues, bone, and soft tissues should be removed. When there is a doubt if a tissue is infected, it must be removed as in tumor surgery. Multiple tissue samples are taken and send for culture, and histopathology and IV antibiotics are administered. At the end of debridement, the joint is irrigated first with hydrogen peroxide and then with pulsatile lavage using normal saline [105,106].

New draping is applied, and the surgeon has to decide what type of temporary cement spacers will be used. Spacers are inserted in order to maintain the length of the leg, to prevent tissue adhesion, and to release antibiotics which were used in the preparation of the spacer (Fig. 7.5). There are two main types of spacers: (1) block spacers which are nonarticulating and (2) articulating spacers.

Block spacers: are hand shaped intraoperatively by the surgeon according to joint space and the bone deficit that resulted after debridement [107]. Antibiotics are added according to sensitivity of the microorganism(s) as given by the antibiogram. Usually 4 g of vancomycin and 2–3 g of gentamicin are added to 80 mg of acrylic cement. Appropriate antibiotics in the desired concentration can be selected during mixing of the cement. This volume of antibiotic-containing cement is usually adequate to fill the joint and is placed between femur and tibia, under the patella and at the medial and lateral gutters. Block spacers are more appropriate for infections with severe bone defects. They can be shaped to fit the unpredictable bone defect and provide better stability to the knee. The block spacer releases antibiotics locally, in high concentrations and prevents shortening of the medial and lateral collateral ligaments and quadriceps tendon [107–109]. There are certain drawbacks of the hand-fashioned block spacers. The main disadvantage is that they do not allow mobility of the knee. Although partial loading of the knee is allowed postoperatively, the knee remains in extension and an orthosis is required for several weeks until the second stage. The quadriceps tendon is contracted, and this causes a certain degree of difficulty during final reimplantation of the prostheses at the second stage [107–109].

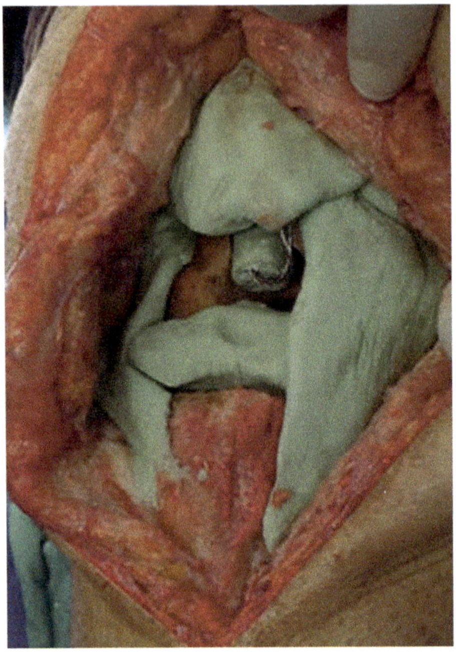

Fig. 7.5 After radical debridement, hand-made cement spacers are filling the supracondylar pouch and medial and lateral gutters. 4 mg of Vancomycin has been added for the preparation of the cement.

Articulating spacers: These are cement spacers that allow mobility of the knee as they are placed to resemble the shapes of the distal femur and proximal tibia [92,110,111]. Mobilization of the knee helps prevent tissue adhesions, contracture (shortening) of the collateral ligaments, and the quadriceps tendon. They can be (1) the implants of the index operation but sterilized and loaded with antibiotic-impregnated cement, (2) molded intraoperatively, and (3) prefabricated commercial spacers. The third type has gained the preference of most surgeons, although it is more expensive than the other two types because it has better mechanical and biological advantages. The sterilized implants of the index operation have the potential risk of bacteria adherence and reinfection, and the molded spacers may fracture causing bone erosion [110,111].

7.7.1 Postoperative course of antibiotics after first stage in two-stage revision

The patient ambulates soon after the first days and is under administration of antibiotics. According to preoperative antibiogram, antibiotics are administered initially intravenously and later orally for 6 weeks. For more effective treatment, preferably two chemotherapeutic agents are administered. Infection indices (CRP and ESR) are evaluated weekly. The patient is considered a candidate for the second stage when these indices are normal after discontinuation of antibiotics; for more than 2 weeks, wound healing has been uneventful and his general condition is good. The type of preoperative antibiotic administered to a patient with prior septic arthritis or PJI should cover the previous infecting organism of the same joint. In these patients, we recommend the use of a cemented component and antibiotic-impregnated cement. The most antibiotic-resistant bacterial strains are found in patients for whom prior antibiotic treatment had failed. Acute postoperative infections have a greater resistance profile than the chronic or hematogenous infections. Bacteria isolated from a hematogenous infection present a high sensitivity to both second-generation cephalosporins and gentamicin and are the recommendation until final cultures are available in the acute hematogenous infections. All chronic and acute postoperative infections with Gram-positive bacteria and all cases in which a gram stain fails to identify bacteria, the use of ciprofloxacin, daptomycin, and vancomycin which have the potential to significantly limit the intracellular SCV emergence are justified with an antibiofilm molecule, such as rifampin. Daptomycin combination with rifampicin in more recent studies displayed the greatest activity and represents a promising combination for the management of biofilm-associated staphylococcal infections. Infections with Gram-negative bacteria should be managed with third- or fourth-generation cephalosporin. Infections with mixed Gram-positive and Gram-negative bacteria should be managed with a combination of vancomycin and third- or fourth-generation cephalosporin.

7.7.2 Second stage in the two-stage revision

The interval between the first stage and the reimplantation of the prosthesis varies from 6 weeks to few months, depending on several factors. Infection markers should be negative without antibiotics and the patients; and local tissue conditions uncompromised. During second stage, the spacers are removed carefully usually piecemeal to prevent additional bone destruction and loss. The joint is then meticulously debrided from scar, tissue reaction to the spacers, and both the femur and the tibia medullary canals are opened. Joint fluid and multiple tissue samples (bone and soft tissues) are taken again and are send for frozen section to evaluate polymorphonucleate cell (PMN) counts per high power field (10×40—hpf) and cultures. Antibiotics which were withheld are administered at this point of surgery [100–102]. If the number of PMNs per high power field is greater than 10, the joint is not considered appropriate for reimplantation and a cement spacer impregnated with antibiotics is inserted again. The joint is closed, and the entire process will be repeated again after a "safe" interval. Special attention will be given to the new results of the culture as a new microorganism may have developed. If the number of PMNs is less than 5 per hpf, the surgeon proceeds with the implantation of a new prosthesis which is usually a constrained or rotating hinge prosthesis [112]. The new implants are inserted with antibiotic-loaded cement. Again as in the "one-stage exchange," a constrained or rotating hinge prosthesis is implanted because the knee may have been unstable after possible excision of the ligaments during the previous debridements (Fig. 7.6A and B). When the count of the PMNs is in the gray zone between 5 and 10, the surgeon will decide based on his experience if the surgery will proceed to implantation of the new arthroplasty or cement spacers will be inserted again. Postoperatively, the patient is mobilized depending on the condition of the quadriceps tendon. The knee may either be mobilized freely or be

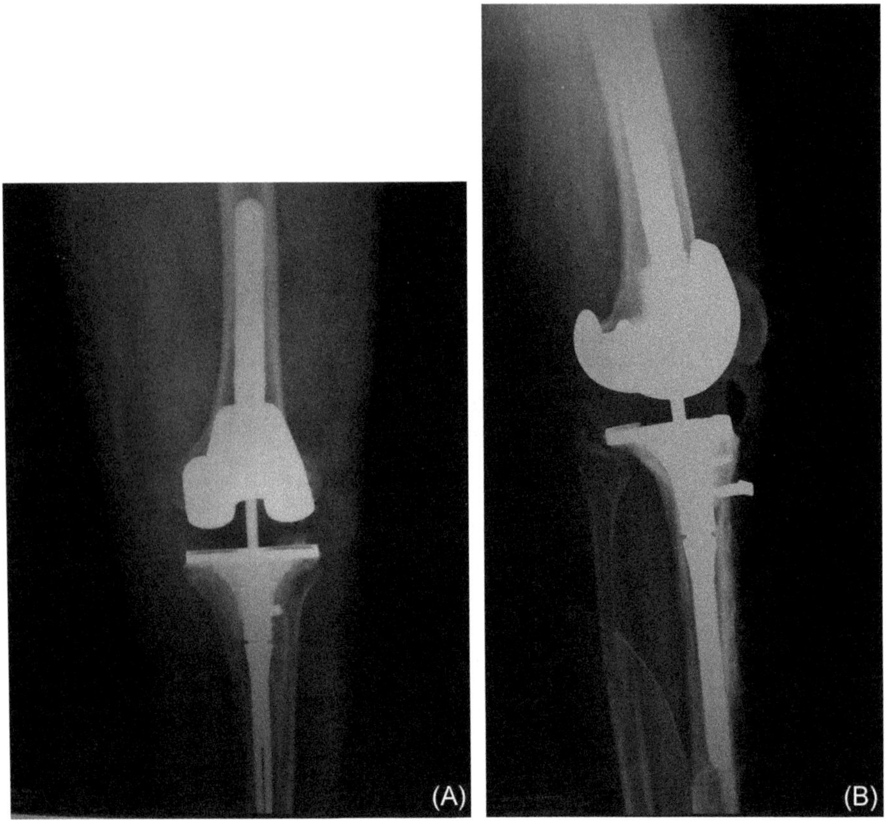

Fig. 7.6 (A, B) Anteroposterior and lateral radiographs of a semiconstrained prosthesis which was implanted at the second stage after eradication of the infection. The radiographs were taken at 3 years after the last procedure.

immobilized in a cast for 3 weeks subsequently followed by rehabilitation. Antibiotics usually are continued for 2 weeks given that the result of the intraoperative tissue cultures will be negative for bacteria [105,106].

Main disadvantages of this approach are: (1) longer hospitalization; (2) higher cost; (3) two additional operative procedures to solve the problem; (4) long period, at least 6 weeks and some months, between the two-stages; (5) pain and instability of the knee in the period between stages that requires wearing of a knee splint; and (6) inferior range of motion, at the end of treatment, than the one-stage exchange [92].

7.8 Outcomes

Very satisfactory results are reported with a single-stage (one-stage) revision when certain criteria are followed for selection of patients. Haddad et al. reported 100% successful results regarding eradication of the infection in their series which included

28 patients but with a relatively short follow up of 3 years [113]. Parkinson et al. [99] reported that all 12 patients (100%) who were treated with one-stage exchange arthroplasty were free of infection at the latest follow up of 2 years. It is interesting that their series included two patients with septic discharging sinus from the index procedure. Gehrke and Breusch [93] reported on 100 consecutive patients who were treated for infected TKA. At a mean follow up of 8.5 years, 90 patients (90%) remained free of infection and with the arthroplasty in place. All authors emphasize the fact that results may deteriorate with time, secondary to mechanical loosening and the possibility of recurrent infection which will require a reintervention [92,95–97,99].

Two-stage revision: remains the gold standard and preferred treatment for the infected TKA [113–116]. When the two stages are followed and executed carefully, the infection will be eradicated in the highest number of cases, leaving a functional joint [89,117–119]. Two-stage approach provides a success rate which can be as high as 100%. Meani and Romano [120] reported on 21 patients who were treated with preformed knee spacers. After 32 months, no patients developed any signs and symptoms of infections or needed reoperation for any reason. Meek et al. [119] accomplished a success rate of 96% in a series with 47 patients with a range of motion measured at 87 degrees. Romano et al [121] analyzed 6 studies reporting on one- and two-stage revision arthroplasties and concluded that the two-stage approach has a success rate of 89.8% in eradication of the infection compared with 81.9% eradication with one-stage exchange. Articulating and static cement spacers seem to provide similar functional outcomes although the articulating spacers facilitate the reimplantation stage [121]. There seems to be no difference in the reinfection rates, complication rates, or reoperation rates between articulating and static spacers, and no particular recommendation can be made regarding the use of a certain type of spacer.

7.9 Knee arthrodesis

When infection recurs or persists after one- or two-stage revision exchange and the joint has been severely deformed due to bone and soft tissue defects and the patient does not wish to undergo any reconstructive procedures knee arthrodesis (fusion of the knee) is an alternative (Fig. 7.7A and B). The procedure was first described by Henry Park for salvage of tuberculous knees. Nelson and Evarts [122] were the first to report arthrodesis for salvage procedure after failed TKA. Main indications include multiple failures and recurrences, very virulent micro organism(s), poor soft tissue coverage, loss of extensor mechanism, knee that is painful after multiple attempts of treatment, and patient with immunosuppression [123–126]. Contraindications are bilateral infection, a fused hip in the same leg, amputation of the contralateral limb, and a nonreconstructable segmental defect [123–126]. Implant removal and aggressive debridement are performed, and antibiotic-loaded cement spacers are inserted. IV antibiotics are administered and infection indices are evaluated. The procedure of arthrodesis is performed when the knee is considered again free of infection. The knee should be in a position with 0–5 degrees of valgus, external rotation of 10 degrees, and a flexion from 0 to 15 degrees. Limbs that are shortened due to severe bone stock

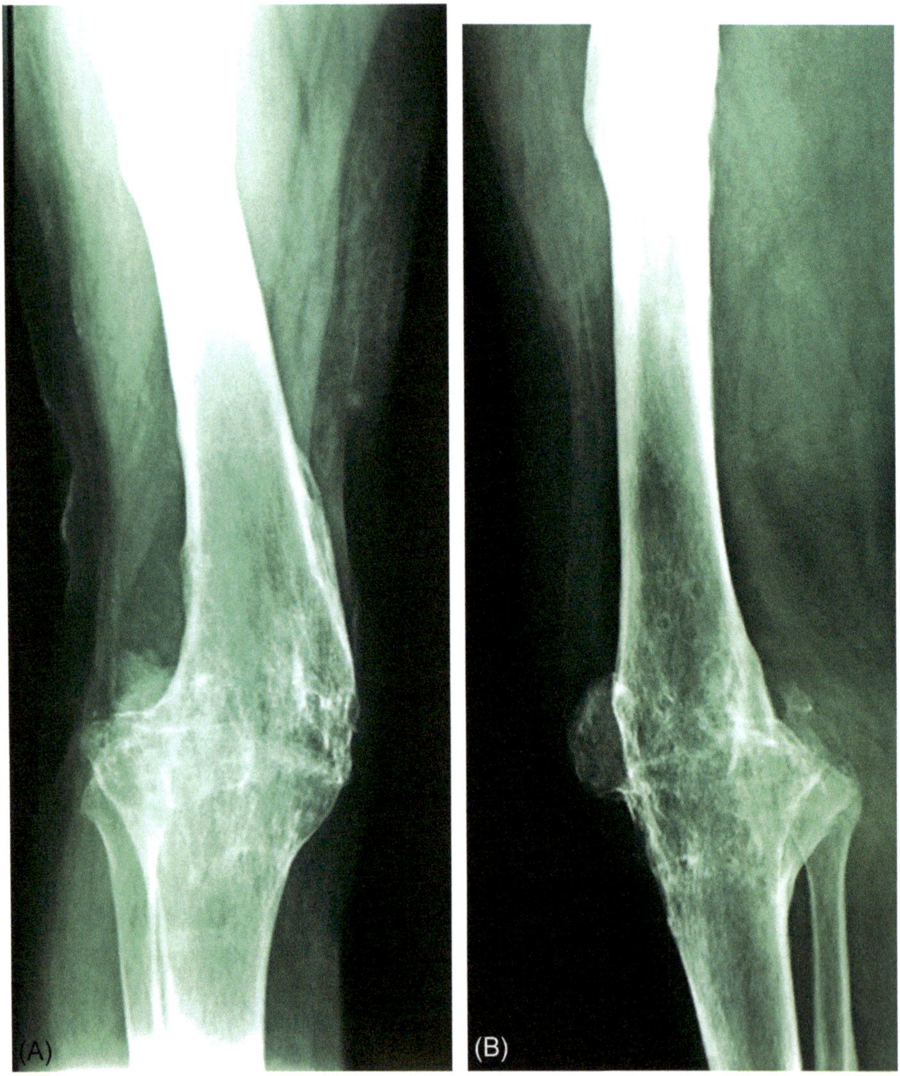

Fig. 7.7 (A, B) Arthrodesis of the knee after repeated failed attempts to eradicate infection of the TKA.

should be arthrodesed in complete extension (0 degree). Methods for performing knee arthrodesis are: (1) intramedullary antegrade nailing, (2) intramedullary fixation with a modular nail, (3) internal fixation with one or two compression plates and screws, and (4) external fixation. Bone autografts and the patella can be used to fill defects and enhance healing [126,127].

IM nailing has generally higher rates of union and usually brings the knee in slight varus (2–4 degrees). External fixation can be performed with unilateral or circular frames. Circular frames have certain advantages and constitute a popular method of treatment for arthrodesis: (1) they are more stable, (2) they have less risk for infection, (3) they can achieve precise anatomic alignment with postoperative modification, (4) they can apply gradual compression to achieve healing, (5) they can allow simultaneous limb lengthening if it is desired, and (6) they can be removed easily in an outpatient basis. Time to union ranges from 4 to 9 months [127–129]. High rates of union have been obtained by all methods of treatment. Circular frames can provide union rates up to 100%, and intramedullary nailing provides union which can approach 90–100% [126–128].

7.10 Complications

Failure of healing (non fusion), recurrence of infection, excessive shortening of the leg, and implant failure are complications that accompany the attempt to fuse the knee. Limb lengthening can be performed in young and active patients. In older patients, a shoe lift may correct in part the length inequality [127–129]. When the infection recurs despite all efforts and surgical interventions (ranging from staged revisions to attempts for arthrodesis), an above knee amputation is indicated. Amputation in this case is not considered a failure of treatment but an effective treatment as it will save the patient from further deterioration of his health and a potential live threat [130].

References

[1] David S. Jevsevar, Treatment of osteoarthritis of the knee: evidence-based guideline, second ed., J. Am. Acad. Orthop. Surg. 21 (2013) 571–576.

[2] E.F. Berbari, D.R. Osmon, B. Lahr, et al., The Mayo prosthetic joint infection risk score: implication for surgical site infection reporting and risk stratification, Infect. Control Hosp. Epidemiol. 33 (8) (2012) 774–781.

[3] D. Ward, Infection control: reducing the psychological effects of isolation, Br. J. Nurs. 9 (2000) 162–170.

[4] C.K. Hebert, R.E. Williams, R.S. Levy, R.L. Barrack, Cost of treating an infected total knee replacement, Clin. Orthop. Relat. Res. 331 (1996) 140–145.

[5] S.A. McGarry, J.J. Engemann, K. Schmader, D.J. Sexton, K.S. Kaye, Surgical-site infection due to *Staphylococcus aureus* among elderly patients: mortality, duration of hospitalization, and cost, Infect. Control Hosp. Epidemiol. 25 (2004) 461–467.

[6] L.A. Poultsides, L.L. Liaropoulos, K.N. Malizos, The socioeconomic impact of musculoskeletal infections, J. Bone Joint Surg. Am. 92 (11) (2010). e13.

[7] J. Chen, Y. Cui, X. Li, et al., Risk factors for deep infection after total knee arthroplasty: a meta-analysis, Arch. Orthop. Trauma Surg. 133 (5) (2013) 675–687.

[8] F.E. Stinchfield, L.U. Bigliani, H.C. Neu, T.P. Goss, C.R. Foster, Late hematogenous infection of total joint replacement, J. Bone Joint Surg. Am. 62 (8) (1980) 1345–1350.

[9] B.J. Thomas, J.R. Moreland, H.C. Amstutz, Infection after total joint arthroplasty from distal extremity sepsis, Clin. Orthop. Relat. Res. 181 (1983) 121–125.

[10] A.D. Hanssen, D.R. Osmon, C.L. Nelson, Prevention of deep periprosthetic joint infection, Instr. Course Lect. 46 (1997) 555–567.

[11] G. Peersman, R. Laskin, J. Davis, M. Peterson, Infection in total knee replacement: a retrospective review of 6489 total knee replacements, Clin. Orthop. Relat. Res. 392 (2001) 15–23.
[12] American Diabetes Association, Standards of medical care in diabetes—2013, Diabetes Care 36 (Suppl. 1) (2013) S11–S66.
[13] E. Jamsen, P. Nevalainen, J. Kalliovalkama, T. Moilanen, Preoperative hyperglycemia predicts infected total knee replacement, Eur. J. Intern. Med. 21 (3) (2010) 196–201.
[14] M.H. Marchant Jr., N.A. Viens, C. Cook, T.P. Vail, M.P. Bolognesi, The impact of glycemic control and diabetes mellitus on perioperative outcomes after total joint arthroplasty, J. Bone Joint Surg. Am. 91 (7) (2009) 1621–1629.
[15] A.L. Adams, E.W. Paxton, J.Q. Wang, et al., Surgical outcomes of total knee replacement according to diabetes status and glycemic control, 2001 to 2009, J. Bone Joint Surg. Am. 95 (6) (2013) 481–487.
[16] G.C. Del Savio, S.B. Zelicof, L.M. Wexler, et al., Preoperative nutritional status and outcome of elective total hip replacement, Clin. Orthop. Relat. Res. 326 (1996) 153–161.
[17] S. Gherini, B.K. Vaughn, A.V. Lombardi Jr., T.H. Mallory, Delayed wound healing and nutritional deficiencies after total hip arthroplasty, Clin. Orthop. Relat. Res. 293 (1993) 188–195.
[18] F.M. Jabari, J. Parvizi, C.T. Haytmanek, A. Joshi, J. Purtill, Procrastination of wound drainage and malnutrition affect the outcome of joint arthroplasty, Clin. Orthop. Relat. Res. 466 (6) (2008) 1368–1371.
[19] C.J. Lavernia, R.J. Sierra, L. Baerga, Nutritional parameters and short term outcome in arthroplasty, J. Am. Coll. Nutr. 18 (3) (1999) 274–278. 38.
[20] J.A. Nicholson, A.S. Dowrick, S.M. Liew, Nutritional status and short-term outcome of hip arthroplasty, J. Orthop. Surg. (Hong Kong) 20 (3) (2012) 331–335.
[21] E. Jamsen, H. Huhtala, T. Puolakka, T. Moilanen, Risk factors for infection after knee arthroplasty. A register-based analysis 43,149 cases, J. Bone Joint Surg. Am. 91 (2009) 38–47.
[22] G.M. Kerkhoffs, E. Servien, W. Dunn, D. Dahm, J.A. Bramer, D. Haverkam, The influence of obesity on the complication rate an outcome of total knee arthroplasty: a meta-analysis and systematic literature review, J. Bone Joint Surg. Am. 94 (2012) 1839–1844.
[23] B. Liabaud, D.A. Patrick Jr., J.A. Geller, Higher body mass index leads to longer operative time in total knee arthroplasty, J. Arthroplasty 28 (2013) 563–565.
[24] A.J. Samson, G.E. Mercer, D.G. Campbell, Total knee replacement in the morbidly obese: a literature review, ANZ J. Surg. 80 (2010) 595–599.
[25] S.M. Kurtz, K.L. Ong, E. Lau, K.J. Bozic, D. Berry, J. Parvizi, Prosthetic joint infection risk after TKA in the medicare population, Clin. Orthop. Relat. Res. 468 (2010) 52–56.
[26] B. Ravi, B. Escott, P.S. Shah, R. Jenkinson, J. Chahal, E. Bogoch, H. Kreder, G. Hawker, A systematic review and meta-analysis comparing complications following total joint arthroplasty for rheumatoid arthritis versus for osteoarthritis, Arthritis Rheum. 64 (2012) 3839–3849.
[27] R.A. Malinzak, M.A. Ritter, M.E. Berend, J.B. Meding, E.M. Olberding, K.E. Davis, Morbidly obese, diabetic, younger, and unilateral joint arthroplasty patients have elevated total joint arthroplasty infection rates, J. Arthroplasty 24 (Suppl. 6) (2009) 84–88.
[28] M.J. McElroy, R. Pivec, K. Issa, S.F. Harwin, M.A. Mont, The effects of obesity and morbid obesity on outcomes in TKA, J. Knee Surg. 26 (2) (2013) 83–88.
[29] J.A. Singh, T.K. Houston, B.A. Ponce, et al., Smoking as a risk factor for short-term outcomes following primary total hip and total knee replacement in veterans, Arthritis Care Res. (Hoboken) 63 (10) (2011) 1365–1374.

[30] E. Mills, O. Eyawo, I. Lockhart, S. Kelly, P. Wu, J.O. Ebbert, Smoking cessation reduces postoperative complications: a systematic review and meta-analysis, Am. J. Med. 124 (2) (2011) 144–154. e148.
[31] K. Myers, P. Hajek, C. Hinds, H. McRobbie, Stopping smoking shortly before surgery and postoperative complications: a systematic review and meta-analysis, Arch. Intern. Med. 171 (11) (2011) 983–989.
[32] L.T. Sorensen, Wound healing and infection in surgery. The clinical impact of smoking and smoking cessation: a systematic review and meta-analysis, Arch. Surg. 147 (4) (2012) 373–383.
[33] D. Lindstrom, O. Sadr Azodi, A. Wladis, et al., Effects of a perioperative smoking cessation intervention on postoperative complications: a randomized trial, Ann. Surg. 248 (5) (2008) 739–745.
[34] K.A. Bradley, A.D. Rubinsky, H. Sun, et al., Alcohol screening and risk of postoperative complications in male VA patients undergoing major non-cardiac surgery, J. Gen. Intern. Med. 26 (2) (2011) 162–169.
[35] R.S. Namba, M.C. Inacio, E.W. Paxton, Risk factors associated with deep surgical site infections after primary total knee arthroplasty: an analysis of 56,216 knees, J. Bone Joint Surg. Am. 95 (9) (2013) 775–782.
[36] J.R. Lieberman, M.D. Fuchs, S.B. Haas, et al., Hip arthroplasty in patients with chronic renal failure, J. Arthroplasty 10 (2) (1995) 191–195.
[37] J.M. Sunday, J.T. Guille, J.S. Torg, Complications of joint arthroplasty in patients with end-stage renal disease on hemodialysis, Clin. Orthop. Relat. Res. 397 (2002) 350–355.
[38] D.P. Sakalkale, W.J. Hozack, R.H. Rothman, Total hip arthroplasty in patients on long-term renal dialysis, J. Arthroplasty 14 (5) (1999) 571–575.
[39] P.H. Hsieh, L.H. Chen, M.S. Lee, C.H. Chen, W.E. Yang, C.H. Shih, Hip arthroplasty in patients with cirrhosis of the liver, J. Bone Joint Surg. (Br.) 85 (6) (2003) 818–821.
[40] S.M. Cohen, H.S. Te, J. Levitsky, Operative risk of total hip and knee arthroplasty in cirrhotic patients, J. Arthroplasty 20 (4) (2005) 460–466.
[41] B. Habermann, C. Eberhardt, A.A. Kurth, Total joint replacement in HIV positive patients, J. Infect. 57 (1) (2008) 41–46.
[42] G. Peersman, R. Laskin, J. Davis, M.G. Peterson, T. Richart, Prolonged operative time correlates with increased infection rate after total knee arthroplasty, HSS J. 2 (1) (2006) 70–72.
[43] R.S. Namba, Y. Chen, E.W. Paxton, T. Slipchenko, D.C. Fithian, Outcomes of routine use of antibiotic-loaded cement in primary total knee arthroplasty, J. Arthroplasty 24 (2009) 44–47.
[44] E.A. Salvati, R.P. Robinson, S.M. Zeno, B.L. Koslin, B.D. Brause, P.D. Wilson Jr., Infection rates after 3175 total hip and total knee replacements performed with and without a horizontal unidirectional filtered air-flow system, J. Bone Joint Surg. Am. 64 (4) (1982) 525–535.
[45] D.W. Bratzler, P.M. Houck, Antimicrobial prophylaxis for surgery: an advisory statement from the National Surgical Infection Prevention Project, Am. J. Surg. 189 (4) (2005) 395–404.
[46] E. Hansen, Perioperative antibiotics, in: Proceedings of the International Concensus Meeting on Periprosthetic Joint Infection, Data Trace Publishing Company, Brooklandville, MD (2007) 29–82.
[47] M.J. Enzler, E. Berbari, D.R. Osmon, Antimicrobial prophylaxis in adults, Mayo Clin. Proc. 86 (7) (2011) 686–701.

[48] A.B. Haynes, T.G. Weiser, W.R. Berry, et al., A surgical safety checklist to reduce morbidity and mortality in a global population, N. Engl. J. Med. 360 (5) (2009) 491–499.
[49] E.N. de Vries, H.A. Prins, R.M. Crolla, et al., Effect of a comprehensive surgical safety system on patient outcomes, N. Engl. J. Med. 363 (20) (2010) 1928–1937.
[50] A.E. Andersson, I. Bergh, J. Karlsson, B.I. Eriksson, K. Nilsson, The application of evidence-based measures to reduce surgical site infections during orthopedic surgery—report of a single-center experience in Sweden, Patient Saf. Surg. 6 (1) (2012) 11.
[51] F.Y. Chiu, C.F. Lin, Antibiotic-impregnated cement in revision total knee arthroplasty. A prospective cohort study of one hundred and eighty-three knees, J. Bone Joint Surg. Am. 91 (3) (2009) 628–633.
[52] R.S. Nickinson, T.N. Board, A.K. Gambhir, M.L. Porter, P.R. Kay, The microbiology of the infected knee arthroplasty, Int. Orthop. 34 (4) (2010) 505–510.
[53] E. Fulkerson, C.J. Valle, B. Wise, M. Walsh, C. Preston, P.E. Di Cesare, Antibiotic susceptibility of bacteria infecting total joint arthroplasty sites, J. Bone Joint Surg. Am. 88 (6) (2006) 1231–1237.
[54] B. Zmistowski, C. Della Valle, T.W. Bauer, K.N. Malizos, A. Alavi, H. Bedair, R.E. Booth, P. Choong, C. Deirmengian, G.D. Ehrlich, A. Gambir, R. Huang, Y. Kissin, H. Kobayashi, N. Kobayashi, V. Krenn, D. Lorenzo, S.B. Marston, G. Meermans, J. Perez, J.J. Ploegmakers, A. Rosenberg, C. Simpfendorfer, P. Thomas, S. Tohtz, J.A. Villafuerte, P. Wahl, F.C. Wagenaar, E. Witzo, Diagnosis of periprosthetic joint infection, J. Orthop. Res. 32 (Suppl. 1) (2014) S98–S107.
[55] A.T. Rasul Jr., D. Tsukayama, R.B. Gustilo, Effect of time of onset and depth of infection on the outcome of total knee arthroplasty infections, Clin. Orthop. Relat. Res. 273 (1991) 98.
[56] R.R. Burger, T. Basch, C.N. Hopson, Implant salvage in infected total knee arthroplasty, Clin. Orthop. Relat. Res. 273 (1991) 105.
[57] B.J. Waldman, E. Hostin, M.A. Mont, et al., Infected total knee arthroplasty treated by arthroscopic irrigation and debridement, J. Arthroplasty 15 (4) (2000) 430.
[58] H. Segawa, D.T. Tsukayama, R.F. Kyle, et al., Infection after total knee arthroplasty. A retrospective study of the treatment of eighty-one infections, J. Bone Joint Surg. Am. 81 (10) (1999) 1434.
[59] F.Y. Chiu, C.M. Chen, Surgical debridement and parenteral antibiotics in infected revision total knee arthroplasty, Clin. Orthop. Relat. Res. 461 (2007) 130.
[60] M.A. Mont, B. Waldman, C. Banerjee, et al., Multiple irrigation, debridement, and retention of components in infected total knee arthroplasty, J. Arthroplasty 12 (4) (1997) 426.
[61] S. Bengtson, G. Blomgren, K. Knutson, et al., Hematogenous infection after knee arthroplasty, Acta Orthop. Scand. 58 (5) (1987) 529.
[62] T.J. Grogan, F. Dorey, J. Rollins, et al., Deep sepsis following total knee arthroplasty. Ten-year experience at the University of California at Los Angeles Medical Center, J. Bone Joint Surg. Am. 68 (2) (1986) 226.
[63] A. Wigren, G. Karlstrom, H. Kaufer, Hematogenous infection of total joint implants: a report of multiple joint infections in three patients, Clin. Orthop. Relat. Res. 152 (1980) 288.
[64] J.L. Cook, R.D. Scott, W.J. Long, Late hematogenous infections after total knee arthroplasty: experience with 3013 consecutive total knees, J. Knee Surg. 20 (1) (2007) 27–33.
[65] E.F. Berbari, A.D. Hanssen, M.C. Duffy, et al., Risk factors for prosthetic joint infection: case–control study, Clin. Infect. Dis. 27 (5) (1998) 1247–1254.
[66] H. Ceri, M.E. Olson, C. Stremick, et al., The Calgary Biofilm Device: new technology for rapid determination of antibiotic susceptibilities of bacterial biofilms, J. Clin. Microbiol. 37 (6) (1999) 1771.

[67] F. Sepandj, H. Ceri, A.P. Gibb, et al., Biofilm infections in peritoneal dialysis-related peritonitis: comparison of standard MIC and MBEC in evaluation of antibiotic sensitivity of coagulase-negative staphylococci, Perit. Dial. Int. 23 (1) (2003) 77.
[68] Beau S. Konigsberg, Craig J. Della Valle, Nicholas T. Ting, Fang Qiu, Scott M. Sporer, Acute hematogenous infection following total hip and knee arthroplasty, J. Arthroplast. 29 (2014) 469–472.
[69] K.J. Bozic, E. Lau, S. Kurtz, K. Ong, D.J. Berry, Patient-related risk factors for postoperative mortality and periprosthetic joint infection in Medicare patients undergoing TKA, Clin. Orthop. Relat. Res. 470 (1) (2012) 130–137.
[70] K.J. Bozic, E. Lau, S. Kurtz, et al., Patient-related risk factors for periprosthetic joint infection and postoperative mortality following total hip arthroplasty in Medicare patients, J. Bone Joint Surg. Am. 94 (9) (2012) 794–800.
[71] L. Pulido, E. Ghanem, A. Joshi, J.J. Purtill, J. Parvizi, Periprosthetic joint infection: the incidence, timing, and predisposing factors, Clin. Orthop. Relat. Res. 466 (7) (2008) 1710–1715.
[72] C. Diaz-Ledezma, P.M. Lichstein, J.H. Dolan, J. Parvizi, What is the best strategy to diagnose hip/knee periprosthetic joint infections in Medicare patients seen in the ambulatory setting? Clin. Orthop. Relat. Res. 11 (2014) 3275–3284.
[73] C.A. Cipriano, N.M. Brown, A.M. Michael, M. Moric, S.M. Sporer, C.J. Della Valle, Serum and synovial fluid analysis for diagnosing chronic periprosthetic infection in patients with inflammatory arthritis, J. Bone Joint Surg. Am. 94 (7) (2012) 594–600.
[74] B. Fink, C. Makowiak, M. Fuerst, I. Berger, P. Schafer, L. Frommelt, The value of synovial biopsy, joint aspiration and C-reactive protein in the diagnosis of late peri-prosthetic infection of total knee replacements, J. Bone Joint Surg. (Br.) 90 (7) (2008) 874–878.
[75] C. Deirmengian, K. Kardos, P. Kilmartin, S. Gulati, P. Citrano, R.E. Booth, The alpha-defensin test for periprosthetic joint infection responds to a wide spectrum of organisms, Clin. Orthop. Relat. Res. 7 (2015) 2229–2235.
[76] S.J. Frangiamore, A. Saleh, M.J. Grosso, M. Farias Kovac, C.A. Higuera, J.P. Iannotti, E.T. Ricchetti, α-Defensin as a predictor of periprosthetic shoulder infection, J. Shoulder Elbow Surg. 24 (2015) 1021–1027, http://dx.doi.org/10.1016/j.jse.2014.12.021. pii: S1058-2746(14)00703-4.
[77] C. Deirmengian, K. Kardos, P. Kilmartin, A. Cameron, K. Schiller, J. Parvizi, Combined measurement of synovial fluid α-Defensin and C-reactive protein levels: highly accurate for diagnosing periprosthetic joint infection, J. Bone Joint Surg. Am. 96 (17) (2014) 1439–1445, http://dx.doi.org/10.2106/JBJS.M.01316.
[78] C. Deirmengian, K. Kardos, P. Kilmartin, A. Cameron, K. Schiller, R.E. Booth Jr., J. Parvizi, The alpha-defensin test for periprosthetic joint infection outperforms the leukocyte esterase test strip, Clin. Orthop. Relat. Res. 473 (1) (2015) 198–203, http://dx.doi.org/10.1007/s11999-014-3722-7.
[79] M. Fuerst, B. Fink, W. Ruther, The value of preoperative knee aspiration and arthroscopic biopsy in revision total knee arthroplasty, Z. Orthop. Ihre Grenzgeb. 143 (1) (2005) 36–41.
[80] G. Meermans, F.S. Haddad, Is there a role for tissue biopsy in the diagnosis of periprosthetic infection? Clin. Orthop. Relat. Res. 468 (5) (2010) 1410–1417.
[81] S. Sadiq, J.R. Wootton, C.A. Morris, M.D. Northmore-Ball, Application of core biopsy in revision arthroplasty for deep infection, J. Arthroplasty 20 (2) (2005) 196–201.
[82] B.L. Atkins, N. Athanasou, J.J. Deeks, et al., Prospective evaluation of criteria for microbiological diagnosis of prosthetic-joint infection at revision arthroplasty. The OSIRIS Collaborative Study Group, J. Clin. Microbiol. 36 (10) (1998) 2932–2939.

[83] A. Trampuz, K.E. Piper, M.J. Jacobson, et al., Sonication of removed hip and knee prostheses for diagnosis of infection, N. Engl. J. Med. 357 (7) (2007) 654–663.
[84] J.H. Lonner, P. Desai, P.E. Decisive, G. Steiner, J.D. Zuckerman, The reliability of analysis of intraoperative frozen sections for identifying active infection during revision hip or knee arthroplasty, J. Bone Joint Surg. Am. 78 (10) (1996) 1553–1558.
[85] R.R. Laffer, P. Graber, P.E. Ochsner, W. Zimmerli, Outcome of prosthetic knee-associated infection: evaluation of 40 consecutive episodes at a single centre, Clin. Microbiol. Infect. 12 (2006) 433–439.
[86] S.S. Lewis, K.V. Dicks, L.F. Chen, M.P. Bolognesi, D.J. Anderson, D.J. Sexton, R. Moehring, Delay in diagnosis of invasive surgical site infections following knee arthroplasty versus hip arthroplasty, Clin. Infect. Dis. 60 (7) (2015) 990–996.
[87] E. Moran, S. Masters, A.R. Berendt, P. McLardy-Smith, I. Byren, B.L. Atkins, Guiding empirical antibiotic therapy in orthopaedics: the microbiology of prosthetic joint infection managed by débridement, irrigation and prosthesis retention, J. Infect. 55 (2007) 1–7.
[88] C.L. Jacovides, R. Kreft, B. Adeli, B. Hozack, G.D. Ehrlich, J. Parvizi, Successful identification of pathogens by polymerase chain reaction (PCR)-based electron spray ionisation time-of-flight mass spectrometry (ESI-TOF-MS) in culture-negative periprosthetic joint infection, J. Bone Joint Surg. Am. 94 (24) (2012) 2247–2254.
[89] F.S. Haddad, B.A. Masri, D. Campbell, The PROSTALAC functional spacer in two-stage revision for infected knee replacements: prosthesis of antibiotic-loaded acrylic cement, J. Bone Joint Surg. (Br.) 82 (2000) 807–812.
[90] S. Oussedik, K. Gould, I. Stockley, F.S. Haddad, Defining periprosthetic infection: do we have a workable gold standard? J. Bone Joint Surg. (Br.) 94 (2012) 1455–1456.
[91] W. Zimmerli, P.E. Ochsner, Management of infection associated with prosthetic joints, Infection 31 (2003) 99–108.
[92] F.S. Haddad, M. Sukeik, S. Alazzawi, Is single-stage revision according to a strict protocol effective in treatment of chronic knee arthroplasty infections? Clin. Orthop. Relat. Res. 473 (2015) 8–14.
[93] T. Gehrke, S.J. Breusch, Management of TKA infection. One-stage exchange, in: Infection and Local Treatment in Orthopaedic Surgery, Springer-Verlag, Berlin Heidelberg, 2007, pp. 272–285.
[94] F.F. Buechel, F.P. Femino, J. D'Alessio, Primary exchange revision arthroplasty for infected total knee replacement: a long-term study, Am. J. Orthop. 33 (2004) 190–198.
[95] S.B. Goksan, M.A. Freeman, One-stage reimplantation for infected total knee arthroplasty, J. Bone Joint Surg. (Br.) 74 (1992) 78–82.
[96] M. Silva, R. Tharani, T.P. Schmalzried, Results of direct exchange or débridement of the infected total knee arthroplasty, Clin. Orthop. Relat. Res. 404 (2002) 125–131.
[97] J. Singer, A. Merz, L. Frommelt, B. Fink, High rate of infection control with one-stage revision of septic knee prostheses excluding MRSA and MRSE, Clin. Orthop. Relat. Res. 470 (2012) 1461–1471.
[98] D. Sofer, B. Regenbrecht, J. Pfeil, Early results of one-stage septic revision arthroplasties with antibiotic-laden cement. A clinical and statistical analysis [in German], Orthopade 34 (2005) 592–602.
[99] R.W. Parkinson, P.R. Kay, A. Rawal, A case for one-stage revision in infected total knee arthroplasty? Knee 18 (2011) 1–4.
[100] R.P. Pitto, Antibiotic loaded bone cement spacers for the two-stage revision in total hip arthroplasty, in: Infection and Local Treatment in Orthopaedic Surgery, Springer-Verlag, Berlin Heidelberg, 2007, pp. 287–293.

[101] A. Abudu, K.A.Z. Sivardeen, R.J. Grimmer, et al., The outcome of perioperative wound infection after total hip and knee arhhroplasty, Int. Orthop. 26 (1) (2002) 40–43.
[102] R.T. Goldman, G.R. Scuderi, J.N. Insall, Two-stage reimplantation for infected total knee replacement, Clin. Orthop. Relat. Res. 331 (1996) 118–124.
[103] J.N. Insall, F.M. Thompson, B.D. Brause, Two-stage reimplantation for the salvage of infected total knee arthroplasty, J. Bone Joint Surg. Am. 65 (8) (1983) 1087–1098.
[104] I.R. Scott, I. Stockley, C.J. Getty, Exchange arthroplasty for infected knee replacements. A new two-stage method, J. Bone Joint Surg. (Br.) 75 (1) (1993) 28–35.
[105] A.H. Wilde, J.T. Ruth, Two-stage reimplantation in infected total knee arthroplasty, Clin. Orthop. Relat. Res. 236 (1988) 23–35.
[106] R.E. Windsor, J.V. Bono, Infected total knee replacements, J. Am. Acad. Orthop. Surg. 2 (1) (1994) 44–53.
[107] R.H. Emerson Jr., M. Muncie, T.R. Tarbox, et al., Comparison of a mobile with a static spacer in total knee infection, Clin. Orthop. Relat. Res. 404 (2002) 132–138.
[108] T.F. Carlton, T.K. Fehring, W.L. Griffin, Bone loss associated with the use of spacer blocks in infected total knee arthroplasty, Clin. Orthop. Relat. Res. 345 (1997) 148–154.
[109] T.K. Fehring, S. Odum, T.F. Calton, J.B. Mason, Articulating versus static spacers in revision total knee arthroplasty for sepsis, Clin. Orthop. Relat. Res. 380 (2000) 9–16.
[110] Garg Parag, Ranjan Rajeev, Bandyopadhyay Utpal, Chouksey Shiv, S.R. Mitra, S.K. Gupta, Antibiotic-impregnated articulating cement spacer for infected total knee arthroplasty, Indian J. Orthop. 45 (6) (2011) 535–545.
[111] C.C. Castelli, R. Ferrari, Key points in two-stage revision for infected knee arthroplasty: bone loss, quality of life between stages and surgical approach at second stage, in: Infection and Local Treatment in Orthopaedic Surgery, Springer-Verlag, Berlin Heidelberg, 2007, pp. 295–301.
[112] S.M. Tintle, J.A. Forsberg, B.K. Potter, R.B. Islinger, R.C. Andersen, Prosthesis retention, serial debridement, and antibiotic bead use for the treatment of infection following total joint arthroplasty, Orthopedics 32 (2009) 87.
[113] I.S. Vanhegan, R. Morgan-Jones, D.S. Barrett, F.S. Haddad, Developing a strategy to treat established infection in total knee replacement. A review of the latest evidence and clinical practice, J. Bone Joint Surg. (Br.) 94 (2012) 875–881.
[114] R.L. Barrack, G. Engh, C. Rorabeck, J. Sawhney, M. Woolfrey, Patient satisfaction and outcome after septic versus aseptic revision total knee arthroplasty, J. Arthroplasty 15 (2000) 990–993.
[115] M.G. Freeman, T.K. Fehring, S.M. Odum, K. Fehring, W.L. Griffin, J.B. Mason, Functional advantage of articulating versus static spacers in 2-stage revision for total knee arthroplasty infection, J. Arthroplasty 22 (2007) 1116–1121.
[116] A.A. Haleem, D.J. Berry, A.D. Hanssen, Mid-term to long-term followup of two-stage reimplantation for infected total knee arthroplasty, Clin. Orthop. Relat. Res. 428 (2004) 35–39.
[117] J.M. Leone, A.D. Hanssen, Management of infection at the site of a total knee arthroplasty, J. Bone Joint Surg. Am. 87 (2005) 2335–2348.
[118] R.M. Meek, B.A. Masri, D. Dunlop, D.S. Garbuz, N.V. Greidanus, R. McGraw, C.P. Duncan, Patient satisfaction and functional status after treatment of infection at the site of a total knee arthroplasty with use of the PROSTALAC articulating spacer, J. Bone Joint Surg. Am. 85 (2003) 1888–1892.
[119] C.L. Romano, L. Gala, N. Loqoluso, D. Romano, L. Drago, Two-stage revision of septic knee prosthesis with articulating knee spacers yields better infection eradication rate than one-stage or two-stage revision with static spacers, Knee Surg. Sports Traumatol. Arthrosc. 20 (12) (2012) 2445–2453.

[120] E. Meani, C.L. Romano, Two-stage revision with preformed knee spacers and modular knee prosthesis, in: Infection and Local Treatment in Orthopaedic Surgery, Springer-Verlag, Berlin Heidelberg, 2007, pp. 302–308.

[121] R. Pivec, Q. Naziri, K. Issa, S. Banerjee, M.A. Mont, Systematic review comparing static and articulating spacers used for revision of infected total knee arthroplasty, J. Arthroplasty 29 (3) (2014) 553–557.

[122] C.L. Nelson, C.M. Evarts, Arthroplasty and arthrodesis at the knee joint, Orthop. Clin. North Am. 2 (1971) 245–264.

[123] J.A. Rand, Alternatives to reimplantation for salvage of the total knee arthroplasty complicated by infection, J. Bone Joint Surg. 75A (1993) 282–289.

[124] S.E. Stulberg, Arthrodesis in failed total knee replacements, Orthop. Clin. North Am. 13 (1982) 213–234.

[125] C. Sandrone, G. Moraga, C. Casalino Finoccio, Arthrodesis for the treatment of severe knee disease, in: Infection and Local Treatment in Orthopaedic Surgery, Springer-Verlag, Berlin Heidelberg, 2007, pp. 309–319.

[126] K. Bargiotas, D. Wohlrab, J.J. Sewecke, G. Lavinge, P.J. DeMeo, N.G. Sotereanos, Arthrodesis of the knee with a long intramedullary nail following the failure of a total knee arthroplasty as the result of infection. Surgical technique, J. Bone Joint Surg. Am. 89 (Suppl. 2, Pt 1) (2007) 103–110.

[127] J.H. MacDonald, S. Agarwal, M.P. Lorei, N.A. Johanson, A.A. Freiberg, Knee arthrodesis, J. Am. Acad. Orthop. Surg. 14 (3) (2006) 154–163.

[128] T.A. Damron, A.A. McBeath, Arthrodesis following failed total knee: comprehensive review and metanalysis of recent literature, Orthopaedics 18 (4) (1995) 361–368.

[129] M.G. Wilson, K. Kelley, T.S. Thornhill, Infection as a complication of total knee replacement arthroplasty: risk factors and treatment in sixty-seven cases, J. Bone Joint Surg. (Am) 72 (6) (1990) 878–883.

[130] R.A. Laun, Arthrodesis and amputation in knee joint infections, in: Infection and Local Treatment in Orthopaedic Surgery, Springer-Verlag, Berlin Heidelberg, 2007, pp. 321–327.

Periprosthetic infection in shoulder and elbow joints

8

M.H. Amini*, P.J. Evans*, E.T. Ricchetti*,†
*Cleveland Clinic, Orthopaedic and Rheumatologic Institute, Cleveland, OH, United States, †Cleveland Clinic Lerner College of Medicine of Case Western Reserve University, Cleveland, OH, United States

8.1 Introduction

The incidence of upper extremity arthroplasty in the United States continues to rise, a trend that increases at a faster pace with time. Shoulder arthroplasty grows each year not only because of the continued growth of hemiarthroplasty and anatomic total shoulder arthroplasty (TSA) but also because of the rapid acceptance of the reverse total shoulder arthroplasty (RSA) and the evolution of its indications. The incidence rose 2.5 fold over a decade [1]. Similarly, the incidence of total elbow arthroplasty (TEA) grew 248% from 1993 to 2007, and the incidence of upper extremity revision arthroplasty grew 500% during the same time period [2]. This rise in surgical volume can lead to a similar rise in volume of surgical complications, including infection. In a systematic review of TSA, Bohsali et al. reported that infection affected 0.7% of all TSAs and that infection made up 4.6% of all complications [3]. More recently, Gonzalez et al. similarly reported an incidence of infection after TSA of 1.1%, which comprised 2.9% of all complications [4]. In 2011, median hospitalization costs for shoulder periprosthetic joint infections (PJI) were $17,163.57, a substantial burden to the healthcare system given the large volume of shoulder arthroplasties performed and the incidence of infection [5]. The risk of infection after TEA is substantially higher than the risk after shoulder arthroplasty, or even hip and knee arthroplasty. Earlier studies reported infection rates near 10%, though more recently this has been reported to be around 3% [6;7]. However, the rate of infection is higher in patients with rheumatoid arthritis at 5% [8,9], and in the revision setting at 8% [10].

Despite the growing incidence of shoulder and elbow arthroplasty, and the difficulty in treating infections in these joints, there is relatively little evidence to guide prognosis, diagnosis, and management in comparison to the large volume of high-quality evidence in the literature on infected hip and knee arthroplasty. Studies are often retrospective, with small case series of a single institution; therefore, results may not be generalizable. Further compounding these issues is the predominance of less virulent organisms in the shoulder and elbow, particularly *Propionibacterium acnes* (*P. acnes*) and *coagulase-negative Staphylococcus*, that may be less clinically evident and more difficult to diagnose, particularly in shoulder arthroplasty [11–13].

We will review the clinical presentation, diagnostic testing, and treatment and outcomes of infection after shoulder and elbow arthroplasty.

Management of Periprosthetic Joint Infections (PJIs). http://dx.doi.org/10.1016/B978-0-08-100205-6.00008-2

8.2 Clinical presentation

Infections in total joint arthroplasty are often classified by chronicity as acute (less than 3 months after surgery), subacute (3–12 months after surgery), and late/chronic (greater than 1 year after surgery) [3,14,15]. A slightly different classification in the hip and knee literature includes four types: type 1 (the presence of positive intraoperative cultures at revision surgery), type 2 (acute postoperative infection within 30 days of arthroplasty), type 3 (acute hematogenous infection occurring at any time), and type 4 (chronic infection) [16]. Infections caused by nonvirulent organisms, such as *P. acnes*, are typically present since the time of primary arthroplasty but are often chronic by the time of diagnosis, as the paucity of clinical signs of infection leads to a delay in diagnosis.

As with any clinical evaluation, the workup should begin with a detailed history of the patient's clinical course and symptoms after the index procedure. The surgeon should specifically ask about the recovery after the initial arthroplasty, whether there was a period of pain relief and functional gain followed by eventual decline or if the patient never experienced improvement in pain and/or motion, and whether there were any wound healing issues, prolonged drainage, or antibiotic use.

Patients with infected shoulder arthroplasty typically present with pain as the primary complaint, frequently noting that the pain never went away postoperatively. While component loosening may cause pain with activity, patients with infection typically report rest pain and constant pain, though the pain may be worse with activity if the infection leads to early component loosening. Stiffness is also commonly reported in patients with infected shoulder arthroplasties [14,17]. They typically never regained full or adequate motion after the arthroplasty despite thorough rehabilitation. The stiffness is often associated with pain, which then further worsens the stiffness.

Patients frequently do not have the typical signs of infection, including wound drainage, fevers, chills, etc., with infections caused by indolent organisms, such as *P. acnes*. A history of wound issues in the immediate postoperative period should be noted, however. A history of hematoma formation at the time of the arthroplasty, particularly one requiring reoperation, is associated with an increased risk of PJI [18]. The surgeon should carefully review the medical history and comorbidities, evaluating for risk factors for PJI. Intrinsic factors include diabetes mellitus, rheumatoid arthritis, systemic lupus erythematosus, prior surgery, other sources of infection, malnutrition, obesity, and older age, while extrinsic factors include repeated steroid injections, oral corticosteroid use, chemotherapy, and local radiation [3,14,17,19,20].

The surgeon should examine the previous incision for healing, erythema, cellulitis, swelling, drainage, or a sinus tract. Sometimes *P. acnes* presents with a diffuse, nonblanching, erythematous rash about the incision; however, most shoulder arthroplasty infections have normal-appearing surgical incisions. Wound complications after elbow arthroplasty, however, are more frequent, particularly in those patients with a poor soft tissue envelope or in immunocompromised hosts. Jeon et al. noted a 5.5% rate (97 of 1749) of wound problems, including delayed healing and hematoma in 34 cases, and hematoma in 33, 9 of which progressed to infection. Of the 97 cases with wound problems, 24 progressed to a deep infection and 11 of them required resection arthroplasty [21]. Palpation of the joint line, along with other relevant structures, will help localize the pain and look for other sources of pain. Active and passive range of motion are commonly restricted in all planes as a result of the scar tissue and ensuing stiffness that form because of an infection. Differences in active and passive motion or deficits in strength testing may suggest rotator cuff pathology in the shoulder or impairment of the biceps and triceps in the elbow. The clinician must also evaluate for other sources of pain and/or dysfunction, especially the cervical spine.

8.3 Diagnostic testing

8.3.1 General

Working up a prosthetic shoulder or elbow joint for infection should begin with a detailed history and physical examination as outlined previously. Further diagnostic testing includes laboratory and imaging studies. Serum white blood cell (WBC) count, erythrocyte sedimentation rate (ESR), C-reactive protein (CRP), radiographs, and joint aspiration are the most common first line tests [3,20]. Advanced imaging, such as CT, US, MRI, or indium In-111-labeled WBC scan may also be obtained. It is important to note that no single test is reliable enough by itself to diagnose PJI [22].

Though ESR and CRP are nonspecific inflammatory markers, they are often elevated in cases of PJI [14,17,23,24]. They both often remain elevated in the acute postoperative period and are less useful during this time, with CRP normalizing by roughly 2 weeks and ESR returning to normal over a more prolonged period [25–27]. It is important to consider that inflammatory conditions, such as rheumatoid arthritis, may elevate inflammatory markers, making the interpretation of ESR and/or CRP in the workup of possible PJI challenging in these patients. If a patient has baseline elevation of ESR and/or CRP from an inflammatory disease process, a further increase in these markers may be a sign of developing PJI. However, interpreting an elevation in this setting may still not always be straight forward, as fluctuations of ESR and/or CRP can occur as a part of the normal disease process for such inflammatory conditions.

Aspiration of synovial fluid from the prosthetic joint can be extremely helpful in the diagnostic work-up. This can often be done in the office, blindly or with US-guidance, or if unsuccessful, this can be performed fluoroscopically by a radiologist to ensure that the needle is within the joint. As with any other aspiration, the needle should not be placed through any areas of cellulitis. Patients should be off antibiotics for at least 2 weeks prior to the aspiration to maximize the yield of the synovial culture [22]. Synovial WBC count and WBC differential have been well reported in the hip and knee literature, and well-established guidelines exist in the diagnosis of hip and knee PJI. Combining the synovial WBC count and differential with ESR and CRP also improves the ability to diagnose a hip or knee PJI [22]. However, in the upper extremity, established cutoff levels do not exist.

Radiographs should be carefully examined for new or progressive radiolucencies, or gross loosening of one or both components. Particularly if there is no identifiable cause for early progressive radiolucencies or loosening, the clinician should have a high index of suspicion for PJI. In addition to lucencies, periosteal new bone formation may suggest a PJI [26]. Technetium Tc-99 bone scan and indium In-111-labeled WBC scan largely play a limited role in the work-up of PJI, though they may be helpful if other testing has been equivocal [22].

For those scheduled to undergo revision surgery, particularly if the work-up was negative but the suspicion for PJI remains, the patient should again be off antibiotics for at least 2 weeks prior to surgery, and intraoperative cultures and pathologic specimens, including frozen sections, should be obtained. The surgeon should obtain samples from the joint capsule/synovial lining, the prosthesis-bone interface, and the medullary canal(s) to adequately sample all potential niduses for infection. Intraoperative frozen sections are often the only point-of-care test available to the surgeon in the operating room. Generally, five or more polymorphonuclear (PMN) leukocytes per high-powered field are considered positive for infection, based on literature from hip and knee arthroplasty [15,23,26,28–30]. Intraoperative gram stain is no longer obtained because of its poor utility and because it has been reported negative in cases of confirmed infection. Even cultures have been reported to be negative in the same scenario. This may be because of insufficient tissue samples, inadequate culture length, or failing to discontinue antibiotics early enough before the operation [14,15,17,31]. Because of the prevalence of slow-growing, low-virulence organisms in upper extremity PJI, cultures should often be held for more prolonged periods of time than usual. We will discuss this in more detail in the following sections.

8.3.2 Shoulder specific

Because of the indolent nature of many of the offending organisms in PJIs of the shoulder, many of the typical diagnostic tests may be negative. Villacis et al., in a study of 14 infected shoulder arthroplasties in 34 patients, evaluated the utility of common serum markers and noted poor sensitivity for all. The sensitivity, specificity, positive predictive value, negative predictive value, and accuracy for WBC were 7%, 95%, 50%, 59%, and 59%, for ESR were 21%, 65%, 30%, 54%, and 47%, and for CRP were 0%, 95%, 0%, 57%, and 56% [32]. Another recent prospective study of 24 infections in 69 patients undergoing revision shoulder arthroplasty by Grosso et al. showed a sensitivity and specificity for ESR of 42% and 82%, and for CRP of 46% and 93% [33]. Piper et al. showed in a systematic review that ESR and CRP have sensitivities of 16% and 42% in shoulder PJIs compared to 75% and 88% in hip and knee PJIs [34].

Serum interleukin-6 (IL-6) has received attention in hip and knee PJI due to increased sensitivity and specificity in diagnosis of PJI [35] and has subsequently been evaluated in the shoulder, as well. The study by Villacis et al. also prospectively evaluated the utility of IL-6 and showed that there was no difference in IL-6 levels between infected and noninfected shoulder arthroplasties, and that the sensitivity, specificity, positive predictive value, negative predictive value, and accuracy were 14%, 95%, 67%, 61%, and 62%, respectively [32]. Similarly, in the study by Grosso et al., the sensitivity was 12% and the specificity was 93%, making it less sensitive than ESR and CRP (42% and 46%, respectively) in that series [33].

Aspiration is an important part of the diagnostic workup in possibly infected shoulder arthroplasties. But again, the indolent nature of many shoulder PJIs leads to a decreased synovial inflammatory response and decreased synovial fluid production relative to infected hip and knee arthroplasties. Frequently, attempts at aspiration are unsuccessful. Sperling et al. noted that aspiration was successful in only 56% of patients who were later confirmed to have a PJI, the majority of which were *coagulase-negative*

Staphylococcus or *P. acnes* [14]. Codd et al. noted that aspiration was successful in only 38.8% of shoulder arthroplasties, and a pathogen was isolated in only 29% [36]. Dilisio et al. recently evaluated fluoroscopically guided aspiration of 19 painful arthroplasties that later went on to revision and found a sensitivity, specificity, positive, and negative predictive values of 16.7%, 100%, 100%, and 58.3% [37].

Even when successful, the volume of aspirate can be limited and may preclude testing of synovial WBC count and differential. As a result, shoulder-specific cutoff values have not been reported. In addition to sending the fluid for aerobic and anaerobic cultures, some newer tests examining synovial fluid biomarkers have shown promise in the detection of prosthetic infection. Synovial IL-6 was prospectively evaluated by Frangiamore et al. in a study of 35 painful shoulder arthroplasties undergoing revision surgery. Using receiver operating characteristic curve analysis, a cutoff value of 359.3 pg/mL led to sensitivity, specificity, positive, and negative likelihood ratios of 87%, 90%, 8.45, and 0.15. Seven patients with negative preoperative workup were later diagnosed with infection based on multiple positive intraoperative cultures, and the synovial IL-6 level was elevated in five of them, with a mean level of 1400 pg/mL. Levels were also significantly elevated in patients with *P. acnes* infections [38]. In a similarly modeled study, Frangiamore et al. evaluated synovial α-defensin in 33 painful shoulder arthroplasties undergoing revision surgery. Sensitivity, specificity, positive, and negative likelihood ratios were 63%, 95%, 12.1, and 0.38, and α-defensin was significantly elevated in the presence of a culture positive for *P. acnes* and moderately correlated with the number of positive intraoperative cultures [39]. Nearly all culture-positive cases in these two studies were *P. acnes* or *coagulase-negative* *Staphylococcus*. Leukocyte esterase is another synovial fluid analysis that has been evaluated in shoulder arthroplasty after showing promising results in hip and knee PJI [40,41]. However, Nelson et al. evaluated 85 primary and revision arthroplasties, including 10 infected revisions, and showed sensitivity, specificity, positive, and negative predictive values of only 30%, 67%, 43%, and 83%. It is important to note that aspirates that contain blood must be centrifuged prior to testing, and 29% of the time, even after centrifuging, the aspirate was too bloody for analysis [42].

If workup of a painful shoulder arthroplasty is negative for infection, but there is no other indication for revision surgery and the concern for PJI remains high, arthroscopic tissue biopsy may be considered. Multiple tissue samples can be taken from around the components as well as from the joint capsule, and other cause of pain can also be evaluated, including component loosening and rotator cuff deficiency. The recent study by Dilisio et al. retrospectively evaluated 19 patients with painful arthroplasties who underwent arthroscopic biopsy prior to revision surgery, 7 (41%) of which later grew *P. acnes*. The sensitivity, specificity, positive, and negative predictive values were all 100%, and all arthroscopic cultures matched cultures taken during the revision surgery [37].

Preoperative workup can often be equivocal, leaving the surgeon to rely on intraoperative findings and testing if revision surgery is performed. Gram staining has previously been a mainstay of point-of-care testing for infection, but because of poor sensitivity noted in the hip and knee literature, is no longer recommended [22,43–45]. In a known infection, however, intraoperative gram stain may guide immediate empiric antibiotic coverage. Frozen section histology has also been evaluated in shoulder arthroplasty. One study noted a 92% rate of negative intraoperative frozen sections in patients later confirmed with culture-positive infection, and another showed no correlation between frozen sections and positive cultures, with the exception of *Staphylococcus aureus* [30]. However, recent data suggest that the use of alternate criteria for intraoperative frozen section analysis may improve the ability to detect *P. acnes* infections in the shoulder. Grosso et al. evaluated 45 patients who underwent frozen section histology during revision shoulder arthroplasty, including 18 *P. acnes* infections and 12 infections from other organisms. Using a standard threshold of 5 PMN cells per high-power field (400×), the sensitivity was 50% for *P. acnes* infections and 67% for other infections, while the specificity was 100%. Using a new threshold of 10 PMNs in each of the 5 densest high-power fields, the sensitivity for *P. acnes* infections improved to 72%, while the specificity remained 100% [46].

Implant sonication is another method that may improve the ability to detect organisms of low virulence. Ultrasonic pressure waves dislodge biofilms from the surface of the implant to increase the sensitivity of detection. Piper et al. showed that sonication improves the sensitivity from 54.5% to 66.7%. However, the workflow and coordination necessary to transmit large, sterile containers for holding the components have limited the widespread use of this technique [12].

Despite advances in the preoperative and intraoperative diagnosis of shoulder PJI, the possibility of an unexpected positive culture remains, and the surgeon may have to rely on other signs to guide intraoperative decision-making. In a study of 193 apparently aseptic revision shoulder arthroplasties, male sex, humeral loosening, membrane formation, and cloudy fluid were independent predictors of postoperative positive culture growth for *P. acnes*. In this study, the rate of unexpected positive cultures was 56% [47]. Another study from the same institution compared *P. acnes* positive revision arthroplasties to aseptic revisions and noted that the *P. acnes* positive patients were more likely to be male; have glenoid erosions, osteolysis, or loosening; and have a humeral membrane [48]. Several other studies have reported lower rates of unexpected positive cultures. In a cohort of 148 apparently aseptic revisions performed 3 years or more after primary arthroplasty, the rate of positive cultures was 9.5% [49]. Another study noted a rate of 15% in a similar cohort of apparently aseptic revision arthroplasties, and of these, only 10% resulted in a symptomatic infection [50]. Grosso et al. found that only 5.9% of patients with unexpected positive cultures experienced a symptomatic infection [51]. Prolonged postoperative antibiotic therapy may not be necessary for patients with unexpected positive cultures, particularly if all previous hardware was removed and antibiotic cement was used at the time of revision, but further data are needed to confirm this.

8.3.3 Propionibacterium acnes

P. acnes is a relatively slow-growing organism that can be difficult to isolate in routine cultures with standard incubation periods and can remain in the soft tissues even after adequate antisepsis. Lee et al. showed that after skin preparation with ChloroPrep (CareFusion, San Diego, California) that punch biopsies of 7 of 10 male volunteers were culture positive for *P. acnes* [52]. Matsen et al. showed that 3 of 10 male patients had *P. acnes* growth from deep tissues during primary arthroplasty after skin preparation and intravenous antibiotics [53]. In a study of revision arthroplasties from the same authors, *P. acnes* grew from dermal specimens in 12 of 21 males and 0 of 20 females, and deep specimens grew from 12 of 20 males and 1 of 20 females [54].

Many have recognized the need to incubate cultures for longer than standard incubation times of five days in order to improve the ability to isolate *P. acnes*. Some studies recommend incubation periods of up to 21–28 days [55–57]. However, the clinical relevance and the risk of reinfection in those with late culture growth remain unknown. Butler-Wu et al. recommended holding cultures for 13 days, as those that grew after this point were considered to be contaminants. They also noted that holding only the anaerobic cultures for prolonged incubation periods would have missed 29.4% of *P. acnes* isolates and suggest holding both aerobic and anaerobic cultures for this time frame [58]. We routinely obtain three to five specimens for culture during revision shoulder arthroplasty from different locations—capsule, periprosthetic humeral and glenoid tissue—and hold each for aerobic and anaerobic culture for a period of 14 days.

P. acnes positive cultures have been reported in several recent studies in patients undergoing first time open shoulder surgery. Levy et al. cultured aspirates and specimens in 55 consecutive patients undergoing primary shoulder arthroplasty and noted that 41.8% of patients were culture positive for *P. acnes*. No patient developed a postoperative infection, though the authors treated culture-positive patients with 4 weeks of oral antibiotics and also suggested that *P. acnes* may be implicated as a possible cause for glenohumeral osteoarthritis based on the high positive culture rate [59]. However, other recent studies using strict specimen collection protocols and/or control specimens suggest that *P. acnes* positive cultures during first time shoulder surgery may more likely represent contaminants rather than true positive results. Maccioni et al. utilized a strict specimen collection protocol in 32 patients undergoing primary shoulder arthroplasty in which 5 capsule/synovium specimens were sent for culture and a sixth was sent for histopathology and noted that only 3 patients (9.4%) grew *P. acnes*, with only 1 showing growth on more than 1 specimen. Histopathology was negative for infection in all positive culture cases [60]. Mook and Garrigues also recently reported a 17.1% (14/82) rate of positive *P. acnes* culture in patients undergoing first time open shoulder surgery, with most cases representing an isolated result (three capsule specimens taken per case) that grew out late. In addition, a sterile gauze sponge was sent as a control culture specimen in all of the prospectively enrolled patients in the study and had a 13.0% (7/54) rate of positive culture (5/7 positive cultures grew *P. acnes*). Taken together, these studies suggest a significant contamination rate with *P. acnes* positive cultures, likely due to the increased incubation times for these specimens and the increased handling of samples as a result of the longer culture times [61].

In the setting of revision shoulder arthroplasty, interpretation of a positive *P. acnes* culture result should be made in the context of the overall clinical picture. This should take into account other positive preoperative and intraoperative markers for infection, including traditional serum markers and intraoperative frozen section findings, as well as newer synovial fluid biomarkers, if available, and the characteristics of the positive culture result(s) themselves, such as the timing of the first positive culture and the number of positive culture results relative to the overall number of cultures taken. Such data taken together can help determine whether a positive culture is likely to represent a false positive result consistent with contamination or a true positive finding. A recent study by Frangiamore et al. highlights this approach. In 46 patients who underwent revision shoulder arthroplasty and had at least one positive *P. acnes* culture, cases were classified into one of two groups based on culture results and other perioperative findings of infection, a probable true positive culture group and a probable contaminant group. Time to *P. acnes* growth in culture was found to be significantly shorter in the probable true positive culture group compared with the probable contaminant group (median of 5 days compared with 9 days). There were also significantly fewer days to *P. acnes* culture growth among cases with a higher number of positive cultures and a higher proportion of positive cultures, regardless of group classification [38,39,62].

8.3.4 Elbow specific

Workup should begin with routine blood tests as in any other possibly infected arthroplasty, including ESR, CRP, and WBC count and differential. However, there is little literature on the utility of these tests in PJI of the elbow, and a large number of patients with TEAs have inflammatory arthritis, limiting the use of inflammatory markers in these patients. Mean ESR in infected TEA has been reported between 30 and 55 mm/h [7,63,64]. Because of the lack of information to guide the use of serum markers, aspiration and culture remains the main diagnostic test to augment clinical suspicion of infection in the elbow.

Similar to the shoulder, we recommend the surgeon should obtain three to five specimens for culture and intraoperative frozen section histology. Ahmadi et al. evaluated 227 revision TEAs and noted a sensitivity, specificity, positive predictive value, negative predictive value, and accuracy of 51.3%, 93.1%, 60.6%, 90.2%, and 85.9% with frozen section histology [65]. As a result, negative frozen section histology does not exclude an infection, but a positive result is highly suggestive of one.

Implant sonication has been evaluated in the diagnosis of an infection. Vergidis et al. reported on a series of 36 patients, including 9 infections, and noted a sensitivity and specificity of 89% and 100% compared to 55% and 93% for standard tissue culture [66]. However, because of limitations previously noted, it is not commonly used.

Just as in the evaluation of a possibly infected shoulder arthroplasty, the perioperative workup may be completely negative in elbow PJI with indolent organisms, but intraoperative cultures may later become positive. Wee et al. evaluated 213 apparently aseptic revision TEAs and noted unexpected positive cultures in 7.5%. One patient was treated for infection because of early loosening, and nine of ten remained infection free at more than 2 years. One patient sustained an infection with a different organism [67].

8.4 Treatment and outcomes—Shoulder

Treatment of shoulder PJI includes several surgical and nonsurgical options and can and should be tailored to each patient's specific clinical scenario, including the virulence and chronicity of infection, the stability of the implants, the quality of the soft tissues, the remaining bone stock and quality, and the patient's physiologic and psychological preparedness to undergo two-stage revision surgery if necessary. Treatment options include antibiotic suppression, irrigation and debridement with implant retention, resection arthroplasty, permanent antibiotic cement spacer placement, one-stage exchange with the use of antibiotic-impregnated cement, two-stage exchange with a temporary antibiotic cement spacer, arthrodesis, or amputation. Regardless of the method of treatment, care should include medical management of the infection with a peripherally inserted central catheter (PICC) line to receive organism-specific IV antibiotics typically for 6 weeks [68].

In patients too medically infirm to undergo surgical treatment of a PJI, or in patients who are unwilling to undergo surgical treatment, long-term antibiotic suppression is an option. This is particularly true if the patient has minimal to no symptoms, is not septic, and has a low-virulence organism. One small series noted a recurrence rate of 60% in shoulder arthroplasty [17]. Other series in knee arthroplasty note rates as low as 24% [69].

Implant retention with irrigation and debridement is most appropriate for acute postoperative infections (type 2) and acute hematogenous infections (type 3). Modular component exchange can be performed as part of the procedure, such as exchange of the modular head component in standard shoulder arthroplasty and exchange of the glenosphere and polyethylene liner in RSA. Modular component exchange also significantly facilitates exposure of the joint for thorough debridement, as much of the posterior joint would be inaccessible with component retention. Postoperatively, the patient receives 4–6 weeks of intravenous antibiotics, potentially followed by a period of oral antibiotic suppression.

Most commonly, treatment of a shoulder PJI includes removal of the infected implant, particularly in chronic infections (type 4), in cases with highly virulent organisms, or in cases that have previously failed less aggressive surgical treatments. Removal of the implant should also include thorough bone and soft tissue debridement and removal of all cement [23,24,70–73]. The surgeon may need fluoroscopy to guide implant and cement removal, and a multitude of instruments should be available, including osteotomes, curettes, reamers, and a saw. A longitudinal unicortical osteotomy made the length of the stem and lateral to the bicipital groove may minimize the risk of unintended humeral fracture during implant removal [74]. The split can be gently hinged open to loosen the stem and remove cement, or if needed, the split can be converted to a cortical window and secured back at the end of the case with a monofilament cerclage wire or suture. As mentioned previously, three to five culture specimens should be obtained from the joint capsule and periprosthetic tissue of the humeral canal and the glenoid vault.

Following implant removal, resection arthroplasty is an option in those with intractable infections, medical illnesses that preclude multiple operations, or soft tissue, bony, or neurologic deficits that preclude reimplantation of a new prosthesis (Fig. 8.1). Typically, resection arthroplasty provides good pain relief but leaves the patient with significant functional limitations. Sperling et al. noted that 11 patients all had moderate to complete pain relief, however, surgery was graded as successful in only three patients because of functional limitations, with mean abduction of 69 degrees and external rotation of 31 degrees [14]. Braman et al. followed seven patients

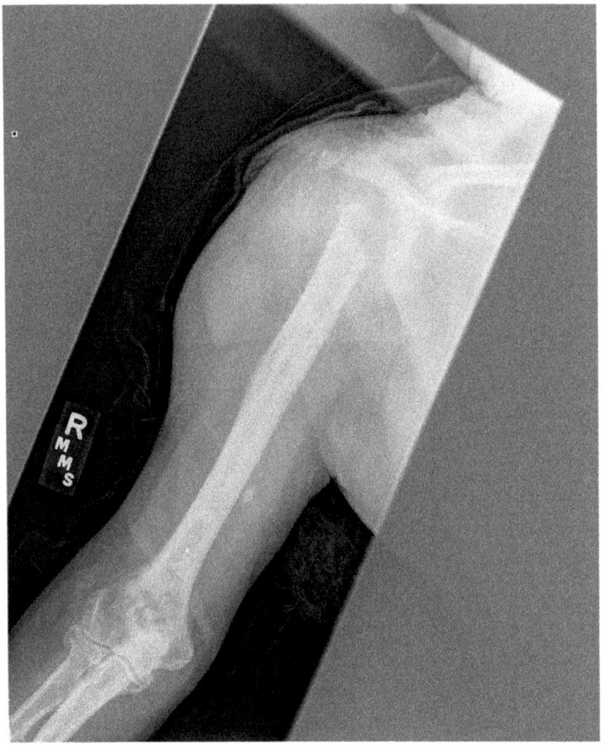

Fig. 8.1 Anteroposterior radiograph demonstrating a resection arthroplasty of the shoulder for infection.

with resection arthroplasty and reported that all patients were satisfied, though none were considered satisfactory by Neer's criteria. All were able to perform activities of daily living (ADLs), but mean flexion was 28 degrees and mean external rotation was 8 degrees [75]. Rispoli followed 18 patients with resection arthroplasty (13 for infected arthroplasty) and reported significant pain relief in all, though five still had moderate to severe pain. Patients had significant functional limitations, with mean elevation of 70 degrees and mean external rotation of 31 degrees, simple shoulder test (SST) score of 3.1, and American Shoulder and Elbow Surgeons (ASES) score of 36 [76]. Muh et al. followed 26 patients and reported significant improvements in visual analog scale pain scores, with mean flexion of 45 degrees and mean external rotation of 9 degrees, and Constant scores of 27.3 [77].

Permanent placement of an antibiotic cement spacer can be performed for the same indications as resection arthroplasty, specifically those with contraindications to multiple surgical interventions (Fig. 8.2). Additionally, patients may be satisfied with the pain relief and function of a spacer initially placed as part of a two-stage protocol and may not wish to undergo reimplantation. A study of nine patients with antibiotic spacers who elected not to undergo reimplantation because of satisfaction with the spacer reported satisfaction in all nine patients, no or mild pain, and adequate performance of

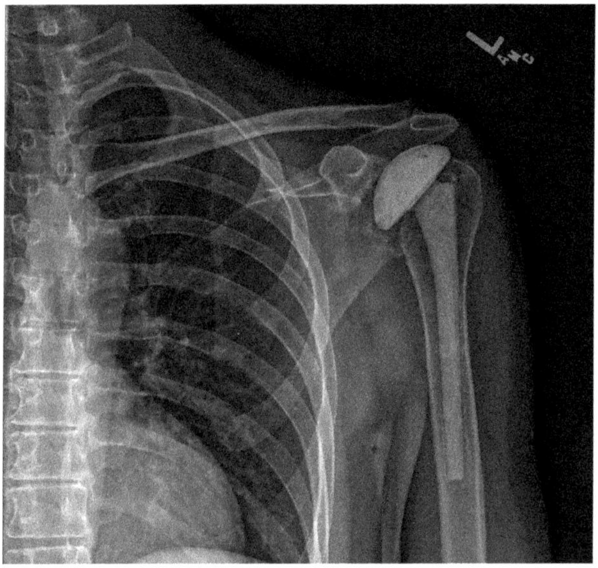

Fig. 8.2 Articulating antibiotic-impregnated cement spacer made with a prefabricated mold.

ADLs. Mean abduction was 75 degrees, mean external rotation was 25 degrees, and QDASH scores were 37.5 [78].

Reimplantation of a prosthesis typically provides the best functional outcome and can be done in one or two stages. One-stage exchange, with placement of a new prosthesis at the same time as the debridement and removal of the infected prosthesis, should be performed with antibiotic-impregnated cement. This approach may not be adequate in cases of highly resistant or highly virulent infections. Ince et al. studied 16 one-stage exchanges (15 to hemiarthroplasty, 1 to RSA) treated with antibiotics until the CRP was trending down, a mean of 8.6 days. There were no recurrent infections, mean abduction was 51.6 degrees, mean Constant scores were 33.6, and mean UCLA scores were 18.3 [79]. Beekman also reported on 11 patients, all of whom were revised to an RSA with antibiotic-impregnated cement and all of whom received at least 3 weeks of antibiotic therapy. There was one recurrent infection, and mean Constant scores were 55 [80]. Klatte et al. evaluated 35 patients treated with single-stage revision to various implants and a mean of 10.6 days of antibiotics. Two patients (5.7%) developed recurrent infection and were treated with resection arthroplasty. Mean Constant scores were 43.3 for hemiarthroplasties, 56.0 for bipolar hemiarthroplasties, and 61 for RSAs [81].

Two-stage exchange involves placement of an antibiotic-impregnated cement spacer followed by delayed reimplantation (Fig. 8.3). The cement spacer delivers a high concentration of antibiotics locally, minimizes soft tissue contractures by maintaining some soft tissue length, and maintains a space for future implant placement. Cement spacers can be made by hand or by a prefabricated mold. Prior to reimplantation, the surgeon should confirm that the infection has been cleared. At least 2 weeks after completion of a 6-week course of IV antibiotics, a new serum WBC, ESR, CRP,

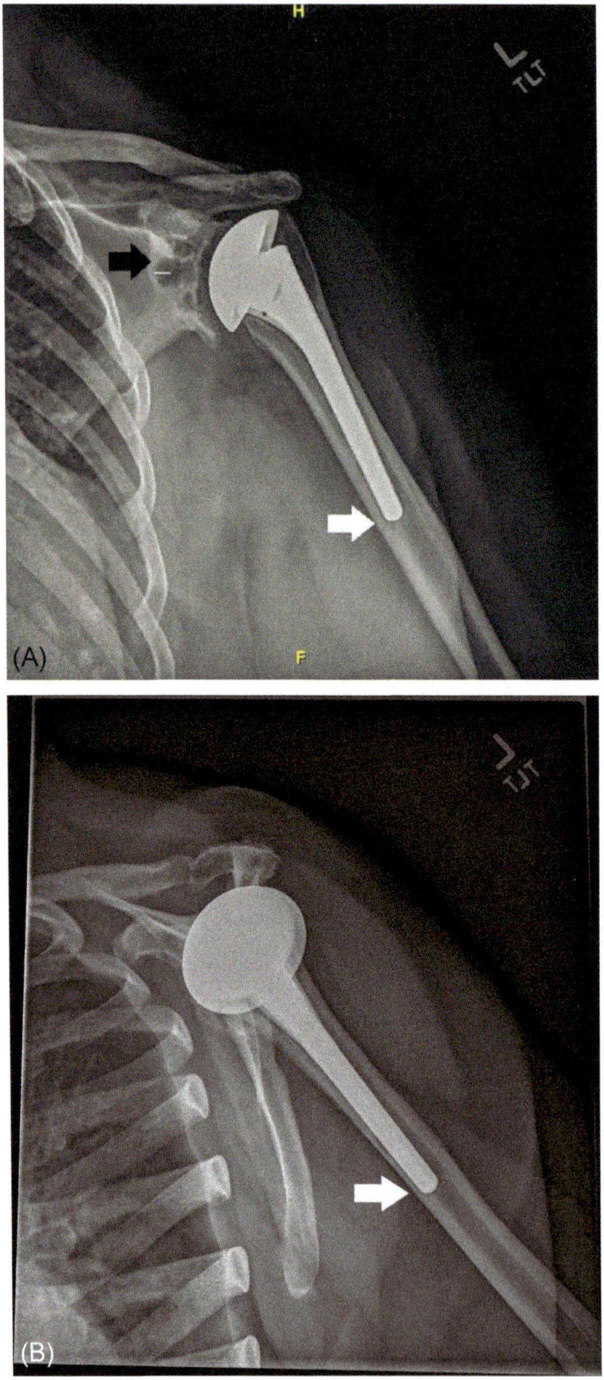

Fig. 8.3 Infected anatomic total shoulder arthroplasty. (A and B) Anteroposterior (AP) and lateral radiograph demonstrating loose glenoid and humeral components. Note the lucencies around the central glenoid peg (black arrow) and the endosteal erosion at the tip of the humeral stem (white arrows).

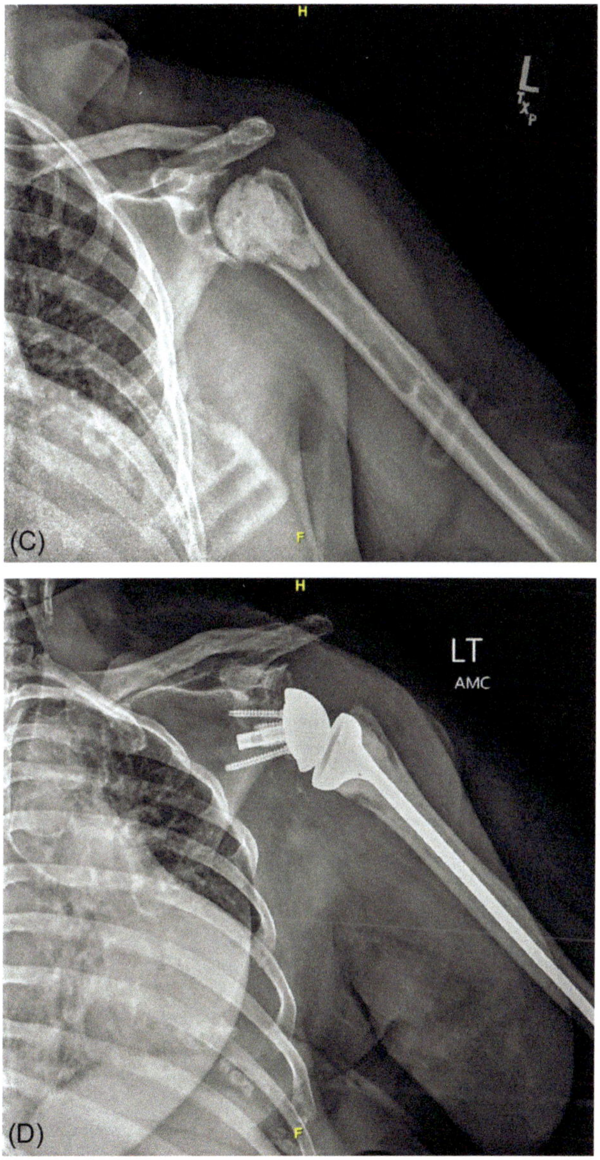

Fig. 8.3 Continued. (C) AP radiograph after removal of the prostheses and cement, and placement of a hand-molded antibiotic-impregnated cement spacer. (D) AP radiograph after revision to a reverse shoulder arthroplasty with a long humeral stem to bypass the endosteal erosion and cancellous allograft to fill the cavitary defects in the glenoid.

and glenohumeral aspiration should be obtained to ensure that serum tests have normalized and that the aspiration is culture negative [73]. If the infection clears after the initial debridement and antibiotic course, reimplantation is performed typically three months later. A computed tomography (CT) scan before reimplantation is helpful to better evaluate glenoid bone loss and morphology. At the time of the reimplantation, the surgeon should obtain multiple specimens for culture and pathology, as outlined previously. If there is still evidence of infection preoperatively or intraoperatively, the surgeon should perform a repeat surgical debridement and should place a new antibiotic spacer. Recently, Zhang et al. reported on 18 patients who underwent open biopsy before reimplantation and noted evidence of persistent infection in 22% of patients, and in 38% of patients with *P. acnes*. These patients underwent repeat debridements and further antibiotic therapy before final reimplantation [82].

Seitz and Damacen first reported on eight patients with infected shoulders (five shoulder arthroplasties) treated with two-stage exchange 6 months apart with 3 months of IV antibiotics. There were no recurrent infections, all patients noted improved pain and function, and mean Penn Shoulder scores were 68.4, though patients had decreased motion and strength compared to their contralateral shoulder [72]. Strickland et al. followed 19 patients who underwent 4–6 weeks of IV antibiotics and reimplantation at 11 weeks and noted that 7 of 19 had recurrent infections based on need for suppressive antibiotics in 6 and a subsequent resection arthroplasty in 1. Pain improved significantly, but outcomes were rated unsatisfactory in 68%, and there were 14 complications in the study group. Mean elevation was 89 degrees, mean external rotation was 43 degrees, and internal rotation was to L5 [15]. Hattrup and Renfree reported on 25 infected shoulders (20 arthroplasties) treated with 6 weeks of IV antibiotics and subsequent reimplantation. There were 3 recurrent infections, mean flexion was 100.9 degrees, abduction was 93.6 degrees, and external rotation was 32.6 degrees [83]. Coffey et al. treated 16 shoulders (11 arthroplasties) with a commercially produced cement spacer, IV antibiotics for 5.6 weeks, and reimplantation at 11.2 weeks. Four patients refused reimplantation because of satisfaction, and one underwent arthrodesis because of deltoid deficiency. There were no recurrent infections, and pain improved in all. Mean flexion was 110 degrees, mean external rotation was 20 degrees, UCLA scores were 26, SST scores were 6.6, ASES scores were 74, and Constant scores were 57 [84]. Sabesan et al. evaluated 17 patients treated with two-stage exchange, 6.3 weeks of IV antibiotics, and revision to RSA at 4.0 months. There was one recurrent infection, and seven complications, including the one recurrence, one hematoma, and five patients with instability. Mean flexion was 123 degrees, mean external rotation was 26 degrees, and Penn Shoulder scores were 66.4 [24].

Several studies have directly compared the previously mentioned treatment methods. Verhelst et al. evaluated 11 patients treated with resection arthroplasty and 10 patients with permanent spacers and noted no difference in recurrence rate or functional outcomes [85]. Codd et al. compared resection arthroplasty in five patients to reimplantation in 13 patients. Pain relief was similar in the two groups, though elevation was 66 degrees compared to 117 degrees, external rotation was 27 degrees compared to 38 degrees, and internal rotation was to the sacrum compared to L2 [36]. Stine compared permanent spacers to two-stage exchange in 30 patients. There were no recurrent infections and no differences in functional outcomes [73]. Cuff et al. compared one-stage exchange in 10 patients to two-stage exchange in 12 patients. There were no recurrent infections and no differences in functional outcomes between groups; however, there were 11 complications in 7 shoulders [23].

Several studies have evaluated outcomes in case with unexpected positive culture results. Topolski et al. reported on 75 cases of revision arthroplasty with unexpected positive cultures. Fifty-four of 75 were treated with standard postoperative antibiotics. Ten patients underwent a second revision surgery, only one of which was for a documented recurrent infection, though 7 of the 10 had positive cultures at the time of the second revision [30]. Kelly and Hobgood evaluated eight patients with unexpected positive cultures and noted that two of eight developed a late infection. They recommended placing all revisions on oral antibiotics until cultures are negative and that culture-positive patients should be treated with 6 weeks of IV antibiotics [28]. We similarly reviewed 17 patients with unexpected positive cultures who were not treated with prolonged antibiotic therapy and noted recurrent infection in 1 of 17. There was no difference in recurrence rate or functional outcomes in these patients compared to one- and two-stage revisions for infection from our institution [51]. In the largest and most recent of these series, Foruria et al. evaluated the results of 107 consecutive cases of revision shoulder arthroplasty without preoperative or intraoperative signs of infection that were found to have at least one positive intraoperative culture. Sixty-eight (64%) of the cases grew *P. acnes*. Following one-stage revision, 53 cases were treated with an extended course of antibiotics, while 54 were not. At mean follow-up of 5.6 years, 11/107 (10%) cases had a subsequent positive culture result either by aspirate or during a second revision surgery that matched the culture result of the original revision surgery. Ten of the cases were *P. acnes* positive. Treatment with antibiotics did not appear to lower the risk of having a second positive culture result [50]. Dodson et al. reported on 11 patients with *P. acnes* infections, all of whom received prolonged antibiotic treatment. In five patients diagnosed preoperatively and treated with two-stage exchange, there were two recurrent infections. Six patients were diagnosed with unexpected positive cultures, though the recurrence rate was not reported [86].

Arthrodesis is an option in those with axillary nerve or brachial plexus injuries, or combined loss of the rotator cuff and deltoid. Functional outcome is typically better than resection arthroplasty as it provides a stable platform for distal function; however, it is a technically demanding procedure given the bone loss typically present after implant removal. Scalise and Iannotti reported on a series of seven patients who underwent arthrodeses after failed arthroplasty and noted the need for a vascularized fibula in three patients and subsequent operations to obtain union in four [87].

As the above information shows, most studies on outcomes of treatment for infected shoulder arthroplasty report results on only a small number of patients, often with varying treatment protocols. This lack of uniformity in treatment approach, as well as in reported outcome measures, makes it difficult to draw definitive conclusions on specific treatment methods. As the most common clinical scenario in shoulder PJI is a chronic infection involving an indolent organism, further data are particularly needed to better define the indications and outcomes in cases of one- and two-stage exchange. Improved diagnostic testing to better identify *P. acnes* and other less virulent organisms preoperatively or intraoperatively may help more clearly define indications for one- versus two-stage exchange, as well as the need for postoperative antibiotic therapy in the setting of a presumed aseptic one-stage revision with unexpected positive culture results. Currently, our preferred management approach for a chronic PJI of the shoulder is a two-stage reimplantation when one or more perioperative signs of infection are present. We also routinely maintain all presumed aseptic revision shoulder arthroplasty cases on oral antibiotics postoperatively until all cultures are negative, due to the possibility of an unexpected positive culture. In this scenario, cases found to have multiple positive intraoperative cultures are treated with 6 weeks of IV antibiotic therapy, with transition to a more extended course of oral antibiotics based on the clinical presentation. If only one intraoperative culture turns positive, no further antibiotic therapy may be needed if the clinical picture is suggestive of a probable contaminant result. This is particularly true if the culture growth is late and all prior components were removed at revision surgery; however, retention of some of the prior components may still be an indication for postoperative antibiotic treatment.

8.5 Treatment and outcomes—Elbow

Choosing a method of treatment for infected elbow arthroplasty depends on patient- and disease-specific factors. The surgeon must consider the duration of symptoms, component fixation, quality of the soft tissue envelope, virulence of the infection, and the patient's health status and comorbidities. Treatment options include long-term antibiotic suppression, irrigation and debridement with implant retention, resection arthroplasty, one- and two-stage exchange, and arthrodesis. Regardless of the method chosen, the patient should receive a PICC line and organism-specific IV antibiotics.

Long-term antibiotic suppression should be reserved for patients unwilling or unable to undergo operative treatment because of end-stage comorbidities. Patients should be appropriately counseled on the risk of continued or recurrent infection, though the exact rate remains unknown.

Irrigation and debridement with component retention is best suited for acute postoperative and acute hematogenous infections with stable components. After exposure of the joint and prosthesis, the components are disarticulated and the bushings are removed. If the components are stable, the joint is thoroughly debrided and irrigated. If repeat debridements are deemed necessary, antibiotic-impregnated cement beads are placed in the joint and removed at the final debridement. Early studies reported poor success rates. Wolfe et al. reported on 11 patients treated with component retention and noted eight failures and intermittent drainage in the remaining three [64]. Cheung et al. reported on nine patients with success in only one of them [88]. Another study used this treatment in infections with durations of 30 days or less and noted a 50% success rate. Of note, all four patients with *Staphylococcus epidermidis* failed treatment, while six of eight with *S. aureus* were treated successfully [7].

Resection arthroplasty is a common treatment method for infected TEA. In the appropriately selected patient, pain relief and satisfaction are high, though functional limitations may persist. Those who are unfit or unwilling to undergo multiple operations and exchange arthroplasty, and those with limited functional demands are ideal candidates. The old components and cement mantle should be completely removed with attention to preserving the remaining bone stock. A unicortical osteotomy or cortical window may be necessary on the humeral and/or ulnar side to prevent accidental fracture during implant and cement removal. Any split or window is then repaired with monofilament cerclage wire or suture. Preservation of both condyles and deepening of the intervening sulcus are necessary to contain the ulna and provide a fulcrum for flexion and extension with a resection arthroplasty. Loss of one or both condyles or significant shortening results in worse functional outcomes [88]. The arm is placed in a cast for three to 4 weeks to obtain soft tissue stability. A hinged elbow brace can be used if there is significant instability. Zarkadas et al. reported on 30 resection arthroplasties with significant pain relief, and eight good results, eleven fair results, and eleven poor results. There were postoperative infections in 47%, intraoperative fractures in 35%, and permanent injury in 18% [89].

There is limited information on the success of one-stage exchange for an infected TEA. The surgeon should perform a thorough debridement of the bone and soft tissues, in addition to complete removal of the components and previous cement mantle. If a unicortical osteotomy or cortical window is necessary for component and cement removal, the new stem should bypass it by two cortical diameters. New components should be cemented with antibiotic-impregnated cement. One small series reported successful treatment in five of six patients with *S. aureus* infections [90].

Two-stage exchange is more commonly used than one-stage exchange for the treatment of infected TEA, particularly if the patient is medically fit to undergo two-stage reimplantation. As in one-stage exchange, the soft tissues should be thoroughly debrided, and the previous components and cement mantle should be carefully removed to prevent further bone loss. The condyles should be preserved in case the final result becomes a resection arthroplasty. Yamaguchi et al. reported an 80% success rate and noted that *S. epidermidis* was a significant risk for failure [7]. Peach et al. reported on a series of 34 infected TEAs that were to undergo two-stage exchange. Of these, 21% refused further surgery because of satisfaction with the resection, one patient did not clear the infection, and 76% underwent reimplantation. In those undergoing reimplantation, the recurrence rate was 11.5%, and mean Mayo elbow performance scores (MEPS) were 81.1 [91]. Cheung et al. reported on reimplantation of a TEA at a mean of 72.5 weeks after resection arthroplasty with a success rate of 72%. There were 52% good to excellent results, 10% fair, 38% poor, and mean MEPS of 66.3. Similar to other reports, *S. epidermidis* was associated with a higher risk of failure [92].

Arthrodesis has been reported in one series by Otto et al., who noted high complication rates and poor success in achieving a fusion. Of five patients, none had complete union, and two had fibrous unions. All patients underwent at least one reoperation, three required revision fixation for hardware failure, and two ultimately underwent resection arthroplasty. As a result, elbow arthrodesis is not recommended in the treatment of infected TEA [93].

Bone loss on the humeral and/or ulnar sides is a difficult problem when dealing with reconstructive efforts in the elbow, and use of allograft-prosthetic composites (APCs) and/or megaprostheses may be required. Morrey et al. reported on 25 patients treated with APCs and noted successful union in 92%. Of the seven patients in the series treated for infected TEAs, five were successfully salvaged with APCs [94].

The extensor mechanism requires special attention in the multiply operated elbow. Duquin et al. importantly noted triceps weakness in 55% of patients after revision surgery for infected TEA, triceps or olecranon deficiency in 41%, and an intact extensor mechanism in only 11% [95]. Patients should be counseled on the risk of triceps insufficiency and wound breakdown because of the frequent need for reoperation and revision arthroplasty, often with allograft that may compromise the volume of the soft tissue envelope. Although very infrequent, soft tissue coverage with a cutaneous or myocutaneous flap could be necessary.

Any surgeon treating PJI of the elbow should be prepared to deal with potentially serious wound complications. The thin soft tissue envelope is often even thinner in many patients because of chronic use of immunosuppressants for inflammatory arthropathies. Even the smallest wound breakdown may expose the effective joint space or the implant itself to the external environment. As a result, the surgeon should have a variety of soft tissue coverage procedures in his or her repertoire or should have a working relationship with a plastic surgeon who does.

8.6 Conclusion and future directions

As the population ages, the volume of upper extremity arthroplasty continues to grow. As a result, upper extremity PJI poses an even greater burden to both patients and global health systems. Despite significant advances in our understanding of the pathogenesis, diagnosis, treatment, and outcomes of PJI, many challenges and questions remain. The frequently indolent nature of the offending organisms makes establishing a firm diagnosis challenging, particularly since patients do not present with classic signs of infection. As in hip and knee PJI, synovial biomarkers have shown promise as more sensitive and specific tests than traditional serum markers. Though not widely available at this point, development of point-of-care use of these synovial biomarkers would be a significant step forward in the pre- and intraoperative identification of shoulder and elbow PJI. We must also establish what culture-growths represent true pathogens as opposed to colonizing bacteria or even contaminants. This chapter highlights the lack of precise algorithms for both diagnosis and treatment of shoulder and elbow PJI currently. Essential to such algorithms is the development of a consensus definition for both shoulder and elbow PJI, based on a combination of preoperative and intraoperative findings and intraoperative culture results. The evaluation and management of the painful shoulder or elbow arthroplasty remains highly variable and needs to be standardized in such areas as preoperative surgical site preparation, choice and timing of intraoperative antibiotics during revision surgery, number and type of intraoperative cultures obtained during revision surgery, culture methods and length of time for culture incubation, and choice and length of postoperative antibiotic therapy. A consensus definition of PJI and a standardized approach to evaluation and management will aid in developing and interpreting future research studies and will ultimately lead to more refined diagnostic algorithms and clinical treatment pathways. Currently, there are many acceptable forms of treatment for shoulder and elbow PJI, as outlined previously. The surgeon must chose a treatment based on the chronicity and virulence of the infection, soft tissue and bony deficiencies, activity level, comorbidities, and the patient's desires. Further studies will clarify the role of each treatment paradigm in specific scenarios and the economic impact of certain treatment methods on both the patients and the health systems.

References

[1] S.H. Kim, B.L. Wise, Y. Zhang, R.M. Szabo, Increasing incidence of shoulder arthroplasty in the United States, J. Bone Joint Surg. Am. 93 (2011) 2249–2254.

[2] J.S. Day, E. Lau, K.L. Ong, G.R. Williams, M.L. Ramsey, S.M. Kurtz, Prevalence and projections of total shoulder and elbow arthroplasty in the United States to 2015, J. Shoulder Elbow Surg. 19 (2010) 1115–1120.

[3] K.I. Bohsali, M.A. Wirth, C.A. Rockwood Jr., Complications of total shoulder arthroplasty, J. Bone Joint Surg. Am. 88 (10) (2006) 2279–2292.

[4] J.F. Gonzalez, G.B. Alami, F. Baque, G. Walch, P. Boileau, Complications of unconstrained shoulder prostheses, J. Shoulder Elbow Surg. 20 (4) (2011) 666–682.

[5] E.M. Padegimas, M. Maltenfort, M.L. Ramsey, G.R. Williams, J. Parvizi, S. Namdari, Periprosthetic shoulder infection in the United States: incidence and economic burden, J. Shoulder Elbow Surg. 24 (5) (2015) 741–746.

[6] I. Voloshin, D.W. Schippert, S. Kakar, E.K. Kaye, B.F. Morrey, Complications of total elbow replacement: a systematic review, J. Shoulder Elbow Surg. 20 (2011) 158–168.

[7] K. Yamaguchi, R.A. Adams, B.F. Morrey, Infection after total elbow arthroplasty, J. Bone Joint Surg. Am. 80 (1998) 481–491.

[8] J.M. Aldridge 3rd., N.R. Lightdale, W.J. Mallon, R.W. Coonrad, Total elbow arthroplasty with the Coonrad/Coonrad-Morrey prosthesis. A 10- to 31-year survival analysis, J. Bone Joint Surg. (Br.) 88 (4) (2006) 509–514.

[9] J.C. Van Der Lugt, R.B. Geskus, P.M. Rozing, Primary Souter-Strathclyde total elbow prosthesis in rheumatoid arthritis, J. Bone Joint Surg. Am. 86-A (2004) 465–473.

[10] S.B. Sneftrup, S.L. Jensen, H.V. Johannsen, J.O. Sojbjerg, Revision of failed total elbow arthroplasty with use of a linked implant, J. Bone Joint Surg. (Br.) 88 (2006) 78–83.

[11] A. Patel, R.P. Calfee, M. Plante, S.A. Fischer, A. Green, *Propionibacterium acnes* colonization of the human shoulder, J. Shoulder Elbow Surg. 18 (6) (2009) 897–902.

[12] K.E. Piper, M.J. Jacobson, R.H. Cofield, J.W. Sperling, J. Sanchez-Sotelo, D.R. Osmon, A. Mcdowell, S. Patrick, J.M. Steckelberg, J.N. Mandrekar, M. Fernandez Sampedro, R. Patel, Microbiologic diagnosis of prosthetic shoulder infection by use of implant sonication, J. Clin. Microbiol. 47 (6) (2009) 1878–1884.

[13] V. Zeller, a. Ghorbani, C. Strady, P. Leonard, P. Mamoudy, N. Desplaces, *Propionibacterium acnes*: an agent of prosthetic joint infection and colonization, J. Infect. 55 (2007) 119–124.

[14] J.W. Sperling, T.K. Kozak, A.D. Hanssen, R.H. Cofield, Infection after shoulder arthroplasty, Clin. Orthop. Relat. Res. (2001) 206–216.

[15] J.P. Strickland, J.W. Sperling, R.H. Cofield, The results of two-stage re-implantation for infected shoulder replacement, J. Bone Joint Surg. (Br.) 90 (2008) 460–465.

[16] H. Segawa, D.T. Tsukayama, R.F. Kyle, D.A. Becker, R.B. Gustilo, Infection after total knee arthroplasty. A retrospective study of the treatment of eighty-one infections, J. Bone Joint Surg. Am. 81 (1999) 1434–1445.

[17] J.S. Coste, S. Reig, C. Trojani, M. Berg, G. Walch, P. Boileau, The management of infection in arthroplasty of the shoulder, J. Bone Joint Surg. (Br.) 86 (2004) 65–69.

[18] E.V. Cheung, J.W. Sperling, R.H. Cofield, Infection associated with hematoma formation after shoulder arthroplasty, Clin. Orthop. Relat. Res. 466 (2008) 1363–1367.

[19] C.J. Linberg, J.W. Sperling, C.D. Schleck, R.H. Cofield, Shoulder arthroplasty in morbidly obese patients, J. Shoulder Elbow Surg. 18 (2009) 903–906.

[20] M.A. Wirth, C.A. Rockwood Jr., Complications of total shoulder-replacement arthroplasty, J. Bone Joint Surg. Am. 78 (1996) 603–616.

[21] I.H. Jeon, B.F. Morrey, O.A. Anakwenze, N.V. Tran, Incidence and implications of early postoperative wound complications after total elbow arthroplasty, J. Shoulder Elbow Surg. 20 (2011) 857–865.

[22] C. Della Valle, J. Parvizi, T.W. Bauer, P.E. Dicesare, R.P. Evans, J. Segreti, M. Spangehl, W.C. Watters 3rd., M. Keith, C.M. Turkelson, J.L. Wies, P. Sluka, K. Hitchcock, Diagnosis of periprosthetic joint infections of the hip and knee, J. Am. Acad. Orthop. Surg. 18 (2) (2010) 760–770.

[23] D.J. Cuff, N.A. Virani, J. Levy, M.A. Frankle, A. Derasari, B. Hines, D.R. Pupello, M. Cancio, M. Mighell, The treatment of deep shoulder infection and glenohumeral instability with debridement, reverse shoulder arthroplasty and postoperative antibiotics, J. Bone Joint Surg. (Br.) 90 (2008) 336–342.

[24] V.J. Sabesan, J.C. Ho, D. Kovacevic, J.P. lannotti, Two-stage reimplantation for treating prosthetic shoulder infections, Clin. Orthop. Relat. Res. 469 (9) (2011) 2538–2543.

[25] R.O. Niskanen, O. Korkala, H. Pammo, Serum C-reactive protein levels after total hip and knee arthroplasty, J. Bone Joint Surg. (Br.) 78 (1996) 431–433.

[26] M.D. Saltzman, G.S. Marecek, S.L. Edwards, D.M. Kalainov, Infection after shoulder surgery, J. Am. Acad. Orthop. Surg. 19 (4) (2011) 208–218.

[27] J. White, M. Kelly, R. Dunsmuir, C-reactive protein level after total hip and total knee replacement, J. Bone Joint Surg. (Br.) 80 (1998) 909–911.

[28] J.D. Kelly, E.R. Hobgood, Positive culture rate in revision shoulder arthroplasty, Clin. Orthop. Relat. Res. 467 (2009) 2343–2348.

[29] J.M. Mirra, H.C. Amstutz, M. Matos, R. Gold, The pathology of the joint tissues and its clinical relevance in prosthesis failure, Clin. Orthop. Relat. Res. (1976) 221–240.

[30] M.S. Topolski, P.Y. Chin, J.W. Sperling, R.H. Cofield, Revision shoulder arthroplasty with positive intraoperative cultures: the value of preoperative studies and intraoperative histology, J. Shoulder Elbow Surg. 15 (2006) 402–406.

[31] J. Jerosch, M. Schneppenheim, Management of infected shoulder replacement, Arch. Orthop. Trauma Surg. 123 (2003) 209–214.

[32] D. Villacis, J.A. Merriman, R. Yalamanchili, R. Omid, J. Itamura, G.F. Rick Hatch 3rd., Serum interleukin-6 as a marker of periprosthetic shoulder infection, J. Bone Joint Surg. Am. 96 (2014) 41–45.

[33] M.J. Grosso, S.J. Frangiamore, A. Saleh, M.F. Kovac, R. Hayashi, E.T. Ricchetti, T.W. Bauer, J.P. Iannotti, Poor utility of serum interleukin-6 levels to predict indolent periprosthetic shoulder infections, J. Shoulder Elbow Surg. 23 (2014) 1277–1281.

[34] K.E. Piper, M. Fernandez-Sampedro, K.E. Steckelberg, J.N. Mandrekar, M.J. Karau, J.M. Steckelberg, E.F. Berbari, D.R. Osmon, A.D. Hanssen, D.G. Lewallen, R.H. Cofield, J.W. Sperling, J. Sanchez-Sotelo, P.M. Huddleston, M.B. Dekutoski, M. Yaszemski, B. Currier, R. Patel, C-reactive protein, erythrocyte sedimentation rate and orthopedic implant infection, PLoS ONE 5 (2010). e9358.

[35] P.E. Di Cesare, E. Change, C.F. Preston, C.J. Liu, Serum interleukin-6 as a marker of periprosthetic infection following total hip and knee arthroplasty, J. Bone Joint Surg. Am. 87 (9) (2005) 1921–1927.

[36] T.P. Codd, K. Yamaguchi, R.G. Pollock, E.L. Flatow, L.U. Bigliani, Infected shoulder arthroplasties: treatment with staged reimplantation vs. resection arthroplasty, J. Shoulder Elbow Surg. 5 (1996) S5.

[37] M.F. Dilisio, L.R. Miller, J.J. Warner, L.D. Higgins, Arthroscopic tissue culture for the evaluation of periprosthetic shoulder infection, J. Bone Joint Surg. Am. 96 (23) (2014) 1952–1958.

[38] S.J. Frangiamore, A. Saleh, M.F. Kovac, M.J. Grosso, X. Zhang, T.W. Bauer, T.M. Daly, E.T. Ricchetti, J.P. Iannotti, Synovial fluid interleukin-6 as a predictor of periprosthetic shoulder infection, J. Bone Joint Surg. Am. 97 (1) (2015) 63–70.

[39] S.J. Frangiamore, A. Saleh, M.J. Grosso, M. Farias Kovac, C.A. Higuera, J.P. Iannotti, E.T. Ricchetti, Alpha-defensin as a predictor of periprosthetic shoulder infection, J. Shoulder Elbow Surg. 24 (7) (2015) 1021–1027.

[40] E.H. Tischler, P.K. Cavanaugh, J. Parvizi, Leukocyte esterase strip test: matched for musculoskeletal infection society criteria, J. Bone Joint Surg. Am. 96 (2014) 1917–1920.

[41] N.G. Wetters, K.R. Berend, A.V. Lombardi, M.J. Morris, T.L. Tucker, C.J. Della Valle, Leukocyte esterase reagent strips for the rapid diagnosis of periprosthetic joint infection, J. Arthroplasty 27 (2012) 8–11.

[42] G.N. Nelson, A.C. Narzikul, E.S. Paxton, M.D. Lazarus, G.R. Williams, M.L. Ramsey, J.A. Abboud, Leukocyte Esterase in the Diagnosis of Shoulder Periprosthetic Joint Infection, American Academy of Orthopaedic Surgeons Annual Meeting, Las Vegas, NV, 2015.

[43] C.J. Della Valle, D.M. Scher, Y.H. Kim, C.M. Oxley, P. Desai, J.D. Zuckerman, P.E. Di Cesare, The role of intraoperative Gram stain in revision total joint arthroplasty, J. Arthroplasty 14 (4) (1999) 500–504.

[44] P.M. Morgan, P. Sharkey, E. Ghanem, J. Parvizi, J.C. Clohisy, R.S. Burnett, R.L. Barrack, The value of intraoperative Gram stain in revision total knee arthroplasty, J. Bone Joint Surg. Am. 91 (2009) 2124–2129.

[45] M. Oethinger, D.K. Warner, S.A. Schindler, H. Kobayashi, T.W. Bauer, Diagnosing periprosthetic infection: false-positive intraoperative Gram stains, Clin. Orthop. Relat. Res. 469 (2011) 954–960.

[46] M.J. Grosso, S.J. Frangiamore, E.T. Ricchetti, T.W. Bauer, J.P. Iannotti, Sensitivity of frozen section histology for identifying *Propionibacterium acnes* infections in revision shoulder arthroplasty, J. Bone Joint Surg. Am. 96 (2014) 442–447.

[47] P. Pottinger, S. Butler-Wu, M.B. Neradilek, A. Merritt, A. Bertelsen, Prognostic factors for bacterial cultures positive or *Propionibacterium acnes* and other organisms in a large series of revision shoulder arthroplasties performed for stiffness, pain, or loosening, J Bone Joint Surg. Am. 94 (22) (2012) 2075–2083.

[48] C. Hou, A. Gupta, M. Chen, F.A. Matsen 3rd., How do revised shoulders that are culture positive for *Propionibacterium* differ from those that are not? J. Shoulder Elbow Surg. 24 (9) (2015) 1427–1432.

[49] E. Mcgoldrick, M.D. Mcelvany, S. Butler-Wu, P.S. Pottinger, F.A. Matsen 3rd., Substantial cultures of *Propionibacterium* can be found in apparently aseptic shoulders revised three years or more after the index arthroplasty, J. Shoulder Elbow Surg. 24 (1) (2015) 31–35.

[50] A.M. Foruria, T.J. Fox, J.W. Sperling, R.H. Cofield, Clinical meaning of unexpected positive cultures (UPC) in revision shoulder arthroplasty, J. Shoulder Elbow Surg. 22 (2013) 620–627.

[51] M.J. Grosso, V.J. Sabesan, J.C. Ho, E.T. Ricchetti, J.P. Iannotti, Reinfection rates after 1-stage revision shoulder arthroplasty for patients with unexpected positive intraoperative cultures, J. Shoulder Elbow Surg. 21 (2012) 754–758.

[52] M.J. Lee, P.S. Pottinger, S. Butler-Wu, R.E. Bumgarner, S.M. Russ, F.A. Matsen 3rd., Propionibacterium persists in the skin despite standard surgical preparation, J. Bone Joint Surg. Am. 96 (17) (2014) 1447–1450.

[53] F.A. Matsen 3rd., S.M. Russ, A. Bertelsen, S. Butler-Wu, P.S. Pottinger, Propionibacterium can be isolated from deep cultures obtained at primary arthroplasty despite intravenous antimicrobial prophylaxis, J. Shoulder Elbow Surg. 24 (6) (2014) 884–887.

[54] F.A. Matsen 3rd., S. Butler-Wu, B.C. Carofino, J.L. Jette, A. Bertelsen, R. Bumgarner, Origin of propionibacterium in surgical wounds and evidence-based approach for culturing propionibacterium from surgical sites, J. Bone Joint Surg. Am. 95 (23) (2013) e1811–e1817.

[55] I. Brook, Soft tissue and muscular infections (including group A streptococci), Baillieres Best Pract. Res. Clin. Rheumatol. 13 (1999) 95–114.

[56] J.H. Lonner, P. Desai, P.E. Dicesare, G. Steiner, J.D. Zuckerman, The reliability of analysis of intraoperative frozen sections for identifying active infection during revision hip or knee arthroplasty, J. Bone Joint Surg. Am. 78 (1996) 1553–1558.

[57] D.A. Stroh, A.J. Johnson, Q. Naziri, M.A. Mont, How do frozen and permanent histopathologic diagnoses compare for staged revision after periprosthetic hip infections? J. Arthroplasty 27 (9) (2012) 1663–1668. e1.

[58] S.M. Butler-Wu, E.M. Burns, P.S. Pottinger, A.S. Magaret, J.L. Rakeman, F.A. Matsen 3rd., B.T. Cookson, Optimization of periprosthetic culture for diagnosis of *Propionibacterium acnes* prosthetic joint infection, J. Clin. Microbiol. 49 (2011) 2490–2495.

[59] O. Levy, S. Iyer, E. Atoun, N. Peter, N. Hous, D. Cash, F. Musa, A.A. Narvani, *Propionibacterium acnes*: an underestimated etiology in the pathogenesis of osteoarthritis? J. Shoulder Elbow Surg. 22 (2013) 505–511.

[60] C.B. Maccioni, A.B. Woodbridge, J.C. Balestro, M.C. Figtree, B.J. Hudson, B. Cass, A.A. Young, Low rate of *Propionibacterium acnes* in arthritic shoulders undergoing primary total shoulder replacement surgery using a strict specimen collection technique, J. Shoulder Elbow Surg. 24 (8) (2015) 1206–1211.

[61] W.R. Mook, M.R. Klement, C.L. Green, K.C. Hazen, G.E. Garrigues, The incidence of *Propionibacterium acnes* in open shoulder surgery: a controlled diagnostic study, J. Bone Joint Surg. Am. 97 (12) (2015) 957–963.

[62] S.J. Frangiamore, A. Saleh, M.J. Grosso, B. Alolabi, T.W. Bauer, J.P. Iannotti, E.T. Ricchetti, Early versus late culture growth of *Propionibacterium acnes* in revision shoulder arthroplasty, J. Bone Joint Surg. Am. 97 (14) (2015) 1149–1158.

[63] B.F. Morrey, R.S. Bryan, Infection after total elbow arthroplasty, J. Bone Joint Surg. Am. 65 (1983) 330–338.

[64] S.W. Wolfe, M.P. Figgie, A.E. Inglis, W.W. Bohn, C.S. Ranawat, Management of infection about total elbow prostheses, J. Bone Joint Surg. Am. 72 (1990) 198–212.

[65] S. Ahmadi, T.M. Lawrence, B.F. Morrey, J. Sanchez-Sotelo, The value of intraoperative histology in predicting infection in patients undergoing revision elbow arthroplasty, J. Bone Joint Surg. Am. 95 (2013) 1976–1979.

[66] P. Vergidis, K.E. Greenwood-Quaintance, J. Sanchez-Sotelo, B.F. Morrey, S.P. Steinmann, M.J. Karau, D.R. Osmon, J.N. Mandrekar, J.M. Steckelberg, R. Patel, Implant sonication for the diagnosis of prosthetic elbow infection, J. Shoulder Elbow Surg. 20 (2011) 1275–1281.

[67] A.T. Wee, B.F. Morrey, J. Sanchez-Sotelo, The fate of elbows with unexpected positive intraoperative cultures during revision elbow arthroplasty, J. Bone Joint Surg. Am. 95 (2013) 109–116.
[68] E.T. Ricchetti, Management of infect shoulder arthroplasty, in: J.P. Iannotti, A. Miniaci, G.R. Williams, J.D. Zuckerman (Eds.), Disorders of the Shoulder. Diagnosis and Management, third ed., Lippincott Williams and Wilkins, Philadelphia, PA, 2014.
[69] A.D. Hanssen, J.A. Rand, Evaluation and treatment of infection at the site of a total hip or knee arthroplasty, Instr. Course Lect. 48 (1999) 111–122.
[70] M.I. Loebenberg, J.D. Zuckerman, An articulating interval spacer in the treatment of an infected total shoulder arthroplasty, J. Shoulder Elbow Surg. 13 (2004) 476–478.
[71] M.L. Ramsey, J.M. Fenlin Jr., Use of an antibiotic-impregnated bone cement block in the revision of an infected shoulder arthroplasty, J. Shoulder Elbow Surg. 5 (6) (1996) 479–482.
[72] W.H. Seitz Jr., H. Damacen, Staged exchange arthroplasty for shoulder sepsis, J. Arthroplasty 17 (2002) 36–40.
[73] I.A. Stine, B. Lee, C.G. Zalavras, G. Hatch 3rd., J.M. Itamura, Management of chronic shoulder infections utilizing a fixed articulating antibiotic-loaded spacer, J. Shoulder Elbow Surg. 19 (5) (2010) 739–748.
[74] J.W. Sperling, R.H. Cofield, Humeral windows in revision shoulder arthroplasty, J. Shoulder Elbow Surg. 14 (2005) 258–263.
[75] J.P. Braman, M. Sprague, J. Bishop, I.K. Lo, E.W. Lee, E.L. Flatow, The outcome of resection shoulder arthroplasty for recalcitrant shoulder infections, J. Shoulder Elbow Surg. 15 (2006) 549–553.
[76] D.M. Rispoli, J.W. Sperling, G.S. Athwal, C.D. Schleck, R.H. Cofield, Pain relief and functional results after resection arthroplasty of the shoulder, J. Bone Joint Surg. (Br.) 89 (2007) 1184–1187.
[77] S.J. Muh, J.J. Streit, C.J. Lenarz, C. Mccrum, J.P. Wanner, Y. Shishani, C. Moraga, R.J. Nowinski, T.B. Edwards, J.J. Warner, G. Walch, R. Gobezie, Resection arthroplasty for failed shoulder arthroplasty, J. Shoulder Elbow Surg. 22 (2013) 247–252.
[78] G. Themistocleous, C. Zalavras, I. Stine, V. Zachos, J. Itamura, Prolonged implantation of an antibiotic cement spacer for management of shoulder sepsis in compromised patients, J. Shoulder Elbow Surg. 16 (2007) 701–705.
[79] A. Ince, K. Seemann, L. Frommelt, A. Katzer, J.F. Loehr, One-stage exchange shoulder arthroplasty for peri-prosthetic infection, J. Bone Joint Surg. (Br.) 87 (2005) 814–818.
[80] P.D. Beekman, D. Katusic, B.M. Berghs, A. Karelse, L. De Wilde, One-stage revision for patients with a chronically infected reverse total shoulder replacement, J. Bone Joint Surg. (Br.) 92 (2010) 817–822.
[81] T.O. Klatte, K. Junghans, H. Al-Khateeb, J.M. Reuger, T. Gehrke, D. Kendoff, J. Neumann, Single-stage revision for peri-prosthetic shoulder infection: outcomes and results, Bone Joint J. 95-B (3) (2013) 391–395.
[82] A.L. Zhang, B.T. Feeley, B.S. Schwartz, T.T. Chung, C.B. Ma, Management of deep postoperative shoulder infections: is there a role for open biopsy during staged treatment? J. Shoulder Elbow Surg. 24 (2015) e15–e20.
[83] S.J. Hattrup, K.J. Renfree, Two-stage shoulder reconstruction for active glenohumeral sepsis, Orthopedics 33 (1) (2010) 20.
[84] M.J. Coffey, E.E. Ely, L.A. Crosby, Treatment of glenohumeral sepsis with a commercially produced antibiotic-impregnated cement spacer, J. Shoulder Elbow Surg. 19 (2010) 868–873.

[85] L. Verhelst, J. Stuyck, J. Bellemans, P. Debeer, Resection arthroplasty of the shoulder as a salvage procedure for deep shoulder infection: does the use of a cement spacer improve outcome? J. Shoulder Elbow Surg. 20 (2011) 1224–1233.

[86] C.C. Dodson, E.V. Craig, F.A. Cordasco, D.M. Dines, J.S. Dines, E. Dicarlo, B.D. Brause, R.F. Warren, *Propionibacterium acnes* infection after shoulder arthroplasty: a diagnostic challenge, J. Shoulder Elbow Surg. 19 (2) (2010) 303–307.

[87] J.J. Scalise, J.P. Iannotti, Glenohumeral arthrodesis after failed prosthetic shoulder arthroplasty, J. Bone Joint Surg. Am. 90 (2008) 70–77.

[88] E.V. Cheung, K. Yamaguchi, B.F. Morrey, Treatment of the infected total elbow arthroplasty, in: B.F. Morrey, J. Sanchez-Sotelo (Eds.), The Elbow and Its Disorders, fourth ed., Saunders, Elsevier, Philadelphia, PA, 2009.

[89] P.C. Zarkadas, B. Cass, T. Throckmorton, R. Adams, J. Sanchez-Sotelo, B.F. Morrey, Long-term outcome of resection arthroplasty for the failed total elbow arthroplasty, J. Bone Joint Surg. Am. 92 (2010) 2576–2582.

[90] J. Gille, A. Ince, O. Gonzalez, A. Katzer, J.F. Loehr, Single-stage revision of periprosthetic infection following total elbow replacement, J. Bone Joint Surg. (Br.) 88 (2006) 1341–1346.

[91] C.A. Peach, S. Nicoletti, T.M. Lawrence, D. Stanley, Two-stage revision for the treatment of the infected total elbow arthroplasty, Bone Joint J. 95-B (2013) 1681–1686.

[92] E.V. Cheung, R.A. Adams, B.F. Morrey, Reimplantation of a total elbow prosthesis following resection arthroplasty for infection, J. Bone Joint Surg. Am. 90 (2008) 589–594.

[93] R.J. Otto, P.J. Mulieri, B.J. Cottrell, M.A. Mighell, Arthrodesis for failed total elbow arthroplasty with deep infection, J. Shoulder Elbow Surg. 23 (2014) 302–307.

[94] M.E. Morrey, J. Sanchez-Sotelo, M.P. Abdel, B.F. Morrey, Allograft-prosthetic composite reconstruction for massive bone loss including catastrophic failure in total elbow arthroplasty, J. Bone Joint Surg. Am. 95 (2013) 1117–1124.

[95] T.R. Duquin, J.A. Jacobson, C.D. Schleck, D.R. Larson, J. Sanchez-Sotelo, B.F. Morrey, Triceps insufficiency after the treatment of deep infection following total elbow replacement, Bone Joint J. 96 (1) (2014) 82–87.

Part Three

Managing and treating periprosthetic joint infections

Practice and guidelines for treating periprosthetic joint infections: Single- and two-stage revision

9

C.L. Romanò, G. Manzi†, S. Scarponi*, N. Logoluso*, D. George‡*
*Orthopaedic Research Institute Galeazzi, Milan, Italy, †CHU Rennes, Rennes, France,
‡University College London Hospitals, London, United Kingdom

9.1 Introduction

Given the ability of bacteria to adhere on implanted biomaterials and to form well structured biofilms, that protect them from both immune-competent host's cells and antibiotics, the removal of a chronically infected prosthesis and a thorough joint debridement is considered the treatment of choice of periprosthetic infections (PJIs).

Following implant removal, in the majority of cases, depending on general and local conditions and on the clinical history of each patient, new components may be implanted.

This can be performed in one or two stages; the two-stage revision, with placement of an interim antibiotic-loaded spacer, is the most often reported procedure and regarded by many as the standard of care.

However, the best surgical modality for treating chronic PJI remains controversial, with a lack of randomized comparative studies and an ongoing discrepancy in the literature between the success rates of a single or one-stage exchange compared to a two-stage exchange arthroplasty [1–5].

Moreover, several variables in the performance of one- or two-stage procedures across different centers, including patient's selection, type of spacer, interval between stages, duration of antibiotic treatment, cemented versus uncemented implant fixation, etc. [6–12], make even pooling data of one-stage or two-stage approach an oversimplification, that should be considered.

With the lack of comparative studies, the operative approach is often determined by a combination of surgeon's experience and host's and microorganism's type, while the best inclusion and exclusion criteria for each treatment modality is still under discussion, with some consensus that two-stage exchange should be undertaken in patients with unknown pathogens or those of high-virulence [13–15].

The aim of this chapter is to provide a general overview of the current techniques of one- and two-stage exchange procedures and to compare the available results of the two treatment modalities through a detailed review of the literature.

Management of Periprosthetic Joint Infections (PJIs). http://dx.doi.org/10.1016/B978-0-08-100205-6.00009-4

9.2 Patient preparation, implant removal, and surgical debridement

Both one- or two-stage procedures do require a preoperative accurate workup, starting with a detailed clinical history and physical examination.

The diagnosis of infection should be established, together with the assessment of the local and general conditions of the patient. The general status of the patient should also be evaluated in order to assess if it is compatible with single or two-staged procedures.

Local signs of acute or chronic infection (redness, warmth, swelling, stiffness, draining sinuses), type and extent of skin scars, bone or implant exposure, mobility of the subcutaneous tissues, residual passive and active range of motion of the affected and adjacent joints, joint stability, peripheral vascular, and neurological function should be carefully evaluated prior to surgery.

Local and systemic risk factors for infection (smoking or alcohol abuse, diabetes, peripheral vasculopathy, renal insufficiency, etc.) must also be recorded, informing the patient of his/her own risk [16].

Laboratory tests and joint aspiration, when possible, should be performed to confirm the diagnosis of infection and to identify the pathogen for culture and sensitivity to antibiotics [17].

Plain long radiographs of the affected and contralateral limb are aimed at assessing joint angles, prosthesis and cement position, degree of periprosthetic osteolysis, and quality and extent of residual bone. Nuclear imaging studies, including leukocyte bone scans and/or PET scan should be reserved for selected cases [18].

Clinical and anaesthesiological assessment of the general operative risk and adequate supply of blood is required. In the case of long exchange operations, preoperative administration of fibrinolysis inhibitors (e.g., tranexamic acid) is recommended.

The surgeon should have experience with the type of implant to be revised and be familiar with its removal and disassembly. Occasionally the use of implant-specific instrumentation will be necessary.

In case of a staged procedure, physical examination is repeated at the time of spacer removal and prior to joint replacement. Persistence or recurrence of clinical signs of infection may warrant further laboratory or imaging tests and delay or prevent joint reimplantation.

In principle, surgical technique concerning implant removal and debridement do not differ in one- or staged procedures, both of which have as the main goal the most complete and accurate mechanical removal of all potentially contaminated foreign bodies and tissues.

Local and systemic risk factors for infection persistence/recurrence should be corrected if possible, including smoking cessation, pre- and peri-operative glycemic control, nutritional support, renal function, etc. Standard precautionary measures should be taken as necessary (septic procedure, for patients with multiresistant strains).

Antibiotic prophylaxis will start at the time of surgery or immediately after intraoperative cultural examination, with targeted antibiotic(s) when the pathogen is known, or with broad-spectrum antibiotics or a combination of antibiotics until the results of the initial sensitivity tests are available.

It is generally recommended that skin incision is made on one of the previous scars, including and excising fistulas, when present. In some cases, a muscular-cutaneous flap may be required to fill the preexisting or the resulting soft tissue defect.

Knee procedures should be started without tourniquet; consequently interfaces between infected tissue, scar, and surrounding healthy bleeding soft tissue can be distinguished more clearly during the debridement. All nonbleeding tissues and related bone need to be excised.

Biopsy material, preferably five or six samples should be taken as a routine measure from all relevant areas of the operation site for combined microbiological and histological evaluation.

Removed implant and cement should be sent for microbiological analysis with a biofilm debonding technique, like sonication or dithiothreitol processing [19] (Fig. 9.1).

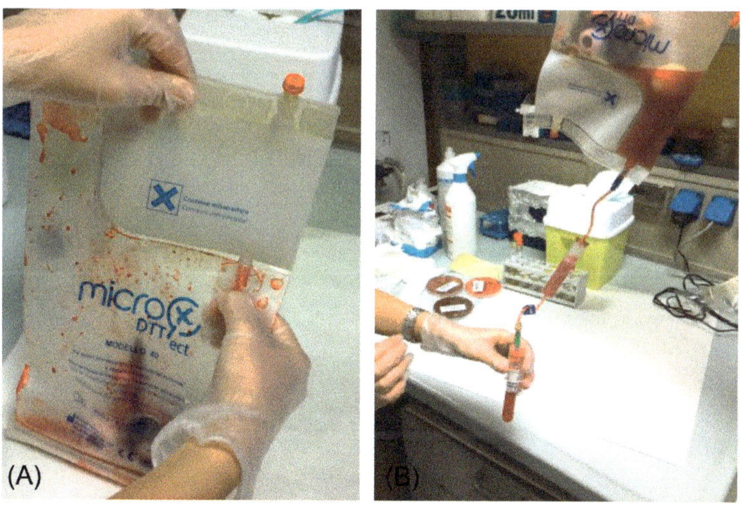

Fig. 9.1 (A and B) Failed removed implants, bone cement and even bioptic specimens should routinely be sent for antibiofilm processing and microbiological examination, since infection may be difficult to recognize preoperatively in all cases. Bacteria adhering on the removed implant are dislodged by means of sonication or through chemical processing, using dithiothreitol, a pure antibiofilm agent that does not interfere with bacterial growth. This can be currently achieved through a disposable kit (microDTTect, 4i Srl, Monza, Italy), that allows to collect samples at surgery and to transport and process them in a completely closed system, avoiding contamination, and increasing sensitivity of cultural examination.

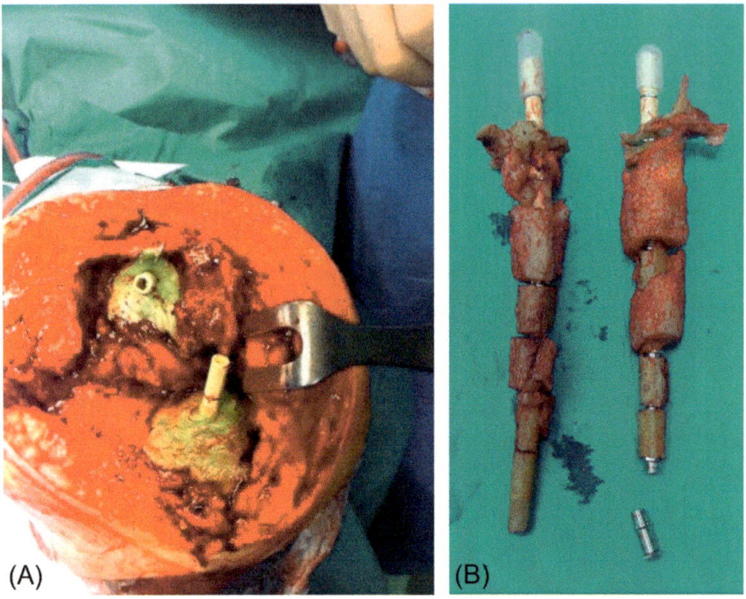

Fig. 9.2 (A) Intra-operative aspect of CEMOVER system (Tecres SpA, Sommacampagna, Italy) to remove well fixed bone cement in failed joint replacements. (B) At the end of the procedure, the removed cement; the bone cement is removed pieces by pieces through a cement-on-cement technique and a specifically designed apparatus, without the need for cortical window opening.

Cortical windows or longitudinal controlled fractures or osteotomies may be required to remove a well fixed prosthesis or cement; high speed burrs, straight osteotomes, curved saw blades, or Gigli's saw can aid removal, however, occasionally significant destruction and related loss of bone stock can occur. Extraction of the implant may necessitate special or universal extraction instruments, otherwise general punches are required. Special curved chisels, long rongeurs, curetting instruments, long drills, cement taps, and special cement removing systems (Fig. 9.2) can be helpful to completely remove the cement. In the hip joint, retrograde chisels can also be useful in many cases.

Debridement must be as radical as possible, including all areas of osteolysis, nonviable bone, and soft tissues. There is no evidence concerning the superiority of pulsatile lavage, local antiseptic, or antibiotic solution compared to simple lavage with normal saline.

9.2.1 One-stage technique

Once debridement is considered complete, swabs are placed over the wound area before redraping the patient. The whole team should change gloves and gowns and new instruments should be obtained for reimplantation. A second dose of antibiotics may be necessary, depending on drug half-life and if blood loss exceeds 1 L.

When choosing a one-stage cemented procedure, antibiotic-loaded bone cement is generally prepared according to the following criteria:

- Appropriate antibiotic choice, according to preoperative antibiogram and to adequate elution properties; whenever possible, bactericidal antibiotics should be preferred. Industrially premanufactured antibiotic-loaded acrylic cement may be appropriate, even if additional antibiotics in powder form can be added intra-operatively. In this case, the antibiotic should be available in powder form. Liquid antibiotic formulations may not be used. Two to three doses of cement (80–120 g) are generally required, with large mixing systems and appropriate cement guns. There are a few studies comparing the different elution properties of antibiotic-loaded bone cement according to the different preparation techniques. In particular, a clear difference between elution properties for low-dose (up to 1–2 g per cement batch) antibiotic-loaded bone cements has not been shown when using four different methods (no mixing, suspension of antimicrobial powder in monomer, hand-stirred, and bowl-mixed) [20,21]; however, if high-dose (10 g per cement batch) antibiotic-loaded bone cements are considered, it has been recently reported that dough-phase mixing leads to greater antimicrobial delivery, but with lower compressive strength, when compared to the hand-stirred or bowl-mixed methods [22].

  On the other hand, there is no clear evidence that vacuum mixing may reduce cement porosity, hence having any detrimental effect on antibiotic release from polymethylmethacrylate [23].

  Preoperative positive bacterial culture with known antibiogram [24] is considered a key point for selecting the appropriate antibiotic to be mixed with the cement used to fix the prosthesis that will be released locally to protect the implant [25].Isolated reports, however, did show good results even without local use of antibiotics in the cement [26] and comparative studies between the use of antibiotic-loaded bone or plain cement are lacking.
- Maximum concentration of antibiotic in bone cement should not exceed 10%/PMMA powder. It should be considered that, depending on the type and concentration, antibiotics may change the polymerization behavior of the cement, often causing an acceleration of cement curing.
- Current principles of modern cementing techniques should be applied. In order to achieve an improved cement bone interface, the tourniquet should be inflated prior to cementing, in knee prosthesis.

Postoperative management generally includes systemic antibiotic administration for 2 weeks [27].

If a cementless one-stage option is preferred, local protection of the implant can be achieved using antibiotic-loaded bone grafts, as described by Winkler and coworkers [11] or, more recently, by means of an antibacterial-hydrogel coating (Fig. 9.3).

It should be noted, however, that even for cementless one-stage exchange, isolated reports, in selected patients, did show satisfactory results also without local antibiotic administration [10,12], while direct scientific clinical evidence for the need of local delivery of antibiotics is still lacking.

In fact, whatever the implant choice is, given the biofilm-related nature of PJIs and the lack of a second debridement chance, the success of a one-stage approach probably mainly relies on the meticulous removal of all hardware material (including cement

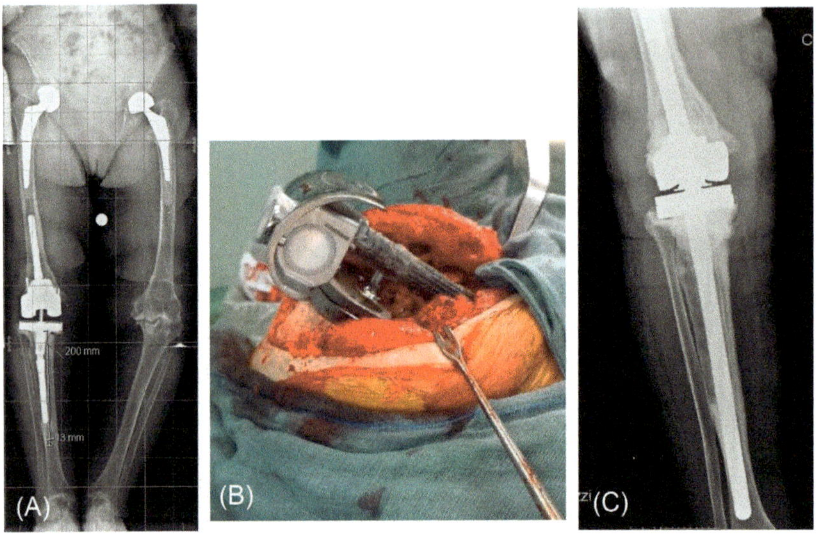

Fig. 9.3 (A) Infected and loosened revision knee prosthesis in female patient affected by severe rheumatoid arthritis; (B) After removing the failed prosthesis a one-stage procedure is performed, using a cementless long stem revision prosthesis, coated intra-operatively with a fast resorbable antibiotic-loaded hydrogel (DAC, Novagenit Srl, Trento, Italy). (C) Radiographic control 1 year after surgery. No clinical or laboratory signs of infection recurrence.

and restrictors) in combination with an aggressive and complete debridement of any infected soft tissue and bone, while local antibacterial protection may only play an adjunctive role.

9.2.2 *Two-stage technique*

Two-stage approach is based on the assumption that debridement should be kept separate from joint replacement and that revision surgery should only be made once the surgical field is considered to be effectively sterile.

Two-stage revision may be performed with or without the use of an interval spacer. In fact, until the 1980s, two-stage reimplantation was done with no interim antibiotic spacer placed. However, while successful eradication of infection was demonstrated in a high percentage of cases, functional results were rather poor [28,29].

In the 1990s, use of static cement spacers in the interim period became widespread. The use of an impregnated antibiotic cement spacer block was introduced to maintain the joint space and stability, prevent soft tissue retraction, and to provide local antibiotic release [30–32].

To overcome the disadvantages of block spacers, including decreased ability to comfortably perform activities of daily life such as sitting, toilet, and stair use, articulating or dynamic spacers were introduced; this new option was developed to

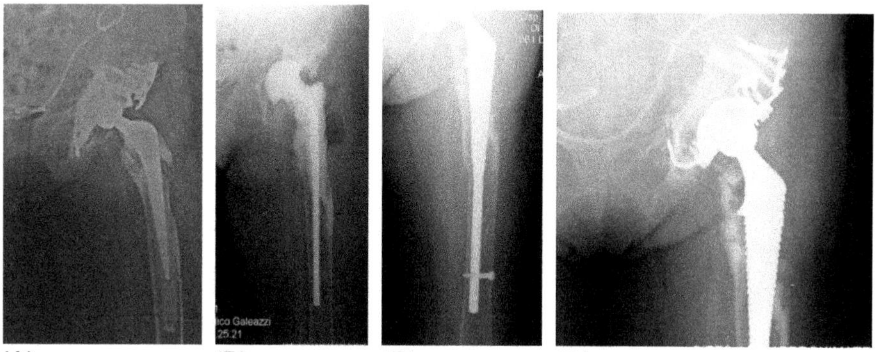

Fig. 9.4 (A) Preoperative X-ray of a 56 years old female, previously treated with a renal transplant, with a failed and septic cemented revision prosthesis. (B) Removal of the implant and of bone cement and implant of a preformed long stem hip spacer (Spacer G, Tecres SpA, Sommacampagna, Italy) to by-pass the large femoral bone defect. (C and D) Radiographic control 2 years after joint revision with a cementless, distally locked, implant.

facilitate exposure at the time of reimplantation, to preserve joint function and to prevent interim bone loss between stages and ultimately provide a better functional outcome.

Different types of articulated spacers have been described, including, for the knee, resterilized prosthetic components or new components (spacer prostheses) [33–36], cement spacer molded during operation with a thin metal-on-polyethylene runner (PROSTALAC system) [37–41]; all cement spacers, molded [42–44] or custom-made [45–47] during the operation and, more recently, preformed cement spacers [48], which are also available for the hip (Fig. 9.4) and shoulder joints [49].

All the described spacers and techniques present distinct advantages and disadvantages and no clinical study has been designed to compare the outcomes of different articulated spacers.

Theoretically, the presence of metallic hardware or polyethylene, like in the resterilized component procedure or in the PROSTALAC system, may be more prone to bacterial recolonization; moreover, the cement spacer molded in the operative room may not have reproducible mechanical and antibiotic elution characteristics and may have an increased risk of component fracture; both, the use of a spacer-prosthesis and of a molded spacer increase operatory times, while, on the other hand, molds and preformed spacers do add some cost to the procedure.

However, intra-operatively molded spacers allow the surgeon to choose during surgery the type and concentration of the antibiotic(s) to be added to the cement, while preformed spacers come with a predetermined concentration of one or two antibiotics (usually gentamicin and/or vancomycin), even if, for special cases, they may be prepared "on order," with any desired antibiotic and size.

In fact, both molded and preformed spacers offer a limited choice of sizes and this may be a problem in large bone defects or to fit extreme anatomic variations.

After a spacer implant has been prepared or selected on the basis of the anticipated bone defect and preoperative antibiogram, when available, it is fixed with antibiotic-loaded cement. To prevent significant interdigitation of the cement into the host's bone, a poor cementing technique should be preferred, also avoiding cementing the intra-medullary part of the spacer. Cement in excess may be left in place if it does not interfere with joint motion or soft tissues. Before soft tissue closure, joint mobility should be checked and any cement remnants within the joint space removed.

Various studies have been documented in the last two decades on how the interim use of an articulating antibiotic-impregnated spacer maintains excellent infection eradication rates and an acceptable function between stages, minimizing bone loss and ultimately improving patient function and satisfaction [49].

Reimplantation after spacer implant, does not differ from that of a one-stage procedure, except by the fact that removal of the spacer is usually easily performed. In this regard, a two-stage procedure may not be seen as a double revision surgery, since both, spacer implant and removal are rather simple parts of the procedure.

Intra-operative cultural examination (usually 4–6 samples) is mandatory at the time of reimplantation, even in the presence of normal laboratory tests and satisfactory clinical aspect. Whenever possible, the whole removed spacer or at least parts of it should also be sent for microbiological analysis using biofilm debonding techniques.

At reimplantation, as already described for the one-stage procedure, implant choice is primarily based on residual bone stock that will result from infected implant removal and debridement.

Inadequate bone stock, possible intra-operative complications as fractures, perforations of the cortex, osseous windows, and more complex fractures must be taken into consideration at the time of surgical planning. In this regard, the advantage over a one-stage procedure is that in a staged revision surgery, the exact amount of bone loss may be assessed intra-operatively at the end of the first stage and hence surgical planning at revision can be more accurate, particularly with regard to the choice of prosthesis or the need for bone grafts or special implants.

In general, one or a combination of two antibiotics, selected on the basis of available antibiograms and eventually under the guidance of an infectious diseases consultant, are administered intravenously and/or orally for 4–6 weeks after the first and the second stage procedure [50]. However, recent studies have reported successful short (1 versus 4–6 weeks) antibiotic course after the second stage [13,51]. Antibiotic choice should take into consideration the possible occurrence of resistance induction. In particular, rifampicin resistance has been reported following inappropriate use and its administration between stages should be carefully evaluated [52].

Care should be taken to control postoperatively any risk factor for infection persistence/recurrence. Smoking and alcohol should be prohibited for at least the first 6 months after surgery, normal glycemia, renal function, and nutritional status should be targeted in any case.

Prophylaxis of thromboembolic complications and adequate pain control is performed in all cases. Monitoring of laboratory tests, including hemochromocytometric and white blood cell count with differential, erythrocyte sedimentation rate and C-reactive protein level, renal and hepatic function is performed periodically after the first and the second stage procedure, to detect any sign of infection recurrence and/or side effects due to the protracted antibiotic therapy.

Plain radiographs of the knee/hip are usually performed immediately after either surgery, prior to spacer removal and then at 2, 6, and 12 months postreimplantation.

The postoperative rehabilitation management of individual patients is dependent on numerous variables, including the general condition, the status of the soft tissue coverage, the type of exposure required, residual ligament apparatus, and the degree of bone defect and implant stability.

After spacer implantation, patients are usually encouraged to actively mobilize the joint immediately after surgery; passive joint motion, walking, and stair climbing with partial weight bearing and two crutches is normally allowed until the second stage surgery. In selected cases with significant bone loss or insufficient soft tissue, a brace may be required and weight bearing restricted, with joint motion postponed or performed at a reduced range.

Gait parameters of patients with a preformed hip or knee spacer have been recently shown to be in the range of a normal reference control population, walking at a reduced speed [49,53].

Rehabilitation after joint reconstruction does not differ from that of aseptic revision surgery.

9.2.3 Indications and patient information

Late chronic periprosthetic knee infections (Type III according to McPherson et al. [16] or Type 2, according to Segawa et al. [54]) are candidates for one- or two-stage revision, provided that the general conditions of the patient allow revision surgery.

Concerning more specifically the indication to one- or two-stage revision, it should be noted that there is no any univocal recommendation and both options have pros and contras, that should be briefly but clearly presented to the patient.

According to Gehrke et al. [24], contraindications to one-stage procedure are the following:

- failure of $\geq$2 previous one-staged procedures;
- infection spreading to the neurovascular bundle;
- unclear preoperative bacterial specification;
- nonavailability of appropriate antibiotics;
- high antibiotic resistance;
- sinus tract with unclear bacterial specification.

In selected cases, one-stage revision may provide excellent results, as recently reported by Singer et al. [55].

Poor or inadequate soft tissue coverage of the affected joint is regarded as a relative contra-indication that can eventually be managed with immediate or early secondary flap reconstruction by a plastic surgeon.

Patient information should include a description of the general risks of surgery and specific risks connected to the one- or two-stage revision procedure; both procedures may in fact expose to a risk of recurrent or new infection ranging from 5% to 20%, depending on the host's type, the local and general conditions, the type of microorganism(s), and the treatment choice; risks of intra- and postoperative fracture, of reoperation for haematoma and for further wound debridement or persistent infection should be made clear, together with possible postoperative peripheral nerve transient or permanent palsy, joint stiffness and loss of function, leg's length discrepancy, increased risk of aseptic loosening, chronic pain, thrombo-haembolism, etc.

In case of two-stage procedure, explanation using simple language of how the temporary implant is aimed at preserving soft tissue balance, some joint mobility and function and delivering antibiotics directly to the site of the infection is critical. Antibiotic resistance and allergies should be noted and the necessity of a second surgical intervention after treatment of the infection made clear. Reduced joint mobility, risk of spacer dislocation or fracture, especially in the case of abnormal weight bearing, ligament laxity, and bone loss should also be pointed out.

Postoperative management, including the need for partial weight bearing with two crutches, eventually coupled with a brace, will be illustrated.

Alternative treatments, including suppressive antibiotic therapy, joint arthroplasty, arthrodesis, or amputation will be discussed as suitable for any given case.

9.3 Analysis of results

We here report a review of published papers with full text or an abstract in English, published from 1974 through Jan. 2015, which reported the success rate of eradication of periprosthetic hip infection after a one-stage or two-stage exchange arthroplasty. The international databases were searched as previously described [1], including EMBASE; PubMed/Medline; Medline Daily Update; Medline In-Process and other nonindexed citations; Google Scholar; SCOPUS; CINAHL; Cochrane Central Register of Controlled Trials and Cochrane Database of Systematic Reviews; NHS Health Technology Assessment; http://www.google.com; and http://www.yahoo.com. The following keywords were entered either alone or in different combinations: hip; knee; infection; arthroplasty; prosthesis; total hip replacement (THR); total knee replacement (TKR); prosthetic hip, knee or infection; periprosthetic hip, knee or infection; exchange arthroplasty; one-stage; one-stage; and two-stage revision.

The inclusion criteria were the following:

(a) Papers written fully or with an abstract in English;
(b) Papers reporting the results of management of infected hip, knee, or arthroplasty with a one-stage or two-stage exchange arthroplasty;
(c) Study reporting results relating to delayed or chronic (6 weeks or later) stages of disease;
(d) Study design classifiable as: randomized controlled trial; comparative prospective study; prospective study with historical controls; prospective case series with no comparison group; comparative retrospective study; retrospective study with historical control group; retrospective study with no control group;

(e) If more than one paper by the same author(s) was retrieved and follow-ups of different papers were found to overlap, only the most recent reference with the longest follow-up and largest patient series was included;

(f) The study population had to be five or more cases treated with a minimum follow-up of 12 months;

(g) At least all of the following variables had to be reported: number of patients treated; type of treatment; number of recurrent infections after treatment.

9.3.1 Hip

The analysis of the available literature data shows 25 original articles on one-stage hip exchange ($n=1454$) and 68 papers on two-stage revisions ($n=3658$) (Tables 9.1 and 9.2).

The database search retrieved no randomized, prospective controlled trials. The majority of the papers described retrospective case series without a control group (level of evidence: IV) and most gave an incomplete description of the study material. In addition, no detailed description was given of the criteria for selecting one- or two-stage revision. Similarly, other relevant variables such as the relative proportion of compromised hosts or multiresistant pathogens were only occasionally reported.

The range in the number of cases varied between 8 and 583 (mean 58 ± 113) and 5 and 294 (mean 53 ± 52), respectively for one- and two-stage revision studies, while the range in the duration of follow-up ranged from 19 to 118 and from 19 to 144 months, respectively.

The infection eradication rate after a one-stage procedure ranged from 50% to 100% (Table 9.1). The eradication rate after a two-stage revision ranged from 74% to 100%.

At a mean follow-up of 65 months, the calculated average success rate in the eradication of infection was 82.3% after one-stage revision, compared to 91.2% after two-stage revision, including all types of spacers and at a mean follow-up of 59 months (Table 9.2).

When considering one-stage procedures performed with *cementless* implants (with or without antibiotic-loaded bone grafts), the average infection eradication rate was 82.4% in a total of 142 patients and at an average follow-up of 72 months (Table 9.3); comparing this value, with one-stage *cemented* hip revision, we found an average infection eradication rate of 82.4% in 1304 patients at an average follow-up of 62 months (Table 9.4). Comparing the same with a rather homogeneous population of patients, all treated with a standard preformed hip spacer (Spacer G/InterSpace Hip—Tecres SpA, Sommacampagna, VR, Italy) and a cementless two-stage revision prosthesis, an average infection eradication rate of 94.6% at 51 months mean follow-up was found (Table 9.5).

9.3.2 Knee

Ten original articles on one-stage knee exchange ($n=392$) and 78 papers on two-stage revisions ($n=3034$) (Tables 9.6 and 9.7) were retrieved. Even for the knee, no randomized, prospective controlled trials were found and all the papers reported retrospective case series without a control group (level of evidence: IV), with an incomplete description of the study material, as previously described for the hip.

Table 9.1 **Infection eradication rate after a one-stage hip exchange**

Author	Year	Number of patients	Number of eradicated infections	Eradication rate (%)	Follow-up (months)		
					Min	Max	Mean
Wilson [56]	1974	19	17	89.5	24	–	–
Hughes [57]	1979	13	12	92.3	32	83	51
Buchholz [58]	1981	583	448	76.8	24	132	–
Miley [59]	1982	46	40	87	32	–	48.5
Wroblewski [60]	1986	102	93	91.2	38.8	–	–
Weber [61]	1986	8	4	50.0	60	96	72
Sanzen [62]	1988	102	77	75.5	24	108	–
Schneider [63]	1989	26	21	80.8	12	108	–
Morscher [64]	1994	47	27	57.4	12	132	84
Raut [65]	1995	57	49	86	24	151	88
Mulcahy [66]	1996	15	15	100	24	84	53
Ure [67]	1998	20	20	100	42	205.2	118.8
Callaghan [68]	1999	12	10	83.3	120	168	109.2

Gao [69]	2008	10	10	100	12	37	19
Winkler [11]	2008	37	34	91.9	24	96	52.8
Rudelli [10]	2008	32	30	93.8	63	183	103
Yoo [12]	2009	12	10	83.3	39.6	135.6	86.4
Oussedik [70]	2010	11	11	100	66	105.7	81.6
De Man [71]	2011	22	21	95.5	17	204	60
Klouche [26]	2012	38	38	100	24	61	35
Choi [72]	2013	17	14	82.4	12	132	61
Wolf [73]	2014	37	21	56.8	24	–	–
Zeller [74]	2014	99	97	98.0	24	–	41.6
Bori [75]	2014	24	23	95.8	25	94	45
Jenny [76]	2014	65	55	84.6	36	72	–
Total		*1454.00*	*1197.00*				
Minimum		8.00	–	50.00	12.00	37.00	19.00
Maximum		583.00	–	100.00	120.00	205.20	118.80
Mean		*58.20*	–	*82.32*	*33.40*	*116.80*	*64.70*
SD		113.04	–	–	23.50	45.60	26.00

Table 9.2 Infection eradication rates after a two-stage hip exchange

Author	Year	Number of patients	Number of eradicated infections	Eradication rate (%)	Follow-up (months)		
					Min	Max	Mean
Hughes [57]	1979	13	12	92.3	32	83	51
Fitzgerald [77]	1985	131	115	87.8	24	108	49
Ketterl [78]	1988	161	144	89.4	24	168	32
McDonald [79]	1989	82	71	86.6	24	163.2	66
Wilson [56]	1989	22	20	90.9	36	120	60.2
Colyer [80]	1994	37	31	83.8	12	88	36
Garvin [81]	1994	40	38	95.0	24	120	60
Lieberman [82]	1994	32	29	90.6	24	74	40
Morscher [64]	1994	27	20	74.1	12	132	84
Nestor [83]	1994	34	28	82.4	24	72	47
Lai [84]	1996	39	34	87.2	30	84	48
Younger [85]	1997	48	45	93.8	24	63	43
Fehring [86]	1999	25	23	92.0	24	98	41
Morales [87]	1999	37	34	91.9	36	156	57.6
Thomas [86]	1999	25	23	92.0	24	98	41
Haddad [88]	2000	50	46	92.0	24	104.4	69.6
Koo [89]	2001	22	21	95.5	24	78	41
Magnan [90]	2001	10	8	80.0	24	48	35
Karpas [91]	2003	18	18	100.0	24	120	42
Yamamoto [92]	2003	17	17	100.0	14	62	38
Evans [93]	2004	23	22	95.7	24	108	48
Hsieh [94]	2004	128	122	95.3	24	96	58.8
Buttaro [95]	2005	30	29	96.7	24	60	32.4
Hofmann [96]	2005	27	26	96.3	28	148	76
Nusem [97]	2006	17	16	94.1	60	168	108
Cabrita [98]	2007	38	34	89.5	24	102	48

Masri [99]	2007	29	26	89.7	24	88	47
Thabe [100]	2007	16	16	100.0	72	120	75.6
Gao [69]	2008	5	5	100.0	12	37	19
Stockley [101]	2008	114	100	87.7	24	175	74
Sudo [102]	2008	7	6	85.7	27.6	73.2	60
Biring [103]	2009	48	43	89.6	120	180	144
Cordero [104]	2009	36	33	91.7	12	144	52.8
Darley [105]	2009	19	19	100.0	24	36	26
Fink [106]	2009	36	36	100.0	24	60	35
Hsieh [51]	2009	99	89	89.9	24	60	43
McKenna [107]	2009	31	31	100.0	24	60	35
Sanchez [108]	2009	168	156	92.9	24	192	84
Seung-Jae [109]	2009	37	30	81.1	24	120	52.8
Toulson [110]	2009	82	78	95.1	24	203	64.8
Whittaker [111]	2009	41	38	92.7	25	83	49
Kent [112]	2010	13	13	100.0	26	60	38
Oussedik [70]	2010	39	37	94.9	66	105.6	81.6
Pignatti [113]	2010	41	40	97.6	60	120	63.6
Takigami [114]	2010	8	8	100.0	24	81	49
D'Angelo [115]	2011	28	27	96.4	18	106	53
De Man [71]	2011	50	49	98.0	17	204	60
Kim [116]	2011	294	288	98.0	60	168	124.8
Leung [117]	2011	38	30	78.9	24	123	58
Neumann [118]	2011	44	43	97.7	36	120	67
Wang [119]	2011	12	11	91.7	36	96	64.8
Babiak [120]	2012	33	31	93.9	36	180	84
Degen [121]	2012	30	28	93.3	24	70	43
Klouche [26]	2012	46	42	91.3	24	68	35
Macheras [122]	2012	35	33	94.3	84	168	139.2

Continued

Table 9.2 Continued

Author	Year	Number of patients	Number of eradicated infections	Eradication rate (%)	Follow-up (months)		
					Min	Max	Mean
Romanò [123]	2012	183	173	94.5	24	104	56
Berend [124]	2013	189	157	83.1	24	180	53
Choi [72]	2013	44	33	75.0	12	132	61
Hsieh [125]	2013	35	26	74.3	48	120	86
Johnson [126]	2013	66	60	90.9	24	105	45
Lee [127]	2013	17	15	88.2	24	96	48
Sabry [128]	2013	78	67	85.9	24.3	135.3	58
van Diemen [129]	2013	136	119	87.5	24	180	72
Wolf [61]	2014	55	52	94.5	24		
Schwarzkopf [130]	2014	56	48	85.7	12		32.4
Ekpo [131]	2014	19	17	89.5	24	132	48
Ibrahim [132]	2014	125	120	96.0	60	75	103.2
Dieckmann [133]	2014	43	39	93.0	24	46.32	139
Total		*3658.00*	*3338.00*		–	–	–
Minimum		5.00	–	74.00	12.00	36.00	19.00
Maximum		294.00	–	100.00	120.00	204.00	144.00
Mean		*53.80*	–	*91.25*	*30.20*	*111.00*	*59.40*
SD		52.70	–	–	18.20	43.40	26.30
Total		*1454.00*	*1197.00*				
Minimum		8.00	–	50.00	12.00	37.00	19.00
Maximum		583.00	–	100.00	120.00	205.20	118.80
Mean		*58.20*	–	*82.32*	*33.40*	*116.80*	*64.70*
SD		113.04	–	–	23.50	45.60	26.00

Table 9.3 Infection eradication rates after a one-stage *cementless* hip exchange

Author	Year	Number of patients	Number of eradicated infections	Eradication rate (%)	Follow-up (months)		
					Min	Max	Mean
Rudelli	2008	32	30	93.8	63	183	103
Winkler	2008	37	34	91.9	24	96	52.8
Yoo	2009	12	10	83.3	39.6	135.6	86.4
Wolf	2014	37	21	56.8	24		
Bori	2014	24	23	95.8	25	94	45
Total		*142*	*118*	–	–	–	–
Minimum		12	–	56.8	24.0	94.0	45.0
Maximum		37	–	95.8	63.0	183.0	103.0
Mean		28.4	–	*83.10*	*35.1*	*127.2*	*71.8*
SD		10.6	–	–	16.9	41.9	27.5

Table 9.4 **Infection eradication rates after a one-stage *cemented* hip exchange**

Author	Year	Number of patients	Number of eradicated infections	Eradication rate (%)	Follow-up (months)		
					Min	Max	Mean
Wilson	1974	19	17	89.5	24	–	–
Hughes	1979	13	12	92.3	32	83	51
Buchholz	1981	583	448	76.8	24	132	–
Miley	1982	46	40	87	32	–	48.5
Wroblewski	1986	102	93	91.2	38.8	–	–
Sanzen	1988	102	77	75.5	24	108	–
Schneider	1989	26	21	80.8	12	108	–
Morscher	1994	47	27	57.4	12	132	84
Raut	1995	57	49	86	24	151	88
Mulcahy	1996	15	15	100	24	84	53
Ure	1998	20	20	100	42	205.2	118.8
Callaghan	1999	12	10	83.3	120	168	109.2
Gao	2008	10	10	100	12	37	19
Oussedik	2010	11	11	100	66	105.7	81.6
De Man	2011	22	21	95.5	17	204	60
Klouche	2012	38	38	100	24	61	35
Choi	2013	17	14	82.4	12	132	61
Zeller	2014	99	97	98.0	24	–	41.6
Jean-Yves	2014	65	55	84.6	36	72	–
Total		*1304.00*	*1075.00*				
Minimum		10.00	–	57.00	12.00	37.00	19.00
Maximum		583.00	–	100.00	120.00	205.20	118.80
Mean		*68.60*	–	*82.44*	*31.60*	*115.40*	*61.80*
SD		128.50	–	–	25.10	49.00	27.20

Table 9.5 **Two-stage exchange for periprosthetic hip infection using preformed hip spacers and *cementless* exchange arthroplasty prosthesis**

Author	Year	Number of patients	Number of eradicated infections	Eradication rate (%)	Follow-up (months)		
					Min	Max	Mean
Magnan	2001	10	8	80	24	48	35
D'Angelo	2011	28	27	96.4	18	106	53
Neumann	2011	44	43	97.7	36	120	67
Degen	2012	30	28	93.3	24	70	43
Romanò	2012	183	173	94.5	24	104	56
Total		*295*	*279*	–	–	–	–
Minimum		10	–	80	18	48	35
Maximum		183	–	98	36	120	67
Mean		59.0	–	*94.58*	*25.2*	*89.6*	*50.8*
SD		70.4	–	–	6.6	29.6	12.3

Table 9.6 **Infection eradication rates after a one-stage knee exchange**

Author	Year	Number of patients	Number of eradicated infections	Eradication rate (%)	Follow-up (months)		
					Min	Max	Mean
von Foerster [134]	1991	104	76	73.1	–	–	76
Goksan [135]	1992	18	16	88.9	–	–	60
Lu [136]	1997	8	8	100	–	–	20
Silva [137]	2002	37	33	89.2	–	–	48
Buechel [138]	2004	22	20	90.	–	–	22
Sofer [139]	2005	15	14	93.3	–	–	18
Singer [55]	2012	63	60	95.2	24	70	36
Jenny 9[76]	2013	47	41	87.0	36	76	–
Tibrewal [140]	2014	50	49	98	24	288	126
Haddad [141]	2015	28	28	100	36	108	78
Total		*392*	*345*	–	–	–	–
Minimum		8	–	73	24	70	18
Maximum		104	–	100	36	288	126
Mean		*39.2*	–	*88.01*	*30.0*	*135.5*	*53.8*
SD		28.6	–	–	6.9	103.0	35.6

Table 9.7 **Infection eradication rates after a two-stage knee exchange**

Author	Year	Number of patients	Number of eradicated infections	Eradication rate (%)	Follow-up (months)		
					Min	Max	Mean
Borden [30]	1987	11	10	90.9	–	–	46
Rosemberg [142]	1988	15	12	80	–	–	29
Wilde [143]	1988	15	12	80	–	–	30
Booth [144]	1989	25	24	96	–	–	25
Windsor [145]	1990	38	34	89.5	–	–	48
Henderson [146]	1991	28	27	96.4	–	–	27
Scott [147]	1993	7	7	100	–	–	24
Hanssen [148]	1994	89	79	88.8	–	–	52
McPherson [149]	1995	21	20	95.2	–	–	17
Whiteside [150]	1995	33	28	84.8	–	–	24
Goldman [151]	1996	64	58	90.6	–	–	90
Gacon [152]	1997	29	24	82.8	–	–	42
Hirakawa [153]	1998	55	41	74.5	–	–	61
Fehring [154]	2000	85	80	94.1			36
Haddad [40]	2000	45	41	91.1	–	–	48
Goldstein [155]	2001	5	5	100	–	–	12
Lonner [156]	2001	53	44	83	–	–	56
Emerson [157]	2002	48	44	91.7	–	–	90
Siebel [158]	2002	10	10	100	–	–	18
Pietsch [159]	2003	24	22	91.7	–	–	15
Durbhakula [160]	2004	24	22	91.7	–	–	33
Evans [93]	2004	31	29	93.5	–	–	36
Haleem [161]	2004	96	87	90.6	–	–	86
Meek [41]	2004	54	52	96.3	–	–	41
Cuckler [162]	2005	44	44	100	–	–	65

Continued

Table 9.7 **Continued**

Author	Year	Number of patients	Number of eradicated infections	Eradication rate (%)	Follow-up (months)		
					Min	Max	Mean
Hofmann [34]	2005	50	44	88	–	–	74
MacAvoy [163]	2005	13	9	69.2	–	–	28
Pitto [48]	2005	19	19	100	–	–	24
Huang [164]	2006	21	20	95.2	–	–	52
Jamsen [165]	2006	34	26	76.5	–	–	32
Vilanueva [47]	2006	30	30	100.0	–	–	36
Freeman [166]	2007	76	69	90.8			71
Hsu [43]	2007	28	25	89.3	–	–	101
Pascale [46]	2007	14	14	100.0	–	–	12
Anderson [167]	2009	25	24	96	–	–	54
Ocguder [168]	2010	17	15	88.2	–	–	20
Chiang [169]	2011	43	40	93.0	24	61	40
Gooding [170]	2011	115	101	87.8	–	–	108
Kohl [171]	2011	16	16	100	24		
Macheras [172]	2011	34	31	91.1	120	168	145.2
Mortazavi [173]	2011	117	84	72	24	113	40.8
Van Thiel [174]	2011	60	53	88.3	–	–	36
Babiak [120]	2012	24	23	95.8	12		
Cai [175]	2012	23	21	91.3	24	60	43
Choi [176]	2012	47	32	68.1	17	69	43
Hwang [177]	2012	30	28	93.4	28	140	52
Kalore [178]	2012	53	47	88.6	12	105	39
Kotwal [179]	2012	37	31	83.8	12.7	73	29.4
Jia [180]	2012	21	21	100	17	54	32.3
Johnson [181]	2012	115	95	82.6	12	121	66

Lee [182]	2012	20	19	95	38	121	64.8
Mahmud [183]	2012	253	237	93.3	12	204	48
Nickinson [184]	2012	9	8	88.8	21	64	32
Wan [185]	2012	33	30	91	24	62	44
Ahmad [186]	2013	75	68	90.6	24	84	43
Brun [187]	2013	81	76	93.8	96	180	144
Brunnekreef [188]	2013	35	35	100	12		
Caulli [189]	2013	9	9	100	48	84	55.2
Kim [190]	2013	20	18	90	14	60	22.3
Schwarkopf [191]	2013	53	48	90.5			
Silvestre [192]	2013	45	43	95.6	60	132	86
Tigani [193]	2013	38	29	76.4	24	139	65
Ueng [194]	2013	9	7	77.7	28	90	41
Castelli [195]	2014	50	46	92	24	156	84
Classen [196]	2014	23	20	86	19	103	47
Corona [197]	2014	22	21	95.4	12.4	64.7	
Nelson [198]	2014	29	21	72.4	19	38	29.9
Prasad [199]	2014	34	32	94	24	96	60
Watts [200]	2014	111	100	90.1	61	130	83
Haddad [141]	2015	74	69	93	36	108	78
Total		*3034*	*2710*	–	–	–	–
Minimum		5	–	67	12	38	12
Maximum		253	–	100	120	204	145
Mean		*38.9*	–	*89.32*	*29.2*	*102.3*	*49.9*
SD		34.6	–	–	23.2	44.8	28.1

Case number ranged between 8 and 104 (mean 39 ± 28) and 5 and 253 (mean 39 ± 35), respectively for one- and two-stage revision studies; follow-up varied from 18 to 126 months in one-stage and from 12 to 145 in two-stage studies.

At a mean follow-up of 54 months, the calculated average success rate in the eradication of infection was 88.0% (range: 73–100%) after one-stage revision, compared to 89.3% (range: 74–100%) after two-stage revision, including all types of spacers and at a mean follow-up of 49.9 months (Table 9.7).

9.4 Conclusions and future trends

Whatever the surgical strategy, one- or two-stage exchange, given the biofilm-related nature of PJIs, the outcome of any revision surgery most probably still relies on the surgeon's ability/possibility to effectively remove all infected biomaterials and tissues and on each host's immunological capability and comorbidities.

Systematic review of available data shows that the number of reported cases treated with a two-stage exchange largely exceeds that of patients treated with a one-stage procedure; however, there is insufficient evidence to drive any definitive conclusion concerning the relative success rate of the two different surgical modalities.

In particular, while the overall success rate in the hip still seems to be relatively higher for the two-stage approach, without any substantial difference between cemented and cementless one-stage revisions in the knee, things are becoming different; in fact, at variance to what we calculated only a few years ago [1], the average success rate of one-stage knee exchange seems currently to be approaching that of the two-stage procedures.

According to our analysis of the data, this finding is substantially related to the highly satisfactory results obtained in the four most recent studies on one-stage, published between 2012 and Jan. 2015, the last of which reports, in selected patients, a 100% success rate [141]; in particular, this latter study points out that, in selected patients, the one-stage option can be a reasonable choice or even the most appropriate.

Different criteria across centers with regard to patient selection and eligibility for a single- or two-stage exchange arthroplasty is just one of the many variables that may bias the results of systematic reviews in this topic.

Other possible bias and limitations include differing pathogen virulence across treatment arms, postoperative management, host's type, surgical technique, etc.

For example, although local antibacterial protection, both in the one- or two-stage procedures, is considered by many as a key factor to the success in infection eradication, and is supported by several in vivo models, clinical evidence is more difficult to achieve. In this regard it is worth noting the study from Klouche and coworkers [26], that recently reported 100% eradication rate after one-stage hip exchange arthroplasty, without using antibiotic-loaded cement.

The limitations and bias of the current literature highlights the need for large, multicenter, prospective, and randomized studies to establish the superiority of one type of surgical treatment over another. However, certain circumstantial limitations such as the low incidence of the disease, relatively small patient cohorts, need for long-term follow-up, and variations in microorganisms, would also make a large controlled prospective study in this field extremely challenging.

In this evolving panorama, the decision to undergo one- or two-stage reimplantation still relies on a number of different variables, such as the host's type, general and local conditions, the correct preoperative identification of the pathogen(s) and its sensitivity to antibiotics, the amount of bone loss and the quality of the soft tissue coverage, the experience of the surgeon, and the requirements of the patient after complete information.

Accurate selection of patients, pharmacological and technological innovations including antibiofilm agents and novel antibacterial coatings, are among the most promising trends of current research and possible clinical applications.

References

[1] C.L. Romanò, L. Gala, N. Logoluso, D. Romanò, L. Drago, Two-stage revision of septic knee prosthesis with articulating knee spacers yields better infection eradication rate than one-stage or two-stage revision with static spacers, Knee Surg. Sports Traumatol. Arthrosc. 20 (2012) 2445–2453.

[2] E. Jamsen, J. Stogiannidis, A. Malmivaara, J. Pajamäki, T. Puolakka, Y.T. Konttinen, Outcome of prosthesis exchange for infected knee arthroplasty: the effect of treatment approach. A systematic review of the literature, Acta Orthop. 80 (1) (2009) 67–77.

[3] K. Steinbrink, L. Frommelt, Treatment of periprosthetic infection of the hip using one-stage exchange surgery, Orthopade 24 (1995) 335–343 (in German).

[4] H.C. Schmitz, B. Schwantes, D. Kendoff, One-stage revision of knee endoprosthesis due to periprosthetic infection and Klippel-Trenaunay syndrome, Orthopade 40 (2011) 624–626, 628–9.

[5] C. Della Valle, J. Parvizi, T.W. Bauer, et al., Diagnosis of periprosthetic joint infections of the hip and knee, J. Am. Acad. Orthop. Surg. 18 (2010) 760–770.

[6] A.D. Hanssen, M.J. Spangehl, Practical applications of antibiotic-loaded bone cement for treatment of infected joint replacements, Clin. Orthop. Relat. Res. (427) (2004) 79–85.

[7] A.A. Haleem, D.J. Berry, A.D. Hanssen, Mid-term to long-term followup of two-stage reimplantation for infected total knee arthroplasty, Clin. Orthop. Relat. Res. (428) (2004) 35–39.

[8] D.A. Hoad-Reddick, C.R. Evans, P. Norman, I. Stockley, Is there a role for extended antibiotic therapy in a two-stage revision of the infected knee arthroplasty? J. Bone Joint Surg. (Br.) 87 (2005) 171–174.

[9] R.T. Goldman, G.R. Scuderi, J.N. Insall, 2-Stage reimplantation for infected total knee replacement, Clin. Orthop. Relat. Res. (331) (1996) 118–124.

[10] S. Rudelli, D. Uip, E. Honda, A.L. Lima, One-stage revision of infected total hip arthroplasty with bone graft, J. Arthroplasty 23 (8) (2008) 1165–1177.

[11] H. Winkler, A. Stoiber, K. Kaudela, F. Winter, F. Menschik, One stage uncemented revision of infected total hip replacement using cancellous allograft bone impregnated with antibiotics, J. Bone Joint Surg. (Br.) 90 (12) (2008) 1580–1584.

[12] J.J. Yoo, Y.S. Kwon, K.H. Koo, K.S. Yoon, Y.M. Kim, H.J. Kim, One-stage cementless revision arthroplasty for infected hip replacements, Int. Orthop. 33 (5) (2009) 1195–1201.
[13] W.J. Hart, R.S. Jones, Two-stage revision of infected total knee replacements using articulating cement spacers and short-term antibiotic therapy, J. Bone Joint Surg. (Br.) 88 (8) (2006) 1011–1015.
[14] M.F. Kurd, E. Ghanem, J. Steinbrecher, J. Parvizi, Two-stage exchange knee arthroplasty: does resistance of the infecting organism influence the outcome? Clin. Orthop. Relat. Res. 468 (8) (2010) 2060–2066.
[15] J. Parvizi, B. Adeli, B. Zmistowski, C. Restrepo, A.S. Greenwald, Management of periprosthetic joint infection: the current knowledge: AAOS exhibit selection, J. Bone Joint Surg. Am. 94 (14) (2012) e104.
[16] E.J. McPherson, C. Woodson, P. Holtom, N. Roidis, C. Shufelt, M. Patzakis, Periprosthetic total hip infection. Outcomes using a staging system, Clin. Orthop. Relat. Res. (403) (2002) 8–15.
[17] M.J. Spangehl, B.A. Masri, J.X. O'Connell, C.P. Duncan, Prospective analysis of preoperative and intraoperative investigations for the diagnosis of infection at the sites of two hundred and two revision total hip arthroplasties, J. Bone Joint Surg. Am. 81 (1999) 672–683.
[18] C.L. Romanò, D. Romanò, C. Bonora, A. Degrate, G. Mineo, "Combined Diagnostic Tool" for joint prosthesis infection, Infez. Med. 17 (3) (2009) 141–150.
[19] L. Drago, V. Signori, E. De Vecchi, C. Vassena, E. Palazzi, L. Cappelletti, D. Romanò, C.L. Romano, Use of dithiothreitol to improve the diagnosis of prosthetic joint infections, J. Orthop. Res. 31 (11) (2013) 1694–1699.
[20] G. Lewis, A. Bhattaram, Influence of a pre-blended antibiotic (gentamicin sulfate powder) on various mechanical, thermal, and physical properties of three acrylic bone cements, J. Biomater. Appl. 20 (4) (2006) 377–408.
[21] A.C. McLaren, M. Nugent, K. Economopoulos, H. Kaul, B.L. Vernon, R. McLemore, Hand-mixed and premixed antibiotic-loaded bone cement have similar homogeneity, Clin. Orthop. Relat. Res. 467 (7) (2009) 1693–1698.
[22] R. Miller, A. McLaren, C. Leon, R. McLemore, Mixing method affects elution and strength of high-dose ALBC: a pilot study, Clin. Orthop. Relat. Res. 470 (10) (2012) 2677–2683.
[23] K.J. Messick, M.A. Miller, L.A. Damron, A. Race, M.T. Clarke, K.A. Mann, Vacuum-mixing cement does not decrease overall porosity in cemented femoral stems: an in vitro laboratory investigation, J. Bone Joint Surg. (Br.) 89 (8) (2007) 1115–1121.
[24] T. Gehrke, A. Zahar, D. Kendoff, One-stage exchange: it all began here, Bone Joint J. 95-B (11 Suppl. A) (2013) 77–83.
[25] L. Frommelt, Periprosthetic infection: bacteria and the interface between prosthesis and bone, in: I.D. Learmonth (Ed.), Interfaces in Total Hip Arthroplasty, Springer Verlag, London, 2000, pp. 153–161.
[26] S. Klouche, P. Leonard, V. Zeller, et al., Infected total hip arthroplasty revision: one- or two-stage procedure? Orthop. Traumatol. Surg. Res. 98 (2) (2012) 144–150.
[27] T. Gehrke, D. Kendoff, Peri-prosthetic hip infections: in favour of one-stage, Hip Int. 22 (Suppl. 8) (2012) S40–S45.
[28] J.N. Insall, F.M. Thompson, B.D. Brause, Two-stage reimplantation for the salvage of the infected total knee arthroplasty, J. Bone Joint Surg. 1087 (1983) 1087–1098.
[29] A.H. Wilde, J.T. Ruth, Two-stage reimplantation in infected total knee arthroplasty, Clin. Orthop. Relat. Res. (1988) 23–35.

[30] L.S. Borden, P.F. Gearen, Infected total knee arthroplasty: a protocol for management, J. Arthroplasty 2 (1) (1987) 27–36.
[31] T.F. Calton, T.K. Fehring, W.L. Griffin, Bone loss associated with the use of spacer blocks in infected total knee arthroplasty, Clin. Orthop. Relat. Res. (345) (1997) 148–154.
[32] R.H. Walker, D.J. Schurman, Management of infected total knee arthroplasties, Clin. Orthop. Relat. Res. (186) (1984) 81–89.
[33] A.A. Hofmann, K.R. Kane, T.K. Tkach, R.L. Plaster, M.P. Camargo, Treatment of infected total knee arthroplasty using an articulating spacer, Clin. Orthop. Relat. Res. (321) (1995) 45–54.
[34] A.A. Hofmann, T. Goldberg, A.M. Tanner, S.M. Kurtin, Treatment of infected total knee arthroplasty using an articulating spacer: 2- to 12-year experience, Clin. Orthop. Relat. Res. (430) (2005) 125–131.
[35] R.H. Emerson Jr., M. Muncie, T.R. Tarbox, et al., Comparison of a static with a mobile spacer in total knee infection, Clin. Orthop. Relat. Res. (2002) 132–138.
[36] I.R. Scott, L. Stockley, C.J.M. Getty, Exchange arthroplasty for infected knee replacements, J. Bone Joint Surg. (Br.) 74 (1) (1993) 28–31.
[37] C.P. Duncan, C. Beauchamp, A temporary antibiotic-loaded joint replacement system for management of complex infections involving the hip, Orthop. Clin. North Am. 24 (4) (1993) 751–759.
[38] E.J. McPherson, K. Lewonoswski, L.D. Dorr, Use of an articulated PMMA spacer in the infected total knee arthroplasty, Clin. Arthroplasty 10 (1) (1995) 87–89.
[39] T.K. Fehring, S. Odum, T.F. Calton, et al., Articulating versus static spacers in revision total knee arthroplasty for sepsis. The Ranawat Award, Clin. Orthop. Relat. Res. (2000) 9–16.
[40] F.S. Haddad, B.A. Masri, D. Campbell, et al., The Prostalac functional spacer in two-stage revision for infected knee replacements, J. Bone Joint Surg. (Br.) 82 (6) (2000) 807–812.
[41] R.M. Meek, D. Dunlop, D.S. Garbuz, et al., Patient satisfaction and functional status after aseptic versus septic revision total knee arthroplasty using the PROSTALAC articulating spacer, J. Arthroplasty 19 (7) (2004) 874–879.
[42] A.V. Lombardi, J.M. Karney, K.R. Berend, A motion maintaining antibiotic delivery system, J. Arthroplasty 22 (2007) 50–55.
[43] Y.C. Hsu, H.C. Cheng, T.P. Ng, K.Y. Chiu, Antibiotic-loaded cement articulating spacer for 2-stage reimplantation in infected total knee arthroplasty. A simple and economic method, J. Arthroplasty 22 (2007) 1060–1066.
[44] H. Shen, X. Zhang, Y. Jiang, Q. Wang, Y. Chen, Q. Wang, J. Shao, Intraoperatively-made cement-on-cement antibiotic-loaded articulating spacer for infected total knee arthroplasty, Knee 17 (6) (2010) 407–411.
[45] M. Villanueva-Martinez, A. Rios-Luna, J. Pereiro, F. Chana, H. Fahandez-Saddi, Hand-made articulating spacers for infected total knee arthroplasty: a technical note, Acta Orthop. 77 (2) (2006) 329–332.
[46] V. Pascale, W. Pascale, Custom-made articulating spacer in two-stage revision total knee arthroplasty. An early follow-up of 14 cases of at least 1 year after surgery, HSS J. 3 (2007) 159–163.
[47] M. Villanueva-Martínez, A. Ríos-Luna, J. Pereiro, H. Fahandez-Saddi, A. Villamor, Hand-made articulating spacers in two-stage revision for infected total knee arthroplasty: good outcome in 30 patients, Acta Orthop. 79 (5) (2008) 674–682.

[48] R.P. Pitto, C.C. Castelli, R. Ferrari, et al., Pre-formed articulating knee spacer in two stage revision for the infected total knee arthroplasty, Int. Orthop. 29 (2005) 305.

[49] E. Meani, C.L. Romanò, L. Crosby, G. Hofmann (Eds.), Infection and Local Treatment in Orthopaedic Infections, Springer-Verlag, Heidelberg, 2007.

[50] L. Bernard, L. Legout, L. Zürcher-Pfund, R. Stern, P. Rohner, R. Peter, M. Assal, D. Lew, P. Hoffmeyer, I. Uçkay, Six weeks of antibiotic treatment is sufficient following surgery for septic arthroplasty, J. Infect. 61 (2) (2010) 125–132.

[51] P.H. Hsieh, K.C. Huang, P.C. Lee, M.S. Lee, Two-stage revision of infected hip arthroplasty using an antibiotic-loaded spacer: retrospective comparison between short-term and prolonged antibiotic therapy, J. Antimicrob. Chemother. 64 (2) (2009) 392–397.

[52] Y. Achermann, K. Eigenmann, B. Ledergerber, L. Derksen, P. Rafeiner, M. Clauss, R. Nüesch, C. Zellweger, M. Vogt, W. Zimmerli, Factors associated with rifampin resistance in staphylococcal periprosthetic joint infections (PJI): a matched case–control study, Infection 41 (2) (2013) 431–437.

[53] N. Logoluso, C.L. Romanò, F. Dell'Oro, A. Elia, L. Drago, Gait analysis in patients with a preformed articulated knee spacer, Knee 19 (4) (2012) 370–372.

[54] H. Segawa, D.T. Tsukayama, et al., Infection after total knee arthroplasty. A retrospective study of the treatment of eighty-one infections, J. Bone Joint Surg. Am. 81 (10) (1999) 1434–1445.

[55] J. Singer, A. Merz, L. Frommelt, B. Fink, High rate of infection control with one-stage revision of septic knee prostheses excluding MRSA and MRSE, Clin. Orthop. Relat. Res. 470 (2012) 1461–1471.

[56] P.D. Wilson, P. Aglietti, E.A. Salvati, Subacute sepsis of the hip treated by antibiotics and cemented prosthesis, J. Bone Joint Surg. 56-A (5) (1974) 879–898.

[57] P.W. Hughes, E.A. Salvati, P.D. Wilson, E.L. Blumenfeld, Treatment of subacute sepsis of the hip by antibiotics and joint replacement criteria for diagnosis with evaluation of twenty-six cases, Clin. Orthop. Relat. Res. (141) (1979) 143–157.

[58] H. Buchholz, R. Elson, E. Engelbrecht, H. Lodenkamper, J. Rottger, A. Siegel, Management of deep infection of total hip replacement, J. Bone Joint Surg. 63-B (3) (1981) 342–353.

[59] G.B. Miley, A.D. Scheller, R.H. Turner, Medical and surgical treatment of the septic hip with one-stage revision arthroplasty, Clin. Orthop. Relat. Res. (170) (1982) 76–82.

[60] B.M. Wroblewski, One-stage revision of infected cemented total hip arthroplasty, Clin. Orthop. Relat. Res. (211) (1986) 103–107.

[61] F.A. Weber, E.E.G. Lautenbach, Revision of infected total hip arthroplasty, Clin. Orthop. Relat. Res. (211) (1986) 108–115.

[62] L. Sanzen, A. Carlsson, G. Josefsson, L.T. Lindberg, Revision operations on infected total hip arthroplasties, Clin. Orthop. Relat. Res. (229) (1988) 165–172.

[63] R. Schneider, The infected total prosthesis, Orthopade 18 (6) (1989) 527–532.

[64] E. Morscher, R. Herzog, R. Bapst, Hip revision surgery in septic loosening, Chir. Organi Mov. 79 (4) (1994) 335–340.

[65] V.V. Raut, P.D. Siney, B.M. Wroblewski, One-stage revision of infected total hip replacements with discharging sinuses, J. Bone Joint Surg. (Br.) 76 (5) (1994) 721–724.

[66] D.M. Mulcahy, J.M. O'Byrne, G.E. Fenelon, One stage surgical management of deep infection of total hip arthroplasty, Ir. J. Med. Sci. 165 (1) (1996) 17–19.

[67] K.J. Ure, H.C. Amstutz, S. Nasser, T.P. Schmalzried, Direct-exchange arthroplasty for the treatment of infection after total hip replacement. An average ten-year follow-up, J. Bone Joint Surg. Am. 80 (1998) 961–968.

[68] J.J. Callaghan, R.P. Katz, R.C. Johnston, One-stage revision surgery of the infected hip. A minimum 10-year followup study, Clin. Orthop. Relat. Res. (369) (1999) 139–143.
[69] H. Gao, H. Lv, One-stage revision operations for infection after hip arthroplasty, Zhongguo Xiu Fu Chong Jian Wai Ke Za Zhi 22 (1) (2008) 5–8.
[70] S.I. Oussedik, M.B. Dodd, F.S. Haddad, Outcomes of revision total hip replacement for infection after grading according to a standard protocol, J. Bone Joint Surg. 92-B (9) (2010) 1222–1226.
[71] F.H. De Man, P. Sendi, W. Zimmerli, T.B. Maurer, P.E. Ochsner, T. Ilchmann, Infectiological, functional, and radiographic outcome after revision for prosthetic hip infection according to a strict algorithm, Acta Orthop. 82 (1) (2011) 27–34.
[72] H.R. Choi, Y.M. Kwon, A.A. Freiberg, H. Malchau, Comparison of one-stage revision with antibiotic cement versus two-stage revision results for infected total hip arthroplasty, J. Arthroplasty 28 (8 Suppl.) (2013) 66–70.
[73] M. Wolf, H. Clar, J. Friesenbichler, et al., Prosthetic joint infection following total hip replacement: results of one-stage versus two-stage exchange, Int. Orthop. 38 (7) (2014) 1363–1368.
[74] V. Zeller, L. Lhotellier, S. Marmor, et al., One-stage exchange arthroplasty for chronic periprosthetic hip infection: results of a large prospective cohort study, J. Bone Joint Surg. Am. 96 (1) (2014) e1.
[75] G. Bori, E. Muñoz-Mahamud, J. Cuñé, X. Gallart, D. Fuster, A. Soriano, One-stage revision arthroplasty using cementless hip arthroplasties, J. Arthroplasty 29 (2014) 1076–1081.
[76] J.Y. Jenny, R. Lengert, Y. Diesinger, J. Gaudias, C. Boeri, J.F. Kempf, Routine one-stage exchange for chronic infection after total hip replacement, Int. Orthop. 38 (12) (2014) 2477–2481.
[77] R.H. Fitzgerald, D.R. Jones, Hip implant infection: treatment with resection arthroplasty and late total hip arthroplasty, Am. J. Med. 78 (6) (1985) 225–228.
[78] R. Ketterl, M.B. Henly, B. Stübinger, T. Beckurts, B. Claudi, Analysis of three operative techniques for infected total hip replacements, Orthop. Trans. 12 (1988) 715.
[79] D.J. McDonald, R.H. Fitzgerald, D.M. Ilstrup, Two-stage reconstruction of a total hip arthroplasty because of infection, J. Bone Joint Surg. 71-A (6) (1989) 828–834.
[80] R.A. Colyer, W.N. Capello, Surgical treatment of the infected hip implant. Two-stage reimplantation with a one-month interval, Clin. Orthop. Relat. Res. (298) (1994) 75–79.
[81] K.L. Garvin, B.G. Evans, E.A. Salvati, B.D. Brause, Palacos gentamicin for the treatment of deep periprosthetic hip infections, Clin. Orthop. Relat. Res. (298) (1994) 97–105.
[82] J.R. Lieberman, G.H. Callaway, E.A. Salvati, P.M. Pellicci, B.D. Brause, Treatment of the infected total hip arthroplasty with a two-stage reimplantation protocol, Clin. Orthop. Relat. Res. (301) (1994) 205–212.
[83] B.J. Nestor, A.D. Hanssen, R. Ferrer-Gonzalez, R.H. Fitzgerald, The use of porous prostheses in delayed reconstruction of total hip replacements that have failed because of infection, J. Bone Joint Surg. Am. 76 (1994) 349–359.
[84] K.A. Lai, W.J. Shen, C.Y. Yang, R.M. Lin, C.J. Lin, I.M. Jou, Two-stage cementless revision THR after infection. 5 recurrences in 40 cases followed 2.5–7 years, Acta Orthop. Scand. 67 (4) (1996) 325–328.
[85] A.S. Younger, C.P. Duncan, B.A. Masri, R.W. McGraw, The outcome of two-stage arthroplasty using a custom-made interval spacer to treat the infected hip, J. Arthroplasty 12 (6) (1997) 615–623.

[86] T.K. Fehring, T.F. Calton, W.L. Griffin, Cementless fixation in 2-stage reimplantation for periprosthetic sepsis, J. Arthroplasty 14 (2) (1999) 175–181.

[87] F.L. Morales, J.F. González, F.B. Ortiz, J.M. Martínez, C.G. Araujo, Treatment of infected hip arthroplasty. Retrospective study, Rev. Ortop. Traumatol. 43 (2) (1999) 84–92.

[88] F.S. Haddad, S.K. Muirhead-Allwood, A.R. Manktelow, I. Bacarese-Hamilton, Two-stage uncemented revision hip arthroplasty for infection, J. Bone Joint Surg. (Br.) 82 (5) (2000) 689–694.

[89] K.H. Koo, J.W. Yang, S.H. Cho, et al., Impregnation of vancomycin, gentamicin, and cefotaxime in a cement spacer for two-stage cementless reconstruction in infected total hip arthroplasty, J. Arthroplasty 16 (2001) 882–892.

[90] B. Magnan, D. Regis, R. Biscaglia, P. Bartolozzi, Preformed acrylic bone cement spacer loaded with antibiotics: use of two-stage procedure in 10 patients because of infected hips after total replacement, Acta Orthop. Scand. 72 (6) (2001) 591–594.

[91] K. Karpas, P. Sponer, Management of the infected hip arthroplasty by two-stage reimplantation, Acta Med. (Hradec Kralove) 46 (3) (2003) 113–115.

[92] K. Yamamoto, N. Miyagawa, T. Masaoka, Y. Katori, T. Shishido, A. Imakiire, Clinical effectiveness of antibiotic-impregnated cement spacers for the treatment of infected implants of the hip joint, J. Orthop. Sci. 8 (6) (2003) 823–828.

[93] R.P. Evans, Successful treatment of total hip and knee infection with articulating antibiotic components: a modified treatment method, Clin. Orthop. Relat. Res. (427) (2004) 37–46.

[94] P.H. Hsieh, C.H. Shih, Y.H. Chang, M.S. Lee, H.N. Shih, W.E. Yang, Two-stage revision hip arthroplasty for infection: comparison between the interim use of antibiotic-loaded cement beads and a spacer prosthesis, J. Bone Joint Surg. Am. 86-A (9) (2004) 1989–1997.

[95] M.A. Buttaro, R. Pusso, F. Piccaluga, Vancomycin-supplemented impacted bone allografts in infected hip arthroplasty. Two-stage revision results, J. Bone Joint Surg. (Br.) 87 (2005) 314–319.

[96] A.A. Hofmann, T.D. Goldberg, A.M. Tanner, T.M. Cook, Ten-year experience using an articulating antibiotic cement hip spacer for the treatment of chronically infected total hip, J. Arthroplasty 20 (2005) 874–879.

[97] I. Nusem, D.A. Morgan, Structural allografts for bone stock reconstruction in two- stage revision for infected total hip arthroplasty: good outcome in 16 of 18 patients followed for 5–14 years, Acta Orthop. 77 (2006) 92–97.

[98] H.B. Cabrita, A.T. Croci, O.P. Camargo, A.L. Lima, Prospective study of the treatment of infected hip arthroplasties with or without the use of an antibiotic-loaded cement spacer, Clinics 62 (2) (2007) 99–108.

[99] B.A. Masri, K.P. Panagiotopoulos, N.V. Greidanus, D.S. Garbuz, C.P. Duncan, Cementless two-stage exchange arthroplasty for infection after total hip arthroplasty, J. Arthroplasty 22 (1) (2007) 72–78.

[100] H. Thabe, S. Schill, Two-stage reimplantation with an application spacer and combined with delivery of antibiotics in the management of prosthetic joint infection, Oper. Orthop. Traumatol. 19 (1) (2007) 78–100.

[101] I. Stockley, B.J. Mockford, A. Hoad-Reddick, P. Norman, The use of two-stage exchange arthroplasty with depot antibiotics in the absence of long-term antibiotic therapy in infected total hip replacement, J. Bone Joint Surg. (Br.) 90 (2008) 145–148.

[102] A. Sudo, M. Hasegawa, A. Fukuda, A. Uchida, Treatment of infected hip arthroplasty with antibiotic-impregnated calcium hydroxyapatite, J. Arthroplasty 23 (1) (2008) 145–150.

[103] G.S. Biring, T. Kostamo, D.S. Garbuz, B.A. Masri, C.P. Duncan, Two-stage revision arthroplasty of the hip for infection using an interim articulated Prostalac hip spacer: a 10- to 15-year follow-up study, J. Bone Joint Surg. (Br.) 91 (11) (2009) 1431–1437.

[104] J. Cordero-Ampuero, J. Esteban, E. Garcia-Cimbrelo, Oral antibiotics are effective for highly resistant hip arthroplasty infections, Clin. Orthop. Relat. Res. 467 (9) (2009) 2335–2342.

[105] E. Darley, G. Bannister, A. Blom, A.P. MacGowan, S. Jacobson, W. Alfouzan, Early intravenous to oral antibiotic switch therapy is effective in the treatment of infected total hip replacement, Clin. Microbiol. Infect. 15 (Suppl. S4) (2009) S139.

[106] B. Fink, A. Grossmann, M. Fuerst, P. Schäfer, L. Frommelt, Two-stage cementless revision of infected hip endoprostheses, Clin. Orthop. Relat. Res. 467 (7) (2009) 1848–1858.

[107] P.B. McKenna, K. O'Shea, E.L. Masterson, Two-stage revision of infected hip arthroplasty using a shortened post-operative course of antibiotics, Arch. Orthop. Trauma Surg. 129 (4) (2009) 489–494.

[108] J. Sanchez-Sotelo, D.J. Berry, A.D. Hanssen, M.E. Cabanela, Midterm to long-term followup of staged reimplantation for infected hip arthroplasty, Clin. Orthop. Relat. Res. 467 (2009) 219–224.

[109] S.-J. Lim, Treatment of periprosthetic hip infection caused by resistant microorganisms using 2-stage reimplantation protocol, J. Arthroplasty 24 (8) (2009) 1264–1269.

[110] C. Toulson, S. Walcott-Sapp, J. Hur, et al., Treatment of infected total hip arthroplasty with a 2-stage reimplantation protocol: update on "our institution's" experience from 1989 to 2003, J. Arthroplasty 24 (7) (2009) 1051–1060.

[111] J.P. Whittaker, R.E. Warren, R.S. Jones, P.A. Gregson, Is prolonged systemic antibiotic treatment essential in two-stage revision hip replacement for chronic Gram-positive infection? J. Bone Joint Surg. (Br.) 91 (1) (2009) 44–51.

[112] M. Kent, R. Rachha, M. Sood, A technique for the fabrication of a reinforced moulded articulating cement spacer in two-stage revision total hip arthroplasty, Int. Orthop. 34 (2010) 949–953.

[113] G. Pignatti, S. Nitta, N. Rani, et al., Two stage hip revision in periprosthetic infection: results of 41 cases, Open Orthop. J. 4 (2010) 193–200.

[114] I. Takigami, Y. Ito, D. Ishimaru, et al., Two-stage revision surgery for hip prosthesis infection using antibiotic-loaded porous hydroxyapatite blocks, Arch. Orthop. Trauma Surg. 130 (10) (2010) 1221–1226.

[115] F. D'Angelo, L. Negri, T. Binda, G. Zatti, P. Cherubino, The use of a preformed spacer in two-stage revision of infected hip arthroplasties, Musculoskelet. Surg. 95 (2) (2011) 115–120.

[116] Y.H. Kim, J.S. Kim, J.W. Park, J.H. Joo, Cementless revision for infected total hip replacements, J. Bone Joint Surg. (Br.) 93 (1) (2011) 19–26.

[117] F. Leung, C.J. Richards, D.S. Garbuz, B.A. Masri, C.P. Duncan, Two-stage total hip arthroplasty: how often does it control methicillin-resistant infection? Clin. Orthop. Relat. Res. 469 (4) (2011) 1009–1015.

[118] D.R. Neumann, T. Hofstaedter, C. List, U. Dorn, Two-stage cementless revision of late total hip arthroplasty infection using a premanufactured spacer, J. Arthroplasty 27 (7) (2012) 1397–1401.

[119] L. Wang, Y. Hu, Z. Dai, J. Zhou, M. Li, K. Li, Mid-term effectiveness of two-stage hip prosthesis revision in treatment of infection after hip arthroplasty, Zhongguo Xiu Fu Chong Jian Wai Ke Za Zhi 25 (6) (2011) 646–649.

[120] I. Babiak, Application of individually performed acrylic cement spacers containing 5% of antibiotic in two-stage revision of hip and knee prosthesis due to infection, Pol. Orthop. Traumatol. 77 (2012) 29–37.
[121] R.M. Degen, J.R. Davey, J.R. Davey, J.L. Howard, R.W. McCalden, D.D. Naudie, Does a prefabricated gentamicin-impregnated, load-bearing spacer control periprosthetic hip infection? Clin. Orthop. Relat. Res. 470 (10) (2012) 2724–2729.
[122] G.A. Macheras, S.D. Koutsostathis, K. Kateros, S. Papadakis, P. Anastasopoulos, A two stage re-implantation protocol for the treatment of deep periprosthetic hip infection. Mid to long-term results, Hip Int. 22 (Suppl. 8) (2012) S54–S61.
[123] C.L. Romanò, D. Romanò, A. Albisetti, E. Meani, Preformed antibiotic-loaded cement spacers for two-stage revision of infected total hip arthroplasty. Long-term results, Hip Int. 22 (Suppl. 8) (2012) S46–S53.
[124] K.R. Berend, A.V. Lombardi Jr., M.J. Morris, A.G. Bergeson, J.B. Adams, M.A. Sneller, Two-stage treatment of hip periprosthetic joint infection is associated with a high rate of infection control but high mortality, Clin. Orthop. Relat. Res. 471 (2) (2013) 510–518.
[125] P.H. Hsieh, K.C. Huang, H.N. Shih, Prosthetic joint infection in patients with rheumatoid arthritis: an outcome analysis compared with controls, PLoS One 8 (8) (2013) e71666.
[126] A.J. Johnson, M.G. Zywiel, L.C. Jones, R.E. Delanois, D.A. Stroh, M.A. Mont, Reduced re-infection rates with postoperative oral antibiotics after two-stage revision hip arthroplasty, BMC Musculoskelet. Disord. 14 (1) (2013) 123.
[127] Y.K. Lee, K.H. Lee, J.H. Nho, Y.C. Ha, K.H. Koo, Retaining well-fixed cementless stem in the treatment of infected hip arthroplasty, Acta Orthop. 84 (3) (2013) 260–264.
[128] F.Y. Sabry, C.R. Szubski, J.J. Stefancin, A.K. Klika, C.A. Higuera, W.K. Barsoum, Comparison of complications associated with commercially available and custom-made articulating spacers in two-stage total hip arthroplasty revision, Curr. Orthop. Pract. 24 (4) (2013) 406–413.
[129] M.P. van Diemen, S. Colen, A.A. Dalemans, J. Stuyck, M. Mulier, Two-stage revision of an infected total hip arthroplasty: a follow-up of 136 patients, Hip Int. 23 (5) (2013) 445–450.
[130] R. Schwarzkopf, B. Mikhael, E. Wright, D.M. Estok, J.N. Katz, Treatment failure among infected periprosthetic total hip arthroplasty patients, Open Orthop. J. 8 (2014) 118–124.
[131] T.E. Ekpo, K.R. Berend, M.J. Morris, J.B. Adams, A.V. Lombardi Jr., Partial two-stage exchange for infected total hip arthroplasty: a preliminary report, Clin. Orthop. Relat. Res. 472 (2) (2014) 437–448.
[132] M.S. Ibrahim, S. Raja, M.A. Khan, F.S. Haddad, A multidisciplinary team approach to two-stage revision for the infected hip replacement: a minimum five-year follow-up study, Bone Joint J. 96-B (2014) 1312–1318.
[133] R. Dieckmann, D. Schulz, G. Gosheger, K. Becker, K. Daniilidis, A. Streitbürger, J. Hardes, S. Hoell, Two-stage hip revision arthroplasty with a hexagonal modular cementless stem in cases of periprosthetic infection, BMC Musculoskelet. Disord. 15 (2014) 398, http://dx.doi.org/10.1186/1471-2474-15-398.
[134] G. von Foerster, D. Kluber, U. Kabler, Mid- to long-term results after treatment of 118 cases of periprosthetic infections after knee joint replacement using one-stage exchange surgery, Orthopade 20 (1991) 244–252.
[135] S.B. Goksan, M.A. Freeman, One-stage reimplantation for infected total knee arthroplasty, J. Bone Joint Surg. (Br.) 74 (1992) 78–82.

[136] H. Lu, B. Kou, J. Lin, One-stage reimplantation for the salvage of total knee arthroplasty complicated by infection, Zhonghua Wai Ke Za Zhi 35 (1997) 456–458.
[137] M. Silva, R. Tharani, T.P. Schmalzried, Results of direct exchange or debridement of the infected total knee arthroplasty, Clin. Orthop. Relat. Res. (404) (2002) 125–131.
[138] F.F. Buechel, F.P. Femino, J. D'Alessio, Primary exchange revision arthroplasty for infected total knee replacement: a long- term study, Am. J. Orthop. 33 (4) (2004) 190–198.
[139] D. Sofer, B. Regenbrecht, J. Pfeil, Early results of one-stage septic revision arthroplasties with antibiotic-laden cement. A clinical and statistical analysis, Orthopade 34 (6) (2005) 592–602.
[140] S. Tibreval, Single-stage revision for the infected total knee replacement: results from a single centre, Bone Joint J. 96-B (6) (2014) 759–764.
[141] F.S. Hadddad, Is single-stage revision according to a strict protocol effective in treatment of chronic knee arthroplasty infections? Clin. Orthop. Relat. Res. 473 (1) (2015) 8–14.
[142] A.G. Rosenberg, B. Haas, R. Barden, D. Marquez, G.C. Landon, J.O. Galante, Salvage of infected total knee arthroplasty, Clin. Orthop. Relat. Res. (226) (1988) 29–33.
[143] A.H. Wilde, J.T. Ruth, Two-stage reimplantation in infected total knee arthroplasty, Clin. Orthop. Relat. Res. (236) (1988) 23–27.
[144] R.E. Booth, P.A. Lotke, The results of spacer block technique in revision of infected total knee arthroplasty, Clin. Orthop. Relat. Res. (248) (1989) 57–60.
[145] R.E. Windsor, J.N. Insall, W.K. Urs, D.V. Miller, B.D. Brause, Two-stage reimplantation for the salvage of total knee arthroplasty complicated by infection. Further follow-up and refinement of indications, J. Bone Joint Surg. Am. 72 (1990) 272–278.
[146] M.H. Henderson Jr., R.E. Booth Jr., The use of an antibiotic- impregnated spacer block for revision of the septic total knee arthroplasty, Semin. Arthroplast. 2 (1) (1991) 34–39.
[147] I.R. Scott, L. Stockley, C.J.M. Getty, Exchange arthroplasty for infected knee replacements, J. Bone Joint Surg. (Br.) 75 (1) (1993) 28–31.
[148] A.D. Hanssen, J.A. Rand, D.R. Osmon, Treatment of the infected total knee arthroplasty with insertion of another prosthesis. The effect of antibiotic-impregnated bone cement, Clin. Orthop. Relat. Res. (309) (1994) 44–55.
[149] E.J. McPherson, M.J. Patzakis, J.E. Gross, P.D. Holtom, M. Song, L.D. Dorr, Infected total knee arthroplasty: two-stage reimplantation with a gastrocnemius rotational flap, Clin. Orthop. Relat. Res. (341) (1997) 73–76.
[150] L.A. Whiteside, Treatment of infected total knee arthroplasty, Clin. Orthop. Relat. Res. (299) (1994) 169–172.
[151] R.T. Goldman, G.R. Scuderi, J.N. Insall, 2-Stage reimplantation for infected total knee replacement, Clin. Orthop. Relat. Res. (331) (1996) 118–124.
[152] G. Gacon, M. Laurencon, D. Van de Velde, D.P. Giudicelli, Two stages reimplantation for infection after knee arthroplasty. Apropos of a series of 29 cases, Rev. Chir. Orthop. Reparatrice Appar. Mot. 83 (1997) 313–323.
[153] K. Hirakawa, B.N. Stulberg, A.H. Wilde, T.W. Bauer, M. Secic, Results of 2-stage reimplantation for infected total knee arthroplasty, J. Arthroplasty 13 (1998) 22–27.
[154] T.K. Fehring, S. Odum, T.F. Calton, J.B. Mason, Articulating versus static spacers in revision total knee arthroplasty for sepsis, Clin. Orthop. Relat. Res. (380) (2000) 9–16.
[155] W.M. Goldstein, M. Kopplin, R. Wall, K. Berland, Temporary articulating methylmethacrylate antibiotic spacer (TAMMAS). A new method of intraoperative manufacturing of a custom articulating spacer, J. Bone Joint Surg. Am. 83 (2001) 92–97.
[156] J.H. Lonner, T.D. Beck Jr., H. Rees, M. Roullet, P.A. Lotke, Results of two-stage revision of the infected total knee arthroplasty, Am. J. Knee Surg. 14 (2001) 65–67.

[157] R.H. Emerson Jr., M. Muncie, T.R. Tarbox, L.L. Higgins, Comparison of a static with a mobile spacer in total knee infection, Clin. Orthop. Relat. Res. (404) (2002) 132–138.

[158] T. Siebel, J. Kelm, M. Porsch, T. Regitz, W.H. Neumann, Two-stage exchange of infected knee arthroplasty with a prosthesis-like interim cement spacer, Acta Orthop. Belg. 68 (2002) 150–156.

[159] M. Pietsch, C. Wenisch, S. Traussing, R. Trnoska, S. Hofmann, Temporary articulating spacer with antibiotic-impregnated cement for an infected knee endoprosthesis, Orthopade 32 (2003) 490–497.

[160] S.M. Durbhakula, J. Czajka, M.D. Fuchs, R.L. Uhl, Antibiotic- loaded articulating cement spacer in the 2-stage exchange of infected total knee arthroplasty, J. Arthroplasty 19 (2004) 768–774.

[161] A.A. Haleem, D.J. Berry, A.D. Hanssen, Mid-term to long- term follow-up of two-stage reimplantation for infected total knee arthroplasty, Clin. Orthop. Relat. Res. 35 (2004) 111–117.

[162] J.M. Cuckler, The infected total knee: management options, J. Arthroplasty 20 (2) (2005) 33–36.

[163] M.C. MacAvoy, M.D. Ries, The ball and socket articulating spacer for infected total knee arthroplasty, J. Arthroplasty 20 (2005) 757–762.

[164] H.T. Huang, J.Y. Su, S.K. Chen, The results of articulating spacer technique for infected total knee arthroplasty, J. Arthroplasty 21 (2006) 1163–1168.

[165] E. Jamsen, P. Sheng, P. Halonen, M.U. Letho, T. Moilanen, J. Pajamäki, T. Puolakka, Y.T. Konttinen, Spacer prostheses in two-stage revision of infected knee arthroplasty, Int. Orthop. 30 (2006) 257–261.

[166] M.G. Freeman, T.K. Fehring, S.M. Odum, K. Fehring, W.L. Griffin, J.B. Mason, Functional advantage of articulating versus static spacers in 2-stage revision for total knee arthroplasty infection, J. Arthroplasty 22 (8) (2007) 1116–1121.

[167] J.A. Anderson, P.K. Sculco, S. Heitkemper, D.J. Mayman, M.P. Bostrom, T.P. Sculco, An articulating spacer to treat and mobilize patients with infected total knee arthroplasty, J. Arthroplasty 24 (2009) 631–635.

[168] A. Ocguder, A. Firat, O. Tecimel, S. Solak, M. Bozkurt, Two-stage total infected knee arthroplasty treatment with articulating cement spacer, Arch. Orthop. Trauma Surg. 130 (2010) 719–725.

[169] E.R. Chiang, Comparison of articulating and static spacers regarding infection with resistant organisms in total knee arthroplasty, Acta Orthop. 82 (4) (2011) 460–464.

[170] C.R. Gooding, B.A. Masri, N.V. Greidanus, D.S. Garbuz, Durable infection control and function with the PROSTALAC spacer in two-stage revision for infected knee arthroplasty, Clin. Orthop. Relat. Res. 469 (2011) 985–993.

[171] S. Kohl, An intraoperatively moulded PMMA prostheses like spacer for two-stage revision of infected total knee arthroplasty, Knee 18 (2011) 464–469.

[172] G.A. Macheras, The long-term results of a two-stage protocol for revision of an infected total knee replacement, J. Bone Joint Surg. (Br.) 93 (11) (2011) 1487–1492.

[173] S.M.J. Mortazavi, Two-stage exchange arthroplasty for infected total knee arthroplasty: predictors of failure, Clin. Orthop. Relat. Res. 469 (2011) 3049–3054.

[174] G.S. Van Thiel, K.R. Berend, G.R. Klein, A.C. Gordon, A.V. Lombardi, C.J. Della Valle, Intraoperative molds to create an articulating spacer for the infected knee arthroplasty, Clin. Orthop. Relat. Res. 469 (2011) 994–1001.

[175] P. Cai, Two-stage revision of infected total knee arthroplasty using antibiotic-impregnated articulating cement spacer, Zhongguo Xiu Fu Chong Jian Wai Ke Za Zhi 26 (10) (2012) 1169–1173.

[176] H.R. Choi, Are prosthetic spacers safe to use in 2-stage treatment for infected total knee arthroplasty? J. Arthroplasty 27 (8) (2012) 1474–1479. e1.
[177] B.H. Hwang, Fungal peri-prosthetic joint infection after primary total knee replacement, J. Bone Joint Surg. (Br.) 94 (5) (2012) 656–659.
[178] N.V. Kalore, Is there a preferred articulating spacer technique for infected knee arthroplasty? Clin. Orthop. Relat. Res. 470 (1) (2012) 228–235.
[179] S.Y. Kotwal, Intramedullary rod and cement static spacer construct in chronically infected total knee arthroplasty, J. Arthroplast. 27 (2) (2012) 253–259. e4.
[180] Y.T. Jia, Antibiotic-loaded articulating cement spacers in two-stage revision for infected total knee arthroplasty: individual antibiotic treatment and early results of 21 cases, Chin. J. Traumatol. 15 (4) (2012) 212–221.
[181] A.J. Johnson, Minimizing dynamic knee spacer complications in infected revision arthroplasty, Clin. Orthop. Relat. Res. 470 (2012) 220–227.
[182] J.K. Lee, Two-stage reimplantation in infected total knee arthroplasty using a re-sterilized tibial polyethylene insert and femoral component, J. Arthroplast. 27 (9) (2012) 1701–1706.
[183] T. Mahmud, Assessing the gold standard: a review of 253 two-stage revisions for infected TKA, Clin. Orthop. Relat. Res. 470 (2012) 2730–2736.
[184] R.S.J. Nickinson, Two stage revision knee arthroplasty for infection with massive bone loss. A technique to achieve spacer stability, Knee 19 (2012) 24–27.
[185] Z. Wan, Preformed articulating knee spacers in 2-stage total knee revision arthroplasty: minimum 2-year follow-up, J. Arthroplasty 27 (8) (2012) 1469–1473.
[186] S.S. Ahmad, The cement prosthesis-like spacer: an intermediate halt on the road to healing, Sci. World J. 2013 (2013) 763434.
[187] D. Brun, Tibial tubercle osteotomy or quadriceps snip in two-stage revision for prosthetic knee infection? A randomized prospective study, Clin. Orthop. Relat. Res. 471 (2013) 1305–1318.
[188] J. Brunnekreef, Recovery of knee mobility after a static or mobile spacer in total knee infection, Acta Orthop. Belg. 79 (1) (2013) 83–89.
[189] C. Carulli, A novel technique to preserve range of motion in two-stage revision of infected total knee arthroplasty, Int. Orthop. 37 (6) (2013) 1069–1074.
[190] Y.S. Kim, Two-stage revision using a modified articulating spacer in infected total knee arthroplasty, Knee Surg. Relat. Res. 25 (4) (2013) 180–185.
[191] R. Schwarzkopf, Treatment failure among infected periprosthetic patients at a highly specialized revision TKA referral practice, Open Orthop. J. 7 (2013) 264–271.
[192] A. Silvestre, Revision of infected total knee arthroplasty: two-stage reimplantation using an antibiotic-impregnated static spacer, Clin. Orthop. Surg. 5 (2013) 180–187.
[193] D. Tigani, Two-stage reimplantation for periprosthetic knee infection: influence of host health status and infecting microorganism, Knee 20 (1) (2013) 9–18.
[194] S.W. Ueng, What is the success of treatment of hip and knee candidal periprosthetic joint infection? Clin. Orthop. Relat. Res. 471 (2013) 3002–3009.
[195] C.C. Castelli, Two-stage treatment of infected total knee arthroplasty: two to thirteen year experience using an articulating preformed spacer, Int. Orthop. 38 (2) (2014) 405–412.
[196] T. Classen, Functional interest of an articulating spacer in two-stage infected total knee arthroplasty revision, Orthop. Traumatol. Surg. Res. 100 (4) (2014) 409–412.
[197] P.S. Corona, Industrially prefabricated cement spacers: do vancomycin- and gentamicin-impregnated spacers offer any advantage? Clin. Orthop. Relat. Res. 472 (3) (2014) 923–932.

[198] C.L. Nelson, Sonication of antibiotic spacers predicts failure during two-stage revision for prosthetic knee and hip infections, Clin. Orthop. Relat. Res. 472 (7) (2014) 2208–2214.
[199] N. Prasad, Two-stage revision for infected total knee arthroplasty: our experience with interval prosthesis, Eur. J. Orthop. Surg. Traumatol. 24 (7) (2014) 1279–1283.
[200] C.D. Watts, Morbid obesity: a significant risk factor for failure of two-stage revision total knee arthroplasty for infection, J. Bone Joint Surg. Am. 96 (18) (2014). e154.

PMMA beads and spacers for local antibiotic administration

10

J.A.P. Geurts, G.H.I.M. Walenkamp
Maastricht University Medical Center, Maastricht, The Netherlands

10.1 History

Historically, bone cement was developed and used for the first time more than 100 years ago by Themistocles Gluck (1870) when he obtained fixation of an ivory total knee replacement by cement made of plaster and colophony [1]. Only several decades later, cement that resembles modern, commercially available bone cement was developed by people as Otto Rohm, Degussa, and Kulzer; the latter two patenting the polymerization mechanism of methyl methacrylate when adding a coinitiator such as a tertiary aromatic amine (1943) [2]. In the orthopaedic field, Sir John Charnley was the first to describe the use of polymethylmethacrylate (PMMA) as a way of fixing an endoprostheses (total hip replacement) to bone [3,4].

Also, as soon as antibiotics were discovered and became more and more available, people looked at ways of administering these antibiotics locally to prevent and eradicate infection.

In infection treatment, one was looking at ways to get a high local concentration of antibiotics in order to kill as much bacteria remaining after the debridement as possible. This has led to development of several different "carriers" like antibiotic-loaded beads, fleeces, pellets, etc. [5].

Bone cement and antibiotics "came together" when Buchholz demonstrated in 1972 that bone cement was capable of releasing various substances (residual monomer, CuS) in a sustained manner. In a pilot study, he mixed four heat stable antibiotic powders with bone cement and showed that all but tetracycline were released for at least 2 weeks in concentrations surpassing the bactericidal threshold [6].

10.2 Rationale and pharmacokinetics of antibiotic-loaded PMMA

PMMA or bone cement is the result of mixing the liquid monomer methylmethacrylate with the powder polymer. In an exothermic reaction, these components form a network of new PMMA chains. Before mixing, all sorts of substances can be added, for instance, antibiotics. These will then incorporate in between the PMMA chains and can be released by diffusion, when the PMMA absorbs water, a process which is determined by the hydrophobicity of the cement [7].

Management of Periprosthetic Joint Infections (PJIs). http://dx.doi.org/10.1016/B978-0-08-100205-6.00010-0

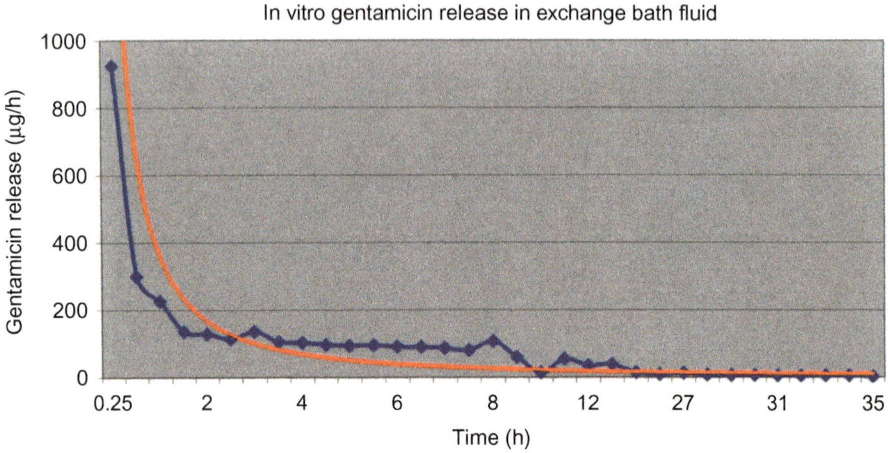

Fig. 10.1 Initial burst release of gentamicin from polylactide, in vitro test (Walenkamp, unpublished data).

Cement properties like porosity and roughness influence the release of antibiotics. Only antibiotics closer than 100 μm to the surface will be released [8]. This accounts to only 3–15% of all the antibiotics in the cement. Initially, there will be a burst release in the first hours (Fig. 10.1). This is a surface phenomenon. Following this initial release is a period of sustained release, a bulk phenomenon, depending on the penetration of fluid in the cement through interconnecting pores [9]. Clinically, both phases have their importance as high local postoperative concentrations at first will result in rapid eradication of sensitive germs in the operation site (without the adverse toxic drug effects). As local concentrations can reach much higher levels than the minimal inhibitory concentration, even resistant germs can be killed. The sustained release phase in the days thereafter prevents biofilm formation on the surface of the cement and colonization with hematogenic germs.

Ways of improving the release of the antibiotic content are hand-mixing cement instead of vacuum mixing. Pores are larger when cement is hand-mixed, and this improves antibiotic release although it can be detrimental for the biomechanical properties of the cement. As beads and spacers are temporary implants, this is seldom an issue.

Also, gel-forming polymeric fillers such as glycin, xylitol, sucrose, or erythritol have been shown to increase antibiotic release from cement by dissolving and leaving behind large pores through which the antibiotics can then be set free from the cement [10–12]. Recently, other research groups have also advocated gelatin sponge and ceramic granules, which increase antibiotic release by a factor 5.65- and 3.75-fold, respectively, [13].

Type of antibiotic, the dose, and the combination of different antibiotics also influence the eventual release characteristics. These antibiotics have to fulfil a certain number of requirements: being heat-stable and chemically inert during polymerization, water-soluble, broad-spectrum, bactericidal (even in low concentrations), available in powder form with low serum binding and at low risk for allergy or delayed hypersensitivity. Most commonly used antibiotics in cement are gentamicin, tobramycin, and

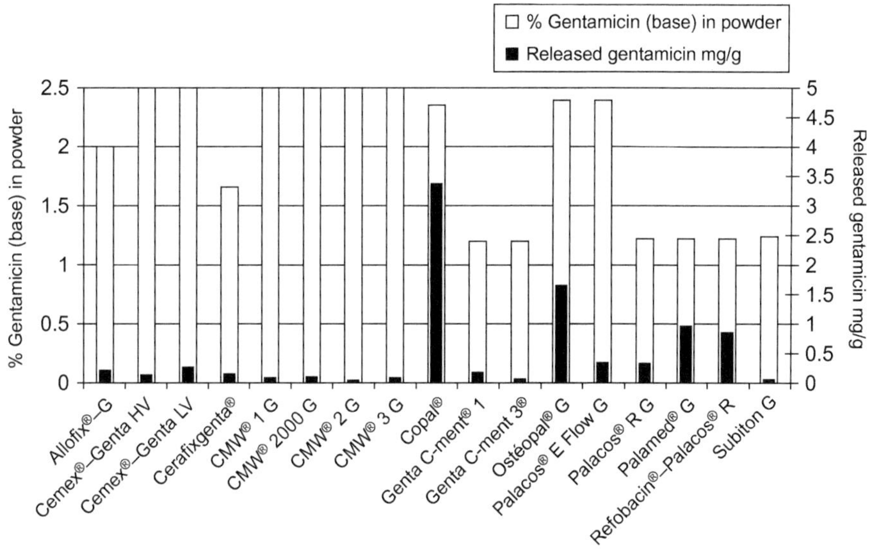

Fig. 10.2 Accumulated release of gentamicin in the first 7 days: comparison of several antibiotic loaded bone cements. G.H.I.M. Walenkamp, Gentamicin-PMMA beads. A clinical, pharmacokinetic and toxicological study, Thesis University Nijmegen, 1983, p. 127. ISBN 90-9000470-X, Fig. 15.8.

erythromycin. Vancomycin is advised for use against methicillin-resistant *Staphylococcus aureus*. Cephalosporins are not heat-stable and generally not effective in treating methicillin-resistant organisms. Potential systemic side effects, notably renal, hepatic, or ototoxicity (gentamicin), were not detected in several in-vivo elution studies. Serum levels remained very low in the postoperative period.

The combination of more kinds of antibiotics in bone cement may increase the release of the individual antibiotics. The best example is the Copal bone cement containing gentamicin as well as clindamycin (Fig. 10.2).

10.3 Beads

To deliver therapeutic tissue levels of *parenteral* antibiotics to the target area, high serum levels of antibiotics must be achieved. These high serum levels, however, may result in an increased incidence of systemic side effects such as nephrotoxicity and ototoxicity. Alternatively, the use of beads as a local adjunct to other treatment offers advantages compared with systemic antibiotics. Systemic levels of antibiotics are low, and the systemic complications are eliminated [14,15], while the local concentration, where it is needed, is extremely high (Fig. 10.3). Antibiotic beads also offer the benefit of management of dead space. They are relatively inexpensive and are easy for the surgeon to insert and the patient to tolerate.

Klemm was one of the first authors describing his results in 1976 with gentamicin-loaded beads in the treatment of bone infection, rather than using large

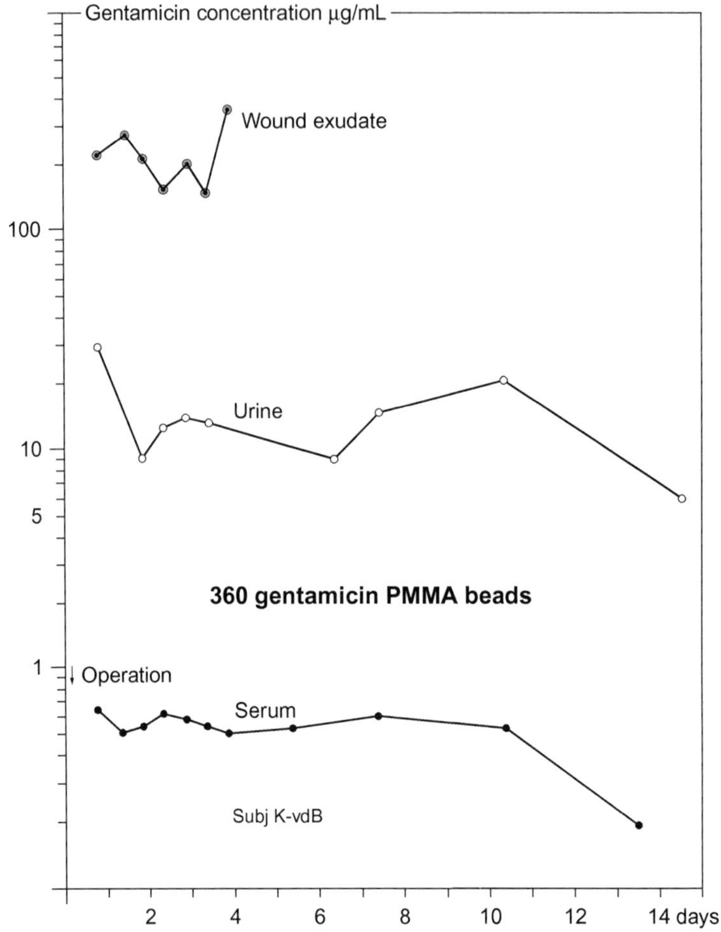

Fig. 10.3 Antibiotic concentrations in exudate, urine and serum, released by beads in vivo [14,15].

deposits of antibiotic impregnated cement [16]. About 128 patients with chronic osteomyelitis were treated with beads, resulting in eradication of infection in 91.4%.

Beads have a large surface area and are capable of releasing a significant amount of the antibiotics they carry, but not all. After 2 weeks in situ, the beads release between 20% and 70% of the total amount of incorporated gentamicin. After that period, release and thus local concentrations drop significantly [15]. This is also the reason why beads should be removed, as they might be colonized and act as a new infected "implant."

If the amount of antibiotic that diffuses to the hematoma and tissues is higher than the amount that is resorbed, then the local concentration of, for example, gentamicin will gradually increase. The maximum local concentration in the exudate in most

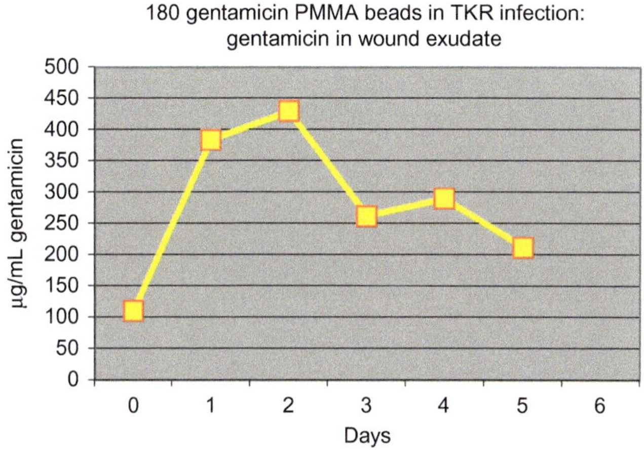

Fig. 10.4 Exudate concentration in a knee joint after removal of a TKR, and implantation of 180 gentamicin beads. The local gentamicin has its maximum value at day 2 of about 430 µg/mL [17].

cases will be achieved with optimal use at day 2 or 3 after implantation (Fig. 10.4) and is in the range of 300–400 µg/mL [17].

This maximum level of concentration depends on the surface of the carrier on one hand, and the volume of the hematoma on the other hand. The much larger surface of beads as compared with spacers explains why the resulting exudate concentration in spacers is much lower. Small gentamicin beads release for the same reason seven times more of their gentamicin content than large beads, resulting in a release of 93% vs 24% of their content in 2 weeks [18].

To achieve these concentrations, as many beads as possible have to fill the cavity remaining after removal of the implant or the bone defect in case of osteomyelitis. In a knee (after removal of a total knee arthroplasty), this can be up to 180 beads and in a hip up to 360 [14].

Debate remains on whether or not to use suction drains in these patients, as this will cause evacuation of the antibiotics from the wound cavity and as a consequence lower local concentrations. However, without the use of a drain, the residual hematoma will likely be bigger and the local antibiotic diluted.

In hand-made beads, the antibiotic release may be much less for several reasons. When the regular gentamicin-containing Refobacin-Palacos is mixed with vancomycin, then the exudate concentrations as depicted in Fig. 10.5 show the maximum concentration as mentioned before on days 2–3. The maximum level of gentamicin, however, is much lower as compared with beads as the bone cement as used has a lower concentration of gentamicin than beads, and the cement does not contain the glycine filler as used in beads.

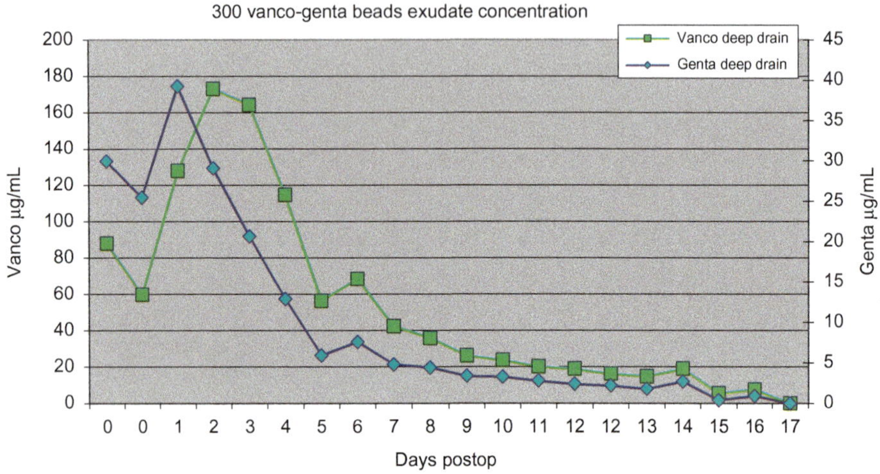

Fig. 10.5 Gentamicin release diminished as compared with genta beads vancomycin, by almost a factor 10, since the cement as a carrier has different properties and a lower gentamicin content (Walenkamp, unpublished data).

The mechanical and elution properties of commercially available premixed antibiotic-loaded bone-cement products are superior to those of hand-mixed preparations.

Figs. 10.6–10.8 show the technique of making beads on the operating room back table when using a metal mold.

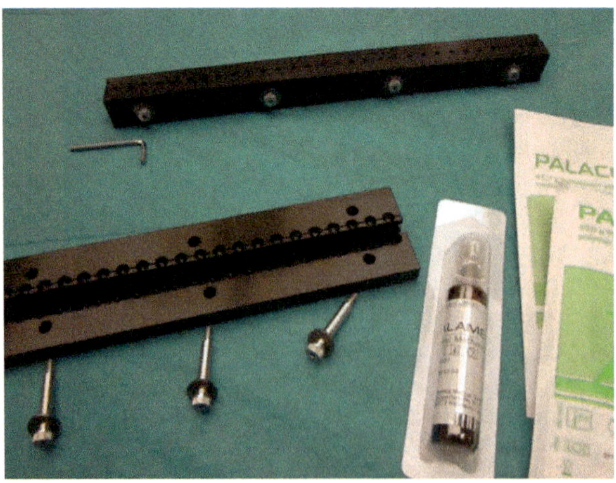

Fig. 10.6 Custom made molds for hand-making beads.

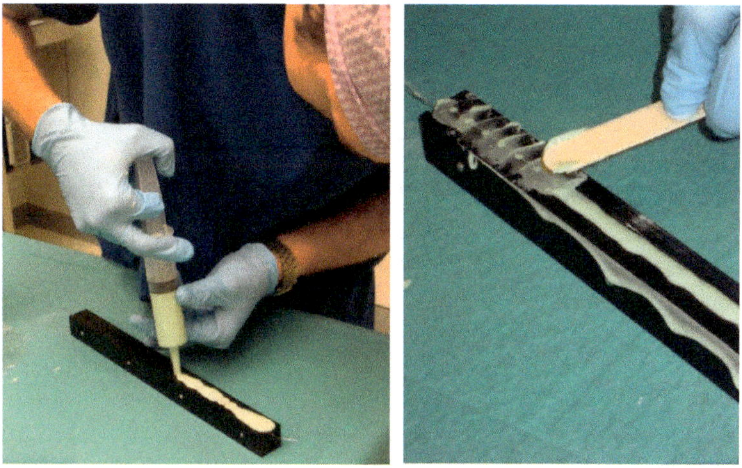

Fig. 10.7 Technique of making beads.

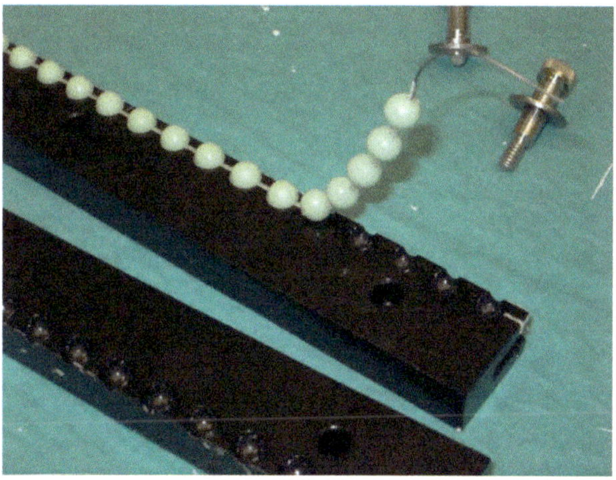

Fig. 10.8 Resulting chain of ready-to-use beads.

10.4 Spacers

In relation to beads, spacers offer the advantage of retainment of length (hip, shoulder) by avoiding soft tissue contracture, joint mobility, possibility of weight-bearing (hip, knee), and sometimes they can serve as a definitive solution in case a patient is no longer operable or refuses another operation.

First-generation spacers were hand molded and adapted to the specific anatomy of the patient, thereby often not allowing adequate motion (block spacers) (Fig. 10.9). Due to incongruence, they could even lead to additional bone loss, extensor

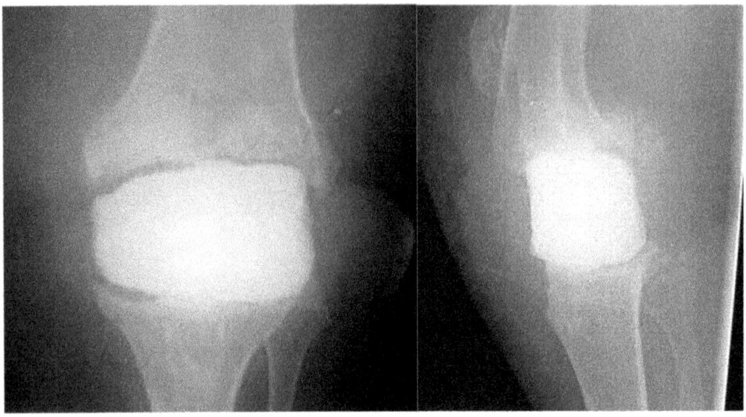

Fig. 10.9 Hand-molded block spacer of the knee.

mechanism erosion, or fracture, making the subsequent reimplantation even more difficult. Advantages were their simplicity, low cost, and possibility of incorporating selective antibiotics.

Second-generation spacers are made with prefabricated molds [19] or ready "off the shelf" (Fig. 10.10). Often these are internally reinforced to prevent fracture or catastrophic failure. Disadvantages include limitations in antibiotic dose and choice and size options. Preformed spacers are commercialised by several companies. Third-generation spacers like the Biomet Stage One Select system (Fig. 10.11) offer a wide range of modularity to fit the patients specific needs. Surface of the spacer is textured in order to increase surface and thus antibiotic release.

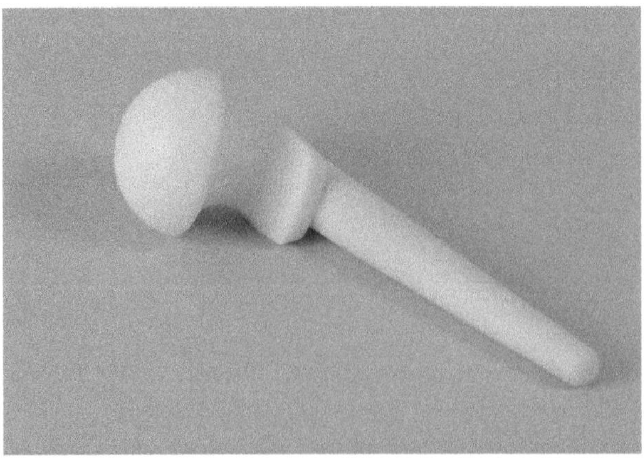

Fig. 10.10 Prefabricated hip spacer.
Courtesy Tecres, Italy.

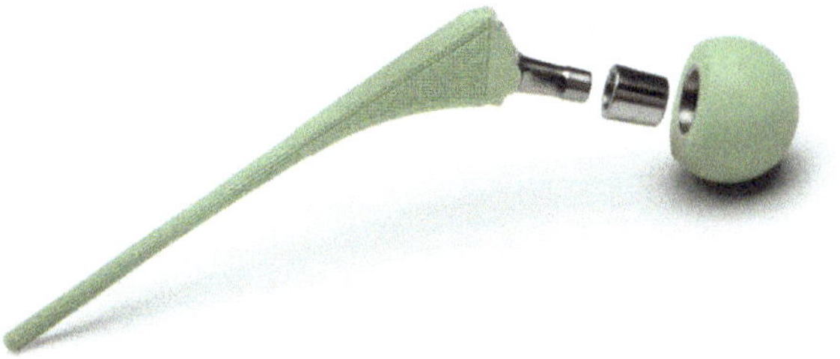

Fig. 10.11 Third-generation hip spacer. (Other than providing permission to use the Zimmer Biomet picture(s), this publication is not financially supported by Zimmer Biomet. Zimmer Biomet is the owner of the copyrights and all other intellectual property rights in relation to the picture(s) used.)

10.4.1 Static vs mobile spacer in the knee [20–23]

Knee spacers in particular are either static or mobile. Static spacers cause a temporary state of arthrodesis, by fusing femur and tibia. This might be advantageous when soft tissues need to settle down and heal properly or when the bony defects on either side of the joint are extensive. However, temporary fusion might be disadvantageous for the function of the extensor mechanism and the knee function after reimplantation, in general [21,24]. As a result, another option is an articulating spacer which preserves mobility of the knee, thereby possibly reducing the chance of arthrofibrosis (Fig. 10.12).

It is important to realize that spacers and beads have significantly different pharmacokinetic properties. Most important is the difference in surface area which is far greater for a comparable amount of cement in the form of beads vs a spacer. Since antibiotic release is a surface phenomenon, beads with the same total as the spacer will release much more of their antibiotic content. As a consequence, spacers cannot reach such high local antibiotic concentrations (Fig. 10.13) [25].

Surface area of spacers can be enlarged by "roughing up" the surface.

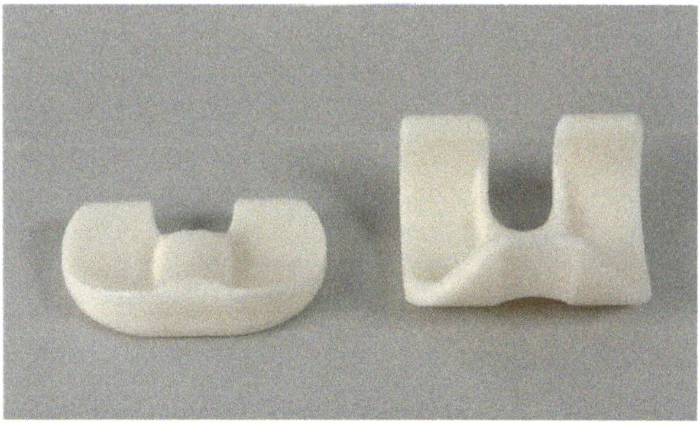

Fig. 10.12 Mobile knee spacer.
Courtesy Tecres, Italy.

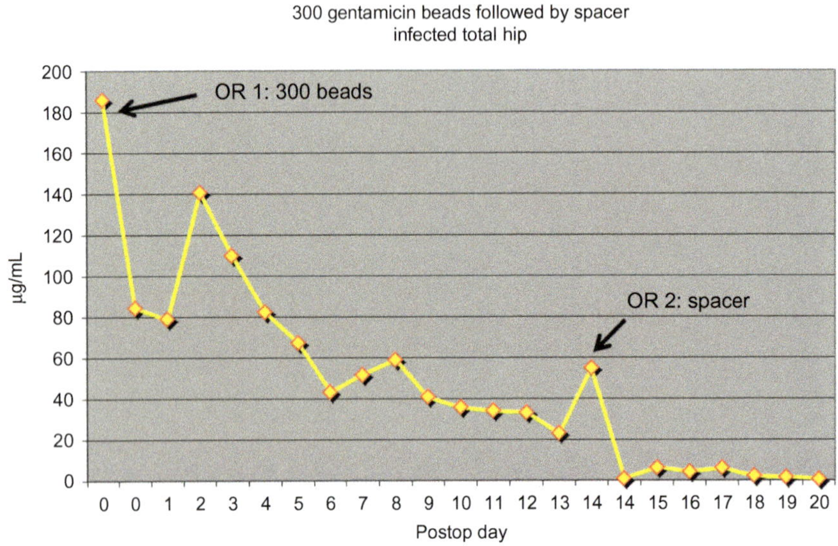

Fig. 10.13 Difference in amount of release of gentamycin, beads vs spacer in the same patient (Walenkamp, unpublished data).

Complications related to spacers are variable in numbers but can be quite substantial. Dislocation, fracture of the spacer, migration (intrapelvine protrusion, back of the knee, etc.), and periprosthetic fracture are mechanical issues, linked to the spacer itself. Certain authors have described complication rates of up to 57% [26,27].

There can also be systemic side-effects like allergic reactions against the antibiotics released from the spacer or acute renal failure [27]. Reports of hepatic failure or ototoxicity are very rare.

General complications may not be directly linked to the use of the spacer and include hematoma formation, wound dehiscence, arthrofibrosis, cardiac decompensation, persisting fistulae, pneumonia, etc. [28].

In the authors opinion, spacers can be considered in infection treatment to keep the space open for a future new implant, retain leg length and mobility of the joint, but not as a genuine antibiotic-delivery device. This is supported by numerous other authors [29].

Studies have shown that even spacers can become contaminated and are covered with biofilm when extracted. This is due to the fact that after a certain time antibiotic elution falls below the minimum inhibitory concentration needed to inhibit bacterial growth.

References

[1] R.A. Brand, M.A. Mont, M.M. Manring, Biographical sketch: Themistocles Gluck (1853–1942), Clin. Orthop. Relat. Res. 469 (6) (2011) 1525–1527.

[2] M. Arora, et al., Polymethylmethacrylate bone cements and additives: a review of the literature, World J. Orthop. 4 (2) (2013) 67–74.

[3] J. Charnley, Anchorage of the femoral head prosthesis to the shaft of the femur, Bone Joint J. 42-B (1) (1960) 28–30.

[4] J. Charnley, The classic: the bonding of prostheses to bone by cement, Clin. Orthop. Relat. Res. 468 (12) (2010) 3149–3159.

[5] D.A. Wininger, R.J. Fass, Antibiotic-impregnated cement and beads for orthopedic infections, Antimicrob. Agents Chemother. 40 (12) (1996) 2675–2679.

[6] H. Buchholz, H. Gartenmann, Infektionsprophylaxe und operative Behandlung des schleichende tiefen Infektion bei der Totalen Endoprothese, Chirurg 43 (1972) 446–453.

[7] K.D. Kuhn, Bone Cements: Up-to-Date Comparison of Physical and Chemical Properties of Commercial Materials, Springer-Verlag, Berlin, Heidelberg, 2000.

[8] D.J. Schurman, et al., Antibiotic-acrylic bone cement composites. Studies of gentamicin and Palacos, J. Bone Joint Surg. 60 (7) (1978) 978–984.

[9] H. van de Belt, et al., Surface roughness, porosity and wettability of gentamicin-loaded bone cements and their antibiotic release, Biomaterials 21 (19) (2000) 1981–1987.

[10] A.C. McLaren, et al., The effect of glycine filler on the elution rate of gentamicin from acrylic bone cement: a pilot study, Clin. Orthop. Relat. Res. 427 (2004) 25–27.

[11] H.N. Rasyid, et al., Concepts for increasing gentamicin release from handmade bone cement beads, Acta Orthop. 80 (5) (2009) 508–513.

[12] A.C. McLaren, et al., Particle size of fillers affects permeability of polymethylmethacrylate, Clin. Orthop. Relat. Res. 461 (2007) 64–67.

[13] K. Wu, et al., Enhancing drug release from antibiotic-loaded bone cement using porogens, J. Am. Acad. Orthop. Surg. 24 (3) (2016) 188–195.

[14] G.H.I.M. Walenkamp, Gentamicin PMMA beads. A clinical, pharmacokinetic and toxicological study. Thesis University Nijmegen, 1983, p. 164.

[15] G.H.I.M. Walenkamp, T.B. Vree, T.J.G. Van Rens, Gentamicin-PMMA beads: pharmacokinetic and nephrotoxicological study, Clin. Orthop. Relat. Res. 205 (1986) 171–183.

[16] K. Klemm, Gentamycin-PMMA-Kugeln in der Behandlung abszedier-ender Knochen- und Weichteilinfektionen, Zentralbl. Chir. 104 (14) (1979) 934–942.

[17] G.H.I.M. Walenkamp, Antibiotic loaded cement: from research to clinical evidence, in: E. Meani, et al. (Eds.), Infection and Local Treatment in Orthopedic Surgery, Springer, Berlin, Heidelberg, 2007, pp. 170–175.

[18] G.H.I.M. Walenkamp, Small PMMA beads improve gentamicin release, Acta Orthop. Scand. 60 (6) (1989) 668–669.

[19] Y.C. Hsu, et al., Antibiotic-loaded cement articulating spacer for 2-stage reimplantation in infected total knee arthroplasty: a simple and economic method, J. Arthroplast. 22 (7) (2007) 1060–1066.

[20] S.-J. Park, et al., Comparison of static and mobile antibiotic-impregnated cement spacers for the treatment of infected total knee arthroplasty, Int. Orthop. 34 (8) (2009) 1181–1186.

[21] R.J. Emerson, et al., Comparison of a static with a mobile spacer in total knee infection, Clin. Orthop. Relat. Res. 404 (2002) 132–138.

[22] E.-R. Chiang, et al., Comparison of articulating and static spacers regarding infection with resistant organisms in total knee arthroplasty, Acta Orthop. 82 (4) (2011) 460–464.

[23] C. Jacobs, C. Christensen, M. Berend, Static and mobile antibiotic-impregnated cement spacers for the management of prosthetic joint infection, J. Am. Acad. Orthop. Surg. 17 (6) (2009) 356–368.

[24] T.K. Fehring, et al., Articulating versus static spacers in revision total knee arthroplasty for sepsis, Clin. Orthop. Relat. Res. 380 (2000) 9–16.

[25] N. Greene, et al., In vitro elution of tobramycin and vancomycin polymethylmethacrylate beads and spacers from Simplex and Palacos, Am. J. Orthop. (Belle Mead, NJ) 27 (3) (1998) 201–205.

[26] B. Struelens, S. Claes, J. Bellemans, Spacer-related problems in two-stage revision knee arthroplasty, Acta Orthop. Belg. 79 (2013) 422–426.
[27] J. Jung, et al., Complications after spacer implantation in the treatment of hip joint infections, Int. J. Med. Sci. 6 (5) (2009) 265–273.
[28] H. Choi, H. Malchau, H. Bedair, Are prosthetic spacers safe to use in 2-stage treatment for infected total knee arthroplasty? J. Arthroplast. 27 (8) (2012) 1474–1479. e1.
[29] W.A. Jiranek, A.D. Hanssen, A.S. Greenwald, Antibiotic-loaded bone cement for infection prophylaxis in total joint replacement, J. Bone Joint Surg. 88 (11) (2006) 2487–2500.

Pathogen-directed antibiotic therapy

11

L.L. Reubsaet, M.B. Ekkelenkamp
University Medical Center Utrecht, Utrecht, The Netherlands

11.1 Introduction

Adequate antibiotic therapy is a major determinant of outcome in treatment of prosthetic joint infection (PJI). However, challenges in the antibiotic treatment of PJI are many: PJI requires agents with sufficient penetration in bone, treatments are prolonged (enhancing the chance of toxic side effects), active agents with sufficient oral resorption are few, and antibiotic resistance may severely limit the therapeutic options. An additional difficulty is the lack of randomized controlled trials comparing antibiotic regimens in PJI, most evidence being therefore derived from observational, retrospective cohort studies.

The optimal antibiotic therapy is determined by (1) the causative microorganism, and its antibiotic susceptibility, and (2) the chosen surgical strategy: will the prosthesis be retained, exchanged in a single-step procedure, exchanged in a two-step procedure, or is a Girdlestone procedure chosen. In case of definitive extraction (or exarticulation) and in two-stage procedures, the antibiotic therapy is practically identical to that of native joint arthritis/osteomyelitis. In case of one-stage revision and when retainment of the prosthesis is attempted with DAIR (debridement, antibiotics, irrigation, retention) the duration of treatment is longer, treatment may require combination therapy and, if available, antibiotics with antibiofilm activity are to be used.

This chapter will discuss the rationale for the choice of antibiotic therapy based on the clinical picture of infection, chosen surgical treatment, patient factors, and culture results.

11.2 Cultures and diagnosis

In the treatment of infectious diseases identification of the pathogen is essential. This applies especially to PJI, due to the required length of antibiotic treatment, the diversity in pathogens, and the sometimes difficult distinction from noninfectious diseases. PJI is seldom acutely life-threatening, leaving ample opportunity to safely collect diagnostic samples before initiating empirical antibiotic therapy. As administration of antibiotics prior to sample recovery will often lead to false-negative cultures, two major mistakes should be avoided:

(1) To start antibiotics (hours or more) prior to the surgical intervention and the recovery of adequate samples. This mistake is often derived from a belief that surgery under an "adequate umbrella of broad-spectrum coverage" will improve the outcome of the operation. In

Management of Periprosthetic Joint Infections (PJIs). http://dx.doi.org/10.1016/B978-0-08-100205-6.00011-2

practice it does not compensate for the therapeutic uncertainties encountered in the absence of a firm microbiological diagnosis.

(2) To hope that immediate initiation of antibiotic therapy in suspected PJI will "catch the infection at an early stage," when it may still be so limited that antibiotics will "suffice to cure." Just like pregnancy, PJI does not occur in intermediate forms: there either *is* an infection or there *isn't*.

The optimal moment to start peri-operative prophylaxis is currently not known. Most centers with a protocollized approach start infusion of antibiotics the moment the cultures have been drawn or as soon as the prosthesis has been removed/the loose parts have been exchanged. Data on whether infusion of antibiotics 30 min prior to the first incision will negatively impact culture results are inconclusive: some smaller studies suggest this does not have an impact on sensitivity [1–3], but one major retrospective study with 2740 cases of orthopedic infection did find a significant and relevant increase in false-negative culture results with an early start [4]. Although no data are available to support this, false-negativity may be more common for pathogens with very low minimal inhibitory concentrations (MICs), such as streptococci and pneumococci.

Culturing the causative microorganism of PJI, even determining whether the prosthesis is infected at all, can be complicated, in particular in late infections. Growth requirements and firm embedding in a biofilm may make pathogens fastidious to culture. Furthermore, some PJI-pathogens, such as coagulase-negative staphylococci (CoNS) and *Propionibacterium acnes*, are also low-pathogenic skin commensals and frequent contaminants.

In order to substantiate the microbiological findings in PJI, preferably 5 or 6 peri-operative tissue samples should be obtained for culture. Also any purulent fluid should be cultured, either on solid media or directly inoculated into blood culture bottles. In general, the diagnosis of infection requires two or three separate cultures yielding identical low-virulent bacteria/known contaminants. One or two positive cultures may suffice to diagnose infection with common pathogens such as *Staphylococcus aureus*, haemolytic streptococci, or Gram-negative rods. Time to positivity of cultures, a positive Gram stain, (semi)quantitative growth, and histopathological findings may be used as additional arguments to distinguish infection from contamination.

Sonication of removed implants with subsequent culture of the sonication fluid can aid identifying the causative microorganism in late infections, where bacteria grow on biofilm, in particular if the patient has recently received antibiotics [5–7]. In early PJI, the added value of sonication of removed materials over adequately taken tissue cultures remains doubtful.

11.3 Prophylaxis and empirical therapy

Peri-operative prophylaxis in orthopedic procedures should be directed at the most frequently encountered (local) causes of surgical site infections, and should always cover *S. aureus*. Short-term therapy, usually no longer than 24 h, with narrow-spectrum beta-lactam antibiotics such as first-generation cephalosporins (cefazolin, cefalotin) is in many hospitals the first choice. In hospitals with significant spread of MRSA (methicillin-resistant *S. aureus*), and for patients known to be colonized with MRSA, this pathogen may also need to be covered in empirical therapy. A high percentage of postoperative infections with Gram-negative bacteria (enterobacteriaceae, nonfermenters) is, or should be, unusual, and it should prompt careful revision of compliance with infection-prevention measures, rather than expansion of the spectrum of antibiotic prophylaxis.

Empirical therapy for suspected infection may be started once collection of adequate samples for culture has been ensured, and it should target either previously identified infectious agents or the most likely pathogens at the institution. *S. aureus* is the most common pathogen in early infections and this bacterium also causes the most severe clinical pictures; therefore, *S. aureus* is the main target of many prophylactic regimens. First-generation cephalosporins are often an adequate choice in the absence of MRSA; in MRSA-epidemic settings vancomycin is more often the drug of choice. The benefits of expanding the coverage towards other microorganisms should be weighed against the possible lower activity of such antibiotics against *S. aureus* and against the risks associated with the use of broad-spectrum antibiotics.

Gram stains may provide an early indication of the bacterial species, and offer guidance into a first adjustment of the therapy. Definitive therapy is based on culture and susceptibility results.

Protocols are divided on the ideal moment to start therapy with rifampicin or fluoroquinolones. Following one-stage revisions and DAIR-procedures, rifampicin (in combination with a second agent) and fluoroquinolones are the two antibiotics of choice for the long-term treatment of staphylococci and Gram-negative bacteria, respectively. Unfortunately, both classes of antibiotics may induce resistance in the targeted bacteria relatively easily, with a few or even a single DNA-mutation [8–10]. We therefore advise to wait with initiation of these drugs until (1) the bacterial load has been lowered by treatment with a beta-lactam antibiotic or vancomycin, (2) susceptibility results are known, and (3) the operation wound is closed and dry. Open and persistently leaking wounds will harbor high bacterial loads, and suboptimal antibiotic concentrations may be achieved at the wound edges; also these patients are at a higher risk to require a reoperation.

Based on recent studies failing to demonstrate any beneficial effects, antimicrobial prophylaxis prior to dental procedures is no longer advised for patients with orthopedic implants [11]. Literature on risks of PJI due to other invasive surgical procedures is limited. Studies evaluating antimicrobial prophylaxis around genitourinary procedures have failed to demonstrate a benefit of its use in patients with orthopedic implants [12].

11.4 The pathogens in PJI

The causing microorganisms determine to a large degree the therapeutic antibiotic options in PJI, and its outcome. Certain pathogens, such as fungi and enterococci, are associated with more comorbidity and worse preconditions, and (post or propter) also with lower cure rates [13,14]. Table 11.1 shows a selection of cohort studies that reported on outcome after DAIR-procedures according to the infecting pathogens. Large variations are observed between centers, probably reflecting differences in patient selections and

Table 11.1 Selection of cohort studies reporting outcome after DAIR-procedure (debridement, antibiotic therapy, irrigation, retention) for specific microorganisms

Study	Time and place	Microorganisms	*N*	Cure rate (%)[a]
Rodríguez-Pardo et al. [15]	2003–2010 16 centers, Spain	Gram-negatives (all) Subgroups: • *P. aeruginosa* • *ESBL-positive*	174 42 15	68 79 53
Hsieh et al. [16]	2000–2006 1 center, Taiwan	Gram-negatives	27	27
Lora-Tamayo et al. [17]	2003–2010 17 centers, Spain	*S. aureus* (MSSA and MRSA)	345	55
Tornero et al. [14]	1999–2012 18 centers, 3 countries	Enterococci (54% polymicrobial)	94	47
Byren et al. [18]	1998–2003 1 center UK	*S. aureus* CoNS	60 31	78 83
Kuiper et al. [13]	2004–2009 3 centers, The Netherlands	*S. aureus* CoNS	50 13	68 31

[a] Cure generally defined as: implant retention, free from infection, and no suppressive antibiotic therapy at 24 months follow-up.

in procedures followed. Resistance, however, certainly influences the chances of cure: rifampicin-resistant staphylococci and ciprofloxacin-resistant Gram-negative bacilli (GNB) show consistently lower cure rates than their susceptible counterparts.

PJI is often classified according to the duration of symptoms (<3 weeks; >3 weeks) or to the time elapsed since implantation ("early" <1 month; "delayed" 1 month to 2 years; "late" >2 years). Furthermore, peri-operative or surgical site infections may be distinguished from haematogenic infections. The clinical presentation also varies, whereby patients with peri-operatively acquired infections more often present with open wounds, sinus tract or abscess compared to haematogenous acquired PJI [19]. The microorganisms causing PJI vary amongst these infection types. Some microorganisms appear to be related to specific comorbidities; haemolytic streptococci may, for instance, be found more often in patients after radiation therapy of the affected anatomical site. The overall incidence of haematogenous acquired PJI in patients with remote infections is reported to be approximately 0.5% [20]. Bacterial seeding and PJI after *S. aureus* bacteraemia may be as high as 39% and is mostly due to community-acquired bacteraemia where diagnosis and treatment are delayed [21].

Overall the most commonly detected microorganisms in cultures of peri-prosthetic tissue or synovial fluid are CoNS and *S. aureus*, but the pathogens differ depending on the timing of infection. Late infections are mostly due to CoNS; *S. aureus* and GNB are more commonly found in early PJI. Streptococci, enterococci, and anaerobes are overall, less frequently encountered. *P. acnes* is particularly found as pathogen in PJI of the shoulder. Table 11.2 shows an aggregate of pathogens identified in cohort studies in PJI [5].

Table 11.2 Pathogens isolated in prosthetic joint infection

Infection	% of patients with prosthetic joint infection					
	Hip and knee					
	All time periods[a]	Early infection[b]	Hip[c]	Knee[c]	Shoulder[d]	Elbow[e]
Staphylococcus aureus	27	38	13	23	18	42
Coagulase-negative *Staphylococcus*	27	22	30	23	41	41
Streptococcus species	8	4	6	6	4	4
Enterococcus species	3	10	2	2	3	0
Aerobic Gram-negative bacilli	9	24	7	5	10	7
Anaerobic bacteria	4	3	9	5		
Propionibacterium acnes					24	1
Other anaerobes					3	0
Culture negative	14	10	7	11	15	5
Polymicrobial	15	31	14	12	16	3
Other	3					

[a] Data aggregated from 2435 joint infections.
[b] Data aggregated from 637 joints.
[c] Data from 1979 hip and 1427 knee PJIs.
[d] Data aggregated from 199 shoulders.
[e] Data aggregated from 110 prosthetic elbow joint infections.
Reproduced with permission from A.J. Tande, R. Patel, Prosthetic joint infection, Clin. Microbiol. Rev. 27 (2) (2014) 302–345.

11.5 The antibiotics

11.5.1 General principles

Only one small randomized trial has ever been performed comparing different antibiotic regimens in PJI. As a consequence, most recommendations are based on analysis of observational cohort studies, on expert opinion, on preclinical data, and on pharmacological considerations. Theoretically, the ideal antibiotic would achieve high concentrations in bone and in biofilms, have bactericidal action (i.e., kill bacteria instead of only inhibiting their growth) also against slowly multiplying bacteria in biofilm, and be tolerated during long treatment periods. Only two antibiotics/antibiotic classes are demonstrated to exert relevant activity in biofilm: rifampicin against staphylococci and fluoroquinolones against Gram-negative bacteria. The relevance of this effect and superiority of these two antibiotics in treatment of PJI is corroborated by clinical studies [15,17,22,23]. Note that rifampicin is commonly referred to as "rifampin," in particular in American literature (but it is one and the same drug).

Antibiotics used for oral treatment of PJI should have high and reliable bioavailability. Beta-lactam antibiotics such as the penicillins and cephalosporins are therefore, in principle, only used for intravenous therapy. Oral dosages of some antibiotics (such as ciprofloxacin and amoxicillin) are often higher than those used for other indications.

11.5.2 Resistance induction and combination therapy

Development of resistance of staphylococci to rifampicin is generally due to point mutations in the chromosomal DNA of the bacteria [8]. One point mutation may suffice, and resistance development under monotherapy with rifampicin is frequent, if not universal. A case–control study of PJI-patients with rifampicin-resistant staphylococci, found the prior use of the drug as monotherapy and inadequate dosage of the second antibiotic to be risk factors for development of this resistance [24]. Considering the clearly higher cure rates when treating with rifampicin, resistance development should be avoided at all cost. Therefore, in our opinion, rifampicin should not be started until: (1) determination and susceptibility results of the pathogens are available; (2) treatment with a second (proven) active intravenous antibiotic has been initiated; (3) the wound is closed and dry, and it is unlikely that a new surgical intervention will follow before the end of the antibiotic treatment.

Resistance to fluoroquinolones (ciprofloxacin, levofloxacin, moxifloxacin) can also be caused by a couple of point mutations [9,10]. It may therefore be prudent to postpone their use in the treatment of PJI until a significant bacterial load reduction has been achieved, for instance by first treating with an intravenous beta-lactam for a minimum of 2 weeks. Alternatively, the fluoroquinolone and a beta-lactam could be combined for the initial treatment.

11.5.3 Duration of treatment and tolerability of antibiotics

The duration of antibiotic therapy following DAIR and one-stage revisions is nowadays 12 weeks, sometimes extended to 6 months in case of PJI of a total knee arthroplasty [5]. These durations are also applied when a prosthetic joint is suspected of aseptic loosening and exchanged, but peri-operative cultures indicate it was actually infected; and when cultures taken at reimplantation in a two-stage procedure demonstrate persisting infection. A treatment period of 12 weeks appears to be highly effective [25]. Longer treatments seem to only postpone relapses, without improving cure [26]. Twelve weeks is actually a shortening of older habits to continue antibiotics up to a year and longer, but even shorter durations of therapy may be possible: in a recent retrospective study of Gram-negative PJI treated by DAIR, the median duration of antibiotic therapy was 70 days, with a 79% success rate if treated with ciprofloxacin (and a 68% overall success rate) [15]. Also, a retrospective study found a 90% cure rate (18/20 patients) in PJI-patients treated with DAIR followed by 6 weeks of antibiotic therapy [27].

Duration of antibiotic therapy in two-stage revisions and explantations is shorter: 4–6 weeks is generally recommended [5], corresponding with the usual duration of osteomyelitis treatment without prosthetic materials. However, high cure rates have been reported in two-stage procedures where patients received only 5 days of systemic antibiotic therapy in combination with an antibiotic-eluting spacer [28] or only peri-operative antibiotics in combination with antibiotic-loaded cement beads [29]. These results suggest that shortening the antibiotic courses in two-stage procedures may very well be completely safe.

Usually, even low-pathogenic bacteria such as CoNS, *P. acnes*, and enterococci are treated with antibiotics, although once the foreign body has been removed the immune system should in principle be able to clear residual infection. For permanent explantations to Girdlestone situations, it is likely justified to strongly limit the duration of treatment for infections caused by low-pathogenic bacteria.

11.5.4 Intravenous to oral switch

Initial treatment should always be intravenous in order to ensure adequate plasma concentrations. After an initial stage of iv-therapy, and once it is certain that no additional operative procedures will take place during the course of the antibiotic treatment, a switch to oral therapy may be made, provided that adequate oral drugs are available. No studies have adequately compared different durations of intravenous therapy, and the recommended minimum (or standard) duration of intravenous therapy differs between experts from 2 to 6 weeks, also depending on the pathogens encountered. The IDSA guidelines recommend using 4–6 weeks of iv-therapy if PJI is due to organisms other than staphylococci or when rifampicin cannot be used [30]. However, studies with short (i.e., 2 weeks) duration of iv-therapy have been reported to accomplish high percentages of cure for different pathogens, including Gram-negatives [17,31,32], and many institutions treat intravenously for approximately 2 weeks regardless of the species and rifampicin susceptibility (Fig. 11.1 and Table 11.3).

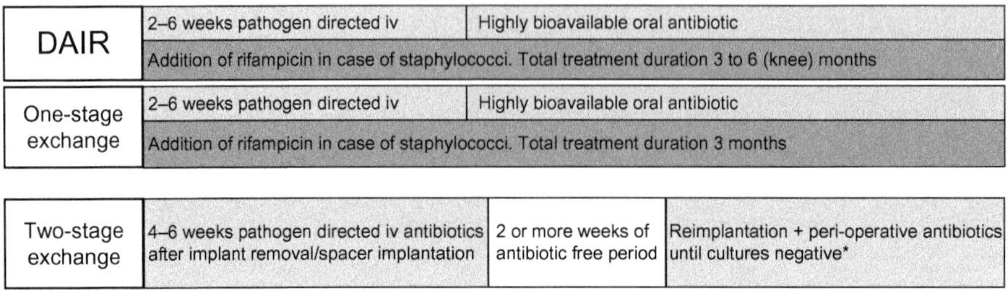

* If peri-operative cultures are positive, continue treatment as in one-stage exchange; adapted from Tande 2014

Fig. 11.1 Treatment algorithms for different surgical strategies.

11.5.5 Suppressive antibiotic therapy

The condition of the bone and the surrounding tissues, and the condition of the patient may make (successful) revision surgery unlikely, unattractive, or even impossible. Also, the patient may simply refuse to undergo another surgical procedure. In such cases, as an alternative to amputation or exarticulation, chronic, indefinite suppression therapy may be an option; long-term successful suppression has been reported between 23% and 83% in cohort studies [30,33,34]. Antibiotic choice should be guided by culture results of representative samples, obtained by either surgical debridement or joint aspiration, and should be tailored to the individual patient, taking into account identification and susceptibility of the pathogen, comorbidity, allergies and toxicity, and interactions of the different drugs. Consultation of a clinical microbiologist or infectious diseases specialist is strongly recommended.

An initial phase with high-dose intravenous therapy for a minimum of 2 weeks is advised, before a switch to suppressive oral therapy is made. Suppressive therapy should in principle be with a single antibiotic. Rifampicin is not recommended, due to its interactions, side effects, and rapid induction of resistance. Clinical evidence on outcome of antibiotic suppression and preferential regimens for suppressive therapy is limited [33,35]. For Gram-positive bacteria (staphylococci, streptococci, enterococci) most experience involves beta-lactam antibiotics, but these are also the antibiotics most often associated with failure of suppressive therapy [33]. Table 11.4 shows a selection of antibiotics used in antibiotic suppression of PJI.

11.5.6 Clinical monitoring of patients receiving long-term antibiotic therapy

The long antibiotic courses used in PJI predispose to toxic side effects of the antibacterial drugs. The most frequent side effects of parenteral therapy are rashes, gastro-intestinal side effects, and fever; however, renal, hepatic, hematologic, and allergic toxicities are also relatively common [36]. Toxicity is an issue in particular for linezolid (symptomatic thrombocytopenia, neuropathy) [37], metronidazole (neuropathy), and aminoglycosides (nephrotoxicity and ototoxicity): these drugs should not be used for long-term therapy if alternatives are available. Less, but still significant toxicity is found with the use of rifampicin, cotrimoxazole, and the peptidoglycans

Table 11.3 **Suggested antimicrobial therapy for treatment of PJI**

Species	Antibiotic	Dosage[a]	Comments
Staphylococci (*S. aureus and* CoNS)			
Methicillin-susceptible	Antistaphylococcal penicillin: Flucloxacillin, cloxacillin, dicloxacillin or nafcillin	12 g/24 h iv, or 6 dd 2 g	+ rifampicin 450 mg bid (oral), if proven susceptible once wound is closed and dry
Methicillin-resistant	Vancomycin	1 g iv tid, or 2 g iv/24 h after 1 g loading dose	+ rifampicin 450 mg bid (oral), if proven susceptible once wound is closed and dry. Monitor serum levels Vancomycin, aim for trough level of 15–20 mg/L or continuous level of 20–25 mg/L. Vancomycin is first alternative for methicillin-susceptible staphylococci
Staphylococci, alternatives initial phase	Daptomycin	8–10 mg/kg once daily	+ rifampicin 450 mg bid (oral), if proven susceptible once wound is closed and dry
	Linezolid	600 mg bid iv or p.o.	+ rifampicin 450 mg bid (oral), if proven susceptible once wound is closed and dry. Oral linezolid therapy only after initial iv-treatment. Do not use for >28 days due to possible toxicity
	Teicoplanin	400 mg iv once daily	+ rifampicin 450 mg bid (oral), if proven susceptible once wound is closed and dry
Staphylococci, first choice long-term treatment after initial therapy	Clindamycin or	600 mg p.o. tid	+ rifampicin 450 mg bid (oral), if proven susceptible
	Ciprofloxacin or Levofloxacin	750 mg p.o. bid 750 mg p.o. once daily	Only use ciprofloxacin and levofloxacin in combination with rifampicin 450 mg bid (if rifampicin-susceptible). Do not use as monotherapy

Continued

Table 11.3 Continued

Species	Antibiotic	Dosage[a]	Comments
Staphylococci, long-term treatment after initial therapy, alternatives	Cotrimoxazole	960 mg p.o. tid	+ rifampicin 450 mg bid (oral), if proven susceptible
	Fusidic acid	500 mg p.o. tid	Only use in combination with rifampicin 450 mg bid (if rifampicin-susceptible). Do not use as monotherapy
N.B.: Other alternatives are iv-therapy with antistaph penicillin, vancomycin, teicoplanin, or daptomycin	Doxycycline or	100 mg p.o. bid	+ rifampicin 450 mg bid (oral), if proven susceptible
	Minocycline	100 mg p.o. bid	
Enterococci (*E. faecalis* and *E. faecium*)	First: Amoxicillin	2 g iv. qid	*E. faecalis* usually amoxicillin-susceptible, *E. faecium* usually amoxicillin-resistant. Oral amoxicillin is not generally recommended because of low oral bioavailability. Piperacillin or piperacillin-tazobactam may be alternative in case of polymicrobial infections
	Second: Vancomycin	1 g iv tid, or 2 g iv/24 h after 1 g loading dose	First choice amoxicillin-resistant isolates. Monitor serum levels Vancomycin, aim for trough level of 15–20 mg/L or continuous level of 20–25 mg/L
Alternatives	Teicoplanin	400 mg iv once daily	
	Daptomycin	8–10 mg/kg once daily	
	Linezolid	600 mg bid iv or p.o.	Oral linezolid therapy only after initial iv-treatment. Do not use for >28 days due to possible toxicity
Streptococci	Penicillin	12×10^6 units iv/24 h	
Alternative	Ceftriaxone	2 g iv once daily	

Streptococci, long-term treatment after initial phase	Clindamycin	600 mg p.o. tid	
N.B.: alternatives are iv-therapy with penicillin or cephalosporin	Amoxicillin	1 g p.o. tid	Oral resorption of amoxicillin is limited
Gram-negative bacilli			
Enterobacteriaceae (*E. coli*, *Klebsiella* spp., *Proteus* spp.)	Ceftriaxone	2 g once daily	Alternative cephalosporins available, according to local practice
Enterobacter spp., *Citrobacter* spp.[b], *Morganella* spp., *Serratia* spp., ESBL-positive enterobacteriaceae	Meropenem or Ertapenem	1000 mg iv. tid 1000 mg iv. once daily	Carbapenems (meropenem, ertapenem) may be used as alternative for enterobacteriaceae. Meropenem (but not ertapenem) may be used as alternative for *P. aeruginosa*
Pseudomonas aeruginosa	Ceftazidime or Cefepime	2 g iv. tid 2 g iv. tid	
Gram-negative bacilli, long-term treatment after initial phase	First: Ciprofloxacin	750 mg p.o. bid or 400 mg iv. bid	Ciprofloxacin proven superior to other antibiotic regimens in Gram-negative PJI. Beware of resistance development
N.B.: alternatives are iv-therapy with third-generation cephalosporin or carbapenem	Cotrimoxazole	960 mg p.o. tid	*P. aeruginosa* intrinsically resistant to cotrimoxazole (ciprofloxacin only oral option)
Anaerobes			
Propionibacterium spp.	Penicillin 12E10^6/24 h	12 × 10^6 units iv/24 h	
	Vancomycin	1 g iv bid, or 2 g iv/24 h after 1 g loading dose	Monitor serum levels Vancomycin, aim for trough level of 15–20 mg/L or continuous level of 20–25 mg/L
	Clindamycin	600 mg iv or p.o. tid	Oral treatment only after initial intravenous therapy

Continued

Table 11.3 Continued

Species	Antibiotic	Dosage[a]	Comments
Other anaerobes	Metronidazole	500 mg iv or p.o. tid	Oral treatment only after initial intravenous therapy. Duration of metronidazole therapy not to exceed 4–6 weeks due to toxicity issues
	Penicillins		Most antibiotics and antibiotic combinations of the penicillin class are active against anaerobes, but susceptibility testing is required. Gram-negative anaerobes may produce beta-lactamase and be resistant
	Clindamycin	600 mg iv. or p.o. tid	Oral treatment only after initial intravenous therapy. Significant proportion of Gram-negative anaerobes resistant to clindamycin

CoNS, coagulase-negative staphylococci.

[a] Where appropriate adjust dosage to renal function.

[b] Excluding *Citrobacter koseri* (which is susceptible to third-generation cephalosporins).

Suggested treatments adapted from W. Zimmerli, A.F. Widmer, M. Blatter, R. Frei, P.E. Ochsner, Role of rifampin for treatment of orthopedic implant-related staphylococcal infections: a randomized controlled trial. Foreign-Body Infection (FBI) Study Group, JAMA 279 (19) (1998) 1537–1541; W. Zimmerli, Orthopaedic implant-associated infections: update of antimicrobial therapy, Orthopade 44 (12) (2015) 961–966; D. Rodríguez-Pardo, C. Pigrau, J. Lora-Tamayo, A. Soriano, M.D. del Toro, J. Cobo, J. Palomino, G. Euba, M. Riera, M. Sánchez-Somolinos, N. Benito, M. Fernández-Sampedro, L. Sorli, L. Guio, J.A. Iribarren, J.M. Baraia-Etxaburu, A. Ramos, A. Bahamonde, X. Flores-Sánchez, P.S. Corona, J. Ariza; REIPI Group for the Study of Prosthetic Infection. Gram-negative prosthetic joint infection: outcome of a debridement, antibiotics and implant retention approach. A large multicentre study, Clin. Microbiol. Infect. 20 (11) (2014) O911–O919; T.N. Peel, K.L. Buising, M.M. Dowsey, C.A. Aboltins, J.R. Daffy, P.A. Stanley, P.F. Choong. Outcome of debridement and retention in prosthetic joint infections by methicillin-resistant staphylococci, with special reference to rifampin and fusidic acid combination therapy, Antimicrob. Agents Chemother. 57 (1) (2013) 350–355; L. Morata, E. Tornero, J.C. Martínez-Pastor, S. García-Ramiro, J. Mensa, A. Soriano. Clinical experience with linezolid for the treatment of orthopaedic implant infections, J. Antimicrob. Chemother. 69 (Suppl. 1) (2014) i47–i52; L. Morata, E. Senneville, L. Bernard, S. Nguyen, R. Buzelé, J. Druon, E. Tornero, J. Mensa, A. Soriano. A retrospective review of the clinical experience of linezolid with or without rifampicin in prosthetic joint infections treated with debridement and implant retention, Infect. Dis. Ther. 2014. Available from: http://link.springer.com/article/10.1007%2Fs40121-0 14-0032-z; J. Cobo, J. Lora-Tamayo, G. Euba, A. Jover-Sáenz, J. Palomino, M.D. del Toro, D. Rodríguez-Pardo, M. Riera, J. Ariza; Red Española para la Investigación en Patología Infecciosa (REIPI). Linezolid in late-chronic prosthetic joint infection caused by gram-positive bacteria, Diagn. Microbiol. Infect. Dis. 76 (1) (2013) 93–98; D.R. Osmon, E.F. Berbari, A.R. Berendt, D. Lew, W. Zimmerli, J.M. Steckelberg, N. Rao, A. Hanssen, W.R. Wilson; Infectious Diseases Society of America. Diagnosis and management of prosthetic joint infection: clinical practice guidelines by the Infectious Diseases Society of America, Clin. Infect. Dis. 56 (1) (2013) e1–e25 and partly based on local practice.

Table 11.4 Reported and suggested antibiotic regimens for long-term suppression of prosthetic joint infections (PJIs)

	Antibiotic	Reported dosages	Suggested regimen	Caveats and recommendations
Staphylococci (MSSA, MRSA, CoNS) and streptococci	Amoxicillin	500–2000 mg tid	1000 mg tid	Staphylococci: only for strains that do not produce penicillinase. Limited oral uptake and tissue penetration. High dose may lead to gastro-intestinal tolerability issues
	Dicloxacillin/cloxacillin/flucloxacillin	500–1000 mg, qid/tid/bid	1000 mg tid	Low resorption, limited tissue penetration
	Cephalexin	500–1000 mg, qid/tid/bid	1000 mg tid	Higher resorption but lower activity than penicillins
	Clindamycin	300 mg bid, up to 600 mg tid	600 mg tid	Infamous for GI-side effects, including *C. difficile* infections
	Macrolides (erythromycin, azithromycin, clarithromycin)	Erythromycin 400 mg bid, azithromycin 500 mg qd	Azithromycin 500 mg qd	Preference for clindamycin over macrolides due to higher expected activity
	Cotrimoxazole	960 mg bid	960 mg bid	Possible hematological side effects, may raise serum creatinine
	Doxycycline/minocycline	100 mg qd or bid	100 mg qd	Oesophageal irritation and photosensitivity common side effects with doxycycline. Skin discolorations with minocycline
	Fluoroquinolones (ciprofloxacin, levofloxacin, moxifloxacin)	Diverse regimens	Moxifloxacin 400 mg qd	Ciprofloxacin not active against streptococci

Continued

Table 11.4 Continued

	Antibiotic	Reported dosages	Suggested regimen	Caveats and recommendations
Enterococci (amoxicillin-sensitive)	Amoxicillin	500–1000 mg tid	1000 mg tid	Limited oral uptake and tissue penetration. High dose may lead to gastro-intestinal tolerability issues. Use of other drugs reported in suppression of enterococci (doxycycline, ciprofloxacin), but unlikely that these would have true clinical effect
Propionibacterium acnes	Amoxicillin	–	1000 mg tid	Limited oral uptake and tissue penetration. High dose may lead to gastro-intestinal tolerability issues
	Clindamycin	–	600 mg tid	Infamous for GI-side effects, including *C. difficile* infections
	Doxycycline	100 mg qd or bid	100 mg qd	Oesophageal irritation and photosensitivity common side effects
Gram-negative bacilli (Enterobacteriaceae, *P. aeruginosa*)	Ciprofloxacin	500–750 mg bid	Ciprofloxacin 750 mg bid	Other fluoroquinolones are possible (levofloxacin, moxifloxacin), but ciprofloxacin is preferred
	Cotrimoxazole	Reports for Gram-positives	960 mg bid	Not for *P. aeruginosa* Possible hematological side effects, may raise serum creatinine
	Amoxicillin-clavulanic acid	–	625 mg tid	Not for *P. aeruginosa*

In all cases, susceptibility of the pathogens should be considered (for instance: beta-lactam antibiotics cannot be used for treatment of MRSA). Some experts advise to start initially with a high dose and reduce dosage for the long-term suppression.

(vancomycin and teicoplanin) [38–40]. But in the end, all antibiotics will cause toxic side effects in a percentage of the patients.

Current guidelines recommend clinical and laboratory monitoring of patients receiving (outpatient) intravenous antibiotic therapy [30], suggesting the laboratory parameters for monitoring during intravenous treatment with antibiotics used in PJI (Table 11.5).

Monitoring of toxicity in patients receiving long-term oral antibiotic therapy is also advised [30], but guidelines or recommendations have not yet been formulated. In practice, laboratory monitoring is far less intensive for oral treatment than for intravenous treatment; considering that drugs such as ciprofloxacin, clindamycin, cotrimoxazole, and linezolid obtain comparable blood and tissue levels by oral and by intravenous administration this policy, of course, lacks some consistency. Since hepatic toxicity, renal toxicity (interstitial nephritis) and, more rarely, hematological toxicity may occur with chronic antibiotic use, periodic laboratory evaluation of these parameters (for instance every 3 months in a stable phase) could be considered prudent.

11.6 Pathogen-directed choice

11.6.1 Staphylococcus aureus *(MSSA (methicillin-sensitive* S. aureus*) and MRSA)*

S. aureus is arguably the most virulent microorganism causing PJI. It is most often seen in early postimplant infection, although also a large percentage of late PJI are caused by *S. aureus*. First choice intravenous therapy for methicillin-susceptible strains is flucloxacillin, dicloxacillin, or cloxacillin. First choice intravenous therapy for methicillin-resistant isolates is vancomycin or teicoplanin; common alternatives are linezolid [41–44] or daptomycin [18]. Vancomycin may be administered as a bolus twice a day, or as continuous infusion. Continuous vancomycin infusion has been shown to be safe and effective [45], and, in particular for longer durations of continuous infusion treatment, it offers the advantage of easier monitoring of blood levels [46]. Continuous infusion through balloon pumps may be used to administer vancomycin in an outpatient setting. Vancomycin trough levels in PJI should be (for all pathogens) 15–20 mg/L, continuous blood levels should be maintained at 20–25 mg/L.

There is a controversy as to whether ceftriaxone is an adequate drug to treat infections by methicillin-susceptible staphylococci; ceftriaxone MICs are 2- to 20-fold higher than those for oxacillin [47], but achieved serum levels are also higher and in clinical settings the two drugs seem equivalent in the treatment of (implant-associated) osteomyelitis [48]. In case of polymicrobial infections or allergy to other beta-lactam antibiotics it may be considered an option.

Combination therapy with rifampicin leads to significantly higher cure in staphylococcal PJI treated by DAIR or one-stage revision [22,23,38], and rifampicin should therefore always be prescribed if the staphylococci are susceptible and the patient is able to tolerate the drug. However, as described in Section 11.5.2, precautions need to be taken to avoid induction of resistance and the antibiotic cannot be used as monotherapy. The added value of rifampicin is such, that even when the drug needs to be

Table 11.5 Suggested laboratory parameters to be monitored during (outpatient) intravenous antimicrobial therapy

Antibiotic	Complete blood count	Renal function tests	Potassium level	Liver enzyme levels	Other
Aminoglycosides (gentamicin, tobramycin, amikacin)	Weekly	2×/week	–	–	Monitor ototoxicity. Serum levels as clinically indicated
Penicillin, cephalosporins	Weekly	Weekly	–	–	
Carbapenems, (flu)cloxacillin, nafcillin	Weekly	Weekly	–	Weekly	
Piperacillin	Weekly	Weekly	Weekly	–	
Fluoroquinolones	–	–	–	Weekly	
Clindamycin	Weekly	Weekly	–	Weekly	
Daptomycin	Weekly	Weekly	–	Weekly	CPK weekly
Linezolid	Weekly	–	–	–	
Cotrimoxazole	Weekly	Weekly	Weekly	–	
Vancomycin	Weekly	Weekly	–	–	Serum levels as clinically indicated

CPK, creatinine phosphokinase.
Adapted from A.D. Tice, S.J. Rehm, J.R. Dalovisio, J.S. Bradley, L.P. Martinelli, D.R. Graham, R.B. Gainer, M.J. Kunkel, R.W. Yancey, D.N. Williams, Practice guidelines for outpatient parenteral antimicrobial therapy. IDSA guidelines, Clin. Infect. Dis. 38 (12) (2004) 1651–1672.

discontinued due to side effects, it is advised to attempt to restart the drug at a later time at a lower dose (300 mg bid, instead of 450 mg bid) if at all possible. Rifabutin may be an alternative in selected cases.

Oral options will always depend on susceptibility test results, but may include: clindamycin, minocycline, cotrimoxazole, fluoroquinolones, and fusidic acid [26,49]. Fluoroquinolones and fusidic acid should not be administered as monotherapy, but only as combination therapy with rifampicin. The use of oral linezolid for PJI has been studied in several trials; the drug is associated with high cure rates but also with high reported toxicity, in particular (symptomatic) thrombocytopenia and anemia. Its use is therefore limited to specific infections with short duration of treatment [41–44].

11.6.2 *Coagulase-negative Staphylococci (CoNS)*

S. epidermidis is the CoNS species most frequently associated with PJI, followed by *S. hominis*, *S. capitis*, and *S. lugdunensis* [50,51]. The majority of CoNS found in PJI are methicillin-resistant. *S. lugdunensis*, however, is mostly susceptible [52]. For methicillin-susceptible isolates (flu)(cl)oxacillin is the intravenous antibiotic of first choice. Generally vancomycin is the treatment of choice in case of methicillin resistance, alternatives being linezolid or daptomycin. The recommendations as to the oral options are identical to those for *S. aureus*, including the recommendations for rifampicin treatment.

11.6.3 *Enterococci:* Enterococcus faecalis *and* Enterococcus faecium

The main enterococcal species causing PJI is *E. faecalis*. *E. faecium* is found less commonly, and other species are truly rare in PJI. Enterococci may occur as monomicrobial PJI, which is usually the case in late PJI, or as part of a polymicrobial infection, as they are usually found in early onset PJI [14,53]. First choice therapy is either intravenous amoxicillin, ampicillin, or vancomycin. *E. faecalis* is generally susceptible to amoxicillin, most infecting strains of *E. faecium* are amoxicillin and ampicillin-resistant (resistance to amoxicillin, ampicillin, and also piperacillin is one and the same).

Due to the resistant nature of enterococci and the high failure-rate of procedures, including procedures in which prostheses are exchanged, combination antibiotic therapy is sometimes attempted. One retrospective study reviewing infections treated with and without addition of an aminoglycoside found only additional toxicity of the combination, without improvement in outcome. However, this study may have been biased towards patients with worse prognosis receiving combination therapy [54]. A noncomparative pilot study treating enterococcal infections with the combination of amoxicillin and ceftriaxone (a common treatment in enterococcal endocarditis) found a promising 90% cure in a very heterogeneous study population [55]. Future comparative studies will need to establish the value of (any) combination therapy.

Alternative intravenous antibiotics for enterococcal PJI include daptomycin and linezolid. Daptomycin is active in vitro against most enterococci, but literature supporting its use in enterococcal PJI is limited to a single case report [40]. Likewise, preclinical data suggests activity of linezolid against enterococci, but cases described in literature are scarce [41,44].

More complex even is oral treatment of enterococci. Amoxicillin and linezolid are the only oral antibiotics with established activity against enterococci. Amoxicillin may fail to reach adequate tissue levels with oral treatment, but is still preferable over exposure to the toxicity of long-term linezolid therapy. For the moment, optimal treatment of enterococcal PJI will probably remain through the intravenous route.

11.6.4 Streptococci

Most cases of streptococcal PJI are caused by haemolytic streptococci of groups A (*Strep. pyogenes*), B (*Strep. agalactiae*), and groups C and G (*Strep. dysgalactiae*). These may act particularly fastidious in case of PJI in a previously irradiated area. More rarely, viridans streptococci are recovered [56,57]. Infections with *Strep. galllolyticus* may be indicative of occult colorectal malignancy and should prompt consultation of a gastro-enterologist [58].

Generally streptococci are susceptible to penicillin and amoxicillin, which are the first choice antibiotics for intravenous treatment. Intravenous alternatives, in case of resistance or allergy are cephalosporins (cefazolin, ceftriaxone), vancomycin, or clindamycin.

First choice antibiotic for oral treatment is clindamycin. Alternatives are amoxicillin, which is very active but has relatively poor oral resorption and tissue penetration. Experience with other drugs potentially active against streptococci, such as tetracyclines, cotrimoxazole, and the new fluoroquinolones (levofloxacin and moxifloxacin), is too scarce to provide any recommendations. Ciprofloxacin is not considered active against streptococci. Although streptococci are usually susceptible to rifampicin, the value of combination therapy has not been established for these pathogens and rifampicin is therefore not currently advised following DAIR and one-stage procedures.

11.6.5 Gram-negative bacilli

GNB comprised 12% of all pathogens in PJI in a large retrospective cohort-study [15]; in this study 45% of such infections were polymicrobial with multiple GNB. Enterobacteriaceae, such as *E. coli*, *Klebsiella*, *Proteus*, and *Enterobacter* species are the most frequently encountered pathogens, followed by the nonfermentative Gram-negative rod *Pseudomonas aeruginosa* [15,27,59].

First choice treatment for enterobacteriaceae is an active beta-lactam antibiotic, such as a third-generation cephalosporin (ceftriaxone, cefotaxime), or, for more resistant isolates, a carbapenem. First-generation cephalosporins have a very limited spectrum of activity against enterobacteriaceae. Depending on local epidemiology, a second-generation cephalosporin may be used, although resistance is more common than against third-generation agents (first and second-generation cephalosporins do exhibit excellent activity against streptococci and methicillin-susceptible *S. aureus*, and are therefore good choices for peri-operative prophylaxis). Alternatively ciprofloxacin or cotrimoxazole could be used, but preferably these drugs are reserved for subsequent oral treatment. Therapy including ciprofloxacin is associated with higher cure rates in PJI caused by GNB [15,59]; if the pathogen is susceptible, a switch should therefore be made to (oral) ciprofloxacin after initial intravenous treatment. An oral alternative may be cotrimoxazole. There is currently no evidence to support combination therapy in the treatment of PJI caused by enterobacteriaceae.

Initial treatment of *P. aeruginosa* PJI is with an antipseudomonal beta-lactam, such as ceftazidime, piperacillin-tazobactam, or a carbapenem (meropenem, imipenem, doripenem, but not ertapenem). If susceptible, the treatment should be switched to oral ciprofloxacin. Ciprofloxacin is the only antibiotic associated with higher cure rates after a DAIR-procedure for *P. aeruginosa*. Furthermore, it is the fluoroquinolones (of which ciprofloxacin is the most active compound) that are the only oral antibiotics available for treatment of *P. aeruginosa*. Particular care should therefore be taken not to induce resistance to the drug. Experts are divided on whether combination therapy with aminoglycosides is indicated for *P. aeruginosa* PJI; its use is in any case not supported by evidence.

11.6.6 Anaerobic microorganisms

Propionibacterium acnes, a Gram-positive bacillus, is the most common anaerobic microorganism isolated in PJI, mainly after shoulder arthroplasty [60]. *P. acnes* is part of the normal skin flora and is also often found as a contaminant in cultures. Clinical presentation of PJI by this low-virulent microorganism is usually insidious and infections generally occur late after implantation. Culture of *P. acnes* may be difficult due to its slow replication, its requirement of anaerobic growth conditions, and its strong embedding in a biofilm. Culture of sonication fluid of explanted prosthetic material may aid in improving the sensitivity of diagnostics.

Recommended first line intravenous therapies for *P. acnes* PJI are penicillin, vancomycin, or clindamycin, with cephalosporins as alternatives (cefazolin, ceftriaxone) [30]. Clindamycin is the oral antibiotic of first choice, if the isolate is proven susceptible; oral amoxicillin may be an alternative. Rifampicin is active in vitro against *P. acnes* biofilms [61], but in an observational study of shoulder PJI with *P. acnes* combination therapy with rifampicin did not lead to better outcomes than monotherapy with an oral antibiotic (mostly clindamycin) [62]. Treatment with rifampicin is therefore not currently recommended for *P. acnes*.

Other anaerobes rarely cause PJI: *Peptostreptococcus*, *Finegoldia magna*, *Bacteroides* species, *Actinomyces*, *Clostridium perfringens*, and *C. difficile* have all been reported [63]. Anaerobes may occur as monomicrobial infections or as part of a mixed infection with intestinal flora [64]. Therapy needs to be tailored to the specific pathogen and its susceptibility. Antibiotics used include penicillins, metronidazole, and clindamycin.

11.6.7 Polymicrobial infections

Polymicrobial infections may have different aetiologies and origins. They may arise by infection with polymicrobial intestinal flora or skin flora, or they may combine different typical nosocomial strains. Superinfection of a PJI may also lead to a complex infecting flora. Overall clinical outcome in polymicrobial infections is worse compared to monomicrobial infections [65]. General treatment strategies cannot be provided for polymicrobial infections. All microorganisms should be evaluated for their contribution to the infection and it should be considered whether any of them are actually contaminants.

11.6.8 Fungi

PJI caused by fungi is extremely rare, with *Candida albicans* as the most frequently cultured microorganism. Retention of the prosthesis in case of fungal PJI is virtually impossible, and a two-stage exchange is advised. For *C. albicans* fluconazole is the treatment of choice. Antifungal treatment can be started based on culture findings and consultation with a clinical microbiologist or infectious disease specialist should always be sought [13,66].

11.6.9 Other microorganisms

A wide array of other rare microorganisms may occasionally cause PJI: *Corynebacterium* species, *Brucella* species, *Haemophilus* species, *Pantoea* species, *Campylobacter* species, *Mycobacterium* species, *Rothia* species, *Pasteurella* species, *Nocardia* species, *Listeria* species, *Weissella* species, etc. have all been reported as pathogens in this disease. Antibiotics should be selected based on available clinical data (if any), antibiotic properties, and susceptibility results.

11.7 Conclusion and areas for further research

Optimal antibiotic therapy is determined by a combination of the planned surgical treatment, and determination and susceptibility of the pathogen(s). Antibiotic treatment courses are long in PJI and pathogens may vary considerably, which makes a firm microbiological diagnosis essential. Currently two biofilm-active antibiotics have demonstrated their superiority: rifampicin in the treatment of staphylococcal PJI, and ciprofloxacin in the treatment of PJI by GNB. Apart from that, most recommendations are based on expert opinion and general microbiological and pharmacological arguments. Trials comparing different antibiotic policies in PJI are therefore very much needed. These may include some alternative antibiotics. Promising in vitro and in vivo results have for instance been found with the combination of fosfomycin and rifampicin [67]. Also, some new antibiotics against Gram-positive bacteria may be of interest for the treatment of PJI, such as telavancin—a long-acting glycopeptide antibiotic—and the fifth-generation cephalosporins: beta-lactam antibiotics active against MRSA and CoNS. Antibiotic therapy is prolonged in the treatment of PJI, but can probably be shortened. Hopefully, in the near future, more evidence will arise to establish the ideal duration of treatment. Finally, a valuable addition to the current protocols would be an evidence-based protocol for laboratory monitoring of long-term antibiotic therapy in PJI: providing adequate follow-up without overshooting into an exaggerated burdening of the patient.

References

[1] K. Bedenčič, M. Kavčič, N. Faganeli, R. Mihalič, B. Mavčič, J. Dolenc, Z. Bajc, R. Trebše, Does preoperative antimicrobial prophylaxis influence the diagnostic potential of periprosthetic tissues in hip or knee infections? Clin. Orthop. Relat. Res. (2015). Available from: http://link.springer.com/article/10.1007%2Fs11999-015-4486-4.

[2] R.S. Burnett, A. Aggarwal, S.A. Givens, J.T. McClure, P.M. Morgan, R.L. Barrack, Prophylactic antibiotics do not affect cultures in the treatment of an infected TKA: a prospective trial, Clin. Orthop. Relat. Res. 468 (1) (2010) 127–134.

[3] E. Ghanem, J. Parvizi, J. Clohisy, S. Burnett, P.F. Sharkey, R. Barrack, Perioperative antibiotics should not be withheld in proven cases of periprosthetic infection, Clin. Orthop. Relat. Res. 461 (2007) 44–47.

[4] M. Al-Mayahi, A. Cian, B.A. Lipsky, D. Suvà, C. Müller, C. Landelle, H.H. Miozzari, I. Uçkay, Administration of antibiotic agents before intraoperative sampling in orthopedic infections alters culture results, J. Infect. 71 (5) (2015) 518–525.

[5] A.J. Tande, R. Patel, Prosthetic joint infection, Clin. Microbiol. Rev. 27 (2) (2014) 302–345.

[6] A. Trampuz, K.E. Piper, M.J. Jacobson, A.D. Hanssen, K.K. Unni, D.R. Osmon, J.N. Mandrekar, F.R. Cockerill, J.M. Steckelberg, J.F. Greenleaf, R. Patel, Sonication of removed hip and knee prostheses for diagnosis of infection, N. Engl. J. Med. 357 (7) (2007) 654–663.

[7] Z. Zhai, H. Li, A. Qin, G. Liu, X. Liu, C. Wu, H. Li, Z. Zhu, X. Qu, K. Dai, Meta-analysis of sonication fluid samples from prosthetic components for diagnosis of infection after total joint arthroplasty, J. Clin. Microbiol. 52 (5) (2014) 1730–1736.

[8] H. Aubry-Damon, C.J. Soussy, P. Courvalin, Characterization of mutations in the rpoB gene that confer rifampin resistance in *Staphylococcus aureus*, Antimicrob. Agents Chemother. 42 (10) (1998) 2590–2594.

[9] G.A. Jacoby, Mechanisms of resistance to quinolones, Clin. Infect. Dis. 41 (Suppl. 2) (2005) S120–S126.

[10] H.J. Tang, C.C. Lai, P.R. Hsueh, C.C. Chen, K.Y. Wu, Y.C. Lin, C.C. Zhang, T.C. Weng, Y.H. Chiu, H.S. Toh, S.R. Chiang, W.L. Yu, W.C. Ko, Y.C. Chuang, RNA polymerase B subunit gene mutations in biofilm-embedded methicillin-resistant *Staphylococcus aureus* following rifampin treatment, J. Microbiol. Immunol. Infect. (2015). pii: S1684-1182(15)00808-7.

[11] T.P. Sollecito, E. Abt, P.B. Lockhart, E. Truelove, T.M. Paumier, S.L. Tracy, M. Tampi, E.D. Beltrán-Aguilar, J. Frantsve-Hawley, The use of prophylactic antibiotics prior to dental procedures in patients with prosthetic joints: evidence-based clinical practice guideline for dental practitioners—a report of the American Dental Association Council on Scientific Affairs, J. Am. Dent. Assoc. 146 (1) (2015). 11.e8–16.e8.

[12] D.J. Mazur, D.J. Fuchs, T.O. Abicht, T.D. Peabody, Update on antibiotic prophylaxis for genitourinary procedures in patients with artificial joint replacement and artificial heart valves, Urol. Clin. North Am. 42 (4) (2015) 441–447.

[13] J.W. Kuiper, M.P. van den Bekerom, J. van der Stappen, P.A. Nolte, S. Colen, 2-Stage revision recommended for treatment of fungal hip and knee prosthetic joint infections, Acta Orthop. 84 (6) (2013) 517–523.

[14] E. Tornero, E. Senneville, G. Euba, S. Petersdorf, D. Rodriguez-Pardo, B. Lakatos, M.C. Ferrari, M. Pilares, A. Bahamonde, R. Trebse, N. Benito, L. Sorli, M.D. del Toro, J.M. Baraiaetxaburu, A. Ramos, M. Riera, A. Jover-Sáenz, J. Palomino, J. Ariza, A. SorianoEuropean Society Group of Infections on Artificial Implants (ESGIAI), Characteristics of prosthetic joint infections due to *Enterococcus* sp. and predictors of failure: a multi-national study, Clin. Microbiol. Infect. 20 (11) (2014) 1219–1224.

[15] D. Rodríguez-Pardo, C. Pigrau, J. Lora-Tamayo, A. Soriano, M.D. del Toro, J. Cobo, J. Palomino, G. Euba, M. Riera, M. Sánchez-Somolinos, N. Benito, M. Fernández-Sampedro, L. Sorli, L. Guio, J.A. Iribarren, J.M. Baraia-Etxaburu, A. Ramos, A. Bahamonde, X. Flores-Sánchez, P.S. Corona, J. Ariza, REIPI Group for the Study of Prosthetic Infection, Gram-negative prosthetic joint infection: outcome of a debridement, antibiotics and implant retention approach. A large multicentre study, Clin. Microbiol. Infect. 20 (11) (2014) O911–O919.

[16] P.H. Hsieh, M.S. Lee, K.Y. Hsu, Y.H. Chang, H.N. Shih, S.W. Ueng, Gram-negative prosthetic joint infections: risk factors and outcome of treatment, Clin. Infect. Dis. 49 (7) (2009) 1036–1043.

[17] J. Lora-Tamayo, J. Parra-Ruiz, D. Rodríguez-Pardo, J. Barberán, A. Ribera, E. Tornero, C. Pigrau, J. Mensa, J. Ariza, A. Soriano, High doses of daptomycin (10 mg/kg/d) plus rifampin for the treatment of staphylococcal prosthetic joint infection managed with implant retention: a comparative study, Diagn. Microbiol. Infect. Dis. 80 (1) (2014) 66–71.

[18] I. Byren, S. Rege, E. Campanaro, S. Yankelev, D. Anastasiou, G. Kuropatkin, R. Evans, Randomized controlled trial of the safety and efficacy of daptomycin versus standard-of-care therapy for management of patients with osteomyelitis associated with prosthetic devices undergoing two-stage revision arthroplasty, Antimicrob. Agents Chemother. 56 (11) (2012) 5626–5632.

[19] P. Sendi, F. Banderet, P. Graber, W. Zimmerli, Periprosthetic joint infection following *Staphylococcus aureus* bacteremia, J. Infect. 63 (1) (2011) 17–22.

[20] I. Uçkay, A. Lübbeke, S. Emonet, L. Tovmirzaeva, R. Stern, T. Ferry, M. Assal, L. Bernard, D. Lew, P. Hoffmeyer, Low incidence of haematogenous seeding to total hip and knee prostheses in patients with remote infections, J. Infect. 59 (5) (2009) 337–345.

[21] P. Sendi, F. Banderet, P. Graber, W. Zimmerli, Clinical comparison between exogenous and haematogenous periprosthetic joint infections caused by *Staphylococcus aureus*, Clin. Microbiol. Infect. 17 (7) (2011) 1098–1100.

[22] A. Holmberg, V.G. Thórhallsdóttir, O. Robertsson, A. W-Dahl, A. Stefánsdóttir, 75% success rate after open debridement, exchange of tibial insert, and antibiotics in knee prosthetic joint infections, Acta Orthop. 86 (4) (2015) 457–462.

[23] W. Zimmerli, A.F. Widmer, M. Blatter, R. Frei, P.E. Ochsner, Role of rifampin for treatment of orthopedic implant-related staphylococcal infections: a randomized controlled trial. Foreign-Body Infection (FBI) Study Group, JAMA 279 (19) (1998) 1537–1541.

[24] Y. Achermann, K. Eigenmann, B. Ledergerber, L. Derksen, P. Rafeiner, M. Clauss, R. Nüesch, C. Zellweger, M. Vogt, W. Zimmerli, Factors associated with rifampin resistance in staphylococcal periprosthetic joint infections (PJI): a matched case–control study, Infection 41 (2) (2013) 431–437.

[25] T. Ilchmann, W. Zimmerli, P.E. Ochsner, B. Kessler, L. Zwicky, P. Graber, M. Clauss, One-stage revision of infected hip arthroplasty: outcome of 39 consecutive hips, Int. Orthop. (2015). Available from: http://link.springer.com/article/10.1007%2Fs00264-015-2833-4.

[26] T.N. Peel, K.L. Buising, M.M. Dowsey, C.A. Aboltins, J.R. Daffy, P.A. Stanley, P.F. Choong, Outcome of debridement and retention in prosthetic joint infections by methicillin-resistant staphylococci, with special reference to rifampin and fusidic acid combination therapy, Antimicrob. Agents Chemother. 57 (1) (2013) 350–355.

[27] L. Bernard, L. Legout, L. Zürcher-Pfund, R. Stern, P. Rohner, R. Peter, M. Assal, D. Lew, P. Hoffmeyer, I. Uçkay, Six weeks of antibiotic treatment is sufficient following surgery for septic arthroplasty, J. Infect. 61 (2) (2010) 125–132.

[28] P.B. McKenna, K. O'Shea, E.L. Masterson, Two-stage revision of infected hip arthroplasty using a shortened post-operative course of antibiotics, Arch. Orthop. Trauma Surg. 129 (4) (2009) 489–494.

[29] I. Stockley, B.J. Mockford, A. Hoad-Reddick, P. Norman, The use of two-stage exchange arthroplasty with depot antibiotics in the absence of long-term antibiotic therapy in infected total hip replacement, J. Bone Joint Surg. (Br.) 90 (2) (2008) 145–148.
[30] D.R. Osmon, E.F. Berbari, A.R. Berendt, D. Lew, W. Zimmerli, J.M. Steckelberg, N. Rao, A. Hanssen, W.R. Wilson, Infectious Diseases Society of America, Diagnosis and management of prosthetic joint infection: clinical practice guidelines by the Infectious Diseases Society of America, Clin. Infect. Dis. 56 (1) (2013) e1–e25.
[31] E.S. Darley, G.C. Bannister, A.W. Blom, A.P. Macgowan, S.K. Jacobson, W. Alfouzan, Role of early intravenous to oral antibiotic switch therapy in the management of prosthetic hip infection treated with one- or two-stage replacement, J. Antimicrob. Chemother. 66 (10) (2011) 2405–2408.
[32] N.G. Daver, S.A. Shelburne, R.L. Atmar, T.P. Giordano, C.E. Stager, C.A. Reitman, A.C. White Jr., Oral step-down therapy is comparable to intravenous therapy for *Staphylococcus aureus* osteomyelitis, J. Infect. 54 (6) (2007) 539–544.
[33] V. Prendki, V. Zeller, D. Passeron, N. Desplaces, P. Mamoudy, J. Stirnemann, S. Marmor, J.M. Ziza, Outcome of patients over 80 years of age on prolonged suppressive antibiotic therapy for at least 6 months for prosthetic joint infection, Int. J. Infect. Dis. 29 (2014) 184–189.
[34] J. Segreti, J.A. Nelson, G.M. Trenholme, Prolonged suppressive antibiotic therapy for infected orthopedic prostheses, Clin. Infect. Dis. 27 (4) (1998) 711–713.
[35] M.B. Siqueira, A. Saleh, A.K. Klika, C. O'Rourke, S. Schmitt, C.A. Higuera, W.K. Barsoum, Chronic suppression of periprosthetic joint infections with oral antibiotics increases infection-free survivorship, J. Bone Joint Surg. Am. 97 (15) (2015) 1220–1232.
[36] A.D. Tice, S.J. Rehm, J.R. Dalovisio, J.S. Bradley, L.P. Martinelli, D.R. Graham, R.B. Gainer, M.J. Kunkel, R.W. Yancey, D.N. Williams, Practice guidelines for outpatient parenteral antimicrobial therapy. IDSA guidelines, Clin. Infect. Dis. 38 (12) (2004) 1651–1672.
[37] J. Cobo, J. Lora-Tamayo, G. Euba, A. Jover-Sáenz, J. Palomino, M.D. del Toro, D. Rodríguez-Pardo, M. Riera, J. Ariza, Red Española para la Investigación en Patología Infecciosa (REIPI), Linezolid in late-chronic prosthetic joint infection caused by gram-positive bacteria, Diagn. Microbiol. Infect. Dis. 76 (1) (2013) 93–98.
[38] O.C. El Helou, E.F. Berbari, B.D. Lahr, J.E. Eckel-Passow, R.R. Razonable, I.G. Sia, A. Virk, R.C. Walker, J.M. Steckelberg, W.R. Wilson, A.D. Hanssen, D.R. Osmon, Efficacy and safety of rifampin containing regimen for staphylococcal prosthetic joint infections treated with debridement and retention, Eur. J. Clin. Microbiol. Infect. Dis. 29 (8) (2010) 961–967.
[39] C. Pulcini, T. Couadau, E. Bernard, A. Lorthat-Jacob, T. Bauer, E. Cua, V. Mondain, R.M. Chichmanian, P. Dellamonica, P.M. Roger, Adverse effects of parenteral antimicrobial therapy for chronic bone infections, Eur. J. Clin. Microbiol. Infect. Dis. 27 (12) (2008) 1227–1232.
[40] J.R. Yuste, M. Quesada, P. Díaz-Rada, J.L. Del Pozo, Daptomycin in the treatment of prosthetic joint infection by *Enterococcus faecalis*: safety and efficacy of high-dose and prolonged therapy, Int. J. Infect. Dis. 27 (2014) 65–66.
[41] M. Bassetti, F. Vitale, G. Melica, E. Righi, A. Di Biagio, L. Molfetta, F. Pipino, M. Cruciani, D. Bassetti, Linezolid in the treatment of Gram-positive prosthetic joint infections, J. Antimicrob. Chemother. 55 (3) (2005) 387–390.
[42] L. Morata, E. Tornero, J.C. Martínez-Pastor, S. García-Ramiro, J. Mensa, A. Soriano, Clinical experience with linezolid for the treatment of orthopaedic implant infections, J. Antimicrob. Chemother. 69 (Suppl. 1) (2014) i47–i52.
[43] L. Morata, E. Senneville, L. Bernard, S. Nguyen, R. Buzelé, J. Druon, E. Tornero, J. Mensa, A. Soriano, A retrospective review of the clinical experience of linezolid with or without rifampicin in prosthetic joint infections treated with debridement and implant retention,

Infect. Dis. Ther. (2014). Available from: http://link.springer.com/article/10.1007%2Fs40121-014-0032-z.

[44] N. Rao, C.W. Hamilton, Efficacy and safety of linezolid for Gram-positive orthopedic infections: a prospective case series, Diagn. Microbiol. Infect. Dis. 59 (2) (2007) 173–179.

[45] V. Dubée, V. Zeller, L. Lhotellier, M.D. Kitzis, J.M. Ziza, P. Mamoudy, N. Desplaces, Continuous high-dose vancomycin combination therapy for methicillin-resistant staphylococcal prosthetic hip infection: a prospective cohort study, Clin. Microbiol. Infect. 19 (2) (2013) E98–E105.

[46] M.F. Waineo, T.C. Kuhn, D.L. Brown, The pharmacokinetic/pharmacodynamic rationale for administering vancomycin via continuous infusion, J. Clin. Pharm. Ther. 40 (3) (2015) 259–265.

[47] K.E. Greenwood-Quaintance, P. Kohner, D.R. Osmon, A. Virk, R. Patel, Ceftriaxone susceptibility of oxacillin-susceptible *Staphylococcus aureus* from patients with prosthetic joint infection, Diagn. Microbiol. Infect. Dis. 82 (2) (2015) 177–178.

[48] B.W. Wieland, J.R. Marcantoni, K.M. Bommarito, D.K. Warren, J. Marschall, A retrospective comparison of ceftriaxone versus oxacillin for osteoarticular infections due to methicillin-susceptible *Staphylococcus aureus*, Clin. Infect. Dis. 54 (5) (2012) 585–590.

[49] C.A. Aboltins, M.A. Page, K.L. Buising, A.W. Jenney, J.R. Daffy, P.F. Choong, P.A. Stanley, Treatment of staphylococcal prosthetic joint infections with debridement, prosthesis retention and oral rifampicin and fusidic acid, Clin. Microbiol. Infect. 13 (6) (2007) 586–591.

[50] L.G. Harris, K. El-Bouri, S. Johnston, E. Rees, L. Frommelt, N. Siemssen, M. Christner, A.P. Davies, H. Rohde, D. Mack, Rapid identification of staphylococci from prosthetic joint infections using MALDI-TOF mass-spectrometry, Int. J. Artif. Organs 33 (9) (2010) 568–574.

[51] D. Teterycz, T. Ferry, D. Lew, R. Stern, M. Assal, P. Hoffmeyer, L. Bernard, I. Uçkay, Outcome of orthopedic implant infections due to different staphylococci, Int. J. Infect. Dis. 14 (10) (2010) e913–e918.

[52] D.J. Diekema, M.A. Pfaller, F.J. Schmitz, J. Smayevsky, J. Bell, R.N. Jones, M. BeachSENTRY Participants Group, Survey of infections due to *Staphylococcus* species: frequency of occurrence and antimicrobial susceptibility of isolates collected in the United States, Canada, Latin America, Europe, and the Western Pacific region for the SENTRY Antimicrobial Surveillance Program, 1997–1999, Clin. Infect. Dis. 32 (Suppl. 2) (2001) S114–S132.

[53] T.N. Peel, A.C. Cheng, K.L. Buising, P.F. Choong, Microbiological aetiology, epidemiology, and clinical profile of prosthetic joint infections: are current antibiotic prophylaxis guidelines effective? Antimicrob. Agents Chemother. 56 (5) (2012) 2386–2391.

[54] O.C. El Helou, E.F. Berbari, C.E. Marculescu, W.I. El Atrouni, R.R. Razonable, J.M. Steckelberg, A.D. Hanssen, D.R. Osmon, Outcome of enterococcal prosthetic joint infection: is combination systemic therapy superior to monotherapy? Clin. Infect. Dis. 47 (7) (2008) 903–909.

[55] G. Euba, J. Lora-Tamayo, O. Murillo, S. Pedrero, J. Cabo, R. Verdaguer, J. Ariza, Pilot study of ampicillin-ceftriaxone combination for treatment of orthopedic infections due to *Enterococcus faecalis*, Antimicrob. Agents Chemother. 53 (10) (2009) 4305–4310.

[56] R.J. Everts, S.T. Chambers, D.R. Murdoch, A.G. Rothwell, J. McKie, Successful antimicrobial therapy and implant retention for streptococcal infection of prosthetic joints, ANZ J. Surg. 74 (4) (2004) 210–214.

[57] A.M. Meehan, D.R. Osmon, M.C. Duffy, A.D. Hanssen, M.R. Keating, Outcome of penicillin-susceptible streptococcal prosthetic joint infection treated with debridement and retention of the prosthesis, Clin. Infect. Dis. 36 (7) (2003) 845–849.

[58] A. Srivastava, N. Walter, P. Atkinson, *Streptococcus bovis* infection of total hip arthroplasty in association with carcinoma of colon, J. Surg. Orthop. Adv. 19 (2) (2010) 125–128.

[59] J.C. Martínez-Pastor, E. Muñoz-Mahamud, F. Vilchez, S. García-Ramiro, G. Bori, J. Sierra, J.A. Martínez, L. Font, J. Mensa, A. Soriano, Outcome of acute prosthetic joint infections due to gram-negative bacilli treated with open debridement and retention of the prosthesis, Antimicrob. Agents Chemother. 53 (11) (2009) 4772–4777.

[60] K.E. Piper, M.J. Jacobson, R.H. Cofield, J.W. Sperling, J. Sanchez-Sotelo, D.R. Osmon, A. McDowell, S. Patrick, J.M. Steckelberg, J.N. Mandrekar, M. Fernandez Sampedro, R. Patel, Microbiologic diagnosis of prosthetic shoulder infection by use of implant sonication, J. Clin. Microbiol. 47 (6) (2009) 1878–1884.

[61] R. Bayston, R. Bayston, B. Nuradeen, W. Ashraf, B.J. Freeman, Antibiotics for the eradication of *Propionibacterium acnes* biofilms in surgical infection, J. Antimicrob. Chemother. 60 (6) (2007) 1298–1301.

[62] A.M. Jacobs, M.L. Van Hooff, J.F. Meis, F. Vos, J.H. Goosen, Treatment of prosthetic joint infections due to *Propionibacterium*, Acta Orthop. 28 (2015) 1–7.

[63] N.B. Shah, A.J. Tande, R. Patel, E.F. Berbari, Anaerobic prosthetic joint infection, Anaerobe 36 (2015) 1–8.

[64] G. Walter, M. Vernier, P.O. Pinelli, M. Million, M. Coulange, P. Seng, A. Stein, Bone and joint infections due to anaerobic bacteria: an analysis of 61 cases and review of the literature, Eur. J. Clin. Microbiol. Infect. Dis. 33 (8) (2014) 1355–1364.

[65] M.D. Wimmer, M.J. Friedrich, T.M. Randau, M.M. Ploeger, J. Schmolders, A.A. Strauss, G.T. Hischebeth, P.H. Pennekamp, P. Vavken, S. Gravius, Polymicrobial infections reduce the cure rate in prosthetic joint infections: outcome analysis with two-stage exchange and follow-up ≥ two years, Int. Orthop. (2015). Available from: http://link.springer.com/article/10.1007%2Fs00264-015-2871-y.

[66] K. Azzam, J. Parvizi, D. Jungkind, A. Hanssen, T. Fehring, B. Springer, K. Bozic, C. Della Valle, L. Pulido, R. Barrack, Microbiological, clinical, and surgical features of fungal prosthetic joint infections: a multi-institutional experience, J. Bone Joint Surg. Am. 91 (Suppl. 6) (2009) 142–149.

[67] R. Mihailescu, U. Furustrand Tafin, S. Corvec, A. Oliva, B. Betrisey, O. Borens, A. Trampuz, High activity of Fosfomycin and Rifampin against methicillin-resistant *staphylococcus aureus* biofilm in vitro and in an experimental foreign-body infection model, Antimicrob. Agents Chemother. 58 (5) (2014) 2547–2553.

Part Four

Case studies

Imaging of prosthetic joint infections

12

S.J. Verberne, O.P.P. Temmerman
NWZ, Centre for Orthopaedic Research (CORAL), Alkmaar, The Netherlands

12.1 Introduction

Imaging of joint prostheses is a specific competence of radiologist and nuclear medicine physicians. However, interpretations are performed in strong collaboration with the orthopedic surgeon who is aware of the type of implant, implantation technique, time after surgery, the use of antibiotics, and clinical status of the patient. The timely identification and localization of a prosthetic joint infection (PJI), including a differentiation from mechanical loosening, is essential to allow the initiation of appropriate medical and surgical therapy. It can make the difference between one-stage or two-stage revision. Unfortunately, assessment of suspected PJI has thus far not been standardized in an universally accepted diagnostic protocol, although efforts are made [1,2]. Despite clinical suspicion, physical examination, multiple diagnostic tests (i.e., laboratory tests, articular biopsy, cultures, and imaging), and complex algorithms, an accurate diagnosis remains challenging. A delay in diagnosing and treatment of a PJI can have a critical impact on loosening or maintaining the prosthesis and joint function.

In the diagnostic work-up, a variety of imaging techniques including radiographs, computed tomography (CT), magnetic resonance imaging (MRI), ultrasound, bone-, and leukocyte scintigraphy (LS) and fluorodeoxyglucose positron emission tomography can contribute in the evaluation of PJI. The use of imaging has been intensively investigated and each imaging modality and radiopharmaceutical is characterized by specific diagnostic accuracy, drawbacks, and added value in diagnosing PJI. The choice of the most accurate imaging technique and optimal radiopharmaceutical depends of its effectiveness in each phase of the diagnostic algorithm of PJI. The objective of this chapter is to illustrate and describe the most frequently used imaging modalities and radiopharmaceuticals in the assessment of PJI. This chapter will mainly focus on imaging of the most prevalent joint replacements, the knee and hip.

12.2 Conventional imaging

Medical imaging has advanced in remarkable ways since the discovery of plain radiography 120 years ago. In the current clinical setting, conventional imaging is capable of imaging the anatomy of the human body in intricate detail using various modalities. Conventional imaging, including radiographs, ultrasound, CT, and MRI, allows improved screening, diagnosis, and monitoring of pathology, mostly in the context of anatomical changes.

Management of Periprosthetic Joint Infections (PJIs). http://dx.doi.org/10.1016/B978-0-08-100205-6.00012-4

12.2.1 Plain radiographs

Radiography is the primary imaging method of prosthetic joints. The initial radiographs serve as baseline and reference films for comparison with all future studies. Initial, sequential radiography are the most valuable methods for detecting complications as dislocations (3%), periprosthetic fracture, proper positioning, and cement extrusion. In the follow-up, radiographs are used in order to detect possible complications including mechanical loosening, particle disease, component wear, dislocation, fracture, heterotopic ossification, and infection (Fig. 12.1).

However, most of the clinical studies confirmed that plain radiography is not reliable in diagnosing the exclusion of PJI [3–5]. Radiographic findings such as radiolucency, osteolysis, and migration could be present in both septic and aseptic loosening of prostheses. Although infrequently present, increased periosteal bone formation or an adjacent soft tissue collection are highly suggestive for periprosthetic infection. Radiographic findings in patients with low-grade periprosthetic infection are possibly unremarkable and could mimic mechanical loosening or small particle disease. In PJI with aggressive micro-organisms, progression could be rapid with bone destruction and sinus tract formation resulting in possible radiographic findings as ill-defined bone resorption and/or endosteal surface, nonconcentric lucency and absence of sclerotic margin. Although neither sensitive nor specific for diagnosing PJI, plain radiographs should be routinely performed in the assessment of suspected PJI [1,2].

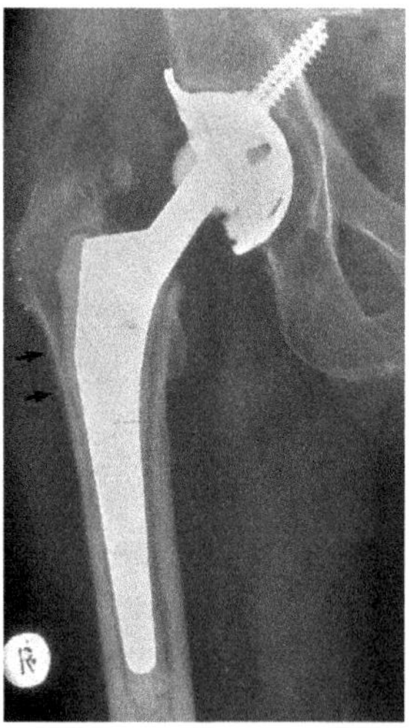

Fig. 12.1 Plain radiograph of a loose total hip prosthesis. This 77-year-old man had right hip replacement 11 months previously. He suffered from increasing pain at rest and on walking 7 months after implantation. Two weeks before the X-ray, bacteremia due to coagulase-negative staphylococci was diagnosed. Besides a loosening line around the stem and a tilting of the cup, a subperiostal new bone formation (➜) suggestive of infection is visible. Biopsies performed during revision grew coagulase-negative staphylococci.

12.2.2 Ultrasound

The value of ultrasound in assessing painful hip prostheses is limited to identifying the presence of hip joint effusion. In the early postoperative phase, sonography can confirm the presence of effusion when PJI is suspected. In low-grade PJI, ultrasound is capable of identifying both the presence and plot of the sinus tract. Clinical investigation reported sensitivity of 93% and specificity of 83% for sonographic detection of hip effusion [6]. The presence of intra-articular fluid, a straightforward ultrasound diagnosis, is nonspecific and more associated with sign of loosening process and its related inflammation [7]. However, ultrasound is an invaluable technique to guide aspiration of an effusion of collection [8]. In the diagnostic work-up of PJI, ultrasound is mostly useful in PJI in which effusion cannot be clinically diagnosed.

12.2.3 Computed tomography and magnetic resonance imaging

Computed tomography (CT) is widely known to be accurate in the depiction of soft tissue abnormalities, including infection. Although details of CT imaging in the assessment of PJI is degraded by metal artifacts, clinical investigations reported that it is possible to depict hip prosthesis [9–11]. PJI could be characterized on CT by aggressive osteolysis with an ill-defined margin, associated with osteomyelitis. However, typically PJI with more chronic indolent changes is possibly difficult to differentiate from other forms of osteolysis. Periostitis, indicated by periosteal new bone, on CT appears an highly sensitive (100%) and not very specific (16%) marker of infection [13]. In PJI, the bone may appear normal on CT in 25% of cases. Investigation of the potential use of CT in differentiating mechanic loosening from PJI reported that CT is accurate in the diagnosis of painful infection at the site of a hip prosthesis on the basis of soft tissue findings, whereas periprosthetic bone abnormalities are not useful [12]. Soft tissue abnormalities including joint distension, fluid filled bursae and fluid collections in muscles and perimuscular fat result in a 83% sensitivity, 96% specificity, and 94% accuracy in the evaluation of painful prostheses. However, there is no further published clinical validation of these results for other type of prostheses. MRI is proposed as better imaging modality than CT for demonstrating periprosthetic soft tissues, mainly due to greater resolution [14]. This technique is more accurate at depicting periprosthetic soft tissues around the femoral stem compared to the region of the acetabulum. The signal intensity of infection on MRI is similar to that of fluid with usually ill-defined margin. Moreover, MRI is capable of depicting bone involvement and assess the extent and route of sinus tracts (Fig. 12.2).

In the evaluation of PJI, CT and MRI are of limited clinical value in the presence of metallic prosthetic implants owing to beam hardening and dephasing artifacts. However, both techniques are useful in detecting soft tissue abnormalities as joint effusion, sinus tracts, soft tissue abscesses, bone erosion, and periprosthetic lucency, especially in the case of total hip arthroplasty.

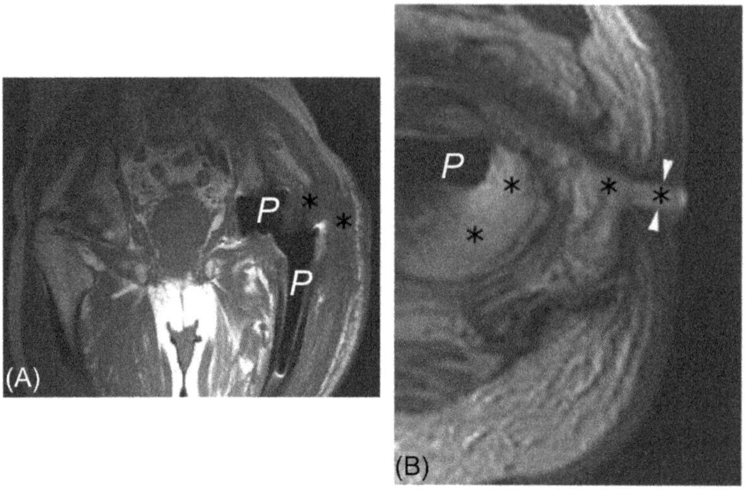

Fig. 12.2 A 70-year-old man who underwent unilateral total hip arthroplasty. (A) Coronal T1-weighted sequence. Hypointense fluid collection (*asterisks*) can be seen around the prosthesis (*P*). Involvement of the skin surface cannot be demonstrated. (B) Axial STIR sequence. The presence of a stula (*arrowheads*) can be easily demonstrated on this image.

12.3 Nuclear imaging

In nuclear medicine procedures, radionuclides are combined with pharmaceutical compounds, to form radiopharmaceuticals. When administered, these radiopharmaceuticals converge in specific organs and in cellular clusters with increased bioactivity. Nuclear techniques image the extent of a pathological process in the body, based on the cellular function and physiology, rather than relying on physical changes in tissue anatomy. Radionuclide imaging is not hampered by metallic artifacts in the evaluation of periprosthetic activity, in contrast to conventional imaging. In the diagnostic algorithm of PJI, the exact role of nuclear imaging remains a subject of debate, although intensively investigated and confirmed diagnostic added value. In the evaluation of painful joint replacements, nuclear imaging is most valuable in differentiating aseptic loosening from PJI. However, some partial analogies between the pathophysiological frameworks of these two conditions make nonspecific tracers of inflammation unreliable. There is no true consensus about the gold standard technique since each radionuclide modality has its drawbacks and limitations.

12.3.1 Bone scintigraphy

The first and most extensively investigated radionuclide procedure used for imaging joint arthroplasty is bone scintigraphy (BS). This technique is widely available at low costs, is easily performed, and rapidly completed (2–4 h after radiopharmaceutical infection). The technique is based on the skeletal uptake of phosphates and labeled phosphates with ^{99m}Tc, the "bone seeking" tracers. Accumulation of these tracers depends on the incorporation with the calcium moiety of the amorphous bone mineral matrix forming calcium-phosphate salts that later mature into hydroxyapatite crystals [15]. Incorporation of the radiopharmaceutical into the bone is related to blood flow and rate of new bone formation [16]. Any cause of accelerated bone turnover induce increased periprosthetic activity on images, including pathological conditions such as fracture, heterotopic ossification, mechanical loosening, and periprosthetic infection (Figs. 12.3 and 12.4). Also, the bone seeking tracers accumulate in areas of sterile inflammation, such as osteolysis induced by polyethylene wear debris.

Most of the clinical investigations reported high sensitivity and low specificity in the assessment of PJI (Table 12.1). The low specificity is caused by increased uptake of the radiopharmaceutical at sites of increased periprosthetic activity in varies conditions. Interpretation of images is performed as three-phase scan or only as delayed scan. The first clinical investigations on delayed scans reported that BS was 85% accurate for diagnosing prosthetic hip loosening with limitation of determining the cause of failure [32–34]. Subsequently, several authors investigated the uptake patterns and suggested that focal periprosthetic uptake indicated loosening and diffuse uptake indicated infection (sensitivity 100%, specificity 54%) [35], while others concluded reversely [37]. In order to improve specificity, BS was performed as triple-phase-BS. This implies evaluation of the flow or perfusion phase, immediately followed by the soft tissue or blood-pool phase and the third (bone) phase after 2 and 4 h. However, most authors reported low sensitivity, low specificity, or both [4,22,28,29]. Although no consensus is reached in the literature regarding interpretation criteria for BS, it is universally accepted that normal findings (no increased perfusion or blood pool, no periprosthetic uptake in the late phase) is considered strong evidence against the presence of PJI. Regardless of the interpretation criteria used, the diagnostic accuracy in the assessment of complicated hip or knee prostheses is approximately 50–70%. Recent metaanalysis showed a sensitivity of 81% and 75% with specificity of 78% and 55% for hip and knee prostheses, respectively [38,39]. The assessment of knee

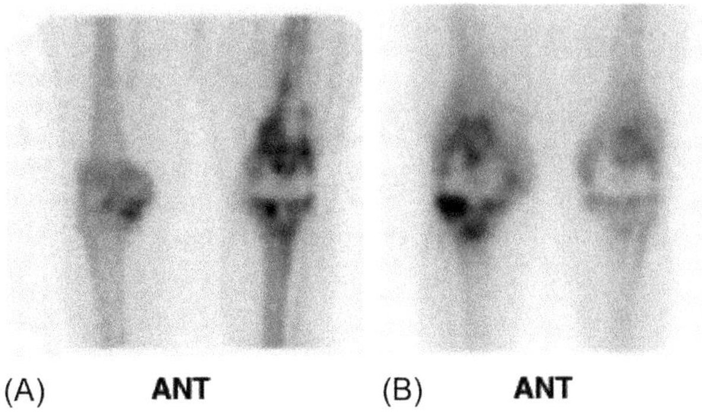

Fig. 12.3 (A) Aseptically loosened 6-year-old knee prostheses. Note the striking bony uptake around the left knee prosthesis, compared to the right. (B) One-year-old infected right knee prosthesis. Intraoperative cultures grew *S. hominis*. Periprosthetic uptake pattern is virtually indistinguishable from that in (A).

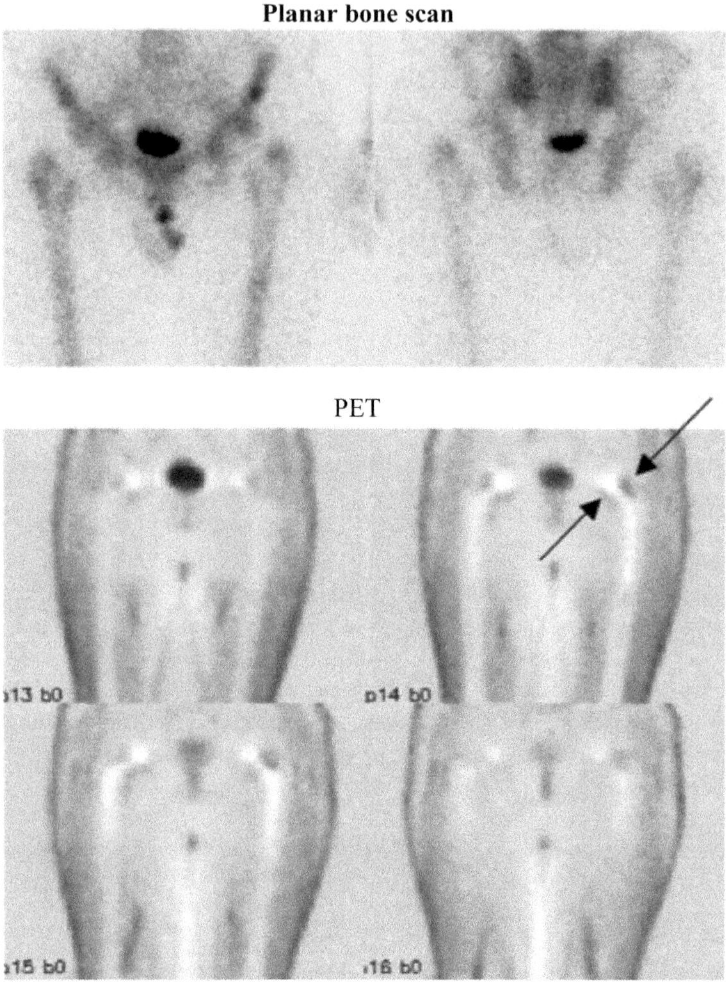

Fig. 12.4 Normal pattern: a clinically asymptomatic 52-year-old patient with a bilateral hip prosthesis from 1999. PET shows slightly increased ^{18}F-DG accumulation around the femoral neck.

prostheses with BS is more complex than that of hip prostheses and several studies confirmed low diagnostic accuracy in diagnosing infection [24].

An important limitation of using BS in PJI is the postoperative physiological bone remodeling after implantation which induces accumulation of the radiopharmaceutical with uptake patterns in asymptomatic prostheses up to 12–24 months after implantation. In case of hip prostheses, this partially depends on the time after implantation and the modality of attachment of prosthetic implant. The accumulation of the radiopharmaceutical in the first postoperative 12 months is independent of the presence of cement. After the first postoperative year, 90% of cemented hip prostheses depict normal uptake pattern while 10% present increased periprosthetic uptake. Porous-coated hip

Table 12.1 Diagnostic accuracy of bone scintigraphy in patients with PJI

Study	Prosthetic joint	Year	No.	Sensitivity (%)	Specificity (%)
Boubaker et al. [17]	Hip	1995	88	92	64
Nijhof et al. [18]	Hip	1997	60	96	15
Itasaka et al. [19]	Hip	2001	48	83	79
Larikka et al. [20]	Hip	2001	64	92	83
Stumpe et al. [4]	Hip	2004	34	56	88
Reinartz et al. [21]	Hip	2005	92	68	76
Mumme et al. [22]	Hip	2005	70	78	70
Nagoya et al. [23]	Hip	2008	46	88	90
Palestro et al. [24]	Knee	1991	23	67	76
Verlooy et al. [25]	Knee	1993	21	100	33
Sacchetti et al. [26]	Knee	1996	13	88	100
Smith et al. [27]	Knee	2001	75	92	76
Magnuson et al. [28]	Hip/Knee	1988	49	100	18
Levitsky et al. [29]	Hip/Knee	1991	72	30	86
Segura et al. [30]	Hip/Knee	2004	77	100	0
Love et al. [31]	Hip/Knee	2008	150	76	51

prostheses present more frequent increased periprosthetic uptake after the first year. Regarding total knee prostheses, physiological bone remodeling after implantation could present periprosthetic uptake up to 5 years, while 60% of the femoral and 90% of the tibial component present increased periprosthetic uptake in the second postoperative year [40]. Investigators proposed serial scans in order to determine increased periprosthetic uptake, however, significant interpatient variability was observed when periprosthetic activity decreased over time in the first 2 years after implantation [41].

The clinical usefulness of BS in the assessment of painful prosthetic joints is limited in differentiating PJI with other causes of prosthetic failure. However, the scan is a sensitive diagnostic tool for detecting periprosthetic bone changes. The clinical advantage is excluding any form of loosening, although of limited value in the 1–2 years after implantation. In the diagnostic algorithm of PJI, investigation with more specificity is often needed.

12.3.2 Gallium scintigraphy

In order to overcome important limitations of BS, several techniques have been investigated over the years. One of the earliest techniques to improve diagnostic accuracy was combined bone and gallium scintigraphy (GS). Gallium-67 has been used for localization of infection over the past decades [42]. After renal excretion (24h), approximately 75% of the injected doses is equally distributed among the liver, soft tissue, and skeletal and bone marrow [43]. Gallium-67 has affinity for polymorphonuclear leukocytes and uptake in area of infection is likely due to several factors including vascular membrane permeability at inflammatory sites and bacterial uptake of gallium. An important advantage is the usefulness in immunocompromised patients, despite the few or lack of circulating leukocytes [44]. Imaging is usually performed 2–3 days after injection.

Since both gallium and diphosponates are bone seeking pharmaceuticals, gallium-67 alone is not accurate enough to diagnose PJI. The combined images are therefore interpreted using standardized criteria comparing congruency of the spatial distribution and intensity [45]. The test is positive for infection when distribution of the two tracers is different or, in case of comparable distribution, the relative intensities of gallium uptake exceeds the BS uptake. The test is equivocal when the distribution of the two radiotracers is the same, both spatially and in intensity. The test is negative when the gallium images are normal, regardless of the bone scan findings or when the distribution of the two tracers is spatially congruent and the relative intensity of gallium uptake is less than that of the diphosphonate. The diagnostic accuracy of PJI in general ranges between 65% and 95% and provides more specificity in the assessment of PJI than BS alone (Table 12.2) [46,51]. Recent metaanalysis for periprosthetic hip infection reported a sensitivity of 59% (38–70%), and specificity of 97% (90–99%) [19,39,46,48]. Although combined interpretation of BS-GS provides more specificity in diagnosing PJI than BS alone, the clinical use is of limited value due to important drawbacks inherent of 67-Ga-citrate. This technique is hampered by high radiation dose and often equivocal cases [19]. Therefore, the use of this technique has declined over the years and often been replaced by other techniques.

12.3.3 Leukocyte scintigraphy

The introduction of in vitro radiolabeling of autologous peripheral leukocytes is one of the most important achievement in the radionuclide diagnoses of infection to date and dramatically improved the accuracy of diagnosing PJI [36]. An important difference with BS-GS is that labeled leukocytes do not accumulate at sites of increased

Table 12.2 Diagnostic accuracy of combined bone and gallium scintigraphy in patients with PJI

Study	Prosthetic joint	Year	No.	Sensitivity (%)	Specificity (%)
Tehranzadeh et al. [3]	Hip	1981	27	100	88
Gómez-Luzuriaga et al. [46]	Hip	1988	40	70	90
Aliabadi et al. [47]	Hip	1989	44	37	100
Kraemer et al. [48]	Hip	1993	43	38	98
Itasaka et al. [19]	Hip	2001	48	67	99
Reing et al. [49]	Hip/Knee	1979	79	95	100
McKillop et al. [50]	Hip/Knee	1984	15	83	78
Merkel et al. [45]	Hip/Knee	1986	154	66	81

bone turnover or remodeling in the absence of infection. Leukocytes are labeled with ^{99m}Tc-hexamethyl propyleneamine oxime (HMPAO) or ^{111}-In-oxine (111In) and migrate to sites of infection and inflammation. The accumulation of in vitro labeled leukocytes depends on several factors including chemotaxis, the quantity of leukocytes labeled, and the primary cellular response of a given inflammatory response [52]. Since circulating neutrophils are predominantly radiolabeled and neutrophils are present at the site of PJI, in theory LS should be accurate in diagnosing PJI and differentiation of inflamed aseptic loosening in which neutrophils are generally absent. No significant differences have been reported regarding the use of leukocytes labeled with ^{99m}Tc-HMPAO or 111In-oxine [24,53–57].

Unfortunately, several investigations reported unsatisfactory results and the value of this technique was a subject of controversy (Table 12.3). Primary clinical investigations reported a sensitivity of 100% and specificity of 89% for infected prosthetic joint replacements [67]. Subsequent studies concluded that this technique was specific and not sensitive for diagnosing PJI, while other authors reported reversely. For lower prosthetic joint replacements, McKillop et al. reported 50% sensitivity and 100% specificity, while other authors reported 100% sensitivity and 45–50% specificity [50,65,66]. Specified for infected hip prostheses, a recent metaanalysis reported pooled sensitivity of 88% (ranges between 81% and 97%) and pooled specificity of 92% (ranges between 57% and 97%). For prosthetic knee infection, Rand et al. reported a sensitivity of 85% and specificity of 83% [68]. In further clinical validations for knee prostheses, sensitivity ranges between 86% and 100% and specificity between 53% and 100% were determined [24,63]. The variability and controversial conclusions in the literature can be attributed to several factors. (1) As for most nuclear imaging techniques, investigated accuracy for diagnosing PJI varies between the different prostheses. This differentiation is often insufficiently investigated or reported. (2) The presence of ectopic bone marrow in the appendicular skeleton is mostly induced by implantation surgery (decrease specificity). (3) Various factors that influence diagnostic accuracy are often not clinically investigated including the assumed negative influence of administrated antibiotics (decrease sensitivity), the type of infecting organism, time after implantation/surgery (decrease specificity) and/or in combination with analyses of acute, delayed low-grade infection, and the presence of biofilm (decrease sensitivity) [20,63]. (4) There is no universally accepted protocol and interpretation criteria for analyzing leukocyte images regarding diagnosing PJI. Clinical studies use different criteria which could explain the heterogeneity of reported diagnostic accuracy.

In order to improve diagnostic accuracy of LS in diagnosing PJI, several attempts have been made including the introduction of delayed imaging and the combined interpretation of LS with other techniques. (1) LS can be performed as routine leukocyte imaging (2–4 h) and delayed imaging (24 h), as illustrated in Fig. 12.5. The importance between both protocols is the difference between the biological basis for recruitment of leukocytes in acute or subacute PJI and chronic, low-grade PJI. In case of early or delayed PJI, bacteria secrete chemotactic factors which continuously recruit leukocytes from peripheral blood to the periprosthetic infected area. In low-grade infections, neutrophils recruitment is less evident due to the biofilm that retards the invasion of labeled leukocytes. Subsequently more time is needed for leukocyte

Table 12.3 Diagnostic accuracy of leukocyte scintigraphy in patients with PJI

Study	Remark	Prosthetic joint	Year	No.	Sensitivity (%)	Specificity (%)
Gómez-Luzuriaga et al. [46]		Hip	1988	40	95	90
Palestro et al. [53]		Hip	1990	92	87	94
Pelosi et al. [58]		Hip	2004	55	95	97
El Espera et al. [56]	Combined with BMS	Hip	2004	45	67	97
Pill et al. [59]	Combined with BMS	Hip	2006	51	50	95
Simonsen et al. [60]		Hip	2007	76	81	94
Fuster et al. [57]		Hip	2008	70	83	57
Fuster et al. [57]	Combined with BMS	Hip	2008	70	92	98
Kim et al. [61]		Hip	2014	71	86	86
Basu et al. [62]	Combined with BMS	Hip	2014	59	39	96
Palestro et al. [24]		Knee	1991	41	89	75
Larikka et al. [63]	Combined with BMS	Knee	2001	30	100	82
El Espera et al. [56]	Combined with BMS	Knee	2004	28	71	95
Pelosi et al. [58]		Knee	2004	40	84	93
Basu et al. [62]	Combined with BMS	Knee	2014	29	33	89
McKillop et al. [50]		Hip/Knee	1984	15	50	100
Pring et al. [64]		Hip/Knee	1986	50	100	89
Wukich et al. [65]		Hip/Knee	1987	50	100	45
Magnuson et al. [28]		Hip/Knee	1988	98	88	73
Johnson et al. [66]		Hip/Knee	1988	28	100	50
El Espera et al. [56]	Combined with BMS	Hip/Knee	2004	73	69	97
Pelosi et al. [58]		Hip/Knee	2004	95	85	71
Love et al. [55]	Combined with BMS	Hip/Knee	2004	59	100	91

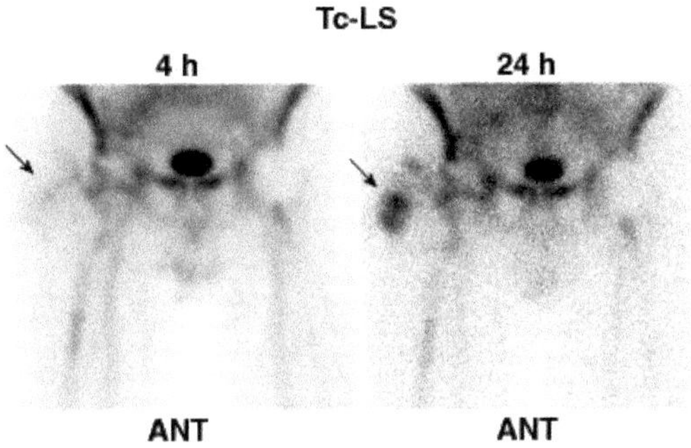

Fig. 12.5 Infected right hip replacement: 4-h *(left)* and 24-h *(right)* labeled leukocyte images. The abnormal labeled leukocyte activity *(arrows)* is seen clearly only on the later images [36].

accumulation [69]. Therefore, several authors reported that a delayed imaging at 24 h is more preferable in the detection of PJI. The authors concluded increased sensitivity (83% vs. 50% and 100% vs. 87%) and specificity (100% vs. 90% and 82% vs. 77%) for infected hip and knee prostheses respectively by comparing early, routine imaging and late images [20,60,63,70]. (2) Besides decreased sensitivity of LS in chronic PJI, LS is less specific in diagnosing PJI due to the accumulation of labeled leukocytes in the bone marrow. The distribution of leukocytes in the bone marrow is characterized by significant interindividual variability. The normal bio-distribution alters in case of localized or generalized marrow expansion, prosthetics implantation, and systematic diseases. During implantation of prostheses, the rasping of the femoral medullary canal and acetabular cup could cause abnormal distribution of normal bone marrow around the prosthesis. This complicates the interpretation of the leukocyte images and makes differentiation between alternation of distribution and infectious uptake difficult. In order to overcome this problem, LS is often combined with bone marrow scintigraphy (BMS). This technique is performed with the tracers ^{99m}Tc-sulfur colloid or ^{99m}Tc-nanocolloid. The principle of combined LS-BMS is the radiotracer accumulation in the reticuloendothelial system of the marrow, with similar physiological distribution of ^{99m}Tc-sulfur colloid and labeled leukocytes in the bone marrow. However, the accumulation of leukocytes in PJI is stimulated and simultaneously suppresses the uptake of sulfur colloid in the bone marrow, resulting in spatially incongruent images, as illustrated in Fig. 12.6 [60]. Consequently, discrimination of infection and implantation induced leukocyte accumulation in the bone marrow is possible. Fuster et al. confirmed improvement in sensitivity of 83% versus 92% and specificity of 57% versus 98% when LS was combined with BMS in the assessment of periprosthetic hip infection. Basu et al. reported lower sensitivity of 39% and 33% in diagnosing periprosthetic hip and knee infection, respectively. Despite some variation in reported diagnostic accuracy (Table 12.3), the combined interpretation of LS-BMS images is highly specific and is the gold standard for diagnosing PJI according to several authors [71]. (3) Another approach in order to improve specificity for diagnosing PJI is the combined interpretation of LS with BS. This did not increase the accuracy of LS for diagnosing knee arthroplasty. For hip prostheses, the added value of this combined technique remains unclear. Very few investigations reported sensitivity of 60–91%

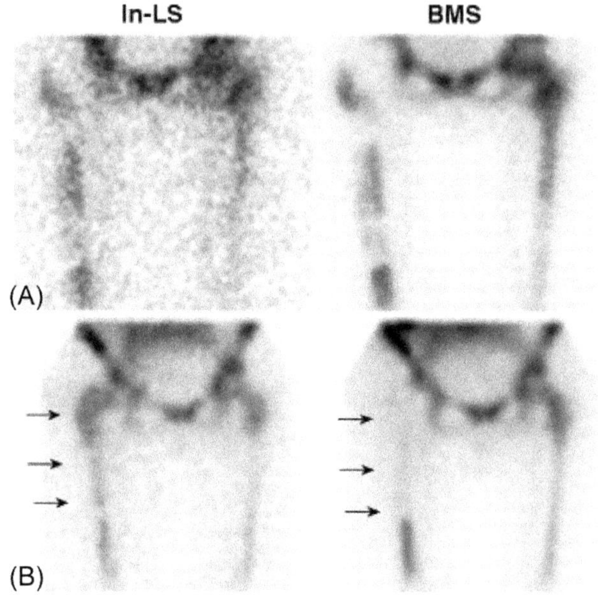

Fig. 12.6 (A) Aseptically loosened right hip replacement. The distribution of activity around the prosthesis on the labeled leukocyte *(left)* and sulfur colloid *(right)* images is spatially congruent, and the combined study is negative for infection. (B) Infected right hip replacement. There is activity around the femoral component on the labeled leukocyte study *(left)*. On the bone marrow image *(right)*, however, activity is present only distal to the tip of the prosthesis. The distribution of activity on the labeled leukocyte and sulfur colloid images is spatially incongruent *(arrows)*, and the combined study is positive for infection [36].

and high specificity of 93–98%. For lower extremity, the largest retrospective study reported lower sensitivity of 64% and moderate high specificity of 78% [72]. In the published investigations, it is proposed that the addition of BS does not improve diagnostic accuracy regarding PJI.

In daily practice, LS is hampered due to important drawbacks. (1) In vitro labeling requires laborious preparation performed in sterile conditions with complicated and time-consuming procedures and appropriate trained staff. An inherent risk is the possible complication of contamination due to in vitro labeling. (2) From patient's perspective, the 3-day imaging procedure is time intensive and inconvenienced. (3) This technique is not routinely available and not widely used in clinical practice. Moreover, there is poor availability of sulfur colloid in Europe. (4) Leukocyte scintigraphy is limited in the use of immunocompromised patients, in contrast to BS-GS. (5) Imaging quality is not ideal. (6) In case of combined imaging with BMS, there is need to perform another imaging protocol.

In conclusion, most recent investigations reported, with few exceptions, diagnostic accuracy that ranges between 86% and 98% regarding the use of LS in diagnosing PJI. LS, especially combined with BMS, improves specificity compared with BS and/or GS [36]. However, the exact role in the diagnostic algorithm of evaluating suspected infected prostheses remains subject of debate, mainly due to the lack of a universal accepted protocol and interpretation criteria for PJI and significant limitations in daily practice.

12.3.4 Antigranulocyte scintigraphy

Radiolabeled antigranulocyte antibodies and antibody fragments have been investigated and explored as alternatives for in vitro labeled leukocytes in order to overcome important limitations of LS. Considerable efforts have been devoted to develop in vivo methods of labeling leukocytes that resulted in antigranulocyte scintigraphy (AGS), performed with different tracers. Besilesomab (/Granuloscint/ScintimunR) is a murine monoclonal G1 immunoglobulin that binds to normal cross-reactive antigen-95 leukocytes (NCA-95), while Sulesomab (/LeukoScanR) is a fragment antigen binding (Fab') portion of a murine monoclonal G1 immunoglobulin that binds nonspecific cross-reactive antigen-90 (NCA-90) on leukocytes [58]. The protocol is based on injection of a prepared agent into the patient, tracing leukocytes in vitro. Both tracers are proposed as promising agents for imaging of inflammations and infectious processes in PJI (Fig. 12.8).

However, the results of clinical validation of AGS have been variable. Boubaker et al. investigated the use of Besilesomab in diagnosing periprosthetic hip infection and reported a sensitivity of 67% and specificity of 75% [17]. Using Sulesomab in diagnosing periprosthetic hip prostheses, they reported sensitivity ranges between 75% and 94% and specificity between 47% and 95% [73–76]. Regardless of which tracer was used, recent metaanalyses reported comparable results for infected hip prostheses (sensitivity 80% and 84%, specificity 72% and 75%) and knee prostheses (83% sensitivity and 83% specificity) [39,77].

In conclusion, in vivo labeling of leukocytes overcomes important limitations of the in vitro labeling procedures. However, based on published studies its accuracy in diagnosing prosthetic infection is less sensitive and less specific compared to that of leukocytes scintigraphy, especially in the assessment of hip prostheses. Despite the proposed promising results in diagnosing PJI, clinical investigations concluded that this technique is not the preferred modality at the moment and therefore has not gained wide acceptance.

12.3.5 Fluorodeoxyglucose positron emission tomography

Fluorodeoxyglucose positron emission tomography (FDG-PET) had been successfully used in the assessment of multiple malignant disorders. Accumulation, however, was also observed in regions of nonmalignant inflammatory and infectious conditions, including in the musculoskeletal system. Therefore, this technique was introduced as potential useful tracer in the assessment of periprosthetic infections. Low dose CT is often added in order to obtain detailed anatomical information.

^{18}F-Fluorodeoxyglucose (FDG) is transported into cells via glucose transporters and phosphorylated to ^{18}F-2-FDG-6 phosphate without being metabolized [78]. Normal distribution of FDG is observed in the brain, myocardium, and genitourinary tract. In areas of inflammatory and infectious processes, uptake of FDG depends on the detection of cells with increased glucose uptake and increased metabolic rate. The amount of expression of glucose transporters increases in activated leukocytes and subsequently results in increased affinity for FDG in the presence of cytokines and growth factor [79]. In order to minimize accumulation of FDG in normal tissue, patients must fast for several hours to reduce competition for glucose transporters and minimize physical activity. The procedure starts with administration of benzodiazepines, 30–60 min before injection of FDG, decreasing muscle and brown adipose tissue activity and is completed within 2–3 h after injection of the tracer FDG.

Compared to conventional radionuclide imaging techniques, FDG-PET has a number of important advantages. (1) The PET technique provides high quality images with improved (high) spatial resolution. (2) Patient convenience; in contrast to LS and or BMS, the imaging procedures is completed within 2–3 h after one injection of FDG, without risk of contaminated blood products and laborious preparations. (3) Low conventional dose CT provides excellent anatomical correlation of areas with increased FDG uptake that improves the accuracy compared to PET alone. (4) In contrast to combined LS-BMS, treatment with antibiotics is assumed not to affect the sensitivity of FDG-PET in delineating sites of infections because FDG does not depend on leukocyte migration [80].

Unfortunately, leukocytes are present in both inflammation and infection and it is assumed that differentiation of accumulation resulting from the inflammatory reaction of mechanical loosening of prostheses and periprosthetic infections is subsequently difficult. However, clinical investigations proposed that site and pattern of FDG uptake is more important than intensity of uptake in the region on interest [81,82]. Standardized uptake values based analysis is assumed as not reliable in differentiating between aseptic loosening and PJI [21,83]. Uptake in bone-prosthesis interface of the middle portion of the hip prosthesis is considered specific, while only increased uptake at the tip is assumed nonspecific for periprosthetic hip infection (Fig. 12.7) [62,81,82]. Similar to conventional tracers, the postoperative remodeling conditions after implantation results in periprosthetic FDG uptake that remains for months to years. Several investigators concluded that this is a nonspecific uptake pattern around the head and neck of asymptomatic hip prosthesis [82,84,85]. Persistently increased nonspecific FDG uptake following knee arthroplasty has also been reported, however, not clinically validated.

Studies that investigated accuracy of FDG-PET in the assessment of PJI reported variable results (Table 12.4). Primary investigation on accuracy using FDG-PET for PJI reported sensitivity of 90%, 91% and specificity of 89%, 72% for hip and knee prostheses, respectively [82]. In subsequently published investigations, sensitivity ranged between 33% and 100% and specificity between 62% and 100%. Recent meta-analyses concluded pooled sensitivity of 86% and 83% and specificity of 93% and 90% for periprosthetic hip infection [39,92]. The most recent, largest, and prospective study reported 82%, 95% sensitivity and 93%, 88% specificity for hip and knee prostheses, respectively [62]. In the assessment of infected knee prostheses, Van Acker et al. reported 100% sensitivity and 73% specificity [90]. Metaanalysis concluded significantly higher specificity for hip prostheses (90%) compared to knee prostheses (75%) [92]. The lower specificity of FDG-PET in knee prostheses may be related to

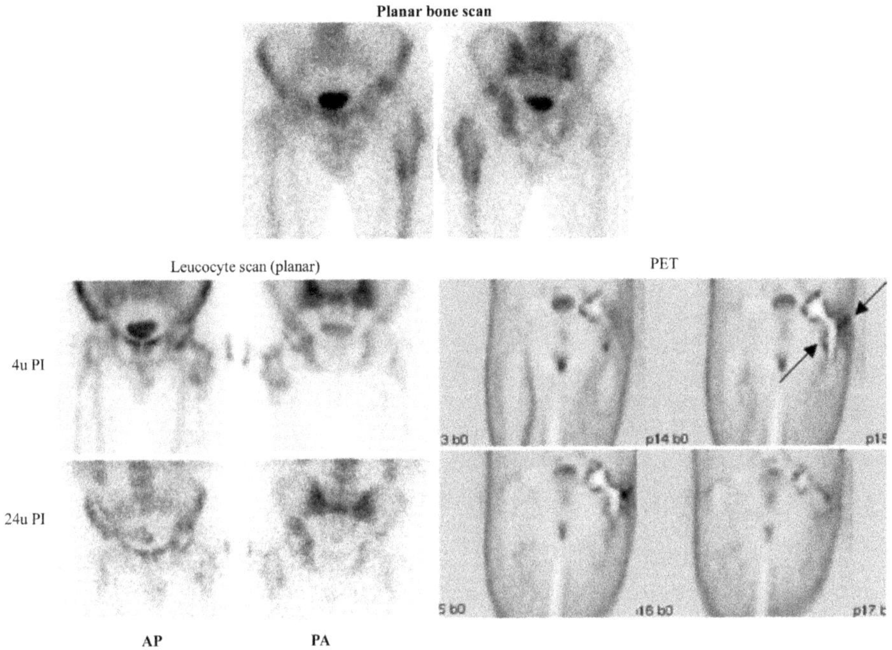

Fig. 12.7 Example of infection. A 77-year-old person who underwent a THP on the left side in 1992 with revision in 1998 and 1999. Pathological tracer accumulation is seen in the trochanteric region on both leukocyte scintigraphy and the the PET scan.

Table 12.4 Diagnostic accuracy of fluorodeoxyglucose positron emission tomography in patients with PJI

Study	Prosthetic Joint	Year	No.	Sensitivity (%)	Specificity (%)
Zhuang et al. [82]	Hip	2001	38	90	89
Chacko et al. [81]	Hip	2003	41	92	97
Vanquickenborne et al. [86]	Hip	2003	17	88	78
Love et al. [55]	Hip	2004	40	43	96
Stumpe et al. [4]	Hip	2004	35	33	81
Mumme et al. [22]	Hip	2005	70	91	92
Reinartz et al. [21]	Hip	2005	92	94	95
Pill et al. [59]	Hip	2006	92	95	93
Garcia-Barrecheguren et al. [87]	Hip	2007	24	64	62
Chryssikos et al. [88]	Hip	2008	127	85	93
Mayer-Wagner et al. [89]	Hip	2010	15	75	71
Basu et al. [62]	Hip	2014	134	82	93
Zhuang et al. [82]	Knee	2001	36	91	72
Van Acker et al. [90]	Knee	2001	21	100	73
Manthey et al. [91]	Knee	2002	14	100	100
Basu et al. [62]	Knee	2014	87	95	88
Love et al. [55]	Hip/Knee	2004	59	36	97

the relatively limited knowledge about the incidence and pattern of nonspecific FDG uptake. Most investigations reported that FDG-PET is a promising imaging tool in the assessment of PJI. In contrast, however, some authors reported that FDG-PET is not reliable in differentiating between mechanical loosening and PJI [4,87,93]. The inconclusive published clinical results on accuracy in the assessment of periprosthetic infections is mainly due to the lack of uniform interpretation criteria used in the clinical investigations of diagnosing PJI. Different FDG uptake patterns and approaches (quantitative versus qualitative approach by means of standardized uptake values) where used as criterion for PJI in the published studies, with subsequent heterogeneity in the results of accuracy. However, in the most recent studies, accumulation of FDG extended along the bone-prosthesis-interface, not limited to only the femoral neck, headzone or periprosthetic soft tissue is the proposed most accurate criterion for periprosthetic hip infection [59,62,88,89]. Further clinical validation, however, is needed in order to confirm the diagnostic accuracy of FDG-PET in the assessment of PJI.

Overall, FDG-PET is considered as potential, accurate imaging technique in the assessment of PJI. This technique enables visualization of hyperglycolytic inflammatory cells during PJI with excellent anatomical correlation and assumed as comparable or even higher diagnostic accuracy than LS. Moreover, this technique has important advantages compared to conventional radionuclide imaging. However, this modality will only gain full clinical acceptance in the assessment of PJI when specific uptake pattern can be developed and its clinical added value is confirmed.

12.4 Imaging in the assessment of PJI

12.4.1 Hybrid imaging and new infection-specific agents

In recent years, the use of nuclear medicine to characterize and diagnose infectious and inflammatory diseases is rapidly increasing. There are some important developments in experimental tracers and imaging techniques which could reflect improvement in current procedures of imaging of PJI. After the introduction of the hybrid imaging modality of PET/CT, preliminary data suggest that single-photon emission computed tomography (SPECT)/electronic computer X-ray tomography technique (CT) potentially contributes to the evaluation of failed joint arthroplasty. In SPECT/CT, which enables three-dimensional information, the CT component provides anatomical details necessary for accumulated radiopharmaceutical localization. Initial clinical investigation of planar, antigranulocyte, scintigraphy compared to fused SPECT/CT images in low-grade PJI (mainly infected knee prostheses) reported improved sensitivity (66% vs. 89%) and specificity (60% vs. 73%) [94]. SPECT/CT made a contribution to the final diagnosis in 23% of the patients, compared to 13% for SPECT alone. Several authors concluded that additional SPECT/CT is capable of accurate anatomical as well as precise information of the extent of the infection, which improves interobserver reliability (Fig. 12.8) [95]. The results regarding hybrid imaging modalities, including PET/CT and SPECT/CT, whether or not combined with new infection agents, is a promising future perspective in diagnosing PJI.

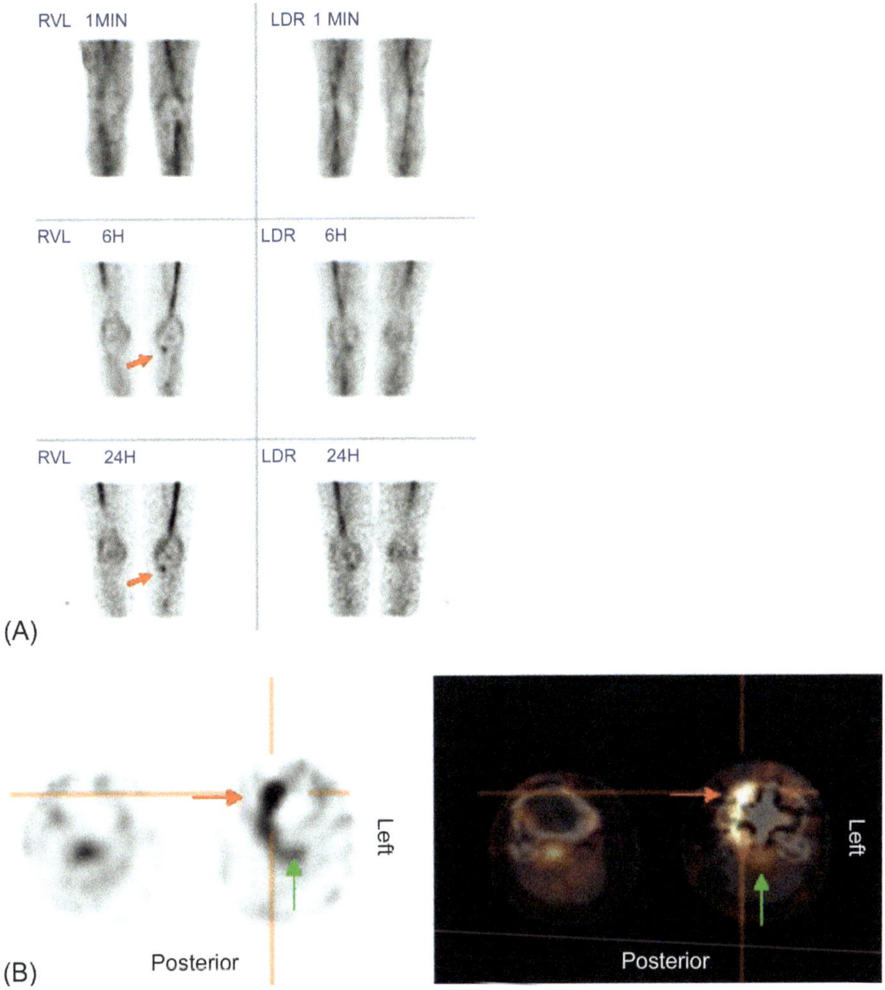

Fig. 12.8 Planar (A) and SPECT, SPECT/CT images (B) from a 61-year-old woman with total knee prosthesis on the left side, presented in ventral and dorsal orientation at 1 min, 6 h, and 24 h for planar imaging and at 6 h after tracer injection for SPECT/CT. Planar imaging revealed tracer accumulation at the tibial component of the prosthesis (*red arrows*). SPECT/CT images additionally demonstrated the extent of the infection in the cortical bone and lack of involvement of the bone marrow. The focal uptake of low intensity lying behind the prosthesis could be attributed to the arteria poplitea (*green arrow*). Hematology laboratory results were completely normal at the time of antigranulocyte scintigraphy (CRP 0.3 mg/dL, leukocytes 5.6×10^9 L^{-1}). However, CRP had increased to 16.7 mg/dL 2 months later. The diagnosis of a low-grade infection was verified by needle aspiration of the left knee, from which *Lactobacillus gasseri* was isolated, such that the scintigraphic result was rated as true-positive. *RVL*, right ventral left; *LDR*, left dorsal right.

Following the developments in hybrid imaging, there is an ongoing search for infection-specific tracers. Among the currently available radionuclide imaging techniques, none of the tracers is truly specific for infection and alternative agents have been explored and investigated [96]. The most important developments in potential tracers, with lack of clinical validation, for diagnosing PJI: (1) Radiolabeled antibiotics have been introduced as potential “infection-specific” tracers with encouraging results in initial reports of ^{99m}Tc-Ciprofloxacin in diagnosing PJI [74,97–99]. Theoretically, the tracers enter and accumulate in infection areas due to metabolic activity of bacteria. However, subsequent investigations did not confirm the clinical validations of previous studies [100,101]. (2) Radiolabeled synthetic fragments of the human antimicrobial peptide ubiquicidin, ^{99m}Tc-UBI 29-41, bind to the bacterial cell membrane. Expression may be constant or induced on contact with microbes and transported via leukocytes [102]. Recent primary data suggest that this agent is able to differentiate between infection and sterile inflammation, also in PJI [103–105]. (3) FDG is not exclusively used as tracer in PET imaging. The bone seeking tracer fluorine-18-fluoride-PET (fluoride-PET) has been investigated in PJI. Kobayashi et al. reported in a prospective study of 65 prosthetic hip joints, a sensitivity of 95% and specificity of 98% using a new uptake pattern classification system for differentiation of aseptic from septic loosening [106]. The authors concluded considerable potential in the assessment of periprosthetic hip infection with this relatively simple classification system. Other experimental alternatives for FDG is the use of ligand PK11195 coupled to a PET tracer, ^{11}C which binds to benzodiazepine receptors on macrophages, ^{11}Choline, ^{68}Ga, ^{64}CU, and ^{18}F-fluorothymidine which are currently been used as markers of cellular proliferation [36,107].

12.4.2 Multimodality diagnostic work-up of PJI

In the assessment of PJI, there is no universally accepted protocol and the algorithm is characterized by multimodality analysis: (1) The combination of elevated C-reactive protein (CRP) and erythrocyte sedimentation rate (ESR) with traditional thresholds appeared a more accurate predictor than isolated elevations, however with poor specificity [108–110]. (2) Elevated synovial fluid white blood cell count and polymorphonuclear neutrophils percentage are accurate predictors of PJI [111]. (3) Positive histological analysis of periprosthetic tissue is considered highly suggestive for PJI [112]. (4) Two positive periprosthetic cultures with phenotypically identical organism is definitive evidence for PJI, while one positive culture should be evaluated in the context of other available evidence [1,2]. (5) The presence of a sinus tract that communicates with the prosthesis is definitive PJI, while the presence of periprosthetic purulence is a subject of debate [1,2]. The multimodality evaluation of PJI reflects the diagnostic challenge and is subject of developments and consensus. The lack of golden standard is a potential source of bias in the reported diagnostic accuracy of imaging in PJI. Although several authors attempted to investigate and report pooled estimated of diagnostic accuracy of imaging PJI in metaanalyses, important variables including the time after surgery, use of antibiotics, sort of implants, and inter- and intraobserver reliability are often insufficiently investigated or reported. Furthermore, mainly due to inconclusive published diagnostic accuracy and high costs, prominent guidelines concluded that nuclear imaging should not be routinely used to diagnose PJI, although there was substantial evidence regarding the effectiveness in diagnosing PJI [1,2].

In recent published investigations with regard to the role of the promising modality of FDG-PET in evaluation of PJI, important developments are reported. In the clinical evaluation of patients suffering from PJI, FDG-PET is the preferred imaging modality in the detection of metastatic infectious foci in complicated gram-positive bacteremia [113]. The diagnosis of complicating infectious foci, including PJI as primarily or metastatic focus, is often difficult due to asymptomatic clinical presentation in up to 32% of the patients [114]. FDG PET/CT detected clinically relevant new complicating infectious foci in 30% of patients, including 30% in joint prostheses. The addition of FDG-PET/CT to the diagnostic work-up led to a decrease of relapse rates overall and infection-related mortality of gram-positive bacteremia by early detection and treatment of metastatic infectious foci [115,116]. CT, MRI, and ultrasonography are less suitable as screening methods when clues for specific sites of infection are absent [117,118]. Furthermore, FDG-PET/CT incorporated in the diagnostic algorithm appeared cost-effective and investigation concluded that it should be standardized in the work-up of complicated gram-positive bacteremia [119]. This sensitive imaging technique appears suitable for effective screening of complicated infectious foci and might identify patients in need of prolonged antibiotic treatment. No analog investigation regarding cost-effectiveness of FDG-PET in primary assessment of PJI has been published. With the introduction of these additional published data and optimal experience in the interpretation of FDG-PET within the setting of complicated arthroplasty, the role of this modality could potentially be firmly established in the literature and clinical evaluation of PJI.

In the context of other diagnostic tests, including clinical information, infectious parameters, and microbiological and histological analysis; nuclear imaging in diagnosing PJI is confirmed valuable. However, each modality has its drawbacks and limitations in each phase of the diagnostic algorithm. Understanding the background, drawbacks, and added value of each technique, improves the effectiveness of imaging in differentiating PJI of other causes of loosening. Of the conventional imaging options, radiographs should be standardized in the diagnostic work-up and follow-up of prosthetic joints. Ultrasound is mostly useful in PJI in which effusion cannot be clinically diagnosed. CT and MRI are of limited value in diagnosing PJI, but could be useful in depicting periprosthetic soft tissue. Nuclear imaging is most valuable in determining whether or not a painful prosthetic joint is infected. BS is a sensitive technique in order to exclude any cause of failure. However, the use of this technique is limited in 12–24 months after implantation. In clinical practice, investigation with more specificity is often needed which could be provided by LS or FDG-PET. Although several authors reported LS-BMS as golden standard in diagnosing PJI, this technique has important drawbacks including laborious preparations and limited availability. FDG-PET is proposed as promising and accurate tool with several advantages in diagnosing PJI, including advanced anatomical images and time-efficient procedures. In conclusion, there is substantial evidence that nuclear imaging is valuable for determining PJI, although further investigation in order to determine the exact role of imaging in the diagnostic algorithm is needed. The future prospects of nuclear imaging of PJI is promising with the ongoing search of "infection-specific" tracers, whether or not combined with improved hybrid imaging modalities.

References

[1] B. Zmistowski, C. Della Valle, T.W. Bauer, et al., Diagnosis of periprosthetic joint infection, J. Arthroplasty 29 (2 suppl.) (2014) 77–83, http://dx.doi.org/10.1016/j.arth.2013.09.040.

[2] D.R. Osmon, E.F. Berbari, A.R. Berendt, et al., Diagnosis and management of prosthetic joint infection: clinical practice guidelines by the infectious diseases Society of America, Clin. Infect. Dis. 56 (1) (2013) e1–e25, http://dx.doi.org/10.1093/cid/cis803.

[3] J. Tehranzadeh, R. Schneider, R.H. Freiberger, Radiological evaluation of painful total hip replacement, Radiology 141 (2) (1981) 355–362. http://www.ncbi.nlm.nih.gov/pubmed/6457316.

[4] K.D.M. Stumpe, H.P. Nötzli, M. Zanetti, et al., FDG PET for differentiation of infection and aseptic loosening in total hip replacements: comparison with conventional radiography and three-phase bone scintigraphy, Radiology 231 (2) (2004) 333–341, http://dx.doi.org/10.1148/radiol.2312021596.

[5] K. Aryana, A. Hootkani, R. Sadeghi, et al., 99mTc-labeled ubiquicidin scintigraphy, Nuklearmedizin 51 (4) (2012) 133–139, http://dx.doi.org/10.3413/Nukmed-0444-11-11.

[6] K. Foldes, M. Gaal, P. Balint, et al., Ultrasonography after hip arthroplasty, Skeletal Radiol. 21 (5) (1992) 297–299.

[7] D.M. Mulcahy, G.C. Fenelon, D.P. McInerney, Aspiration arthrography of the hip joint. Its uses and limitations in revision hip surgery, J. Arthroplasty 11 (1) (1996) 64–68, http://dx.doi.org/10.1016/s0883-5403(96)80162-8.

[8] H. Douis, D.J. Dunlop, A.M. Pearson, J.N. O'Hara, S.L.J. James, The role of ultrasound in the assessment of post-operative complications following hip arthroplasty, Skeletal Radiol. 41 (9) (2012) 1035–1046, http://dx.doi.org/10.1007/s00256-012-1390-9.

[9] W.R. Reinus, K.C. Merkel, J.J. Gilden, K.L. Berger, Evaluation of femoral prosthetic loosening using CT imaging, AJR Am. J. Roentgenol. 166 (6) (1996) 1439–1442, http://dx.doi.org/10.2214/ajr.166.6.8633459.

[10] E. Barmeir, B. Dubowitz, M. Roffman, Computed tomography in the assessment and planning of complicated total hip replacement, Acta Orthop. Scand. 53 (4) (1982) 597–604. http://www.ncbi.nlm.nih.gov/pubmed/7102278 (accessed 01.02.15).

[11] S.W. Mian, G. Truchly, F.A. Pflum, Computed tomography measurement of acetabular cup anteversion and retroversion in total hip arthroplasty, Clin. Orthop. Relat. Res. 276 (1992) 206–209. http://www.ncbi.nlm.nih.gov/pubmed/1537154 (accessed 01.02.15).

[12] C. Cyteval, V. Hamm, M.P. Sarrabère, F.M. Lopez, P. Maury, P. Taourel, Painful infection at the site of hip prosthesis: CT imaging, Radiology 224 (2) (2002) 477–483, http://dx.doi.org/10.1148/radiol.2242010989.

[13] J.G. Cahir, A.P. Toms, T.J. Marshall, J. Wimhurst, J. Nolan, CT and MRI of hip arthroplasty, Clin. Radiol. 62 (12) (2007) 1163–1171, http://dx.doi.org/10.1016/j.crad.2007.04.018 discussion 1172–1173.

[14] J.G. Cahir, A.P. Toms, CT and MRI of hip replacements, Orthop. Trauma 23 (2009) 101–108, http://dx.doi.org/10.1016/j.mporth.2009.03.003.

[15] H.K. Genant, G.J. Bautovich, M. Singh, K.A. Lathrop, P.V. Harper, Bone-seeking radionuclides: an in vivo study of factors affecting skeletal uptake, Radiology 113 (2) (1974) 373–382, http://dx.doi.org/10.1148/113.2.373.

[16] C.S. Galasko, The pathological basis for skeletal scintigraphy, J. Bone Joint Surg. (Br.) 57 (3) (1975) 353–359. http://www.ncbi.nlm.nih.gov/pubmed/1158946 (accessed 30.11.15).

[17] A. Boubaker, A. Bischof Delaloye, C.H. Blanc, M. Dutoit, P.F. Leyvraz, B. Delaloye, Immunoscintigraphy with antigranulocyte monoclonal antibodies for the diagnosis of septic loosening of hip prostheses, Eur. J. Nucl. Med. 22 (2) (1995) 139–147.
[18] M.W. Nijhof, W.J. Oyen, A. van Kampen, R.A. Claessens, J.W. van der Meer, F.H. Corstens, Evaluation of infections of the locomotor system with indium-111-labeled human IgG scintigraphy, J. Nucl. Med. 38 (8) (1997) 1300–1305.
[19] T. Itasaka, A. Kawai, T. Sato, S. Mitani, H. Inoue, Diagnosis of infection after total hip arthroplasty, J. Orthop. Sci. 6 (4) (2001) 320–326. http://www.ncbi.nlm.nih.gov/pubmed/11479760.
[20] M.J. Larikka, A.K. Ahonen, J.A. Junila, O. Niemela, M.M. Hamalainen, H.P. Syrjala, Extended combined 99mTc-white blood cell and bone imaging improves the diagnostic accuracy in the detection of hip replacement infections, Eur. J. Nucl. Med. 28 (3) (2001) 288–293, http://dx.doi.org/10.1007/s002590000463.
[21] P. Reinartz, Radionuclide imaging of the painful hip arthroplasty: positron-emission tomography versus triple-phase bone scanning, J. Bone Jt. Surg. Br. 87-B (4) (2005) 465–470, http://dx.doi.org/10.1302/0301-620X.87B4.14954.
[22] T. Mumme, P. Reinartz, J. Alfer, R. Müller-Rath, U. Buell, D.C. Wirtz, Diagnostic values of positron emission tomography versus triple-phase bone scan in hip arthroplasty loosening, Arch. Orthop. Trauma Surg. 125 (2005) 322–329, http://dx.doi.org/10.1007/s00402-005-0810-x.
[23] S. Nagoya, M. Kaya, M. Sasaki, K. Tateda, T. Yamashita, Diagnosis of peri-prosthetic infection at the hip using triple-phase bone scintigraphy, J. Bone Joint Surg. (Br.) 90 (2) (2008) 140–144, http://dx.doi.org/10.1302/0301-620X.90B2.19436.
[24] C.J. Palestro, A.J. Swyer, C.K. Kim, S.J. Goldsmith, Infected knee prosthesis: diagnosis with In-111 leukocyte, Tc-99m sulfur colloid, and Tc-99m MDP imaging, Radiology 179 (3) (1991) 645–648, http://dx.doi.org/10.1148/radiology.179.3.2027967.
[25] H. Verlooy, J. Victor, L. Renson, et al., Limitations of quantitative radionuclide bone scanning in the evaluation of total knee replacement, Clin. Nucl. Med. 18 (8) (1993) 671–674, http://dx.doi.org/10.1097/00003072-199308000-00010.
[26] G.M. Sacchetti, F. Ghisellini, M. Brambilla, et al., Quantitative scintigraphic evaluation of total knee arthroplasties: a feasibility study, Clin. Orthop. Relat. Res. 325 (1996) 181–189. http://www.ncbi.nlm.nih.gov/pubmed/8998872 (accessed 07.12.15).
[27] S.L. Smith, M.L. Wastie, I. Forster, Radionuclide bone scintigraphy in the detection of significant complications after total knee joint replacement, Clin. Radiol. 56 (3) (2001) 221–224, http://dx.doi.org/10.1053/crad.2000.0620.
[28] E. Magnuson, H. Berquist, L. Brown, H. Fitzgerald, M.F. Hauser, G. Klee, In-111-labeled leukocyte scintigraphy in suspected orthopedic prosthesis infection: comparison with other imaging modalities, Radiology 168 (1988) 235–239.
[29] K.A. Levitsky, W.J. Hozack, R.A. Balderston, et al., Evaluation of the painful prosthetic joint. Relative value of bone scan, sedimentation rate, and joint aspiration, J. Arthroplasty 6 (3) (1991) 237–244. http://www.ncbi.nlm.nih.gov/pubmed/1940929.
[30] A.B. Segura, A. Muñoz, Y.R. Brulles, et al., What is the role of bone scintigraphy in the diagnosis of infected joint prostheses? Nucl. Med. Commun. 25 (5) (2004) 527–532, http://dx.doi.org/10.1097/01.mnm.0000121145.07955.60.
[31] C. Love, G. Tronco, A. Yu, S. Marwin, K. Nichols, C. Palestro, Diagnosing lower extremity (LE) prosthetic joint infection: bone, gallium & labeled leukocyte imaging, Soc. Nucl. Med. Annu. Meet. Abstr. 49 (Suppl. 1) (2008) 133. http://jnumedmtg.snmjournals.org/cgi/content/short/49/MeetingAbstracts_1/133P-a (accessed 07.12.15).

[32] M.I. Gelman, R.E. Coleman, P.M. Stevens, B.W. Davey, Radiography, radionuclide imaging, and asthrography in the evaluation of total hip and knee replacement [/sup 99m/ Tc-phosphate], Radiology (United States) 28 (1978) 3. http://www.osti.gov/scitech/biblio/6737268 (accessed 04.12.15).

[33] P.E. Weiss, J.C. Mall, P.B. Hoffer, W.R. Murray, J.J. Rodrigo, H.K. Genant, 99mTc-methylene diphosphonate bone imaging in the evaluation of total hip prostheses, Radiology 133 (3 Pt 1) (1979) 727–729, http://dx.doi.org/10.1148/133.3.727.

[34] D.P. McInerney, I.D. Hyde, Technetium 99Tcm pyrophosphate scanning in the assessment of the painful hip prosthesis, Clin. Radiol. 29 (5) (1978) 513–517. http://www.ncbi.nlm.nih.gov/pubmed/710036, (accessed 04.12.15).

[35] B. Williamson, R. McLaughlin, Radionuclide bone imaging as a means of differentiating loosening and infection in patients with a painful total hip prosthesis, Radialogy 133 (1979) 723–726. http://radiology.rsna.org/content/133/3/723.short.

[36] F. Gemmel, H. Van den Wyngaert, C. Love, M.M. Welling, P. Gemmel, C.J. Palestro, Prosthetic joint infections: radionuclide state-of-the-art imaging, Eur. J. Nucl. Med. Mol. Imaging 39 (5) (2012) 892–909, http://dx.doi.org/10.1007/s00259-012-2062-7.

[37] P.J. Mountford, A.J. Coakley, S.M. Drive, Role of technetium-99m phosphonate bone and indium-111 leukocyte scanning for detecting the infected hip prosthesis, J. Nuclear Med. 30 (1989) 562–563.

[38] Z. Ouyang, H. Li, X. Liu, Z. Zhai, X. Li, Prosthesis infection: diagnosis after total joint arthroplasty with three-phase bone scintigraphy, Ann. Nucl. Med. 28 (10) (2014) 994–1003, http://dx.doi.org/10.1007/s12149-014-0899-5.

[39] S.J. Verberne, P.G. Raijmakers, O.P. Temmerman, The accuracy of imaging techniques in the assessment of periprosthetic hip infection: a systematic review and meta-analysis (accepted).

[40] L. Rosenthall, L. Lepanto, F. Raymond, Radiophosphate uptake in asymptomatic knee arthroplasty, J. Nucl. Med. 28 (10) (1987) 1546–1549. http://www.ncbi.nlm.nih.gov/pubmed/3655908, (accessed 04.12.15).

[41] A.A. Hofmann, R.W.B. Wyatt, A.U. Daniels, L. Armstrong, N. Alazraki, A. Taylor, Bone scans after total knee arthroplasty in asymptomatic patients, Clin. Orthop. Relat. Res. 251 (1990) 183–188, http://dx.doi.org/10.1097/00003086-199002000-00031.

[42] J.P. Lavender, J. Lowe, J.R. Barker, J.I. Burn, M.A. Chaudhri, Gallium 67 citrate scanning in neoplastic and inflammatory lesions, Br. J. Radiol. 44 (521) (1971) 361–366, http://dx.doi.org/10.1259/0007-1285-44-521-361.

[43] P. Hoffer, Gallium and infection, J. Nucl. Med. 21 (5) (1980) 484–488. http://www.ncbi.nlm.nih.gov/pubmed/6989964 (accessed 30.11.15).

[44] C.J. Palestro, The current role of gallium imaging in infection, Semin. Nucl. Med. 24 (2) (1994) 128–141. http://www.ncbi.nlm.nih.gov/pubmed/8023169 (accessed 30.11.15).

[45] K.D. Merkel, M.L. Brown, R.H. Fitzgerald, Sequential technetium-99m HMDP-gallium-67 citrate imaging for the evaluation of infection in the painful prosthesis, J. Nucl. Med. 27 (9) (1986) 1413–1417. http://www.ncbi.nlm.nih.gov/pubmed/3462352.

[46] M.A. Gomez-Luzuriaga, V. Galan, J.M. Villar, Scintigraphy with Tc, Ga and In in painful total hip prostheses, Int. Orthop. 12 (2) (1988) 163–167. http://www.ncbi.nlm.nih.gov/pubmed/3410621.

[47] P. Aliabadi, S.S. Tumeh, B.N. Weissman, B.J. McNeil, Cemented total hip prosthesis: radiographic and scintigraphic evaluation, Radiology 173 (1) (1989) 203–206, http://dx.doi.org/10.1148/radiology.173.1.2675184.

[48] W.J. Kraemer, R. Saplys, J.P. Waddell, J. Morton, Bone scan, gallium scan, and hip aspiration in the diagnosis of infected total hip arthroplasty, J. Arthroplasty 8 (6) (1993) 611–616. http://www.ncbi.nlm.nih.gov/pubmed/8301279.

[49] C.M. Reing, P.F. Richin, P.I. Kenmore, Differential bone-scanning in the evaluation of a painful total joint replacement, J. Bone Joint Surg. Am. 61 (6A) (1979) 933–936. http://jbjs.org/content/61/6/933.abstract (accessed 07.12.15).
[50] J.H. Mckillop, I. Mckay, G.F. Cuthbert, I. Fogelman, H.W. Gray, R.D. Sturrock, Scintigraphic evaluation of the painful prosthetic joint: a comparison of gallium-67 citrate and indium-111 labelled leucocyte imaging, Clin. Radiol. 35 (3) (1984) 239–241.
[51] Tehranzadeh J, RefID 2103—Tehranzadeh—prop seq tech-99m phosphate and gallium imaging painTHA.pdf, 1987.
[52] C. Love, S.E. Marwin, C.J. Palestro, Nuclear medicine and the infected joint replacement, Semin. Nucl. Med. 39 (1) (2009) 66–78, http://dx.doi.org/10.1053/j.semnuclmed.2008.08.007.
[53] J. Palestro, C.K. Kim, A.J. Swyer, R.W. Solomon, J.D. Capozzi, S.J. Goldsmith, Total-hip arthroplasty: periprosthetic indium-111-labeled leukocyte activity and complementary technetium-99m-sulfur colloid imaging in suspected infection, J. Nucl. Med. 31 (12) (1990) 1950–1955.
[54] T.N. Joseph, M. Mujtaba, A.L. Chen, et al., Efficacy of combined technetium-99m sulfur colloid/indium-111 leukocyte scans to detect infected total hip and knee arthroplasties, J. Arthroplasty 16 (6) (2001) 753–758, http://dx.doi.org/10.1054/arth.2001.24446.
[55] C. Love, S.E. Marwin, M.B. Tomas, et al., Diagnosing infection in the failed joint replacement: a comparison of coincidence detection 18F-FDG and 111In-labeled leukocyte/99mTc-sulfur colloid marrow imaging, J. Nucl. Med. 45 (11) (2004) 1864–1871.
[56] I. El Espera, C. Blondet, V. Moullart, et al., The usefulness of 99mTc sulfur colloid bone marrow scintigraphy combined with 111In leucocyte scintigraphy in prosthetic joint infection, Nucl. Med. Commun. 25 (2) (2004) 171–175, http://dx.doi.org/10.1097/01.mnm.0000109830.10116.61.
[57] D. Fuster, J. Duch, A. Soriano, et al., Potential use of bone marrow scintigraphy in suspected prosthetic hip infection evaluated with 99mTc-HMPAO-leukocytes, Rev. Esp. Med. Nucl. 27 (6) (2008) 430–435, http://dx.doi.org/10.1016/S1578-200X(08)70046-9.
[58] E. Pelosi, C. Baiocco, M. Pennone, et al., 99mTc-HMPAO-leukocyte scintigraphy in patients with symptomatic total hip or knee arthroplasty: improved diagnostic accuracy by means of semiquantitative evaluation, J. Nucl. Med. 45 (3) (2004) 438–444.
[59] S.G. Pill, J. Parvizi, P.H. Tang, et al., Comparison of fluorodeoxyglucose positron emission tomography and (111)indium-white blood cell imaging in the diagnosis of periprosthetic infection of the hip, J. Arthroplasty 21 (6 suppl.) (2006) 91–97, http://dx.doi.org/10.1016/j.arth.2006.05.021.
[60] L. Simonsen, A. Buhl, T. Oersnes, B. Duus, White blood cell scintigraphy for differentiation of infection and aseptic loosening: a retrospective study of 76 painful hip prostheses, Acta Orthop. 78 (5) (2007) 640–647, http://dx.doi.org/10.1080/17453670710014338.
[61] H.O. Kim, S.J. Na, S.J. Oh, B.S. Jung, S. Lee, Usefulness of adding SPECT/CT to 99mTc-hexamethylpropylene amine oxime (HMPAO)-labeled leukocyte imaging for diagnosing prosthetic joint infection, J. Comput. Assist. Tomogr. 38 (2) (2014) 313–319.
[62] S. Basu, M. Hons, T.C. Kwee, et al., FDG PET for diagnosing infection in hip and knee prostheses: prospective study in 221 prostheses and subgroup comparison with combined (111)In-labeled leukocyte/(99m)Tc-sulfur colloid bone marrow imaging in 88 prostheses, Clin. Nucl. Med. 39 (7) (2014) 609–615, http://dx.doi.org/10.1097/RLU.0000000000000464.
[63] M.J. Larikka, A.K. Ahonen, J.A. Junila, O. Niemelä, M.M. Hämäläinen, H.P. Syrjälä, Improved method for detecting knee replacement infections based on extended combined 99mTc-white blood cell/bone imaging, Nucl. Med. Commun. 22 (10) (2001) 1145–1150. http://www.ncbi.nlm.nih.gov/pubmed/11567190 (accessed 10.11.15).

[64] D.J. Pring, Autologous granulocyte scanning of painful prosthetic joints, J. Bone Joint Surg. (Br.) 68 (4) (1986) 647–652.

[65] D.K. Wukich, S.H. Abreu, J.J. Callaghan, et al., Diagnosis of infection by preoperative scintigraphy with indium-labeled white blood cells, J. Bone Joint Surg. Am. 69 (9) (1987) 1353–1360. http://www.ncbi.nlm.nih.gov/pubmed/3126189 (accessed 18.12.15).

[66] J.A. Johnson, M.J. Christie, M.P. Sandler, P.F. Parks, L. Homra, J.J. Kaye, Detection of occult infection following total joint arthroplasty using sequential technetium-99m HDP bone scintigraphy and indium-111 WBC imaging, J. Nucl. Med. 29 (8) (1988) 1347–1353. http://www.ncbi.nlm.nih.gov/pubmed/3404252.

[67] D.J. Pring, R.G. Henderson, A. Keshavarzian, et al., Indium-granulocyte scanning in the painful prosthetic joint, AJR Am. J. Roentgenol. 147 (1) (1986) 167–172, http://dx.doi.org/10.2214/ajr.147.1.167.

[68] J.A. Rand, M.L. Brown, The value of indium 111 leukocyte scanning in the evaluation of painful or infected total knee arthroplasties, Clin. Orthop. Relat. Res. 259 (1990) 179–182. http://www.ncbi.nlm.nih.gov/pubmed/2208853 (accessed 10.11.15).

[69] A.W.J.M. Glaudemans, F. Galli, M. Pacilio, A. Signore, Leukocyte and bacteria imaging in prosthetic joint infection, Eur. Cells Mater. 25 (2012) 61–77.

[70] P. Fernandez, A. Monet, C. Matei, et al., 99mTc-HMPAO labelled white blood cell scintigraphy in patients with osteoarticular infection: the value of late images for diagnostic accuracy and interobserver reproducibility, Eur. J. Clin. Microbiol. Infect. Dis. 27 (12) (2008) 1239–1244, http://dx.doi.org/10.1007/s10096-008-0563-x.

[71] R. Sousa, M. Massada, A. Pereira, F. Fontes, I. Amorim, A. Oliveira, Diagnostic accuracy of combined 99mTc-sulesomab and 99mTc-nanocolloid bone marrow imaging in detecting prosthetic joint infection, Nucl. Med. Commun. 32 (9) (2011) 834–839, http://dx.doi.org/10.1097/MNM.0b013e3283496695.

[72] R.E. Teller, M.J. Christie, W. Martin, E.P. Nance, D.W. Haas, Sequential indium-labeled leukocyte and bone scans to diagnose prosthetic joint infection, Clin. Orthop. Relat. Res. 373 (2000) 241–247. http://www.ncbi.nlm.nih.gov/pubmed/10810483.

[73] V. Ivanćević, C. Perka, O. Hasart, D. Sandrock, D.L. Munz, V. Ivanèeviae, Imaging of low-grade bone infection with a technetium-99m labelled monoclonal anti-NCA-90 Fab' fragment in patients with previous joint surgery, Eur. J. Nucl. Med. Mol. Imaging 29 (4) (2002) 547–551, http://dx.doi.org/10.1007/s00259-001-0744-7.

[74] M.J. Larikka, M.J. Larikka, A.K. Ahonen, O. Niemelä, et al., Comparison of 99mTc ciprofloxacin, 99mTc white blood cell and three-phase bone imaging in the diagnosis of hip prosthesis infections: improved diagnostic accuracy with extended imaging time, Nucl. Med. Commun. 23 (7) (2002) 655–661. http://www.ncbi.nlm.nih.gov/pubmed/12089488.

[75] A. Fabj, T. Von Rothenburg, M. Schoellhammer, J. Schaffstein, O. Koester, Imaging of infected total arthroplasty with Tc-99m, Clin. Nucl. Med. 29 (9) (2004) 548–551.

[76] K.P. Iyengar, S. Vinjamuri, Role of 99m Tc Sulesomab in the diagnosis of prosthetic joint infections, Nucl. Med. Commun. 15–17 (2005).

[77] E.E. Pakos, T.A. Trikalinos, A.D. Fotopoulos, J.P. Ioannidis, Prosthesis infection: diagnosis after total joint arthroplasty with antigranulocyte scintigraphy with 99mTc-labeled monoclonal antibodies—a meta-analysis, Radiology 242 (1) (2007) 101–108, http://dx.doi.org/10.1148/radiol.2421052011.

[78] B.M. Gallagher, J.S. Fowler, N.I. Gutterson, R.R. MacGregor, C.N. Wan, A.P. Wolf, Metabolic trapping as a principle of oradiopharmaceutical design: some factors resposible for the biodistribution of [18F] 2-deoxy-2-fluoro-D-glucose, J. Nucl. Med. 19 (10) (1978) 1154–1161. http://www.ncbi.nlm.nih.gov/pubmed/214528 (accessed 27.11.15).

[79] T. Mochizuki, E. Tsukamoto, Y. Kuge, et al., FDG uptake and glucose transporter subtype expressions in experimental tumor and inflammation models, J. Nucl. Med. 42 (10) (2001) 1551–1555. http://www.ncbi.nlm.nih.gov/pubmed/11585872 (accessed 27.11.15).
[80] S. Basu, T. Chryssikos, S. Moghadam-Kia, H. Zhuang, D.A. Torigian, A. Alavi, Positron emission tomography as a diagnostic tool in infection: present role and future possibilities, Semin. Nucl. Med. 39 (1) (2009) 36–51, http://dx.doi.org/10.1053/j.semnuclmed.2008.08.004.
[81] T.K. Chacko, H. Zhuang, K. Stevenson, B. Moussavian, A. Alavi, The importance of the location of fluorodeoxyglucose uptake in periprosthetic infection in painful hip prostheses, Nucl. Med. Commun. 23 (2002) 851–855, http://dx.doi.org/10.1097/00006231-200209000-00008.
[82] H. Zhuang, P.S. Duarte, M. Pourdehnad, et al., The promising role of 18F-FDG PET in detecting infected lower limb prosthesis implants, J. Nucl. Med. 42 (9) (2001) 44–48, http://dx.doi.org/10.1097/01.rlu.0000252457.54929.b7.
[83] W. van der Bruggen, C.P. Bleeker-Rovers, O.C. Boerman, M. Gotthardt, W.J.G. Oyen, PET and SPECT in osteomyelitis and prosthetic bone and joint infections: a systematic review, Semin. Nucl. Med. 40 (1) (2010) 3–15, http://dx.doi.org/10.1053/j.semnuclmed.2009.08.005.
[84] H. Zhuang, T.K. Chacko, M. Hickeson, et al., Persistent non-specific FDG uptake on PET imaging following hip arthroplasty, Eur. J. Nucl. Med. Mol. Imaging 29 (10) (2002) 1328–1333, http://dx.doi.org/10.1007/s00259-002-0886-2.
[85] N. Beslic, D. Heber, R. Walter Lipp, C. Sonneck-Koenne, P. Knoll, S. Mirzaei, Metabolic pattern of asymptomatic hip-prosthesis by 18F-FDG-positron-emission-tomography, Iran. J. Radiol. 12 (1) (2015), e11204. http://dx.doi.org/10.5812/iranjradiol.11204.
[86] B. Vanquickenborne, A. Maes, J. Nuyts, et al., The value of (18)FDG-PET for the detection of infected hip prosthesis, Eur. J. Nucl. Med. Mol. Imaging 30 (5) (2003) 705–715, http://dx.doi.org/10.1007/s00259-002-1109-6.
[87] E. Garcia-Barrecheguren, FDG-PET: a new diagnostic approach in hip prosthesis infection. 26 (4) (2007) 9–14, http://dx.doi.org/10.1016/S1578-200X(07)70059-1.
[88] T. Chryssikos, J. Parvizi, E. Ghanem, A. Newberg, H. Zhuang, A. Alavi, FDG-PET imaging can diagnose periprosthetic infection of the hip, Clin. Orthop. Relat. Res. 466 (6) (2008) 1338–1342, http://dx.doi.org/10.1007/s11999-008-0237-0.
[89] S. Mayer-Wagner, W. Mayer, S. Maegerlein, R. Linke, V. Jansson, P.E. Müller, Use of 18F-FDG-PET in the diagnosis of endoprosthetic loosening of knee and hip implants, Arch. Orthop. Trauma Surg. 130 (10) (2010) 1231–1238, http://dx.doi.org/10.1007/s00402-009-1000-z.
[90] F. Van Acker, J. Nuyts, A. Maes, et al., FDG-PET, 99mtc-HMPAO white blood cell SPET and bone scintigraphy in the evaluation of painful total knee arthroplasties, Eur. J. Nucl. Med. 28 (10) (2001) 1496–1504, http://dx.doi.org/10.1007/s002590100603.
[91] N. Manthey, P. Reinhard, F. Moog, P. Knesewitsch, K. Hahn, K. Tatsch, The use of [18F] fluorodeoxyglucose positron emission tomography to differentiate between synovitis, loosening and infection of hip and knee prostheses, Nucl. Med. Commun. 23 (7) (2002) 645–653. http://www.ncbi.nlm.nih.gov/pubmed/12089487.
[92] T.C. Kwee, R.M. Kwee, A. Alavi, FDG-PET for diagnosing prosthetic joint infection: systematic review and metaanalysis, Eur. J. Nucl. Med. Mol. Imaging 35 (11) (2008) 2122–2132, http://dx.doi.org/10.1007/s00259-008-0887-x.
[93] K.-S. Delank, M. Schmidt, J.W.-P. Michael, M. Dietlein, H. Schicha, P. Eysel, The implications of 18F-FDG PET for the diagnosis of endoprosthetic loosening and

infection in hip and knee arthroplasty: results from a prospective, blinded study, BMC Musculoskelet. Disord. 7 (2006) 20, http://dx.doi.org/10.1186/1471-2474-7-20.

[94] V. Graute, M. Feist, S. Lehner, et al., Detection of low-grade prosthetic joint infections using 99mTc-antigranulocyte SPECT/CT: initial clinical results, Eur. J. Nucl. Med. Mol. Imaging 37 (9) (2010) 1751–1759, http://dx.doi.org/10.1007/s00259-010-1431-3.

[95] O. Dobrindt, H. Amthauer, A. Krueger, et al., Hybrid SPECT/CT for the assessment of a painful hip after uncemented total hip arthroplasty, BMC Med. Imaging 15 (1) (2015) 18, http://dx.doi.org/10.1186/s12880-015-0056-1.

[96] J.R. Buscombe, The quarterly journal of nuclear medicine and molecular imaging: official publication of the Italian Association of Nuclear Medicine (AIMN) [and] the International Association of Radiopharmacology (IAR), [and] Section of the Society of Radiopharmaceutical Chemistry and Biology, Q. J. Nucl. Med. Mol. Imaging 50 (2) (2006) 99–103. http://www.ncbi.nlm.nih.gov/pubmed/16770299 (accessed 09.12.15).

[97] K. Sonmezoglu, M. Sonmezoglu, M. Halac, et al., Usefulness of 99mTc-ciprofloxacin (infecton) scan in diagnosis of chronic orthopedic infections: comparative study with 99mTc-HMPAO leukocyte scintigraphy, J. Nucl. Med. 42 (4) (2001) 567–574. https://www.researchgate.net/publication/11998708_Usefulness_of_99mTc-Ciprofloxacin_(Infecton)_scan_in_diagnosis_of_chronic_orthopedic_infections_comparative_study_with_99mTc-HMPAO_leukocyte_scintigraphy._J_Nucl_Med (accessed 09.12.15).

[98] M.J. Larikka, A.K. Ahonen, O. Niemelä, et al., 99mTc-ciprofloxacin (Infecton) imaging in the diagnosis of knee prosthesis infections, Nucl. Med. Commun. 23 (2) (2002) 167–170, http://dx.doi.org/10.1097/00006231-200202000-00009.

[99] S. Vinjamuri, K. Solanki, J. Bomanji, et al., Comparison of 99mTc Infecton imaging with radiolabelled white-cell imaging in the evaluation of bacterial infection, Lancet 347 (8996) (1996) 233–235, http://dx.doi.org/10.1016/S0140-6736(96)90407-9.

[100] K.E. Britton, D.W. Wareham, S.S. Das, et al., Imaging bacterial infection with (99m) Tc-ciprofloxacin (Infecton), J. Clin. Pathol. 55 (11) (2002) 817–823. http://www.pubmedcentral.nih.gov/articlerender.fcgi?artid=1769796&tool=pmcentrez&rendertype=abstract (accessed 09.11.15).

[101] L. Sarda, A.-C. Crémieux, Y. Lebellec, et al., Inability of 99mTc-ciprofloxacin scintigraphy to discriminate between septic and sterile osteoarticular diseases, J. Nucl. Med. 44 (6) (2003) 920–926. http://www.ncbi.nlm.nih.gov/pubmed/12791820.

[102] A. Lupetti, E.K.J. Pauwels, P.H. Nibbering, M.M. Welling, 99mTc-antimicrobial peptides: promising candidates for infection imaging, Q. J. Nucl. Med. 47 (4) (2003) 238–245. http://www.ncbi.nlm.nih.gov/pubmed/14973416 (accessed 09.12.15).

[103] A. Lupetti, M.M. Welling, U. Mazzi, P.H. Nibbering, E.K.J. Pauwels, Technetium-99m labelled fluconazole and antimicrobial peptides for imaging of Candida albicans and Aspergillus fumigatus infections, Eur. J. Nucl. Med. Mol. Imaging 29 (5) (2002) 674–679, http://dx.doi.org/10.1007/s00259-001-0760-7.

[104] L. Sarda-Mantel, A. Saleh-Mghir, M.M. Welling, et al., Evaluation of 99mTc-UBI 29-41 scintigraphy for specific detection of experimental Staphylococcus aureus prosthetic joint infections, Eur. J. Nucl. Med. Mol. Imaging 34 (8) (2007) 1302–1309, http://dx.doi.org/10.1007/s00259-007-0368-7.

[105] C. Arteaga de Murphy, F. Gemmel, J. Balter, Clinical trial of specific imaging of infections, Nucl. Med. Commun. 31 (8) (2010) 726–733, http://dx.doi.org/10.1097/MNM.0b013e32833a3d7f.

[106] N. Kobayashi, Y. Inaba, H. Choe, et al., Use of F-18 fluoride PET to differentiate septic from aseptic loosening in total hip arthroplasty patients, Clin. Nucl. Med. 36 (11) (2011) e156–e161, http://dx.doi.org/10.1097/RLU.0b013e3182291ae7.

[107] M.A. Kropholler, R. Boellaard, E.H. Elzinga, et al., Quantification of (R)-[11C]PK11195 binding in rheumatoid arthritis, Eur. J. Nucl. Med. Mol. Imaging 36 (4) (2009) 624–631, http://dx.doi.org/10.1007/s00259-008-0987-7.

[108] E. Ghanem, V. Antoci, L. Pulido, A. Joshi, W. Hozack, J. Parvizi, The use of receiver operating characteristics analysis in determining erythrocyte sedimentation rate and C-reactive protein levels in diagnosing periprosthetic infection prior to revision total hip arthroplasty, Int. J. Infect. Dis. 13 (6) (2009) e444–e449, http://dx.doi.org/10.1016/j.ijid.2009.02.017.

[109] N.V. Greidanus, B.A. Masri, D.S. Garbuz, et al., Use of erythrocyte sedimentation rate and C-reactive protein level to diagnose infection before revision total knee arthroplasty: a prospective evaluation, J. Bone Joint Surg. Am. 89 (7) (2007) 1409–1416, http://dx.doi.org/10.2106/JBJS.D.02602.

[110] J.S. Olshaker, D.A. Jerrard, The erythrocyte sedimentation rate, J. Emerg. Med. 15 (6) (1997) 869–874, http://dx.doi.org/10.1016/S0736-4679(97)00197-2.

[111] B. Zmistowski, C. Restrepo, R. Huang, W.J. Hozack, J. Parvizi, Periprosthetic joint infection diagnosis: a complete understanding of white blood cell count and differential, J. Arthroplasty 27 (9) (2012) 1589–1593, http://dx.doi.org/10.1016/j.arth.2012.03.059.

[112] G. Tsaras, A. Maduka-Ezeh, C.Y. Inwards, et al., Utility of intraoperative frozen section histopathology in the diagnosis of periprosthetic joint infection: a systematic review and meta-analysis, J. Bone Joint Surg. Am. 94 (18) (2012) 1700–1711, http://dx.doi.org/10.2106/JBJS.J.00756.

[113] F.J. Vos, C.P. Bleeker-Rovers, P.D. Sturm, et al., 18F-FDG PET/CT for detection of metastatic infection in gram-positive bacteremia, J. Nucl. Med. 51 (8) (2010) 1234–1240, http://dx.doi.org/10.2967/jnumed.109.072371.

[114] M.L.H. Cuijpers, F.J. Vos, C.P. Bleeker-Rovers, et al., Complicating infectious foci in patients with Staphylococcus aureus or Streptococcus species bacteraemia, Eur. J. Clin. Microbiol. Infect. Dis. 26 (2) (2007) 105–113, http://dx.doi.org/10.1007/s10096-006-0238-4.

[115] A.G. Jensen, C.H. Wachmann, F. Espersen, J. Scheibel, P. Skinhøj, N. Frimodt-Møller, Treatment and outcome of Staphylococcus aureus bacteremia: a prospective study of 278 cases, Arch. Intern. Med. 162 (1) (2002) 25–32. http://www.ncbi.nlm.nih.gov/pubmed/11784216 (accessed 02.12.15).

[116] D.W.M. Verhagen, J.T.M. van der Meer, T. Hamming, M.D. de Jong, P. Speelman, Management of patients with Staphylococcus aureus bacteraemia in a university hospital: a retrospective study, Scand. J. Infect. Dis. 35 (8) (2003) 459–463. http://www.ncbi.nlm.nih.gov/pubmed/14514144 (accessed 02.12.15).

[117] C.P. Bleeker-Rovers, F.J. Vos, E.M.H.A. de Kleijn, et al., A prospective multicenter study on fever of unknown origin: the yield of a structured diagnostic protocol, Medicine (Baltimore) 86 (1) (2007) 26–38. http://dx.doi.org/10.1097/MD.0b013e31802fe858.

[118] E.M. de Kleijn, J.P. Vandenbroucke, J.W. van der Meer, Fever of unknown origin (FUO). I A. prospective multicenter study of 167 patients with FUO, using fixed epidemiologic entry criteria. The Netherlands FUO Study Group, Medicine (Baltimore) 76 (6) (1997) 392–400. http://www.ncbi.nlm.nih.gov/pubmed/9413425 (accessed 11.11.15).

[119] F.J. Vos, C.P. Bleeker-Rovers, B.J. Kullberg, E.M.M. Adang, W.J.G. Oyen, Cost-effectiveness of routine 18F-FDG PET/CT in high-risk patients with gram-positive bacteremia, J. Nucl. Med. 52 (11) (2011) 1673–1678, http://dx.doi.org/10.2967/jnumed.111.089714.

Index

Note: Page numbers followed by *f* indicate figures and *t* indicate tables.

U

X

CPI Antony Rowe
Chippenham, UK
2017-02-01 18:52

Introduction

Takfīr,[1] the act of a Muslim's declaring another Muslim infidel or unbeliever, is not a recent phenomenon. The act occurred for centuries; however, it has never before achieved the force and popularity that it currently enjoys. Today, *takfīr* has become a household topic in Muslim societies, a subject for debate among Muslim theologians and legal scholars, and a field of study in academia.[2] The issue of *takfīr*, writes Ghālib Ḥasan al-Shābandar, has reached an unprecedented level that the Muslim world has never experienced before.[3] This frightening phenomenon, he adds, is gaining popularity not only in Muslim societies but also all over the world. This trend of making the other infidel, al-Shābandar tells us, is as ancient as the human race itself.[4] In recent years, according to al-Shābandar, this phenomenon has taken a new turn in that it has been transformed into a tyrannical ideology that is evaluative of the other.[5]

For this reason, a work on *takfīr* in the English language is long overdue and necessary for understanding the origin, history, and meaning of the precept as well as the groups that have adopted it. This theological concept, which has been widely discussed in books written mainly in the Arabic language,[6] has yet to make its way to Western audiences. This book attempts to fill this gap and will contribute to the fields of Islamic and Middle Eastern studies.

The ongoing debate surrounding *takfīr* is none but proof that the various Islamic schools of law, theological schools, and sects have offered a variety of interpretations on the topic. Throughout the centuries, Muslim theologians, jurists, and others[7] have written volumes explaining the theological meanings of *takfīr*. In their writings, they also have warned against independent interpretations of the concept, especially the interpretations against whom *takfīr* should be declared. Although there had been a wide opposition to *takfīr* in early and medieval Islam, the concept, rather than dying out, has spread wide and far in the Muslim world to the extent that it has become a common tool among some modern and contemporary Muslim groups to declare the other infidel.

Muslim extremist groups and militant leaders across the Muslim world, and in some Western societies as well, have called for the "institutionalization and politicization" of this phenomenon.[8] These calls have led to greater divisions and violence among Islamic sects and groups, because *takfīr*, which some groups today use as a justification for the elimination of the other socially, religiously, and physically, is based on "the holy books and the holy heritage in general, and justified by some scriptural verses and passages, which are used in arbitrary ways for the benefit of the few."[9] *Takfīr*, writes al-Shābandar, is only a figure of speech used in the scripture to highlight deviations from the Islamic path and never meant to be taken literally, that is, making practicing Muslims infidels. For this reason, "historical reports never registered cases where prophets and messengers practiced *takfīr* against the other."[10]

Takfīr, when practiced against another Muslim, is one of the most dangerous and extreme practices in Islam, for it strips the one who is declared a *kāfir* of all legal and civil rights (i.e., inheritance, child custody, and marriage to a Muslim) and justifies the shedding of his or her blood.[11] *Takfīr* was first practiced during the first Muslim Civil War (656–657),[12] when a group of Muslims expressed dissatisfaction with the fourth caliph ʿAlī b. Abī Ṭālib (d. 661).[13] After they seceded from the caliph's army, these Muslims—known as the Kharijites—denounced as *kafir* not only ʿAli but also every Muslim who agreed with his decision and supported him.[14] ʿUmar ʿAbd Allāh Kāmil calls the act of declaring a group of Muslims as *kuffār* "*takfīr al-mujtamaʿ*" (declaring a Muslim society a *kāfir* society), which has serious implications.[15] The Kharijites, for instance, who declared other Muslims *kuffār*, cut all ties with the Muslim community and established their own community after they lived in isolation, away from the "unbelievers."[16]

This book provides an overview of the theological views on *takfīr* in classical and medieval Islamic thought, explores the Islamic context of the concept, and considers the following questions: in what context does the term appear in the Qur'ān and *ḥadīths* (the sayings of the Prophet Muḥammad)? In what circumstances did Muslim theologians engage in *takfīr*, and what were the social, religious, theological, and political implications on the society? How did the meaning of *takfīr* evolve in classical and later Islamic theology, and what justifications did prominent medieval Muslim theologians and jurists provide for declaring other Muslims *kuffār*?

Chapter 1 discusses the meaning and origin of *takfīr* and the early Islamic attitude to the concept. This chapter attempts to establish the historical roots of the concept and how the Prophet Muḥammad understood it and used it.

Chapters 2 and 3 provide a survey of *takfīr* in early and later Islamic theology and demonstrate how and why the concept became an important issue to be discussed in theological schools. These chapters also demonstrate how the various interpretations of *takfīr* were the product of a classical theological debate on the differing views of the concepts of *īmān* (faith) and *islām* (submission to the will of God). Finally, chapter 4 addresses the views of two prominent medieval theologians, Abū Ḥāmid Muḥammad al-Ghazālī (d. 1111) and Taqī al-Dīn Aḥmad ibn Taymiyyah (d. 1328). This chapter demonstrates how al-Ghazālī and Ibn Taymiyyah validated the practice of *takfīr* against other Muslim sects and groups and how they differed in their interpretations of the concept.

NOTES

1. The idea to write about *takfīr* emerged after presenting a paper on the *takfīr* in the Kharijite thought at an international conference entitled "*Takfir*: a Diachronic Perspective," in Madrid, Spain in 2011.
2. See, for example, Ghālib Ḥasan al-Shābandar, *Lā Takfīr fī al-Qur'ān* (Beirut, Lebanon: Dār al-Hudā, 2009).
3. Ibid., 5.
4. Ibid.
5. Ibid.
6. See, for example, 'Abd al-Mun'im Muṣṭafā Ḥalīmah, *Qawā'id fī al-Takfīr* (Amman, Jordan: Dār al-Bashīr, 1994); Mālik Muṣṭafā Wahbī al-'Āmilī, *Ẓāhirat al-Takfīr fī al-Fikr al-Islāmī* (Beirut, Lebanon: Dār al-Hādī, 2007); Ismā'īl b. Aḥmad al-Bustī, *Kitāb al-Baḥth 'an Adillat al-Takfīr wa al-Tafsīq* (Koln., Germany: Al-Kamel Verlag, 2009); Ḍuḥā al-Khatīb, *Ḍawābit al-Takfīr* (Beirut, Lebanon: Dār al-Burāq, 2007); Ghālib Ḥasan al-Shābandar, *Lā Takfīr fī al-Qur'ān* (Beirut, Lebanon: Dār al-Hudā, 2009); Maḥmūd Salīm 'Ubaydāt, *Qaḍiyyat al-Īmān wa al-Takfīr fī ārā' al-Muslimīn* (Amman, Jordan: Dār al-Bashīr, 1996); Rifqī Zāhir, *Qaḍiyyat al-Takfīr 'inda al-Ghazālī* (Cairo, Egypt: Maktabat al-Azhar, 1977).
7. Abū 'Ubayd al-Qāsim b. Sallām (d. 838), Muḥammad b. Saḥnūn (d. 854-55), Abū al-Ḥasan al-Ash'arī (d. 935), Abū Manṣūr al-Māturīdī (d. 944), Qādi 'Abd al-Jabbār (d. 1024), Abū Muḥammad 'Alī ibn Aḥmad ibn Sa'īd ibn Ḥazm (d. 1064), Abū Ḥāmid Muḥammad b. Muḥammad al-Ṭūsī al-Ghazālī (d. 1111), Tāj al-Dīn Abū al-Fatḥ Muḥammad ibn 'Abd al-Karīm al-Shahrastānī (d. 1153), Abū Zakariyyah Yaḥya ibn Sharaf al-Nawawī (d. 1277), Mas'ūd b. 'Umar al-Taftazānī (d. 1389 or 1390), and Taqī al-Dīn Aḥmad ibn Taymiyyah (d. 1328).
8. Al-Shābandar, *Lā Takfīr*, 5.
9. Ibid., 6.
10. Ibid.
11. See, for example, al-Bustī, *Kitāb al-Baḥth,* 6.
12. 'Abd al-Qāhir al-Baghdādī, *Al-Farq Bayna al-Firaq* (Beirut: Dār al-Kutub al-'Ilmiyyah, n.d.), 49.
13. Ibid.
14. Ibid.; see also Muḥammad Nāsir al-Dīn al-Albānī, *Fitnat al-Takfīr* (n.c.: Dār ibn Khazīmah, 1997), 14.
15. 'Umar 'Abd Allāh Kāmil, *al-Khawārij al-Judud: Al-Mutaṭariffūn* (Cairo, Egypt: Maktabat al-Turāth al-Islamī, 1998), 131.
16. Hussam S. Timani, *Modern Intellectual Readings of the Kharijites* (New York: Peter Lang Publishing, 2008), 103.

Chapter One

Takfīr: Meaning and Origin

This chapter provides an overview of the meanings, origins, and types of *takfīr*. It begins by exploring the lexical meaning of *takfīr* then moves to discuss the evolution of the theological connotations of the term as it appeared in Qur'ānic verses. The chapter also provides an overview of the types of *takfīr* and the variety of actions that trigger the accusations of *takfīr* against an individual and concludes with a discussion of the historical, theological, and linguistic contexts of the usage of the term in the Qur'ān and *ḥadīths* as well as in the early years of Islam.

The terms *takfīr*, *kufr* (disbelief or denial), and *kāfir* (an unbeliever) are derivatives of the Arabic root *k-f-r*. According to *Mu'jam Matn al-Lugha*, the original meaning of the term *kufr* is to conceal something, to cover up (*kaffara*), to disown, or to humiliate.[1] According to *Lisān al-'Arab*, the term also refers to those who carry weapons, for the weapons conceal their entire bodies; to peasants, for they sow seeds in the ground and cover them with soil;[2] to the night, for darkness conceals all objects; and to the sea, for the water covers everything in it.[3] According to the definitions above, an ignorant person is also a *kāfir,* for *kufr* has concealed his entire heart and has shielded him from obtaining the necessary knowledge.

Kufr also means to reject, to disobey, or to deny. For instance, when God called the people to believe in his oneness, the *kāfir* refused to do so. Thus, rejecting God means concealing (covering) the self from his guidance and blessing. Hence, the unbelievers—referred to in the Qur'ān as the *kāfirūn* or *kuffār*, plural of *kāfir*—committed disbelief (*kafarū*) for they rejected, denied, and disobeyed God's commandments: "They know God's blessings, but refuse to recognize them: Most of them are ungrateful [*wa aktharuhum al-kāfirūn*]" (Q. 16:83). According to Taqī al-Dīn Aḥmad ibn Taymiyyah (d. 1328)—a prominent jurist and theologian—*kufr* is "The disbelief in God and

His Messenger, whether that is accompanied by denial or it is not accompanied by denial but rather doubt, or turning away from all of this out of jealousy or arrogance, or because of desires that prevent one from following the message."[4]

The concept of *takfīr* is closely connected with the Islamic conception of *īmān* (faith). The lexical meaning of *īmān*, according to *al-Qāmūs al-Muḥīṭ*, is "confidence, truthfulness, obeisance, and the acceptance of the *sharīʿah* (Islamic law)."[5] Thus, *īmān* means having confidence in, surrendering to, and accepting the word of God. If *īmān* indicates a belief in the oneness of God and in Muhammad as his messenger, *islām*, or submission to God's will, means the practice of *īmān*. Possessing *īmān*, thus, becomes essential to affirming the Islamic identity in the individual or the community, for a lack of *īmān* is tantamount to unbelief or *kufr*. Therefore, a Muslim who lacks *īmān* is a *kāfir*, and the lack of belief or its antidote—the denial of God's existence—is defined as major *kufr*. Minor *kufr*,[6] however, is a lesser crime and is considered more or less a minor offense.[7]

The theological debate on the concepts of *kufr* and *īmān* emerged very early in Islamic history.[8] Muslim scholars[9] have argued that although the question of *īmān* had generated theological debates in the ninth century and thereafter, debates on *īmān*, *kufr*, and apostasy in recent times resurfaced due to a variety of factors, chief among them is the rise in the number of Islamic sects and the concurrent development of ideological trends. Scholars also attribute the recent surge in the number of Muslim groups adopting the concept of *takfīr* to the decline of Islamic values and the loss of solidarity among the people after centuries of colonialism and foreign domination of Muslim societies.[10] The deterioration of Islamic values, argues Ḍuḥā al-Khatīb, has prompted Muslim leaders of modern revivalist and fundamentalist Islamic movements to raise the issue of legitimacy in Islam (i.e., what acts are legitimate or illegitimate?), an issue that was never raised in classical Islam.[11] The question of legitimacy, for instance, came to the fore after the dismantling of the Islamic caliphate at the beginning of the twentieth century and the establishment of a new socio-political order. The Islamists,[12] therefore, considered the new order illegitimate, for it impacted the *īmān* of the people by imposing on them laws alien to Islam.[13]

Of concern to the Islamists are 1) issues relating to the Islamic dress, behavior and character of the individual, and the religious teachings; 2) issues relating to the social order and its variants, and the extent of its compatibility with the early Muslims (*al-salaf al-ṣāliḥ*); and 3) issues relating to the socio-political order and whether the foundations of that order are legitimate or illegitimate, that is, based on *īmān* or *kufr*, and the alternative to that order as exemplified in an Islamic order that implements the *sharīʿah*.[14] Thus, one may argue that the factors mentioned above have contributed to the surge of *takfīrī* groups in recent times.

Before discussing the concept of *kufr* and *takfīr*, this chapter first delves into the meanings and interpretations of the terms *īmān* and *islām* and their implications and importance in Islam. A discussion of these two terms may help answer some questions as to why the practice of *takfīr* has become so central and detrimental to various Muslim groups past and present. According to the Qur'ān, a person with *īmān* is someone who increases his obedience to the teachings of the religion; his utterances are compatible with religious doctrines; and his works are dedicated to promoting the common good (Q. 2:143; 12:17; 3:53; 20:112). A belief in the message of God is a proof of *īmān*. The term *īmān*, which appears in 729 Qur'ānic verses,[15] is used in the Qur'ān in various forms to indicate all that is required from the people in terms of belief and right conduct, such as believing in God, his messengers, and his books, and performing the required religious duties. For instance, the Qur'ān uses the term *īmān* to refer to prayer: "God would never let your faith [*īmānakum*, i.e., your prayers][16] go to waste" (Q. 2:143).[17] The Qur'ān also uses the term as a reference to belief in God's message: "You will not believe us [*wa mā anta bi mu'min lanā*]" (Q. 12:17); "Lot believed him [*fa amana lahu Lūṭ*]" (Q. 29:26); and "Lord, we believe [*āmannā*] in what You have revealed" (Q. 3:53). God used the term when He commanded all Muslims to confess their belief in His revelation: "say, 'We believe [*āmannā*] in God and in what was sent down to us'" (Q. 2:136). *Īmān* is also used as a condition for right conduct: "but whoever has done righteous deeds and believed [*wa huwa mu'minun*] need have no fear of injustice . . ." (Q. 20:112). An individual may also increase or decrease his *īmān* depending on his deeds and words that express, test, or measure his *īmān* (Q. 3:173; 8:2; 9:124–125; 33:22; 48:4; 74:31). *Īmān* also means a state of security, when the soul is at peace and is devoid of fear and chaos.[18]

The proof of *īmān*, according to Ibn Taymiyyah, is the acceptance of what the prophets have brought down (*taṣdīq*[19] *al-rusul fīmā akhbarū*).[20] According to Mālik Muṣṭafa Wahbī al-'Āmilī, *taṣdīq* is evaluative of the degree of *īmān*,[21] meaning that works and deeds are proof of *taṣdīq*, which in turn is used to evaluate the degree of *īmān*. He asserts that most Islamic sects agree that *taṣdīq* is necessary for *īmān*.[22] The sects that define *īmān* based on works, he adds, were accused of being fixated on outward belief. He supports this claim with Qur'ānic verses like 2:134: "What they earned belongs to them, and what you earn belongs to you." Al-'Āmilī reads this verse to mean that work is an act of *īmān* and argues that those who attach work to *īmān* want to include work under the definition of *īmān*, not to make work a condition for *īmān*.[23]

According to 'Abd al-Mun'im Muṣṭafā Ḥalīmah, the majority of Muslim theologians and scholars have argued that practicing Islam publicly (i.e., fulfilling all religious duties and rituals) without accepting (*taṣdīq*) the message of the religion is not *īmān* but *kufr*, hypocrisy, and deviation from

Islam.[24] For instance, the Qur'ān speaks against hypocrisy in the following verse: "The hypocrites will be in the lowest depths of Hell, and you will find no one to help them" (Q. 4:145). Similarly, the Prophet Muḥammad expressed resentment against those who proclaim membership in the Islamic community without accepting the message of Islam. For instance, Muḥammad once said: "works are by intentions only,"[25] meaning that works without *taṣdīq* is not *īmān*.

Another trait of *īmān* is the acceptance in the heart—after testifying publicly (confessing by the tongue)—of the two monotheistic testimonies (*shahādatay al-tawḥīd*) that there is no god but God and that Muḥammad is His Messenger and Servant. This is the most important pillar of Islam, followed by praying (*ṣalāt*), fasting (*ṣawm*), obligatory charity (*zakāt*), and making the pilgrimage (*ḥajj*) to Mecca. It is reported that Muḥammad used to visit with his relatives and companions who were on their deathbeds and required them to utter the testimony, "There is no god but God," with passionate belief so they could die the death of a believer.[26] It is also reported that early theologians and legal scholars agreed that true believers are only those who believe in their hearts without any doubt.[27]

According to the Prophet Muḥammad, *īmān* is both belief and works. When asked, "What is the best deed?" He replied, "To believe in Allāh and His Apostle (Muḥammad)."[28] It is reported that the Prophet and his companions, as well as early Muslims, accepted the Islam of those who uttered the two testimonies without inquiring about their inner belief (*al-taṣdīq al-qalbī*) or their deeds. They judged belief based on the two testimonies.[29] Thus, Muḥammad considered *īmān* a work of piety and regarded it as the best work for a human being. The Prophet also declared: "None of you will have faith [*īmān*] till he wishes for his (Muslim) brother what he likes for himself."[30] He also said: "An adulterer, at the time he is committing illegal sexual intercourse, is not a believer; and a person, at the time of drinking an alcoholic drink, is not a believer; and a thief, at the time of stealing, is not a believer."[31] For Muḥammad, *īmān* (having faith in God) is by itself an act of piety, and whoever has *īmān* is considered a believer (*muslim*).[32]

Classical and medieval Muslim theologians[33] debated whether *īmān* is faith alone or both faith and works. For instance, the Mu'tazilites asserted that *īmān* is the belief in the Islamic doctrine, the utterance of the monotheistic testimonies, as well as works.[34] In contrast, the Murji'ites have defined *īmān* as belief with the utterance of the testimonies only to the exclusion of works.[35] Those who argued that *īmān* is both faith and works also defined the acts that would decrease *īmān* in the person. Among these acts are worshipping a deity other than God, bowing to idols, visiting the dead, mocking the religion of God and His beautiful names, insulting God or His Messenger, collaborating with a tyrant or with the infidels against the Muslims, and implementing laws other than the laws sent from God. Committing any of these acts strips the Muslim from his monotheism and *īmān* and renders him a *kāfir*.[36]

The Qur'ān admonishes those who abandon *īmān* after declaring belief. Verse 16:106 says: "Those who reject God after believing . . . will have the wrath of God upon them and grievous punishment awaiting them." The Qur'ān also says: "'You have gone from belief to disbelief [*kafartum ba'da īmānikum*].' They swear by God that they did not, but they certainly did speak words of defiance [*kalimat al-kufr*] and became defiant [*wa kafarū*] after having submitted" (Q. 9:66, 74). The verses demonstrate that abandoning *īmān* renders the person a *kāfir*.

Likewise, correct action (acting wisely) is as equally important as *īmān* because those who do not act according to what God has sent down are indeed unbelievers [*kāfirūn*] (Q. 5:44). The Arabic term for the correct or wise action is *yaḥkum*, meaning to make a decision according to the laws God sent in the Qur'ān. To act based on what God has sent down is an act aimed at enforcing God's word, message, and laws. Therefore, every individual has the ability to act wisely, and to commit an act based on what God has sent down is an act that springs from *īmān*; abandoning that act means abandoning *īmān*, which means *kufr*.

Like *īmān*, the term *islām*, a derivative from the Arabic root *s-l-m*, means peace, security, or harmony. It means the relationship between the people and their Creator as demonstrated in the following verse in reference to the patriarch Abraham: "His Lord said to him, 'Devote [*aslim*] yourself to Me.' Abraham replied, 'I devote [*aslamtu*] myself to the Lord of the Universe'" (Q. 2:131). Thus, in the Qur'ān, the term *muslim* means submission (*taslīm*) to God only, not to anything or anyone. The term *muslim* also meant the complete submission to the *sharī'ah*. If the person declares submission and follows the *sharī'ah*, he becomes entitled to all rights allotted to him by Islam, the religious institution.[37]

The declaration of submission takes effect when the person who is not originally a Muslim (i.e., a member of the Muslim community) utters the two testimonies. This, however, is not required from a person who was born to Muslim parents.[38] Islam is of two types: the Islam of the tongue and the Islam of the heart.[39] The Islam of the tongue is called the outer Islam (*al-islām al-ẓāhirī*). Muslims who show outward belief enjoy all Islamic rights as full Muslim citizens. The Islam of the heart, on the other hand, is referred to as the inner Islam (*al-islām al-ḥaqīqī al-wāqi'ī*), which guarantees its adherents a place in heaven.[40]

When applied to a society, group, or individual, the term "Islamic" means that the society, the group, or the individual follows the right path or the path of Islam, practicing Islam based on the Qur'ān and the examples of Muḥammad in matters of marriage, divorce, inheritance, and so on.[41] The question remains: Which Islam is meant here? Is it the inward Islam or the outward one? If the right path or the path of Islam is *the* inner Islam, and if we assume that there is only one unified inner Islam, then we are faced with a crisis, for it is impossible to have a uniformed inner Islam when, in the same sect, there are many interpretations of what inner Islam means.[42] Thus, to keep all the Muslims who practice different types of inner Islam within the realm of the religion, Muslim theologians and jurists have dismissed the idea that inner Islam is one and unified; rather, they have argued for the existence of multiple paths to inner Islam.[43] This approach is meant to protect people from being declared non-Muslim or *kuffār* if their form of inner Islam is incompatible with the inner Islam of another sect.[44] Imām Muḥammad ibn Idrīs Al-Shāfi'ī (d. 820)—the founder of the Sunnī Shāfi'ī school of law—argued that complete faith begins with the belief in God and His Messenger. The belief in God without the belief in His Messenger is not a complete faith, and no one has a complete faith until he believes in God's Messenger.[45] Al-Shāfi'ī adds that "Muḥammad followed this formula when he tested the people's faith."[46] Verse 24:62 supports the above views: "The true believers are those who believe in God and His Messenger . . . those who ask your permission [Prophet] are the ones who truly believe in God and His Messenger." Thus, one may argue that this outer Islam (i.e., uttering the two testimonies) is enough to brand a society Islamic or an individual a Muslim. However, despite the scholars' and the theologians' attempts to avoid the crises generated by the various interpretations of what Islam means and what makes a person a Muslim or a *kāfir*, it is evident today how some Islamic groups, like the self-declared Islamic State of Iraq and Syria, denounce members of another sect as *kuffār* based on the differences in opinion.

Pre-modern prolific writers and scholars opined that any statement contradicting the two testimonies of monotheism is considered an act of *kufr*, regardless of whether the person is aware of the contradiction or not. For instance, the eighth-century Muslim jurist Abū Ḥanīfah al-Nu'mān (d. 767) contends that *islām* is submission and obedience to God's commands, whereas *īmān* is the acknowledgement and belief that there is no god but God and that Muḥammad is His messenger, that God has no associates, that God has His angels, books, and messengers, and His heaven, hell, and the Day of Judgment.[47] *Islām*, Abū Ḥanīfah adds, is what appears on the outside, and *īmān* is what in the heart, and both *islām* and *īmān* are one.[48] The twelfth-century Muslim theologian Abū Ḥāmid Muḥammad al-Ghazālī (d. 1111), on the other hand, asserts that a person is a Muslim if he declares his *islām* by the tongue without believing in the heart, and should be judged on what appears of him, for it is not known what is in his heart.[49] The fourteenth-century Muslim scholar Badr al-Dīn al-Zarkashī (d. 1392) asserts that the two testimonies are a condition for *Islām*; any statement or act that undermines them, even unintentionally, is a departure from *Islām*.[50] Ibn Taymiyyah writes that a person cannot be a *mu'min* unless he is a *muslim*.[51] *Īmān* is an aspect of *islām* but it is not the whole of *islām*. *Islām* is the submission to God and the obedience to Him in every matter. The name *islām* covers all aspects of the religion: Obedience to God (*ṭa'at Allāh*), *īmān*, *taṣdīq*, and the like.[52]

Based on the exegetical literature, there are several types of *kufr*.[53] The first is *Kufr al-'inād* (disbelief out of stubbornness). This *kufr* applies to someone who knows the truth and admits to knowing the truth and admits to knowing it with his tongue, but refuses to accept it and refrains from making a declaration. God says, "Hurl every obstinate disbeliever [*kaffār 'anīd*] into Hell" (Q. 50:24). The second type is *Kufr al-inkār* (disbelief out of denial). This *kufr* applies to someone who denies with both the heart and the tongue. God says, "They know God's blessings, but refuse to recognize them: Most of them are ungrateful [*wa aktharuhum al-kāfirūn*]" (Q. 16:83). The third is *Kufr al-kibr* (disbelief out of arrogance and pride). The disbelief by the devils (e.g., Iblīs) is an example of this type of *kufr*. The fourth is *Kufr al-juḥūd* (disbelief out of rejection). This *kufr* applies to someone who acknowledges the truth in his heart but rejects it with his tongue. This type of *kufr* is applicable to those who call themselves Muslims but who reject any necessary and accepted norms of Islam such as prayer and almsgiving. God says, "They denied them [*wa jaḥadū*], in their wickedness and their pride, even though their souls acknowledged them as true" (Q. 27:14). The fifth is *Kufr al-nifāq* (disbelief out of hypocrisy). This type of *kufr* applies to someone who pretends to be a believer but conceals his disbelief. Such a person is called a *munāfiq* or hypocrite. God says, "The hypocrites [*al-munāfiqīna*] will be in the lowest depths of Hell, and you will find no one to help them" (Q. 4:145). The sixth type of *kufr* is *Kufr al-istiḥlāl* (disbelief out of trying to permit what is forbidden). This *kufr* applies to someone who accepts as lawful (*ḥalāl*) that which God has made unlawful (*ḥarām*) like alcohol or adultery. Only God has the prerogative to make things *ḥalāl* and *ḥarām* and those who seek to interfere with His rights are like rivals to Him and therefore fall outside the boundaries of faith. The seventh type is *Kufr al-kurh* (disbelief out of detesting any of God's Commands). God says, "how wretched will be their state! God has brought their deeds to nothing. It is because they hate [*karihū*] what God has sent down that He has caused their deeds to go to waste" (Q. 47:8–9). The eighth type is *Kufr al-istihzā'* (disbelief due to mockery and derision). God says, "Say, 'Were you making jokes [*tastahzi'ūna*] about God, His Revelations, and His Messenger? . . . [Y]ou have gone from belief to disbelief'" (Q. 9:65–66). The ninth type is *Kufr al-i'rāḍ* (disbelief due to avoidance). This *kufr* applies to those who turn away and avoid the truth. God says, "Who could be more wrong than the person who is reminded of his Lord's messages and turns his back on them [*fa 'a'raḍa 'anha*], ignoring what his hands are storing up for him [in the Hereafter]?" (Q. 18:57); and the last type is *Kufr al-istibdāl* (disbelief through trying to substitute God's Laws). This type of *kufr* could take the form of (a) rejecting God's Law (*sharī'ah*) without denying it, (b) denying God's law and therefore rejecting it, or (c) substituting God's laws with man-made laws. The Qur'ān says, "How can they take protectors other than Him? God alone is the Protector" (Q. 42:9). God also says, "Do not say falsely, 'This is lawful and that is forbidden,' inventing a lie about God: those who invent lies about God will not prosper" (Q. 16:116).[54]

The major *kufr*, or *al-kufr al-akbar*, is defined as the straightforward disbelief in God and His message. This *kufr* renders the person non-Muslim or *kāfir*, where he loses all rights and dignity allotted to him by Islam. He becomes subject to laws concerning disbelief (*aḥkām al-kufr*) if his *kufr* were due to blasphemies, or to laws concerning apostasy if his *kufr* were the rejection of Islam after accepting its message.[55] The following Qur'ānic verses demonstrate the context in which the major *Kufr* appears: "[*kafara*, they defied God] Those who say, 'God is the Messiah, son of Mary . . . ' [*kafara*, they defied God] Those people who say that God is the third of three" (Q. 5:72–73). Also, "those who disbelieve [*kafarū*] and deny Our messages shall be the inhabitants of the Fire" (Q. 2:39). Based on these verses, not only Muslims but also non-Muslims are subject to accusation of *kufr* since verses 5:72–73 reference the Christians as committing *kufr* for believing in the Trinity and the divinity of Jesus; the Muslims are referenced in verse 2:39 for rejecting Islam after accepting its message.

In the *ḥadīths*, there is also reference to major *kufr*. It is reported that the Prophet Muḥammad once asked the people to be believers in God alone by testifying that none has the right to be worshipped but Allāh and Muḥammad is Allāh's Apostle, to offer prayers, pay the *zakāt* (obligatory charity), and to observe fast during the month of Ramadan.[56] Thus, according to the above *ḥadīth*, refusing to do so was considered an act of major *kufr*. This type of *kufr*, which results in the exclusion of the person from the community, is categorized as a *kufr* due to stubbornness, denial, arrogance, hypocrisy, hatred, and "collapse before the passions and caprices that blind the eyes to understanding and lead to mockery of God's revealed message."[57]

The minor *kufr*, or *al-kufr al-aṣghar*, is a *kufr* without a *Kufr*, meaning that a person committing a minor *kufr* will not be subject to the laws of the major *kufr*, that is, the exclusion from the community. The person who is accused of a minor *kufr* does not lose the rights and dignity allotted to him by Islam. In this case, his punishment is left to God on Judgment Day. God may punish him for a while, but at the end He will bestow His mercy on him. An example of the minor *kufr* is the denial of or the failing to acknowledge God's blessing.[58] The Qur'ān has several references to this type of *kufr*. For instance, verse 27:40 says: "But one of them who had some knowledge of the Scripture said: 'I will bring it to you in the twinkling of an eye.' When Solomon saw it set before him, he said, 'This is a [favor] from my Lord, to test me whether I am grateful [*ashkuru*] or not [*akfuru*].'" Thus, here the Arabic term *akfuru* (to commit unbelief or *kufr*) is a reference to the act of being ungrateful to God for Solomon seems to be saying that if he is not grateful then he has committed *kufr*. Nonetheless, the act is a minor *kufr* because the person here is rejecting God's blessing rather than God Himself or His message.

According to Islamic legal concepts and understanding, there are eleven basic rules of *takfīr*.[59] The first basic rule states that the general laws regarding *kufr* do not always lead to the conviction of the person who is accused of committing *kufr*.[60] In other words, the acts that are deemed acts of *kufr* may be dismissed and the charges may be dropped depending on the characteristics of the individual who is accused of committing such acts.[61] Possessing certain characteristics may shield the accused from being declared a *kāfir*. In theory, the laws consider certain deeds, sayings, or behavior as grounds for *takfīr*; however, in practice the laws are not always enforced because of the person's background and past behavior. According to the Prophet Muḥammad, although God has condemned those who produce, sell, consume, and transport alcohol, this offense does not always earn the offender the accusations of *takfīr* as long as religious piety is apparent in that person.[62] For instance, when Muḥammad ordered the lashing of a man for drinking alcohol, some people suggested that the man be declared a *kāfir*. Upon hearing this suggestion, Muḥammad came to the defense of the man, saying that he had always showed utmost love and respect for the messenger of God.[63] Thus, the man possessed some characteristics (i.e., love and respect for Muḥammad) that shielded him from being declared a *kāfir*. Also, during the time of Muḥammad, the accusations of *kufr* did not materialize against people who drank alcohol because they were not aware of the newly revealed Qur'ānic injunctions that prohibited this practice.[64] Thus, the grounds for dismissing the charges of minor *kufr* pertaining to drinking alcohol are expressed in the following Qur'ānic verse: "Those who believe and do good deeds will not be blamed for what they may have consumed [in the past] as long as they are mindful of God, believe and do good deeds" (Q. 5:93).

The second basic rule of *takfīr* is called "the fulfillment of the promise, and the postponement of the judgment."[65] This rule states that those who believe in God's message and obey His teachings will be rewarded in this world; as for those who commit minor *kufr*, their fate will be at the hands of God on Judgment Day: Either He will punish them or He will bestow His mercy upon them.[66] The ground for this rule is the following verse: "God does not forgive the joining of partners with Him: anything less than that He forgives . . . anyone who joins partners with God has concocted a tremendous sin" (Q. 4:48). Based on this verse, it is clear that associating partners with God (*shirk*) is an act of *kufr*, which earns the *kāfir* God's ultimate punishment.

The third basic rule of *takfīr* proclaims that accepting *kufr* is *kufr* in and of itself.[67] This rule maintains that Muslims who refrain from taking action when someone commits *kufr*, they themselves commit *kufr*. Inaction against *kufr* means the acknowledgement of the legitimacy of the act that is considered an act of *kufr*. Inaction against acts of *kufr* means satisfaction in and acceptance of what God has condemned.[68] This basic rule is based on the following verse: "As He has already revealed to you [believers] in the Scripture, if you hear people denying and ridiculing God's revelation [*yukfaru bihā*], do not sit with them unless they start to talk of other things, or else you yourselves will become like them: God will gather all the hypocrites and disbelievers [*wa al-kāfirīna*] together into Hell" (Q. 4:140). Based on the above verse, a Muslim ruler ordered the punishment of a group of Muslims who were drinking alcohol, including one person who was fasting. When asked about punishing a man who was fasting and never drank a drop of alcohol, he cited the above verse and added that the man's guilt was greater and he was the first to be punished.[69]

The fourth basic rule of *takfīr* states that the apparent behavior of a person can be used as an indicator of either *kufr* or *īmān*.[70] Based on the person's behavior, whatever appears to be *kufr* is indeed *kufr* and whatever appears to be *īmān* is indeed *īmān*. Ibn Taymiyyah explains this rule thus: An individual can be accused of either *kufr* or *īmān* based on his apparent behavior or what appears to be his actions. If his actions appear to be acts of *kufr*, then he is declared a *kāfir*; if his actions appear to be acts of *īmān*, then he is declared a man of faith without even investigating what the person holds in the inside.[71] For the scholars who uphold this basic rule, what the person holds in his heart does not matter, for only God knows what is in the hearts of the people.[72] This rule draws its support from the following *ḥadīth*: "I was ordered to fight the people until they testify that there is no god but God and Muḥammad is His messenger, perform the prayers, and pay the almsgiving. If they obey, they are spared my punishment and God is their judge."[73] Thus, actions obvious to the eye speak for what is in the heart. In other words, if religious duties are fulfilled, then a person appears to be a man of faith.

In the situation where both *kufr* and *īmān* appear simultaneously in the person (i.e., showing signs of *īmān* but committing acts of *kufr*), some theologians such as Ibn Taymiyyah argued that the person is a *kāfir* notwithstanding the simultaneous appearance of *īmān*.[74] The latter does not benefit him or change his status; he is a *kāfir* and an outcast until he rejects his *kufr*. Verse 12:106 supports Ibn Taymiyyah's claims: "Most of them will only believe in God while also joining others with Him." According to Ibn Taymiyyah, both *kufr* and *īmān* were evident in the Mongols who invaded Muslim territories in the thirteenth century and declared their Islam afterwards. For Ibn Taymiyyah, the Mongol leaders were considered *kuffār* although many of them were Muslim and testified that there is no god but God. For Ibn Taymiyyah, although these Mongols uttered the testimony, performed the prayers, and fasted during the month of Ramadan, most of them neglected the path of Islam. They lacked *īmān*, Ibn Taymiyyah contends, because they did not punish those who abandoned Islam, they praised those who supported their rule regardless of whether they have committed *kufr* or not, and they killed God-fearing Muslims just because they rose against Mongol rule. It is important to mention that Ibn Taymiyyah issued a *fatwā* (religious ruling) declaring the Mongols *kuffār* because he believed they governed with a mix of Islamic and non-Islamic legal codes.[75]

The accusations of *kufr* may also be applied to those who believe in Muḥammad's message but fail to acknowledge the oneness of God by declaring the monotheistic testimony: There is no god but God.[76] For instance, it is reported that the Prophet Muḥammad once asked his dying uncle Abī Ṭālib (d. 619) to utter the testimony so that Muḥammad would be able to testify on his uncle's behalf on Judgment Day. Abī Ṭālib was hesitant and died without saying it.[77] Also, the following verse is related to the *ḥadīth* above: "It is not for you [Prophet] to guide them; it is God who guides whoever He will" (Q. 2:272). Thus, Muḥammad's uncle died as a *kāfir* although he protected Muḥammad against the *kuffār* of his own tribes, and at times he expressed belief in Muḥammad's message when he said: "I am aware that Muḥammad's religion is one of the best religions."[78] Thus, he died a *kāfir* because he refused to acknowledge the oneness of God when he had the chance to do so.

Now, defending Muḥammad and believing that his religion is a true faith were not enough to make a person a believer without the testimony. According to a *ḥadīth*, Muḥammad asked a slave shepherd: "'Where is God?' She replied: 'In heaven.' He then said: 'Who am I?' She replied: 'You are the messenger of God.' He then turned to her master and said: 'Free her, for she is a believer.'"[79] Thus, the slave was a believer for acknowledging the existence of one single God although her testimony took the form of answering two simple questions.

The fifth basic rule of *takfīr* states that an act of disbelief or mere words expressing lack of faith are sufficient to render the person a *kāfir*, unless he is compelled to commit such an act or utter such words.[80] This rule is justified in the following Qur'ānic verse: "With the exception of those who are forced to say they do not believe, although their hearts remain firm in faith, those who reject God [*man kafara billāh*] after believing in Him and open their hearts to disbelief [*kufr*] will have the wrath of God upon them" (Q. 16:106). The statement "open their hearts to disbelief" ("*sharaḥa bil kufr ṣadran*") means that those who commit *kufr* other than by compulsion are deemed *kuffār*. According to the Qur'ān commentator and founder of the Saudi Wahhābī school of thought Muḥammad b. 'Abd al-Wahhāb (d. 1792), God does not excuse those who commit *kufr* unless they are forced to do so.[81]

Committing *kufr* willingly means committing *kufr* out of fear (of losing property, belongings), out of persuasion, out of passion to either a country, a family, or a tribe, love for money, or simply due to a candid joke. *Kufr* committed due to compulsion, however, as the verse above states, is not *kufr*.[82]

The sixth basic rule deals with a *kufr* called the minor act of *kufr*. This type of *kufr*, according to this rule, may not be valid without a legal clause that points directly to the act.[83] In other words, if the people consider a certain deed an act of *kufr*, the *kufr* is not valid unless the laws clearly define the deed. Committing a minor *kufr* does not make the person an unbeliever because the act committed is a minor disobedience rather than a disbelief or lack of *īmān*. According to one *ḥadīth*, "a fornicator who fornicates is not a believer as long as he commits fornication, and no one who steals is a believer as long as he commits theft, and no one who drinks wine is a believer as long as he drinks it, and repentance may be accepted after that."[84] The acceptance of repentance, according to this *ḥadīth*, is an indication that a minor *kufr* has been committed, for the *ḥadīth* says that a person commits unbelief only at the time when he is committing these acts. Belief will return when the person is no longer committing such acts. This is in stark contrast to someone who does not believe in what God has sent down. Here, the person is an unbeliever because of lack of faith rather than acts of disobedience.

Muslim theologians justify the above rule by arguing that no one should decide on matters of the law unless God is clear about it.[85] This rule goes against assumptions or analogies, and it protects the person if there is a doubt about whether the deed is an act of *kufr*. They also explain this rule by arguing that it is unlawful to declare people as either believers or unbelievers unless God has already defined them as such.[86] For instance, if the Qur'an defines the believers as people who "do not neglect their Prayers" (Q. 6:92), then the community will be justified to make the *kufr* accusations against those who abandon the prayers. The question remains: What if the person abandons the prayers for reasons other than disbelief in God? The theologians argued that in this case the accusations of minor *kufr* will be made against that person since abandoning the prayer is caused by anything other than disbelief in the message of God.[87] This rationale is supported by the following verse: "If they turn to God, keep up the prayer, and pay the prescribed alms, then they are your brothers in faith" (Q. 9:11). This verse indicates that those who resume the prayers ("turn to God") after abandoning them have committed a minor *kufr*. They were believers (having *īmān*) even when they did not pray, and by resuming the regular prayers, they acknowledge their *īmān*. If the prayers were abandoned due to polytheism (*shirk*),[88] then it is a major *kufr*. God does not forgive those who commit *shirk*: "anything less than that He forgives" (Q. 4:48).

The seventh basic rule of *takfīr* states that permitting religious prohibitions or prohibiting permitted things is an act of *kufr*.[89] The Qur'ān calls on the Muslims to promote righteousness and prohibit vice (*al-amr bi al-ma'rūf wa al-nahī 'an al-munkar*) (Q. 3:110; 9:71; 31:17). Going against this rule is *kufr*. Verses like the following ones support this rule: "Those Children of Israel who defied [God] (*kafarū*) were rejected . . . because they disobeyed, they persistently overstepped the limits. They did not forbid each other to do wrong (*kānū lā yatanāhawna 'an munkarin fa'alūh*)" (Q. 5:78–79). Thus, it is against God's teachings to permit what God has prohibited or vice versa. According to the Qur'ān, only what God commands will prevail, and "He does not allow anyone to share His rule" (Q. 18:26). For instance, verse 9:31 reproaches some Christians and Jews for taking "their rabbis and their monks as lords." Qur'ān and *ḥadīth* scholar Abū Muḥammad Ḥusayn b. Mas'ūd al-Baghawī (d. 1122) interpreted the statement "taking their priests to be their lords" to mean that they obeyed rules established by their priests even though the rules were contrary to God's command.[90] Thus, a mere acknowledgement of a law, method, or rule that was the work of someone other than God is *shirk*: "Most of them will only believe in God while also joining others with Him (*illā wa hum mushrikūna*)" (Q. 12:106).[91]

The eighth basic rule of *kufr* states that when a person converts to *islām*,[92] it will be unlawful to expel that person from the sphere of *islām* unless he commits an obvious act of *kufr* that is not subject to interpretations.[93] In other words, if there were disagreements about whether or not the deed committed by a person is an obvious act of *kufr*, then *kufr* is not committed.[94] There are Qur'ānic verses that require proof and evidence of *kufr* before issuing a declaration of *takfīr*. These injunctions prohibit people from declaring *takfīr* based on mere assumptions, doubt, or appearance. For example, the Qur'ān states: "So, you who believe, be careful [*fatabayyanū*] when you go to fight in God's way, and do not say to someone who offers you a greeting of peace, 'You are not a believer'" (Q. 4:94). Also, the Prophet Muḥammad said: "and not to fight against him unless we noticed him having open *Kufr* for which we would have a proof with us from Allāh."[95]

According to Muḥammad ibn Jarīr al-Ṭabarī (d. 923), the term "*fatabayyanū*" means do not rush to judgment if you are in doubt about the person's belief or *kufr*, and do not rush to issue a death sentence against those whom you are in doubt regarding their status, unless their *kufr* is clear to you beyond any reasonable doubt.[96] In the Islamic tradition, declaring an innocent person a *kāfir* is one of the most heinous crimes, as stated earlier, for the accused could suffer grave and dangerous consequences such as the shedding of his blood, the loss of his rights to inheritance, and the annulment of his marriage, among other things.[97]

Declaring an innocent believer a *kāfir* is a graver sin than bestowing belief on someone who in reality is an unbeliever. To make a Muslim a *kāfir* is equal to killing him because he will be stripped of all his rights and dignity. One *ḥadīth* says: "and cursing a believer is like murdering him; and whoever accuses a believer of disbelief, then it is as if he had killed him."[98] Thus, it is a graver sin to declare a Muslim a *kāfir* than to bestow belief on an unbeliever. According to al-Ghazālī, "those who rush to declare other Muslims *kuffār* are ignorant of the teachings of Muḥammad.[99]

The ninth basic rule of *takfīr* concerns "the status of the person at the time of death."[100] If the person's status were that of *īmān*, then he is a believer and will be destined for heaven; if his status were that of *kufr*, then he is a *kāfir* and will be destined to hellfire. This rule draws its support from the following verses: "As for those who disbelief [*inna al-ladhīna kafarū*], and die as disbelievers, God rejects them. If any of you revoke your faith and die as disbelievers [*fa yamūt wa huwa kāfirun*], your deeds will come to nothing in this world and the Hereafter, and you will be inhabitants of the Fire . . . even if they offer enough gold to fill the entire earth" (Q. 2:161, 2:217, 3:91). According to a *ḥadīth*: "whoever dies knowing that there is no god but God will enter heaven."[101] Thus, entering Heaven is conditional on acknowledging God's oneness upon death.[102] According to Muslim theologians, "he who testifies to the oneness of God before dying will escape eternal hellfire, regardless of the sins he has committed in his life, and he who dies as a *kāfir* will never enter Heaven, regardless of the good deeds he has done throughout his life."[103]

The benefit of the rule above is that it protects the dead from being declared *kuffār* if, throughout their lives, they had committed acts of *kufr* but while dying showed signs of repentance and *īmān*. This rule paves the way for the dead person to be judged based on the signs he shows during the final moments of his life. For instance, those who are known for their *kufr* throughout their lives and do not show signs of repentance or *īmān* while they are dying are judged as *kuffār*. Ḥalīmah writes that Muslim theologians and legal scholars have debated the status of a dying person who repents and believes inwardly without showing any signs of belief: Is he a *kāfir* or a believer? Since no human is able to know the inside of another person except by the signs he shows, the theologians concluded, according to Ḥalīmah, then it is permitted to judge the inner belief of another based only on the outward signs he shows, for the signs reflect the truth of the inner side.[104] But, if the person repents and becomes an inward believer without showing any signs of his belief, he may still appear as a *kāfir*, for the people do not know the hidden belief. Now, if the people were wrong, and the person was truly a believer, then God, who is able to see his inner belief, will be his final judge.[105] Thus, the benefit of this rule is that it gives the dying person a last chance to repent and die as a believer—although this is not shown by outward signs—and it gives the family of the dead person assurances that their dead relative has died a believer, a good reputation the family will enjoy, plus a guarantee that he or she will be spared eternal hellfire.

The tenth basic rule states that declaring a fellow Muslim a *kāfir* is in itself an act of *kufr*.[106] This rule draws its validity from the following *ḥadīth*: "If a person declares his brother a *kāfir*, one of them ends up a *kāfir*."[107] A Muslim may declare another Muslim a *kāfir* only in two instances: 1) for denying the truth of God's words, and 2) for mocking God, His message, or His messenger. In the first instance, denying the truth of God's words is a clear *kufr*. The guilt is that those who deny the truth that was sent from God have allowed themselves the authority to understand this world and man's nature in opposition to what God has said in the scripture.[108] The second instance is affirmed in the Qur'ān in the following verse: "Say: 'Were you making jokes about God, His Revelations, and His Messenger? Do not try to justify yourselves; you have gone from belief to disbelief [*qad kafartum*]'" (Q. 9:65–66). Thus, in these two instances only will a Muslim not be at fault in declaring another Muslim a *kāfir*.

The eleventh and final basic rule of *takfīr* states that failing to declare a person who has committed an act of *kufr* a *kāfir* is in and of itself an act of *kufr*, and doubting the *kufr* of a person who has committed an act of *kufr* is also an act of *Kufr*.[109] This rule draws its support from the *sharī'ah*[110] as well as from the writings of prominent theologians like Muḥammad b. 'Abd al-Wahhāb and others. For instance, Ibn 'Abd al-Wahhāb writes: "He who does not declare the *kufr* of the polytheists or doubt their *kufr* or correct their beliefs, he himself is considered a *kāfir*."[111] He adds: "A person cannot be a monotheist (*muwaḥḥidan*) except by denying that God has partners (*shirk*) and by declaring all those who commit *shirk* as *kuffār*."[112] Muḥammad b. Saḥnūn (d. 854–55), a jurist in the Maliki school of law, claims that "the religious scholars agree that anyone who reviles the Prophet and disparages him is an unbeliever and the threat of Allāh's punishment is on him. The community's judgment on him is that he be killed. Anyone who has any doubts about such a person's disbelief and punishment is also an unbeliever."[113] Thus, Ibn 'Abd al-Wahhāb and Ibn Saḥnūn do not leave room here for indifference or for postponement of judgment (i.e., deferring the person's fate to God on Judgment Day). The statements of the above theologians can be taken to mean that Muslims are required to take action against *kufr* now; failing to do so is considered an act of *kufr*. This position may explain the reasons for the dramatic rise in the number of *takfīrī* groups, especially in a time when globalization, secularization, and Westernization are affecting Muslim countries, and when turmoil and upheavals have wreaked havoc in the Middle East in recent years. These *takfīrī* groups attribute the deterioration of Muslim countries to the collective *shirk* of Muslim societies and governments. For them, Muslims who have embraced secularism and globalization have committed *shirk* because they have replaced God's laws with

Western and civic laws. For them, this is a clear violation of monotheism, which requires declaring an entire society as *kāfir* (*takfīr al-mujtama'*).[114]

Modern Muslim scholars like 'Abd al-Salām b. Salīm al-Sāhimī tells us that every human being is guaranteed five necessities under Islam. These necessities are religion, life, sound reason, honor and dignity, and financial stability.[115] He adds that Muslim jurists have also established laws to protect these necessities and have defined punishments for those who violate these necessities. Under these laws, not only are the Muslims entitled to these necessities, but also non-Muslims and even the *kuffār* who are granted the pledge of protection by the Muslims.[116] A *kāfir* granted the pledge of protection is known as *mu'āhid* and is entitled therefore to all the laws allotted to Muslims.[117] This rule draws its support from the following *ḥadīth*: "Whoever killed a *mu'āhid* shall not smell the fragrance of Paradise."[118] The Qur'ān also refers to this matter in the following verse: "If any one of idolaters should seek your protection [Prophet], grant it to him so that he may hear the word of God, then take him to a place safe for him" (Q. 9:6).

According to al-Sāhimī, the *kuffār* who are granted protection by the Muslims are of three types: The first type comprises those who are granted temporary protection during their stay in Muslim territories. They are called *al-musta'minūn* (the receivers of security). The second type includes those who enter into a peace treaty with the Muslims. They are called *al-Mu'āhidūn* (plural of *mu'āhid*) and are granted protection until the peace treaty has expired. The third type refers to those who are required to pay tribute tax (*al-jizyah*) and are referred to as the people of *dhimmah*, or *ahl al-dhimmah*.[119] Al-Sāhimī writes that Islam guarantees these types of *kuffār* the security of their blood, money, and family, for Islam makes it a religious duty to extend justice to the *kuffār* even if they are not granted protection.[120] Al-Sāhimī writes, "a Muslim committing aggression against peaceful *kuffār* is a betrayal of Islam, a crime subject to severe punishment under Islamic law."[121]

TAKFĪR IN THE QUR'ĀN

The words *kufr*, *kāfir*, and *kuffār* appear hundreds of times in the Qur'ān as the paragraphs below will demonstrate. It is important to understand these terms based on their lexical, historical, and theological contexts. The Qur'ān, for example, makes the accusations of *kufr* against those who express disbelief in God's words. The Qur'ān says: "Those who disbelieve [*kafarū*] and deny [*kadhdhabū*] Our messages" (Q. 2:39). According to this verse, those who deny the messages of God (i.e., the Qur'ān) are the *kuffār*. Thus, denying the word of God makes the person a *kāfir*. In Q. 2:89 God says, "When a Scripture came to them from God confirming what they already had . . . they disbelieved [*kafarū*] in it." In this verse, the *kuffār* are those who deny God, His angels, and His messengers. This verse states that the Qur'ān came down to confirm the belief of those who had already received the word of God but refused to believe in the Qur'ān because of enmity with Muḥammad and his followers. Thus, according to this verse, enmity with the Muslims, rather than disbelief, triggered the status of *kufr*.

In some verses, God addresses the *kuffār* or refers to *kufr* using terms other than *kufr* or *kāfirūn*. For instance, the term *al-fāsiqīn* is also used in the Qur'ān to describe those who rejected faith. Verse 2:26 states:

> [God] does not shy from drawing comparisons even with something as small as a gnat, or larger: the believers know it is the truth from their Lord, but the disbelievers say, "What does God mean by such a comparison?" Through it He makes many go astray and leads many to the right path. But it is only the rebels [*al-fāsiqīn*] He makes go astray.

The term *fāsiqīn* (plural of *fāsiq*) is a reference to the unbelievers who derided God for using similes and parables in the Qur'ān.[122] The verse above was revealed as a reply to the unbelievers who said, "God disdains to use the lowest creatures, such as a spider, for simile."[123] Thus, a *fāsiq* in the context of verse 2:26 is someone who ridicules God's words. Thus, the act of mocking God's words is called *fusūq*. The act of mocking God's words is also an act of *kufr*, according to the Qur'ān: "Say: 'Were you making jokes about God, His Revelations, and His Messenger? Do not try to justify yourselves; you have gone from belief to disbelief [*qad kafartum*]'" (Q. 9:65–66). According to al-Ṭabarī, the literal meaning of *fusūq* is "going out," "seceding," or "deviation."[124] For instance, verse 18:50 describes the act of the angelic figure Iblīs as *fusūq* when he broke the command of God: "He was one of the jinn and he disobeyed his Lord's command [*fafasaqa*, i.e., committed *fusūq*]." For this reason, writes al-Ṭabarī, the *kuffār* and the hypocrites are called *fāsiqīn* for their deviation from the right path, that is, disobeying God's commands.[125]

Some verses refer to *kufr* and *kuffār* without any mention of the terms, as verse 5:21 demonstrates: "My people, go into the holy land which God has ordained for you—do not turn back or you will be the losers [*khāsirīna*]." The term *al-khāsirīna* (the losers) in this context is synonymous to *al-kāfirīn* or *al-kuffār* (disbelievers). The Qur'ān tells stories about nations that committed acts of defiance and ended up among the losers. For instance, the nations of 'Ād and Thamūd had committed acts of defiance, and because of this defiance, they ended up among the losers or *kuffār* (Q. 14:9; 22:42; 29:38; 41:17; 53:51). Thus, in some verses, the term *al-khāsirīn* is used as a reference to disbelief or *kufr*.

Modern Muslim scholars provided varied analyses of the Qur'ānic origin of the term *kufr*. For instance, Ghālib Ḥasan al-Shābandar begins by arguing that the notion of major *kufr* is lacking in the Qur'ān. He states that nowhere in the Qur'ān does God accuse people of *kufr* for denying His existence.[126] Rather, what is mentioned in the text, al-Shābandar continues, is God's criticism of those who skew the image of God due to the misunderstanding of His nature. God does not chastise the people for rejecting or denying His existence, argues al-Shābandar, but for not understanding Him.[127] Even those who worship idols, he adds, they do it because they long to be closer to God. God, therefore, replied to those who misunderstood Him by criticizing them and correcting their views.[128] For example, the criticisms and the corrections the Qur'ān makes in some verses were directed at some Christians who understood the Trinity to be three separate and different gods,[129] at the unbelievers who rejected the Qur'ān as the word of God, at those who disbelieved in God's prophets, angels, and heaven and hell, and at those who refused to believe in the necessity of key Islamic pillars, such as prayer (*ṣalāt*); fasting (*ṣawm*); charity tax (*zakāt*); and pilgrimage (*ḥajj*). Thus, the Qur'ān chastises these unbelievers not for denying God's existence, which is a major *kufr*, but for committing a minor *kufr*, that is, deviating from the right path.[130]

Contemporary Qur'ān scholar Fāyiz Salhab tells us that the Qur'ān defines *kufr* based on different acts and beliefs, chief among them is the denial of God, God's signs, God's grace, God's messenger, the Day of Judgment, the angels, the revelations sent before the Qur'ān, and the prophets.[131] The *kufr* of God is described in the following verse: "those who reject God [*kafara billāhi*] after believing in Him and open their hearts to disbelief will have the wrath of God upon them and a grievous punishment awaiting them" (Q. 16:106). The denial of God's *āyāt* or signs (Qur'ānic verses, miracles) is referred to in Q. 2:39: "Those who disbelieve [*kafarū*] and deny [*kadhdhabū*] Our messages." The Qur'ān condemns those who reject God's grace: "how can they believe in what is false and deny God's blessing [*wa bi ni'matillāhi yakfurūna*]?" (Q. 29:67). The *kuffār* are the unbelievers who rejected Muḥammad as the Messenger of God: "Do not hold prayers for any of them if they die, and do not stand by their graves: they disbelieved in God and His Messenger [*kafarū billāhi wa rasūlihi*]" (Q. 9:84). And "anyone who does not believe in God [*wa man yakfur bi*] Allāh, His angels, His Scriptures, His Messengers, and the Last Day has gone far, far astray" (Q. 4:136). Also condemned in the Qur'ān as types of *kufr* are the following traits: hypocrisy and lying (Q. 4:156; 9:68, 73; 33:48), rebellion (Q. 49:7; 21:74), *fitnah*[132] (Q. 2:191; 8:25; 9:49), arrogance (Q. 2:34; 25:21; 29:39; 34:32; 40:47), treachery and betrayal (Q. 22:38; 4:107; 8:58), wickedness (Q. 80:42; 82:14; 83:7), and shameful and unjust deeds (Q. 6:151; 7:33; 16:90; 24:21; 29:45).

The Qur'ān also defines apostasy (*irtidād*) as *kufr* and refers to the *kāfir* as *murtad* or apostate. The lexical meaning of *irtidād* is "to turn back."[133] In religious terms, it denotes turning back from, abandoning, or changing one's religion, which means committing *kufr* after belief (*īmān*).[134] A believer in any of the revealed religions may become a *murtad* upon leaving that religion to a religion other than those God has revealed (i.e, leaving monotheism).[135] The Qur'ānic text, which condemns apostasy or *irtidād* as unbelief, warns believers not to turn back (i.e., commit *irtidād*) from their faith:

> If any of you revoke your faith [*yartadid*] and die as disbelievers [*kāfirūn*] . . . you will be inhabitants of the Fire. . . . You who believe, if any of you go back on your faith [*yartadda*]. . . . [And] those who turn on their heels [*irtaddū*] after being shown guidance are duped and tempted by Satan (Q. 2:217; 5:54; 47:25).

TAKFĪR IN THE *ḤADĪTHS*

The *ḥadīth* literature throws the accusations of *kufr* on Muslims and non-Muslims alike, and it condemns acts of disbelief as major *kufr* and minor offenses as minor *kufr*. According to the *ḥadīths*, the Prophet Muḥammad issued statements against Jews and Christians for deviating from monotheism. For instance, in one *ḥadīth*, Muḥammad condemns "Jews and Christians[136] for they turned the graves of their Prophets into places of worship."[137] This *ḥadīth*, according to Fāyiz Salhab, is directed against *some* Jews and Christians who, like some followers of Muḥammad, showed *īmān* outwardly yet committed *kufr* inwardly. Thus, this *ḥadīth* is an example of declaring non-Muslims *kuffār* for committing a major *kufr*. Also, some *ḥadīths* refer to neglecting the prayers as *kufr*: "Verily between man and between polytheism and unbelief is the negligence of prayer"[138] and "it is the utmost apathy and sign of *kufr* and hypocrisy that one who hears the call of Allāh to salvation does not respond."[139] These *ḥadiths* may seem to indicate that Muslims who do not pray are *kuffār*. Imām Abū Zakariyyah Yaḥya ibn Sharaf al-Nawawī (d. 1277), a Sunni Muslim author on jurisprudence and *ḥadīths*, explains the preceding *ḥadīths*. He contends that the accusations of *kufr* were indeed directed at the Muslims who neglected the prayers not because of their inability to pray but as a result of their own belief that this practice is not mandated (or unnecessary) in Islam.[140] Thus, for al-Nawawī, the mere neglect of prayer is not ground for *kufr*.

But, al-Nawawī's interpretation of the preceding *ḥadīths* may be contradicted by a *ḥadīth* that indicates that a mere neglect of the prayers is equivalent to rejecting God, which is an act of *kufr*. For instance, in that *ḥadīth*, Muḥammad said: "I have been commanded to fight against people till they testify that there is no god but Allāh, that Muḥammad is the messenger of Allāh, and they establish prayer, and pay *zakāt* and if they do it, their blood and property are guaranteed protection on my behalf except when justified by law, and their affairs rest with Allāh."[141] The phrase "establish prayer . . . and pay *zakāt*" in the quote above, one may argue, is a clear reference to *any* person who failed to pray or give the *zakāt*.

On the other hand, Muḥammad vehemently condemned the killing of a Muslim by another Muslim, and declared such a deed an act of *kufr*. The Prophet said, "[d]o not (become infidels) revert to disbelief after me [*lā tarji'ū ba'dī kuffāran*] by striking the necks (cutting the throats) of one another (killing each other)"[142] for "[a]busing a Muslim is *Fusūq* (evil doing) and killing him is *Kufr*."[143] Imām Muslim ibn al-Ḥajjāj (d. 875), a *muḥaddith* (a scholar of *ḥadīths*), explains the phrase "do not become infidels (*kuffār*)" in the *ḥadīth* above to mean that it is *kufr* to kill a Muslim unjustly or to fight one another. Some *ḥadīth* commentators, however, argue that the term "*kuffāran*" in the phrase "*lā tarji'ū ba'dī kuffāran*" is a prohibition against carrying weapons or arming oneself with weapons since the term *kaffara*, as mentioned previously, means to cover the body with weapons.[144] Thus, the command "do not become *kuffār*," in the context of this *ḥadīth*, has two meanings: it either says "do not become infidels" or "do not carry weapons," for the latter might lead one to killing another if two are in dispute among themselves on matters of faith.[145] Thus, what is meant by the above-*ḥadīths*, according to Muḥammad Salīm 'Ubadydāt, is that the accusation of *kufr* is justified against those who kill Muslims unjustly, for they have full knowledge that this act is condemned in the Qur'ān.[146] 'Ubaydāt supports his claims with the following verses: "If anyone kills a person—unless in retribution for murder or spreading corruption in the land—it is as if he kills all mankind" (Q. 5:32) and "[i]f anyone kills a believer deliberately, the punishment for him is Hell" (Q. 4:93).

While the *ḥadīth* literature condemns the acts mentioned above as major *kufr*, it also speaks against minor offenses and describes them as minor *kufr*. Minor offenses include defaming the ancestry of other families, wailing over the dead,[147] declaring an innocent Muslim a *kāfir*,[148] and the fleeing of the slave from his master.[149] Based on the aforementioned *hadiths*, minor offenses warrant minor *kufr*, and the term *kufr* is used on such offenses only to make the people aware of their dangers rather than declaring them infidels.

The term *kāfir*, in its early usage, was used on Muslims who deviated from the core teachings of the religion. The usage of the term was meant only to warn these Muslims rather than to strip them of their Islamic identity or exclude them from the community. The concept of *takfīr* evolved throughout the centuries, especially in the formative periods of Islam, and became a theological issue that stirred theological debates among major theological schools in classical Islam. As the following chapter will demonstrate, emerging theological schools provided their own definitions and interpretations of *takfīr* to assert their *īmān* in Islam, on the one hand, and to affirm the legitimacy of their own religious views against the religious views of the other, on the other hand.

NOTES

1. See, for example, Aḥmad Riḍā, *Mu'jam Matn al-Lugha*, vol. 5 (Beirut, Lebanon: Dār al-Ḥayāt, 1960), 82–83; see also Abū Muḥammad 'Alī ibn Aḥmad ibn Sa'īd ibn Ḥazm, *Al-Iḥkām fī Uṣūl al-Aḥkām* (Beirut, Lebanon: Dār al-Āfāq al-Jadidah, 1983), 1:49.
2. In Q. 57:20, the term *kuffār* is used of tillers or husbandman because they sow the seed and cover it up with soil.
3. Jamāl al-Dīn b. Manẓūr, *Lisān al-'Arab*, v. 5 (Cairo, Egypt: Dār al-Ma'ārif, n.d.), 145.
4. See, for example, Taqī al-Dīn Aḥmad ibn Taymiyyah, *Majmū' Fatāwā* (Al-Madina, Saudi Arabia: Mujamma' al-Malik Fahd, 2004), 12:335.
5. Majd al-Dīn Muḥammad b. Ya'qūb al-Fayrūz Ābādī, *Al-Qāmūs al-Muḥīṭ*, ed. Muḥammad Na'īm al-'Arqasūsī, 8th ed. (Beirut, Lebanon: Mu'asasat al-Risālah, 2005), 1176.
6. Major and minor *kufrs* will be discussed at length later in this chapter.
7. See, for example, 'Abd al-Mun'im Muṣṭafā Ḥalīmah, *Qawā'id fī al-Takfīr* (Amman, Jordan: Dār al-Bashīr, 1994).
8. See, for example, 'Abd al-Qāhir al-Baghdādī, *Al-Farq Bayna al-Firaq* (Beirut, Lebanon: Dār al-Kutub al-'Ilmiyyah, n.d.), 50.
9. See, for example, Nu'mān 'Abd al-Razzāq al-Sāmarrā'ī, *Al-Takfīr fī al-Qur'ān wa al-Sunna Qadīman wa Ḥadīthan* (Riyadh, Saudi Arabia: Markaz al-Malik Fayṣal lil Buḥūth wa al-Dirāsāt al-Islāmiyyah, 2007) and al-Shābandar, *Lā Takfīr fī al-Qur'ān.*
10. Ḍuḥā al-Khaṭīb, *Ḍawābiṭ al-Takfīr* (Beirut, Lebanon: Dār al-Burāq, 2007), II-III.
11. Ibid.
12. Political Islam or Islamism "refers to a set of worldviews and practices that places the Islamic doctrine at the center of any organization of social order. Islamism describes the position of those individuals and institutions believing that the message of Islam is of the utmost importance and that it ought to be implemented as fully as possible, as soon as possible" (Frederic Volpi, ed., *Political Islam: A Reader* (New York: Routledge, 2011), 9.
13. Al-Khaṭīb, *Ḍawābiṭ al-Takfīr*, II-III.
14. Ibid.
15. See, for example, Muḥammad Fu'ād 'Abd al-Bāqī, *al-Mu'jam al-Mufahras li Alfāẓ al-Qur'ān al-Karīm* (n.c.: Dār al-Fikr, n.d.), 81, 93.
16. Abū Ja'far Muḥammad ibn Jarīr al-Ṭabarī explains the verse to mean that prayers toward Jerusalem had not been wasted. Abū Ja'far Muḥammad ibn Jarīr al-Ṭabarī, *Jāmi' al-Bayān 'an Ta'wīl āy al-Qur'ān*, eds. Bashār 'Awaḍ Ma'rūf and 'Iṣām Fāres al-Ḥirstāni (Beirut, Lebanon: Mu'asasat al-Risālah, 1994), 1:417.
17. The translation of M. A. S. Abdel Haleem, *The Qur'an* (Oxford: Oxford University Press, 2005), will be used throughout this book.
18. 'Abd al-Razzāq b. 'Abd al-Muḥsin al-Badr, *Ziyādat al-Īmān wa Nuqṣānuh* (Riyadh, Saudi Arabia: Dār Kunūz Ishbīlyah, 2006), 33.
19. *Taṣdīq* is an Arabic term from the three letters s-d-q. According to Aḥmad Riḍā, *Matn al-Lugha* (Beirut, Lebanon: Dār al-Ḥayāt, 1959), 434, the term *ṣadaq* means, "conforming to the truth or reality (*ṭābiq al-wāqi'*)." In the Arabic scholarship on *īmān* and *kufr*, the term *taṣdīq* means the acceptance of the message of the Qur'ān. See, for example, Mālik Muṣṭafā Wahbī al-'Āmilī, *Ẓāhirat al-Takfīr fī al-Fikr al-Islāmī* (Beirut, Lebanon: Dār al-Hādī, 2007).
20. Ibn Taymiyyah, *Majmū' Fatāwā*, vol. 7, 59.
21. Al-'Āmilī, *Ẓāhirat al-Takfīr*, 92.
22. Ibid.
23. Ibid.
24. Ḥalīmah, *Qawā'id fī al-Takfīr*, 40.
25. Ibid.
26. Ibid., 40–41.

27. Ibid.
28. *Ṣaḥīḥ Bukhārī* Volume 001, Book 002, Hadith Number 025. Retrieved from www.hadithcollection.com. Accessed on December 9, 2013.
29. Al-ʿĀmilī, *Ẓāhirat al-Takfīr*, 79.
30. *Ṣaḥīḥ Bukhārī* Volume 001, Book 002, Hadith Number 012. Retrieved from www.hadithcollection.com. Accessed on December 9, 2013.
31. *Ṣaḥīḥ Bukhārī* Volume 007, Book 069, Hadith Number 484. Retrieved from www.hadithcollection.com. Accessed on March 11, 2017.
32. The term *muslim* with the lower case "m" means a submitter to one God or a believer in the oneness of God (*tawḥīd*), and the term Muslim with the upper case "M" means a member in the Muslim community who is practicing Islam. The term *islām* with the lower case "i" means submission to the One God, and the term Islam with the upper case "I" means the organized religion or the Islamic community. In the Qur'ān, it is hard to tell whether the text means monotheism or the Muslim community when referring to Islam (*al-islām*), and whether it means the believers or members of the Muslim community when referring to the muslims (*al-muslimūn*) since in Arabic there are no lower or upper case letters; see Mahmut Aydin, "Islam in a World of Diverse Faiths: A Muslim View," *Islam and Inter-Faith Relations*, edited by Perry Schmidt-Leukel and Lloyd Ridgeon (London: SCM Press, 2007), 41–47.
33. See, for example, Abū ʿUbayd al-Qāsim b. Sallām, *Kitāb al-Īmān*, ed. Muḥammad Nāṣir al-Dīn al-Albānī (Beirut, Lebanon: Al-Maktab al-Islāmī, 1983) and Qāḍī ʿAbd al-Jabbār, *Sharḥ al-Uṣūl al-Khamsa* (Beirut, Lebanon: Dār Iḥyā' al-Turāth al-ʿArabī, 2001).
34. See, for example, al-Jabbār, *Sharḥ*.
35. See, for example, Abū Ḥanīfa al-Nuʿmān, *al-ʿĀlim wa al-Mutaʿallim* (Cairo, Egypt: Maṭbaʿat al-Anwār, 1949).
36. Ḥalīmah, *Qawāʿid al-Takfīr*, 44.
37. Quoted in al-ʿĀmilī, *Ẓāhirat al-Takfīr*, 17–18.
38. Ibid., 18.
39. See, for example, Muḥammad Ibn Yaʿqūb al-Kulaynī, *Al-Kāfī* (Beirut, Lebanon: Manshūrāt al-Fajr, 2007), 2:235.
40. al-ʿĀmilī, *Ẓāhirat al-Takfīr*, 18.
41. Ibid., 24.
42. Ibid.
43. Ibid.
44. Ibid.
45. Muḥammad ibn Idrīs al-Shāfiʿī, *Kitāb al-Umm*, vol. 1, ed. Rifʿat Fawzī ʿAbd al-Muṭalib (Cairo, Egypt: Dār al-Wafā,' 2001), 33.
46. Ibid.
47. Abū Ḥanīfah al-Nuʿmān, *Al-Sharḥ al-Muyassar*, ed. Muḥammad b. ʿAbd al-Raḥmān al-Khamīs (Ajman, UAE: Maktabat al-Furqān, n.d.), 96.
48. Ibid.
49. Abū Ḥāmid Muḥammad al-Ghazālī, *Qawāʿid al-ʿAqā'id*, ed. Mūsa Muḥammad ʿAlī (Beirut, Lebanon: ʿĀlam al-Kutub, 1985), 250.
50. Badr al-Dīn al-Zarkashī, *Al-Manthūr fī al-Qawāʿid*, vol. 2, ed. Muḥammad Ḥasan Ismāʿīl (Beirut, Lebanon: Dār al-Kutub al-ʿIlmiyyah, 2000), 208.
51. Ibn Taymiyyah, *Majmū*,' vol. 7, 157.
52. Ibid.
53. Sunnahonline. Retrieved from http://sunnahonline.com/library/beliefs-and-methodology/87-types-of-kufr-disbelief. Accessed on October 28, 2013. See also Ismāʿīl ibn Kathīr, *Tafsīr al-Qur'ān*, ed. Sāmī b. Muḥammad al-Salāmah, 2nd ed. (Riyadh, Saudi Arabia: Dār Ṭaybah, 1999), 7:402, 4:592, 2:6, 2:32, 2:441, 4:171, 7:307, 5:172, 7:193, 4:609.
54. Ibid.
55. Ḥalīmah, *Qawāʿid*, 12.
56. *Ṣaḥīḥ Bukhārī* Volume 001, Book 002, Hadith Number 050. Retrieved from www.hadithcollection.com. Accessed on March 12, 2017.
57. Lenn E. Goodman, *Islamic Humanism* (Oxford: Oxford University Press, 2003), 89.
58. Ḥalīmah, *Qawāʿid*, 14.

59. Ibid., 53
60. Ibid.
61. Ibid., 54–55.
62. Quoted in ibid.
63. *Ṣaḥīḥ Bukhārī* Volume 008, Book 081, Hadith Number 771. Retrieved from www.hadithcollection.com. Accessed on March 12, 2017.
64. Ibid. Volume 008, Book 081, Hadith Number 769. Retrieved from www.hadithcollection.com. Accessed on March 12, 2017.
65. Ḥalīmah, *Qawā'id*, 73.
66. Ibid., 73–85.
67. Ibid., 86.
68. Ibid.
69. Ibid., 86–87.
70. Ibid., 113.
71. Taqī al-Dīn Aḥmad ibn Taymiyyah, *Kitāb al-Īmān, Book of Faith*, trans. Salman Ḥassan al-Ani and Shadia Ahmad Tel (Bloomington, Ind.: Iman Publishing House, 1999), 56–57.
72. Ḥalīmah, *Qawā'id*, 113–114.
73. Ibid.
74. Ibn Tyamiyyah, *Kitāb al-Īmān*, 138.
75. Ibn Taymiyyah, *Majmū,'* vol. 28, 520–525. See also John Kelsay, *Arguing the Just War in Islam* (Cambridge, Mass.: Harvard University Press, 2007), 122.
76. Ḥalīmah, *Qawā'id*, 124.
77. *Ṣaḥīḥ Bukhārī* Volume 006, Book 060, Hadith Number 295. Retrieved from www.hadithcollection.com. Accessed on March 16, 2017.
78. Ḥalīmah, *Qawā'id*, 124.
79. *Maliks Muwatta* Book 038, Hadith Number 008. Retrieved from www.hadithcollection.com. Accessed on March 16, 2017.
80. Ḥalīmah, *Qawā'id*, 127.
81. Quoted in ibid., 128.
82. Ibid.
83. Ibid., 145.
84. *Ṣaḥīḥ Muslim* Book 001, Hadith Number 0109. Retrieved from www.hadithcollection.com. Accessed on March 19, 2017.
85. See, for example, Abū Muḥammad 'Alī ibn Ḥazm, *Al-Fiṣal fī al-Milal wa al-Ahwā' wa al-Niḥal*, vol. 3 (Beirut, Lebanon: Dār al-Ṣādir, n.d.), 191.
86. See, for example, Abū Ḥāmid Muḥammad al-Ghazālī, *Fayṣal al-Tafriqah Bayna al-Islām wa al-Zandaqah* (n.c.: Dār Ihyā' al-Kutub al-'Arabiyyah, 1961), 202.
87. Ibid.
88. *Kufr* and *shirk* are terms used in the Qur'ān interchangeably to refer to infidelity and disbelief (Q. 31:13; 23:117). Also, both terms are mentioned in a single *ḥadīth* to mean one thing: "Verily between man and between polytheism (*shirk*) and unbelief (*kufr*) is the negligence of prayer" (*Ṣaḥīḥ Muslim* Book 001, Hadith Number 0146. Retrieved from www.hadithcollection.com. Accessed on March 18, 2017). However, according to the former Grand Muftī of Saudi Arabia 'Abd al-'Azīz b. 'Abd Allāh b. Bāz, *shirk* is different from *kufr*. *Kufr*, he writes, is "denial and concealment of the truth, such as one who denies that the prayer is obligatory." Shirk, on the other hand, "is devoting acts of worship to something or someone other than God. A *kāfir* may be called a *mushrik* and a *mushrik* may be called a *kāfir*" (see, for example, 'Abd al-'Azīz b. 'Abd Allāh b. Bāz, *Majmū' Fatāwā*, vol. 9 (Riyadh, Saudi Arabia: Al-Buḥūth al-'Ilmiyyah lil Iftā,' 1999), 174.
89. Ḥalīmah, *Qawā'id*, 157.
90. Abū Muḥammad Ḥusayn b. Mas'ūd al-Baghawī, *Tafsīr al-Baghawī*, vol. 4, ed. Muḥammad 'Abd Allāh al-Nimr et al. (Riyadh, Saudi Arabia: Dār Ṭaybah, 1989), 38–39; see also Fakhr al-Dīn al-Rāzī, *Tafsīr al-Rāzī*, vol. 16 (Beirut, Lebanon: Dār al-Fikr, 1981), 38.
91. Qur'ān commentators explained the verse to say that even believers associate partners with God (*yushrikūna*), or commit polytheism, by "worshipping idols, the dead in the grave, statues, trees, stones, angels, or prophets," see, for example, Ibn Bāz, *Majmu'*, 22.

92. It is unclear whether the theologians have meant *islām* with the small "i," meaning monotheism, or Islam with the capital "I," meaning the religious institution.
93. Ḥalīmah, *Qawā'id*, 177.
94. Ibid.
95. *Ṣaḥīḥ Bukhārī* Volume 009, Book 088, Hadith Number 178. Retrieved from www.hadithcollection.com. Accessed on December 9, 2013.
96. Bashār 'Awaḍ Ma'rūf and 'Iṣām Fāris al-Ḥirstānī, eds., *Tafsīr al-Ṭabarī*, vol. 2 (Beirut, Lebanon: Mu'asasat al-Risālah, 1994), 533.
97. 'Abd al-Salām b. Salīm al-Sāhimī, *Fikr al-Takfīr Qadīman wa Ḥadīthan* (Cairo, Egypt: Dār al-Imām Aḥmad, 2005), 71.
98. *Ṣaḥīḥ Bukhārī* Volume 008, Book 078, Hadith Number 647. Retrieved from www.hadithcollection.com. Accessed on March 25, 2017.
99. Abū Ḥāmid Muḥammad al-Ghazālī, *Fayṣal al-Tafriqa Bayna al-Īmān wa al-Zandaqa*, ed. Maḥmūd Bījū (n.c.: n.p., 1993), 66.
100. Ḥalīmah, *Qawā'id*, 187.
101. Muslim b. al-Ḥajjāj, *Ṣaḥīḥ Muslim*, vol. 1, trans. Nasiruddin al-Khattab (Riyadh, Saudi Arabia: Darussalam, 2007), 122.
102. For more on death and dying in Islam, see Hussam S. Timani, "Death and Dying in Islam," in *Ultimate Journey: Death and Dying in the World's Major Religions*, ed. Steven J. Rosen (Westport, Conn.: Praeger, 2008).
103. Ḥalīmah, *Qawā'id*, 189.
104. Ibid., 190.
105. Ibid.
106. Ibid., 203.
107. Ibid.
108. Ibid.
109. Ibid., 207.
110. Ibid.
111. Muḥammad b. 'Abd al-Wahhāb, *Al-Rasā'il al-Shakhṣiyyah*, 213. Retrieved from https://d1.islamhouse.com/data/ar/ih_books/single2/ar_Personal_Message.pdf. Accessed on March 26, 2017.
112. Muḥammad b. 'Abd al-Wahhāb, *Majmū'at al-Tawḥīd* (n.c.: Dār Ihyā' al-Turāth, n.d.), 34–35.
113. Al-Qāḍī al-Iyād, *Kitāb al-Shifā bi Ta'rīf Ḥuqūq al-Muṣṭafā*, trans. Aisha Abdarrahman Bewley. Retrieved from Islamicsystem.blogspot.com/2012/09/imam-al-qadi-iyad-judgements-concerning.html. Accessed on March 26, 2017; see also Ḥalīmah, *Qawā'id*, 208.
114. See, for example, Sayed Khatab, *The Political Thought of Sayyid Qutb: The Theory of Jahiliyyah* (London and New York: Routledge, 2006).
115. Al-Sāhimī, *Fikr al-Takfīr*, 64.
116. Ibid., 64–65.
117. Ibid., 65.
118. *Ṣaḥīḥ Bukhārī* Volume 009, Book 083, Hadith Number 049. Retrieved from www.hadithcollection.com. Accessed on December 9, 2013.
119. The Arabic term *dhimmah* means protection. It appeared in the Qur'ān as a reference to Christians and Jews.
120. Al-Sāhimī, *Fikr al-Takfīr*, 65.
121. Ibid., 64–65.
122. Al-Khaṭīb, *Ḍawābiṭ*, 15.
123. Muḥammad Ḥasan al-Ḥumṣī, ed., *al-Qur'ān al-Karīm* (Beirut, Lebanon: Dār al-Rashīd, n.d.), 5.
124. Muḥammad b. Jarīr al-Ṭabarī, *Tafsīr al-Qur'ān*, 3rd ed. (Beirut, Lebanon: Dār al-Ma'rifah, 1978), 142.
125. Ibid.
126. Al-Shābandar, *Lā-Takfīr*, 38.
127. Ibid.
128. Ibid.

129. See, for example, C. Jonn Block, "Philoponian Monophysitism in South Arabia at the Advent of Islam with Implications for the English Translation of '*Thalatha*' in Qur'ān 4. 171 and 5. 73," in *Journal of Islamic Studies* 23:1 (2012): 50–75.

130. Al-Shābandar, *Lā-Takfīr*, 38.

131. See, for example, Fāyiz Salhab, *al-Takfīr wa al-Takfīr al-Muḍāḍ: Dirāsah* (Damascus, Syria: Dār al-Farqad, 2005), 126–129.

132. *Fitnah* may mean "either trial or temptation, or else tumult, turmoil, or sedition," see, for example, 'Abdullah Yūsuf 'Alī, *The Meaning of the Holy Qur'ān*, 11th ed. (Beltsville, Md.: Amana Publications, 2008), 453.

133. Riḍā, *Mu'jam Matn al-Lugha*, vol. 2, 571; see also Salhab, *al-Takfīr*, 177.

134. Salhab, *al-Takfīr*, 177.

135. Ibid.

136. Although the *ḥadīth* refers to Jews and Christians without exception, the general understanding of the *ḥadīth* in the circles of Muslim theologians and scholars is that the reference was made to minority groups of Jews and Christians who deviated from the path of monotheism during the time of Muhammad. If this *ḥadīth* in fact refers to all Jews and Christians, as some radical Muslims claim today, then Muhammad, one may argue, had committed a clear violation of key Qur'ānic teachings that command the Muslims to respect Jews and Christians for they believe in God and have nothing to fear (Q. 2:62).

137. *Fiqh-us-Sunnah* Volume 004, Funerals and Dhikr, Fiqh 4.069A. Retrieved from www.hadithcollection.com. Accessed on November 14, 2013.

138. *Ṣaḥīḥ Muslim* Book 001, Hadith Number 0146. Retrieved from www.hadithcollection.com. Accessed on November 14, 2013.

139. *Fiqh-us-Sunnah* Volume 001, Purification and Prayer, Fiqh 1.103B. Retrieved from www.hadithcollection.com. Accessed on November 14, 2013.

140. Abū Zakariyyah Yaḥya b. Sharaf al-Nawawī, *Sharḥ Ṣaḥīḥ Muslim* (Cairo: Al-Maṭba'a al-Maṣriyyah, 1929), 2:70–71.

141. *Ṣaḥīḥ Muslim* Book 001, Hadith Number 0033. Retrieved from www.hadithcolletion.com. Accessed on November 17, 2013.

142. *Ṣaḥīḥ Bukhārī* Volume 001, Book 003, Hadith Number 122. Retrieved from www.hadithcollection.com. Accessed on November 17, 2013.

143. *Ṣaḥīḥ Bukhārī* Volume 009, Book 088, Hadith Number 197. Retrieved from www.hadithcollection.com. Accessed on November 17, 2013.

144. Al-Nawawī, *Sharḥ Ṣaḥīḥ Muslim*, 2:55.

145. Ibid.

146. 'Ubaydāt, *Qaḍiyat al-Īmān wa al-Takfīr*, 52.

147. *Ṣaḥīḥ Bukhārī* Volume 005, Book 058, Hadith Number 189. Retrieved from www.hadithcollection.com. Accessed on November 17, 2013.

148. *Ṣaḥīḥ Bukhārī* Volume 008, Book 073, Hadith Number 125A. Retrieved from www.hadithcollection.com. Accessed on November 17, 2013.

149. 'Ubaydāt, *Qaḍiyyat al-Īmān*, 53.

Chapter Two

Takfīr in Early Islamic Theology

This chapter discusses the concept of *takfīr* in classical Islamic theology and demonstrates how the theological debate that prevailed in classical Islam on a variety of topics such as the nature of God, the creation of the Qur'ān, free will, and *takfīr* led to the development of later Islamic theology and to the rise of a number of theological schools, which have shaped and influenced Islamic beliefs throughout the centuries and up to the modern period. The concept of *takfīr* is a topic that contributed to the development of a large corpus of medieval writings on the meaning of *takfīr* and *īmān*. The theological schools posed several questions that will be discussed in this chapter such as: Who should declare others as *kuffār*? Can someone declare a Muslim ruler a *kāfir*? What makes a person *kāfir*?

This chapter discusses the concept of *takfīr* in the theological contexts of the Kharijites, the Murji'ites, and the Karramites and demonstrates how each school provided a different but interesting interpretation of *takfīr*. The chapter also demonstrates how the responses of each school to one another not only contributed to the debate but also further developed the concept of *takfīr* to the extent that it became a major theological issue not only in medieval Islam but also in the twenty-first century.

TAKFĪR IN THE KHARIJITE THOUGHT

The first group to be discussed in this chapter is the Kharijites (*al-Khawārij*). The term *Khawārij* is a derivative from the Arabic verb *kharaja*, meaning "to go out." According to the Muslim historiographer Abū al-Fatḥ Muḥammad b. 'Abd al-Karīm al-Shahrastānī (d. 1153), the term "khawārij" applies on anyone who rebels (*kharaja*) against the imām (the Muslim leader).[1] Some scholars[2] have argued that the term first appeared in al-Ṭabarī's *Ta'rīkh al-*

Umam wa al-Mulūk. Al-Ṭabarī writes that 'Alī b. Abī Ṭālib (d. 661) referred to those who rebelled against him during the Battle of Ṣiffīn (657–658) as *khawārij*.[3] The term Kharijites also points to the group's migration (*hijrah*) from the community of "unbelievers" (*kuffār*) to God and his messenger. "But those who have believed, migrated, and striven for God's cause, it is they who can look forward to God's mercy" (Q. 2:218). They were also known as the people of *al-Ḥarūriyyah* in reference to a place called Ḥarūrā',[4] where 'Alī b. Abī Ṭālib held a fight against them.[5] They are also referred to as *al-muḥakkimah*,[6] those who say *"lā ḥukma illā lillāh"* (judgment belongs to God alone), a Qur'ānic term, in response to 'Alī's acceptance of the arbitration of men to end the Battle of Ṣiffīn[7]: "Mankind was a single community, then God sent Prophets to bring good news and warning, and with them He sent down the Scripture with the Truth, to judge [Yaḥkuma] between people in disagreements" (Q. 2:213). *Al-shurāt*,[8] or those who sell themselves to God, is another Qur'anic term given to the group: "But there is also a kind of man who gives his life away [*yashrī nafsahu*] to please God" (Q. 2:207).

The Kharijites first emerged as an opposition force to the arbitration during the Battle of Ṣiffīn and later adopted a political doctrine, which concerned itself with the caliphate.[9] Toward the end of the Umayyads' rule (r. 661–750), the Kharijites split into five major groups and some twenty subgroups. The five major factions were *al-Muḥakkimah al-Ūla*, *al-Azāriqah*, *al-Najdāt*, *al-'Ajāridah*, *al-Ibāḍiyyah*, and *al-Ṣufriyyah*.[10] In spite of their ideological, political, and tribal differences, the Khariji splinter groups shared similar views regarding the unbelief (*ikfār*) of 'Alī (r. 656–661), 'Uthmān b. 'Affān (r. 644–656), the arbitrators of the Battle of Ṣiffīn, the People of the Camel,[11] and those who were satisfied with the arbitration. According to Ibn Taymiyyah, they declared 'Uthmān and 'Alī as well as non-Kharijite Muslims as a community of *kuffār*.[12]

According to the Shāfi'ī scholar 'Abd al-Qāhir al-Baghdādī (d. 1037), all the Khariji splinter groups proclaimed the unbelief (*ikfār*) of 'Alī, 'Uthmān, the arbitrators, the People of the Camel, and whoever agreed to the appointment of the arbitrators.[13] They also declared the *ikfār* of fellow Muslims who committed a grave sin, and advocated revolts against a tyrant *imām*.[14] Al-Baghdādī argues that Abū al-Ḥasan al-Ash'arī (d. 935), the founder of the Ash'arite theological school, disagrees that the Kharijites proclaimed the *ikfār* of Muslims who committed a grave sin.[15] The omission of "grave sin" in al-Ash'arī's account may be explained to say, in the words of Toshihiko Izutsu, that al-Ash'arī "gives here a more faithful picture of the early Kharijite thought which was so characteristically bound to the actual political conditions"[16] whereas the less favorite accounts of the group reflect the later "Kharijite thinking, at which people began to think about their problems on a more theoretical level."[17]

The Kharijites perhaps were the first *takfīr*-ists in Islam. They were the first to declare Islamic rulers as *kuffār*, a notion that was unprecedented in early Islam. The Muslim historian Aḥmad b. Yaḥya al-Balādhurī (d. 892) writes that Sahm b. Ghālib al-Hujaymī, a Kharijite affiliate, "was the first person to give the term *kufr* on the people of the *qiblah* (*ahl al-qibla*, i.e., the Muslims).[18] Some scholars, however, have alleged that *takfīr* as a concept may have started to take shape among some Muslims prior to the rise of the Kharijites, and the evidence was in the killing of the third caliph 'Uthmān.[19] Seeing the caliph's blunders and nepotism as a deviation from Islam was perhaps a motive to believe that 'Uthmān is no longer a Muslim, thus justifying his killing.[20] In 657–58, during the Battle of Ṣiffīn, the group that had its roots in the insurgency against 'Uthmān revolted against the fourth caliph 'Alī b. Abī Ṭālib, openly declared him a *kāfir* for objecting to their demands not to make peace with the other side, and assassinated him as a result.[21]

The Egyptian scholar 'Āmir al-Najjār writes that the Kharijite school of thought was short lived because of the group's deviation from Islamic teachings, especially in matters relating to belief and unbelief. Al-Najjār argues that the Kharijites did not hesitate to accuse Muslims committing lesser sins of infidelity regardless of their testimony to the oneness of God.[22] He states that the Kharijites justified their *takfīr* of the others who commit lesser sins based on Qur'ānic verses that declare God-fearing people as unbelievers: "[Prophet], have you not considered how those who were given a share of the Scripture purchase misguidance and want you [believers], too, to lose the right path? . . . But God has spurned them for their defiance; they believe very little" (Q. 4:44, 46). Also, according to al-Najjār, the Kharijites supported their *takfīr* of those who committed lesser sins with verse 5:44: "those who do not judge according to what God has sent down are [*al-kāfirūn*]" (disbelievers of a lesser degree as they do not act on God's laws.[23]

Some Kharijite splinter groups, like the Ṣufriyyah, the followers of Ziyād b. al-Aṣfar, developed the concept of *takfīr* to include those who commit sins that do not have a form of punishment in the Qur'ān. They justified their interpretation by arguing that acts that are mentioned in the Qur'ān, such as adultery and theft, do not earn the doers the title *kāfir* because God already labeled them as adulterer (*zānī*) and thief (*sāriq*).[24] Therefore, it is against God's will to call them *kuffār*. Other Karijite groups, like the Ḥafṣiyyah, a sub-division of the Ibāḍiyyah, make a distinction between *kufr* and *shirk* (associating partners with God).[25] For the Ḥafṣiyyah, *shirk* but not *kufr* is the ultimate sin in Islam. While some Kharijites considered the *kāfir* to be automatically *a mushrik* (committing the gravest sin), the Ḥafṣiyyah asserted that a Muslim is only a *kāfir*, but not a *mushrik*, as long as he attests to the oneness of God while committing forbidden acts. All the Kharijites, however, agree that belief in God alone does not make a person immune to *kufr*. In their views, a mere deviation from the path renders the believer a *kāfir*.[26]

The Kharijites developed their concept of *takfīr* based not on their definition of *īmān* or *islām* but on their understanding of who is *not* a believer or a Muslim. As Izutsu rightly puts it: "They did not ask 'What is belief?' or even 'What is unbelief?' Rather they asked, '*Who* is an unbeliever or infidel?'"[27] Asking "who is a *kāfir*?" rather than asking "what is belief?" indicates that the Kharijites were concerned with the people's beliefs rather than with the questions about faith and belief in general.[28] "Instead of trying to define *mu'min* [faithful, possessing faith]" Izutsu writes, "they tried to determine rigorously those who should be driven out of the Muslim community."[29] Thus, by declaring other Muslims infidel, the Kharijites legitimized and authenticated their own religious devotion, which they used as a measuring stick to evaluate (and consequently eliminate) those who challenged them religiously, socially, and politically. Their skewed interpretation of *takfīr* is a clear indication that the Kharijites stripped the concept of its theological connotations as presented in the Qur'ān and the *Sunnah* and transformed it into a political ideology concerned with political matters.

Izutsu's analysis of *takfīr* in the Kharijite thought demonstrates how the Kharijite approach to the concept led to "a very profound change of values"[30] from the Qur'ānic view of *takfīr*. In the Qur'ān, for instance, there is a clear distinction between the abode of *kufr* and the abode of *īmān* (Q. 6:127; 10:25). The abode of *īmān* is referred to in Q. 6:127 and 10:25 as the "House of Peace" (*dār al-salām*), meaning the abode of those who are on the straight path.[31] In more practical terms, those who lived within the boundaries of the House of Peace are the believers, while non-Muslims inhabited territories outside the realm of the House of Peace, also referred to as the abode of Islam (*dār al-Islām*). Those who lived within the House of Peace, or *dār al-Islām,* were the believers who took Muḥammad as the Prophet of God and believed in the oneness of God. The unbelievers did not belong to the House of Allāh. The change of values that the Kharijites introduced and Izutsu highlights is the dismantling of the Qur'ānic separation between Muslims (who lived within) and non-Muslims (who lived without). In other words, the change of values that the Kharijites introduced meant that a mere residence within the abode of Islam does not automatically bestow on the person the title *mu'min* nor does it shield him from being defined as a *kāfir*. For the first time since the creation of the Islamic community, the abode of Islam can contain Muslims who are deemed *kuffār.* With this modification of the concept of *takfīr*, the Kharijites brought *kufr* into the abode of Islam, and this meant that no one was safe from being stripped of his *īmān* and hence declared *kāfir*. As Izutsu nicely put it: "Formerly, people were safe, so to speak, within the wall. Once they accepted Islam formally, they were treated as authentic members of the new community."[32]

This distorted interpretation of *takfīr* by the Kharijites is well noted in modern intellectual readings of the group. For instance, al-Najjār has argued that the Kharijites constructed a theology incompatible with the Islamic teachings.[33] He demonstrates their deviation by comparing them to the Mu'tazilites who argued that even those who commit a grave sin remain Muslims, because in the Qur'ān, according to the Mu'tazilites, disobedience to God (*fisq*) is in a place between belief and unbelief.[34] So they rejected completely the Khariji notion of making disobedience (or committing a grave sin) equal to *kufr*.

The concept of grave sin (*al-kabīrah*) also was central in the Kharijite thought. In classical Islamic theology, different opinions dominated the debate on the issue of the grave sinner (*murtakib al-kabīrah*). While the Kharijites rendered the grave sinner a downright *kāfir*, the Murji'ites argued that a Muslim who has committed a grave sin remains a believer.[35] The Mu'tazilites, on the other hand, placed the grave sinner on a position between the two: He is neither a Muslim nor a *kāfir*.[36] The Kharijites were the most prolific on this issue as the various Kharijite groups had different opinions on the formula that makes a grave sinner a *kāfir*. Although it is handed down to us that the Kharijites consider a sin, light or grave, an act of *kufr*, some Kharijites had different opinions. For instance, the Azraqites adopted the position that a grave sinner is both a *kāfir* and a *mushrik* (rejecting God and associating partners with God); the Najdites, however, make that person a *kāfir* but not a *mushrik*, and he is rendered a *kāfir* not because of his rejection of God but because of being ungrateful for God's blessings. The Ṣufriyyah, on the other hand, distinguished between a grave sin and a light sin and only the former is *kufr*. The Ibāḍiyyah put the grave sinner in a position where he is neither a pure believer nor a pure *kāfir*.[37]

Although the Qur'ān refers to greater sin only on three occasions, the text is not clear about which acts do constitute a greater sin. For instance, in 4:31, 42:37, and 53:32, the Qur'ān promises reward for those who "avoid the greater sins (*yajtanibūna kabā'ira al-ithimi*)." The Qur'ān also assures the believers throughout the text that God will forgive their sins if they pursue the right path (Q. 2:182; 2:199; 2:218). Although some Islamic views hold that disobeying God's commands is equal to committing lesser sins, the general consensus is that rejecting God or associating partners with Him (*shirk*) is a grave sin. In theory, since all Muslims believe in one single God, it is almost impossible for someone to accuse another of committing a grave sin, unless that person openly declares his or her unbelief. The Kharijites, however, changed the rules and provided their own interpretations of the term. They regarded as a *kāfir* a person who commits a grave sin even though that person acknowledges God and His Apostle, for a person is either a *kāfir* or a believer—nothing in between. They justify their views, according to the Shāfi'ī scholar Muḥammad b. Aḥmad al-Malaṭī (d. 987 or 988), on Qur'ānic verses: "The deeds of anyone who rejects faith will come to nothing" (Q. 5:5), and "some of you disbelieve and some believe" (Q. 64:2). Therefore, for them, the Qur'ān does not mention a third category between *kufr* and *īmān.* But, al-Malaṭī challenges this claim and argues that the Qur'ān uses the term *fāsiq* for those who have committed a grave sin:[38] "As for those who accuse chaste women of fornication, and then fail to provide four witnesses . . . they are the lawbreakers [*wa 'ūlā'ika hum al-fāsiqūna*]" (Q. 24:4). Thus, for al-Malaṭī, *fisq* is an intermediate category between *kufr* and *īmān*. Al-Malaṭī also criticizes the Khraijites for lumping together grave sin (*kabīrah*) and lesser sin (*ṣaghīrah*) in one category when the Qur'ān does make a distinction between the two: "But if you avoid the great sins [*kabā'ira*] you have been forbidden, We shall wipe out your minor misdeeds [*sayyi'ātikum*, or *ṣaghā'ir*, plural of *ṣaghīrah*]" (Q. 4:31).[39]

For the Kharijites, a grave sin was determined based not only on belief or disbelief in God but also on acts and practices. Some extremists among the Kharijites declared all those who did not share their views and ideology guilty of committing *shirk.*[40] With this new way of thinking, no one was now immune to the accusations of committing a grave sin since the Kharijites broadened the definition of the term to include many acts that some people normally commit on a daily basis. For the Kharijites, a mere disobedience to the commands of God, such as the consumption of alcohol, is considered a grave sin and a ground for excluding members from the Muslim community.[41] Their exclusion from the community, however, would render them *kuffār*. "From this thesis," writes A. J. Wensinck, "a practical conclusion and a theoretical inference were drawn. The practical conclusion was, that Muslims who were guilty of grave sins must be combated with the sword as being infidels. The theoretical inference was that faith is not a constant entity, but that it is impaired by sin.[42]

Nāṣir b. Sulaymān b. Sa'īd al-Sābi'ī discusses in detail the concept of *takfīr* in the Kharijite thought. He writes that the early Kharijites, or the people of al-Nahrawān (*ahl al-Nahrawān*) as he refers to them, are distinguished from other Khariji splinter groups in "their belief in 'the most dangerous' religio-political issues that no other Islamic group had ever ventured to adopt or even consider."[43] These two issues are: *Isti'rāḍ* (shedding the blood of innocent Muslims) and *takfīr*. Early Muslims were frustrated with their practice of *isti'rāḍ*, Wensinck tells us, for the practice was uncommon at that time, on the one hand, and no other Muslim groups legitimized it, on the other.[44] Wensinck writes that some Kharijites questioned other Muslims whether or not they shared their views and "this question was often asked at the point of the sword and it even gave rise to the term *isti'rāḍ*, which literally means 'inviting someone to speak his opinion.'"[45] The term also became associated with violence as the Kharijites shed the blood of those whose opinions disagreed with their views. According to Wensinck, in practice, the term also meant "killing for the sake of religion."[46] This practice prompted 'Alī b. Abī Ṭālib to declare: "I did not prevent them [the Kharijites] from entering the mosques, nor did I provoke them except when they shed innocent blood."[47]

Al-Sābi'ī also discusses how the Kharijites developed and used the concept of *takfīr* for political gains. He writes that for the Kharijites only 'Alī was a *kāfir* for agreeing to the arbitration but then they declared as *kuffār* some of their own people who did no agree with their ideology. However, al-Sābi'ī does not believe that the Kharijites took the concept of *kufr* literally to mean infidel. He justifies his views by arguing that if one gives the term *kufr* a serious study, it becomes clear that it is a synonym for sinning. In other words, he contends that when the Kharijites used the term *kufr* against other Muslims, they only meant to chastise them for disobeying God's commands rather than to declare them *kuffār*.[48]

The Kharijites' accusations of *takfīr* were not confined only to those who opposed them or disagreed with their views but they were also extended to the children of some Kharijites. For instance, in a book entitled *Ḥarakāt al-Khawārij*, Laṭīfah al-Bakkāy argues that the Kharijites engaged in intellectual debates on a variety of issues, among them whether to declare as *kuffār* children of Kharijite parents who deviated from the Kharijite principles.[49] The Kharijite practice of "combating as infidels those who did not share their views," Wensinck writes, "culminated in the killing of children."[50] The most fanatical among the Kharijites opined that children who are born to infidel parents are in the state of unbelief because the child follows the parents' state in this world.[51] Other Kharijites held the view that children, "whether born of the Faithful or of infidels, were in a state of neutrality" until they reach adulthood and are invited to embrace Islam.[52] Al-Bakkāy tells us that disagreements among the Kharijites on this issue of whether children belong to the state of unbelief or neutrality not only divided the group but also led to the emergence of new splinter groups that contributed to the ongoing debate, which led to further disagreements and divisions within the movement.[53]

The development of the concept of *takfīr* in the Kharijite thought may have been the product of their views on Islamic theocracy. The theocratic nature of the classical Islamic *umma* allowed various views to flourish on matters of authority and rule. For the Kharijites, for instance, an authority is legitimate only if it rules according to the *sharī'ah*. They believed that an authority must be subordinate to the religion because the former draws its laws from the latter.[54] The German theologian Julius Wellhausen (d. 1918) defines theocracy as a state made of two opposing elements: the Religion (*al-dīn*) and the Community (*al-jamā'ah*). One is heavenly (the rule of God) and the other is worldly (the rule of the imām).[55] For the Kharijites, only *one* must be obeyed and followed, which means the other must be rejected. Since the Kharijites took the side of the *dīn*, Wellhausen argues, their rejection of *al-jamā'ah* justified their rebellion against the imāms who ruled the community contrary to God's laws.[56] Because *al-jamā'ah* followed the rule of imāms whom the Kharijites declared as *kuffār*, the Kharijites never hesitated to tear the community apart. And of course, this position requires declaring the *jamā'ah* a community of *kufr*. Their declaration of the community as a community of *kufr* (*dār al-kufr*), Wellhausen contends, obliged them to withdraw from this community—which they also deemed as the abode of war (*dār al-ḥarb*)—to the abode of emigration (*dār al-hijrah*).[57]

As *takfīr* became widely practiced by the Kharijites as a tool to legitimize their own religious and political doctrines, on the one hand, and to condemn and justify the elimination of those who challenged them, on the other, more groups and even individuals utilized *takfīr* against others to enhance and solidify their own agendas. Thus *takfīr* became a free practice. According to Izutsu, the door was open to *takfīr*, and "there could hardly be any limitation set as to how far one should go and where should stop."[58]

With the concept of *takfīr* becoming a free practice in society, even prophets from the past were declared as *kuffār*. According to al-Malaṭī, some Jahmites[59] believed that Moses committed *Kufr* when he asked his God to show Himself to him.[60] For the Jahmites, this is *kufr* because Moses asked for what was impossible. In the Jahmites' view, al-Malaṭī adds, nobody could see God, not in this world, nor in the Hereafter.[61] Jesus also committed *Kufr*, the Jahmites went on to say, when he said to God "you know what is in my soul, but I do not know what is in your soul (*nafs*)."[62] For the Jahmites, Jesus committed *kufr* because he thought that God has a *nafs*. Their fanaticism, al-Malaṭī writes, led them to declare the *kufr* of the Prophets.[63] The Kharijites' free practice of *takfīr* and the danger that resulted from that practice prompted some Muslims to challenge these views. The response to the Kharijites first came from those who believed that a judgment on *īmān* and *kufr* should be deferred to God, hence the rise of the Murji'ite school of theology.

TAKFĪR IN THE MURJI'ITE THOUGHT

The Murji'ites, who emerged after the Kharijites, contributed to the debate on *īmān* and *kufr* as well, but they took a different approach to the issue of grave sin and what makes a Muslim a *kāfir*. While the Kharijites insisted that a grave sinner loses his belief and must be dealt with accordingly, the Murji'ites preferred to defer such a judgment to God. That is how they earned their Arabic name *murji'ah*, from *irjā'*, meaning "to postpone," "to defer," or to "shrink from judging the conduct of man and its consequences, leaving this exclusively to God."[64] The theologians of this school of thought—and this can be said of all theologians in early Islam—had to grapple with political issues at that time, which shaped their thinking and theology. They were reluctant to make judgments on rulers who, by Kharijite standards, were *kuffār* for being irreligious and unjust. The Murji'ites were simply noncommittal to making fatwas in regard to *takfīr* or the belief of a ruler.

The Murji'ites criticized the Kharijites for making judgments against people who openly confess a belief in God. They also believed that men cannot and should not issue judgments on other people, for the final judgment rests on God. The Murji'ites were adamant about basing their concept of *īmān* on a solid religious basis (i.e., confessing belief in the oneness of God), especially after taking accusations from fellow Muslims that they introduced a religious ideology to serve unjust rulers. Thus, as their concept of *irjā'* "was marked with a singularly political coloring,"[65] the Murji'ites set the stage for the beginning of a theological debate on *īmān* and *kufr* and marked the beginning of Islamic theology as well.

The Murji'ite central doctrine of *irjā'* appeared in a major treatise entitled "Kitāb al-Irjā',"[66] which is traditionally ascribed to Ḥasan b. Muḥammad b. al-Ḥanafiyyah (d. 718). This treatise was written in reference to the first *fitnah*, or Civil War (656–657), in early Islam;[67] more specifically, it was a blueprint on how one is to relate to the actors involved in the *fitnah*, that is, Uthmān b. 'Affān and 'Alī b. Abī Ṭālib.[68] Based on "K. al-Irjā'," the Murji'ites, who preferred to refrain from approving one leader at the expense of another, suspended their judgment or deferred it to God.[69] This was reminiscent of the Kharijite call to defer the arbitration between the Syrians and the Iraqis to God when 'Alī agreed to the arbitration of men. One may argue that the Murji'ites deferred matters to God because they preferred inaction, whereas the Kharijites saw God's arbitration as a validation for their activism. The modern historical sources do not elaborate on the rationale behind the concept of *Irjā'* but some suggest that it is politically driven.[70] For instance, "earlier generations of Western scholars," according to Daniel Lav, "tended to view these early factional disputes over the Caliphate as essentially political, with these factions only in time evolving into religious sects."[71]

The concept of *irjā'* in the theological views of the Murji'ites evolved from its original meaning of "suspension of judgment" to "postponement." The difference between the two is that the former meant to refrain from making a judgment on the grave sinner; the latter, however, indicates, as Izutsu puts it, a postponement "not in terms of time but in terms of ranking,"[72] to degrade the importance of work vis-à-vis faith. Degrading the importance of work meant that acts and deeds come second to faith. For example, a Muslim who has committed a grave sin will remain a believer because his belief will shield him from becoming a *kāfir*. Therefore, for the Murji'ites, works—whether good or bad—do not count or are of minor importance. 'Abd al-Qāhir al-Baghdādī writes: "They are called *Murji'ah* simply because they put 'work' after 'belief' in importance."[73] Thus, for the Murji'ites, a believer does not commit *kufr* based on acts of disobedience.

Izutsu put it concretely: “sin does not do any injury where there is *īmān*, just as acts of obedience are of no use where there is *kufr*.”[74]

Some scholars of Islamic history have established a connection between the Murji’ite doctrine of the exclusion of acts from faith and the anti-Umayyad revolt in the East, namely Khurāsān and Transoxania.[75] The revolt was sparked by new converts to Islam who were discontent with the Umayyad rule that continued to require from them the non-Muslim *jizyah* tax despite their conversion to Islam. The Umayyads believed that new converts were not fully Muslim because they believed in the message of religion but had not practiced Islam, and that was faith without works.[76] The Doctrine of *Irjā’*, therefore, served as a blueprint to refute this definition and to assert that faith is not contingent on acts. Wilferd Madelung, for instance, establishes the connection between these Murji’ite views and the revolt of Khurāsān:

> Doctrinally this effort on behalf of the new converts was based on the Murji’ite thesis that the status of faith depended on the mere confession of belief in Islam to the exclusion of all works, i.e., the actual performance of the ritual and legal obligations of Islam. This meant that the mass converts in central Asia could not be denied the full status by the government on the pretext that they continued to ignore many of their duties as Muslims.[77]

According to Lav, the classical doctrine of the Murji’ites was attributed to the theology of Abū Ḥanīfah al-Nu‘mān (d. 767)—the founder of the Ḥanafī school of jurisprudence—whose works *Risālah ilā ‘Uthmān al-Baṭṭī* was a defense of those “labeled Murji’ites.”[78] In this work, as well as in other works of his circle,[79] Abū Ḥanīfah lays out his doctrine of the exclusion of works from faith based on the distinction between religion (*dīn*) and obligation (*farīḍah*). For Abū Ḥanīfah, religions, which all prophets brought, are one and the same, but religious laws and obligations differ. For this reason, works are not part of faith.

The views of the Murji’ites are well expressed in *Kitāb al-‘Ālim wa al-Muta‘allim*, one of the earliest sources on Murji’ite theology. *Kitāb al-‘Ālim*, which some sources ascribe to either Abū Ḥanīfah al-Nu‘mān (the main proponent of the Murji’ite doctrine),[80] to someone from his circle,[81] or to Abū Muqātil al-Samarqandī (d. 823),[82] is a dialogue between master and disciple by the name of Abū Muṭī‘.[83] *Kitāb al-‘Ālim* expresses the doctrines of the Murji’ites on the issue of faith and is a blueprint to “determine the attitude which the Muslim should take towards other Muslims.”[84] *Kitāb al-‘Ālim*, which reflects theological debates among Muslims from several schools of thought, as well as polemical discussions between Muslims and non-Muslims, does not seem to be addressed to specialists, like other treatises by Abū Ḥanīfa such as *Risālah ‘ilā ‘Uthmān al-Baṭṭī.* The *Kitāb al-‘Ālim*, in the opinion of Joseph Schacht, “concentrates on popular, sometimes even somewhat naïve problems . . . [and] introduces us to the popular aspect" of the Murji'ites.[85]

In *Kitāb al-'Ālim* the master answers questions on the exclusion of acts from faith. When the disciple (*al-muta'alim*) asks for an argument to refute the views that God's religion (*dīn Allāh*) is the doing all that is obligatory and the avoidance of all that is forbidden, the master (*al-'ālim*) answers by saying that the obligations (*farā'iḍ*) are not part of faith (*īmān*).[86] When the disciple asks, "What is *īmān*?" the answer was that *īmān* is *taṣdīq* (assent in the heart), *ma'rifah* (recognizing God), *yaqīn* (knowing God), *iqrār* (verbal profession), and *islām* (submitting to God). The master explains that in regard to *taṣdīq*, three types of people can be sorted out: those who have it in their hearts and on their tongues; those who have it in their hearts but not on their tongues; and those who have it on their tongues but not in their hearts. The first are believers in the eyes of God and the community; the second are believers in the eyes of God but unbelievers in the eyes of the community; and the third are believers in the eyes of the community but unbelievers in the eyes of God. The master adds that these are different terms with one meaning: *īmān*.[87] Although this definition of *taṣdīq* confirms the *īmān* of the first and third groups, it does not necessarily make the second group unbelievers or *kuffār* because they only seem as such to others and only God knows what is in their hearts. Without a doubt, this definition shielded *everyone* from being branded a *kāfir*.

In another part of *Kitāb al-'Ālim*, the master responds to several questions relating to grave sins (*al-kabā'ir*), *kufr* (unbelief), *tawḥīd* (oneness of God), hypocrisy (*nifāq*), the concept of *irjā'* (postponement), the unbelievers (*kuffār*), and the meaning of worship (*'ibādah*), among other issues. With regard to grave sins, the master believes that a Muslim who has committed a grave sin remains a believer "as long as he does not abandon *tawḥīd*."[88] He explains the term *kufr* by saying that *kufr* in Arabic means *inkār* or *juḥūd* (denying or rejecting faith). He elaborates by arguing that people should distinguish between the Muslims who omit a duty (*farīḍah*) and those who deny it. The former are believers for acknowledging (the *ma'rifah* of) the obligation; while the latter are *kuffār* for denying (*inkār*) it.[89]

In *Kitāb. al-'Ālim*, the master explains the term *nifāq* (hypocrisy) as the denial in the heart but an outward confession of the tongue. For God, such people are unbelievers; for the Muslims in the time of the Prophet, they were never *kuffār* but hypocritical believers (*munāfiqūn*). The master continues by saying that God calls some people believers and others unbelievers on the basis of what is in their hearts, whereas human beings identify them as believers or unbelievers based on what is on their tongues and on what they appear to be, assuming that what is on their tongues reflects what is in their hearts. The master reminds his disciple that even the recording angels[90] have no knowledge of what is in the hearts of people for they write down only what appears outwardly. Thus, a person who pretends to know what is in the heart of another human being commits an act of "*kufr* punishable by Hell" because that person "arrogates to himself knowledge which only Allāh has."[91] These views of deferring judgment to God shaped the central doctrine of *irjā'* among the Murji'ites.

The origin of *irjā'* goes back to the angels and the Prophet, according to the master in *K. al-Irjā'*.[92] When God asked the angels to inform Him of the names of the creatures in the heavens, earth, and the seas (Q. 2:31, 32), the master writes, they declined (i.e., postponing their answers) because they had no knowledge of them. Likewise, God did not permit Muḥammad to speak or take action against anyone without certain knowledge.[93] The master explains *irja'* by saying that it means to suspend one's judgment (*al-wuqūf*) and "to utter the words only God knows (*Allāhu a'lam*) with regard to things permitted or forbidden, which one does not know and cannot verify by experience or analogies (*al-tajārub wa al-maqāyīs*)."[94]

On the matter of a person declaring a fellow Muslim an unbeliever, the master replies by saying that that person lies but does not commit *kufr*.[95] He is not a *kāfir* because his insult was directed toward another human being rather than toward God and the Prophet, according to the master. Conversely, this accusation of *kufr* does not entitle the accused to brand his accuser as *kāfir* for he will lie about the person who lies about him.[96] This Murji'ite doctrine of *irjā'*, in regard to making judgment on *īmān* and *kufr*, is a clear refutation of the Kharijite views on declaring the believers as *kuffār* for a mere act of disobedience to God's rules. The Murji'ite doctrine of *irjā'*, however, prompted other Muslim traditionalists to refute their theology of the exclusion of acts from faith. Chief among these are the Ḥanbalī and the Ash'arī schools of theology, which will be discussed at length in the following paragraphs.

One of the earliest treatises providing an opposing theology to a traditionalist point of view,[97] and in particular to the Murji'ite doctrine of *īmān*,[98] is *Kitāb al-Īmān*, attributed to Abū 'Ubayd al-Qāsim b. Sallām (d. 839), one of the well-known traditionists and scholars of law and Qur'ānic exegesis.[99] In the first chapter of *Kitāb al-Īmān*, Abū 'Ubayd answers a question concerning the disagreement among Muslims regarding faith and the doctrine of the traditionists, or *Ahl al-Sunnah*. He responds by stating that the community is of two opinions: One supports the views that faith is sincere devotion to God in the hearts, confession of the tongues, and work of the limbs. The other opinion, he adds, maintains that faith "is in the hearts and tongues" to the exclusion of works from faith.[100] While Abū 'Ubayd accords the confession of the tongues, or *shahādah*, "a place of prominence," according to J. Meric Pessagno, he does "not accept it as solely constitutive of *īmān*."[101] For Abū 'Ubayd, the inclusion of works in faith is proven in the Qur'ān.[102] He argues that in early Islam, when the message was revealed to Muḥammad in Makkah, faith was only by hearts and tongues.[103] Religious duties, like almsgiving and fasting, were not incumbent upon the early believers in Makkah because the chief concern at that time was to change the hearts of the people, to make them believers, while they were accustomed to the ways and customs of the pre-Islamic society (*jāhiliyyah*).[104] Works did not become part of the faith until Muḥammad migrated to Madīnah, where God imposed new duties on the believers such as the change of the direction of the prayer toward the Ka'bah.[105] Those who refused to perform this new religious duty—even though confessing *īmān* in their hearts and on their tongues—have nullified their previous confession of *īmān* and this would have availed them nothing.[106]

In response to a Murji'ite argument that God called the Muslims "believers" when they were in Makkah before imposing on them religious duties, Abū 'Ubayd argues that God introduced new duties and made them part of *īmān*. Each new duty was added to the previous ones and these newly imposed obligations constituted *īmān*.[107] Here Abū 'Ubayd establishes stages of *īmān*: initial stage and perfected stage. Confession of the hearts and the tongues can be considered initial *īmān*, whereas works perfected or completed the confessed *īmān*, hence the increase or decrease (*ziyādah* or *nuqṣān*) of *īmān*. While works do increase *īmān*, neglecting religious duties decreases it. Pessagno writes that Abū 'Ubayd does not consider the Murji'ites *kuffār* or outside the community of believers since their teachings do not reflect *bida'* (inventions); rather, they "hold to an erroneous doctrine."[108]

In *Kitāb al-Īmān*, Abū 'Ubayd discusses the question of increase and decrease of faith and argues that good works lead to increase in faith.[109] Abū 'Ubayd shows that the majority of the traditionists opposed the Murji'ite view on the increase in faith. The Murji'ites maintained that the affirmation of the religious duties (i.e., that the prayers consist of five daily prayers and a number of prostrations) leads to an increase in faith. For Abū 'Ubayd, this explanation is awkward. He rejects it on the basis that if affirmation leads to *increase* in faith, this would mean that Muslims who are ignorant of religious duties or deny them all together are faithful and a mere affirmation can lead to extra faith.[110]

Another argument Abū 'Ubayd makes against the Murji'ites is in his reference to Iblīs, the fallen angel who is called infidel in the Qur'ān.[111] Here, Abū 'Ubayd argues that Iblīs affirmed God's lordship but rejected God's order to bow to Ādam. Thus, according to the Murji'ite view, Abū 'Ubayd maintains, Iblīs is a believer, and bowing to Ādam could have increased his faith.[112] Unlike the Kharijites, Abū 'Ubayd agrees with the Murji'ites that sins and acts of disobedience do not nullify *īmān* in the person nor do they make the person a *kāfir*; however, they weaken the *īmān* in the person.[113] Abū 'Ubayd's "main point of contention with the Murji'a," according to Pessagno, "is their exclusion of works from the essential of faith . . . because this exclusion . . . can lead to increasing subjectivity in *īmān*."[114]

On the issue of whether a Muslim who commits sins and acts of disobedience remains a believer, Abū 'Ubayd quotes several well-known *ḥadīths* that support the Murji'ite doctrine that no Muslim should be declared a *kāfir* or a *mushrik*.[115] For instance, Abu 'Ubayd cites the following *ḥadīth*: "When an adulterer commits illegal sexual intercourse, then he is not a believer at the time he is doing it; and when somebody drinks an alcoholic drink, then he is not believer at the time of drinking."[116] These *ḥadīths*, according to Madelung, "were not intended originally to formulate a doctrinal or legal rule. Yet . . . they could be used . . . to bolster [the views] that the Muslim sinner loses name and status of a believer."[117] Traditionalists as well as theologians in Murji'ite and Ash'arite circles faced the dilemma of accepting these *ḥadīths* at their face value without interpretation. For instance, the Sunni Muslim jurist Aḥmad b. Ḥanbal (d. 855) says of these *ḥadīths*: "We accept them, even though we do not know their explanation (*wa in lam na'lam tafsīrahā*). We do not talk nor argue about them, nor do we explain them, but we narrate them as they have come down and we believe in them and we know that they are true."[118]

The traditionalists' reluctance to take these *ḥadīths* at face value or to provide some explanations to them was perhaps a way to avoid the Kharijite experience, that is, using these *ḥadīths* as a pretext to eliminate rivals and cause bloodshed. The traditionalists preferred to be silent about them in order to keep the community integrated.[119] This method of looking at some controversial and disturbing sacred literature with a blind eye can be used by contemporary Muslims as a precedent to argue against extremist groups that cite this literature as a motive for their views and actions.

Four interpretations have been provided regarding these *ḥadīths*, according to Abū 'Ubayd.[120] One position states that the *kufr* meant in these *ḥadīths* is the rejection of God's graces (*kufr al-ni'am*); a second position contends that these *ḥadīths* were meant to underscore the severity (*al-ṭaghlīdh wa al-tarhīb*) of the acts of disobedience in them; a third opinion states that the term *kufr* in these *ḥadīths* points to apostasy (*kufr ahl al-riddah*); and the fourth position rejects these *ḥadīths* altogether.[121] Abū 'Ubayd explains that none of the above positions was acceptable.[122] According to Madelung, Abū 'Ubayd explains these *ḥadīths* based on his views that sins and acts of disobedience neither deprive a person of faith nor do they make him a *kāfir*. Rather, "they deprive faith of its full reality and sincerity," which the Qur'ān describes in numerous verses.[123]

In regard to the *ḥadīths* in which the Prophet severed his ties with certain sinners, Abū 'Ubayd argues against the complete severance of the bond between sinners and the community and opposes its literal implementation.[124] Abū 'Ubayd, for instance, challenges some interpretations that explained the Prophet's statement "he [a person committing an act of disobedience] is not from us (*laysa minnā*)" as meaning "he is not like us (*laysa mithlanā*)." The scholars, according to Abū 'Ubayd, explained the statement "*laysa mithlanā*" to mean severing the bond between the sinners and Muḥammad. Abū 'Ubayd disagrees with these scholars and contends that the Prophet's statement *laysa minnā* was meant only to highlight the seriousness of their disobedience to him and his examples. Abū 'Ubayd adds that the statement *laysa minnā* (he is not from us) cannot be explained to mean *laysa mithlanā* (he is not like us) because no one can be like Muḥammad and he has no equals.[125]

With respect to the *ḥadīths* that designate some acts of disobedience as *kufr* or *shirk*, Abū 'Ubayd argues against the general understanding of this *ḥadīth*.[126] He contends that committing acts of disobedience is not a proof of *shirk* or *kufr* nor does it remove *īmān* from the person. He rather believes that the *ḥadīths* meant to point out that these acts are "characteristic of the infidels and polytheists (*al-kuffār wa al-mushrikūn*)."[127] Abū 'Ubayd supports his claims by a *ḥadīth* that says, "What I fear most for my community is the minor *shirk*."[128] He explains the above *ḥadīth* by saying that committing this kind of *shirk* does not deprive the Muslim from the status of a believer.[129]

The structure of Abū 'Ubayd's *Kitāb al-Īmān* corresponds directly to the Murji'ite response to three questions: What is *īmān*, what qualities distinguish it from *kufr*, and what are the objects of *īmān*? According to Pessagno, Abū 'Ubayd's main problem with the Murji'ites is their "distinction between *īmān* and its objects."[130] For Abū 'Ubayd, Pessagno adds, "*īmān* is a composite of *niya* [intention or devotion], *iqrār* [confession of the tongues], and *a'māl* [works]. No one aspect is more important than the other."[131] For the Murji'ites, *īmān* is defined as "that which preserves (one) from unbelief."[132] Here, the Murji'ites, according to Pessagno, speak of *īmān* as a "name (*ism*) for a group of qualities," such as *ma'rifah* (knowledge of God), *maḥabbah* (love of God), *khuḍū'* (humility before God), *tark al-istikbār* (giving up exalting oneself), *lā istikhfāf* (not making light of God's claim), *khawf* (fear of God), *ta'dhīm* (praise to God), and *ikhlās* (purity of intention).[133] All the Murji'ites, except the Jahmites, Pessagno tells us, added *iqrār* (formal verbal expression) as an important precept directed toward the objects of faith.[134]

For the Murji'ites, because *īmān* is composed of several qualities or characteristics (*khiṣāl*, plural of *khaṣlah*) and is not an entity with independent parts, then *īmān* cannot be eliminated from the heart by a single act of sin.[135] Since the *khiṣāl*, like *maḥabbah*, *khuḍū'*, and the other qualities, are dependent on *īmān*, Pessagno argues, the "force of their existence and their value as religiously good come only through the existence of *īmān* in the heart."[136] Therefore, *īmān*, which God has instilled in the heart of people, neither roots the will in good nor does prevent the believer from committing sin. Hence the Murji'ite dictum: "A sin (*ma'ṣiyah*) will not harm the believer, just as a good deed (*ṭā'ah*) will not be of any use to the unbeliever."[137] The direction of the will, writes Pessagno, determines the nature of the human act, which means that unless the believer commits "an act of the will contradictory of . . . the act of belief, any sin committed cannot harm the essence of *īmān*."[138]

In respect to increase and decrease in *īmān*, the Murji'ite position rejects the possibility of increase and decrease in *īmān* since the latter has *khiṣāl* (i.e., *ma'rifah*, *maḥabbah*, etc.) as its essential characteristics.[139] For the Murji'ites, once *īmān* is "possessed, only a formal act of *kufr* [i.e., denying the existence of God] could affect the essence of *īmān* in the believer,"[140] according to Pessagno. Because *īmān* is from God, no work can affect what God has put in the heart of the believer except a formal rejection of God. For the Murji'ites, *īmān* instills value in work and for this reason they do not consider work a pillar of *īmān*. For Abū 'Ubayd, however, it is sin that affects *īmān*, and for this reason he situates works behind faith. As the Kharijites equated faith with works and the Murji'ites insisted that faith alone determined membership in the Islamic community, the Karramites espoused a position proclaiming that *īmān* is only a confession by the tongue.

TAKFĪR IN THE KARRAMITE THOUGHT

The Karramites (*al-Karrāmiyyah*), followers of Muḥammad b. Karrām (d. 869), claim that *īmān* is by the confession of the tongue alone, the pronouncement of the *shahādah*-formula. Abū Muḥammad 'Alī ibn Aḥmad ibn Sa'īd ibn Ḥazm (d. 1064) writes that for the Karramites, "*īmān* is the confession by the tongue despite the *kufr* in the heart. This person is a believer in God's sight, a friend of God (*waliyun lahu*), and will be among the people of the Garden."[141]

The Muslim historiographer Tāj al-Dīn al-Shahrastānī (d. 1153) tells us that the Karramites believed that *īmān* is the confession by the tongue alone.[142] Confession by the tongue, as understood by many people, did not include *taṣdīq* by the heart nor any external works. Thus, how does someone without *taṣdīq* in his heart be a believer in the sight of God and in the Hereafter be in the Garden forever? Al-Shahrastānī explains the views of the Karramites on this issue by saying that the group makes a distinction between the problem of calling a man a believer in this world and the final reward and punishment.[143] For the Karramites, al-Shahrastānī writes, a man who confesses by the tongue without having *taṣdīq* in the heart is a believer in the present world, but he is a *kāfir* and is doomed to an eternal punishment in the fire in the Hereafter.[144] According to Izutsu, the Karramites are the "real representatives of the thesis that *īmān* is nothing but 'saying' *qawl*."[145]

Ibn Taymiyyah believes that the Karramite thesis on the distinction between calling a hypocrite a Believer in the present world and a *kāfir* in the Hereafter is an innovation (*bid'ah*) and a false view.[146] He writes that the Karramites assert that, "whoever displayed belief outwardly is described by them as being a true believer. But if the individual is a hypocrite, he will abide eternally in Hell. . . . Thus, whoever claims that the Karramites said that the hypocrite enters paradise has misrepresented their views."[147] Ibn Taymiyyah is somewhat satisfied with the Karramite thesis that a hypocrite is at the end doomed to the eternal fire in the Hereafter. He explains the position to clarify the misunderstanding that when the Karramites say a hypocrite is a Believer, they do not mean that a hypocrite will go to heaven. He asserts that the Karramite theory of *īmān* is "fundamentally a mistaken view and an innovation which nobody before them has ever asserted."[148]

The Karramites assert, according to 'Abd al-Qāhir al-Baghdādī (d. 1037), that *īmān* is only the single confession (*iqrār fard*) made at the very beginning (when a man becomes a Muslim by confessing his belief in God and the Prophet).[149] This first confession invalidates and makes unnecessary all subsequent acts of confession, and its repetition after that does not produce *īmān* except in the apostate who does the same confession for the first time after his apostasy.[150] The question remains: If *īmān* is a matter of a single confession made at the time when a man becomes a Muslim, what logic do the Karramites draw from the Qur'ānic verse 7:172 that all humans had already made their confession by the tongue at the beginning of human history? Does this first verbal confession make all subsequent confessions by individuals unnecessary? The Karramite theory of *īmān* as a single confession at the very beginning, Izutsu writes, "would simply destroy the Karramite thesis of *īmān* itself, for everyone, according to this interpretation, is a Believer."[151] According to al-Baghdādī, the Karramites made the first confession effective in everyone "even if he be silent or dumb" and this confession is valid until the Day of Resurrection except in the case of apostasy.[152] The Karramite thesis of first confession seems to be a departure from all the schools that preceded them, and their theory of *īmān* focused more on what defines *īmān* rather than on what defines *kufr*. Their broad definition of *īmān* narrowed the realm of *kufr* in this world and made it a matter to be decided by God in the Hereafter.

The concept of *takfīr* was one of the few topics that were largely debated among the theologians of early Islam. The debate on *takfīr* not only contributed to the development of the meaning of *takfīr* but also provided the theological schools with ideas on how to define and understand *īmān*. It seems that theologians had to define the meaning of *kufr* in order to understand what *īmān* really means. This method of understanding *īmān* through defining *kufr* is explored further in the next chapter on later medieval theological schools.

NOTES

1. Tāj al-Dīn al-Shahrastānī, *al-Milal wa al-Niḥal*, vol. 1, eds. 'Āmir 'Alī Muhannā and 'Alī Ḥasan Fā'ūr (Beirut, Lebanon: Dār al-Ma'rifah, 1996), 132.
2. Nāsir b. Sulaymān b. Sa'īd al-Sābi'ī, *al-Khawārij wa al-Ḥaqiqa al-Ghā'iba*, 2nd ed. (Muscat, Oman: Maktabat al-Jīl al-Wa'īd, 2003), 141–142.
3. Muḥammad bin Jarīr al-Ṭabarī, *Ta'rīkh al-Umam wa al-Mulūk*, vol. 3 (Beirut, Lebanon: Dār al-Kutub al-'Ilmiyyah, 1997), 116.
4. Muḥammad b. Aḥmad al-Malaṭī, *al-Tanbīh wa-al-Radd 'ala Ahl al-Ahwā' wa al-Bida'*, ed. Muḥammad Zāhid b. al-Ḥasan al-Kawtharī (Beirut, Lebanon: Maktabat al-Ma'ārif, 1968), xiii.
5. Taqī al-Dīn Aḥmad ibn Taymiyyah, *Majmū' Fatāwā*, vol. 28 (Riyadh, Saudi Arabia: Mujamma' al-Malik Fahd, 2004), 473.
6. Al-Malaṭī, *al-Tanbīh*, xiii.
7. Al-Shahrastānī, *al-Milal*, vol. 1, 132–133.
8. Al-Malaṭī, *al-Tanbīh*, 2.
9. Al-Shahrastānī, *al-Milal*, vol. 1, 134.
10. 'Abd al-Qāhir al-Baghdādī, *al-Farq Bayna al-Firaq* (Beirut, Lebanon: Dār al-Kutub al-'Ilmiyyah, n.d.), 49.
11. The People of the Camel (*aṣhāb al-jamal*) were those who fought against 'Alī in the battle of the Camel in 656 in Baṣra. See al-Tabarī, *Ta'rīkh*, vol. 3, 53.
12. Ibn Taymiyyah, *Majmū'*, 473.
13. Al-Baghdādī, *Al-Farq*, 50.
14. Ibid.
15. Ibid.
16. Toshihiko Izutsu, *The Concept of Belief in Islamic Theology* (New York: Books for Libraries, 1980), 5.
17. Ibid.
18. Aḥmad b. Yaḥya al-Balādhurī, *Ansāb al-Ashrāf*, vol. 5, eds. Suhayl Zakkār and Riyāḍ Zarkalī (n.c.: Dār al-Fikr, 1997), 179.
19. See Hussam S. Timani, *Modern Intellectual Readings of the Kharijites* (New York: Peter Lang Publishing, 2008), 6–9.
20. Ibid., 26–27.
21. Al-Ṭabarī, *Ta'rīkh*, vol. 3, 110, 156.
22. 'Āmir al-Najjār, *Al-Khawārij: 'Aqīdatan wa Fikran wa Falsafatan* (Beirut, Lebanon: 'Ālam al-Kutub, 1986), 103.
23. Ibid., 108.
24. Al-Baghdādī, *al-Farq*, 71.
25. Ibid.
26. Ibid.; See also Izutsu, *The Concept*, 13–15.
27. Ibid., 4.
28. Ibid.
29. Ibid., 11.
30. Izutsu, *The Concept*, 7.
31. The mentioning of the abode of *īmān* is referred to in verses 6:127 and 10:25: "They shall have the Home of Peace with their Lord. God invites [everyone] to the Home of Peace." According to Fakhr al-Dīn al-Rāzī (d. 1209), the term Peace is used interchangeably with the word Allāh, hence the Home of Allāh. This term is also used for al-Ka'ba as the House of Allāh (*Bayt Allāh*) and for Paradise (*al-Jannah*). The inhabitants of this Home are only the believers. Fakhr al-Dīn al-Rāzī, *al-Tafsīr al-Kabīr*, vol. 13 (Beirut, Lebanon: Dār al-Fikr, 1981), 198.
32. Izutsu, *The Concept*, 8.
33. Al-Najjār, *al-Khawārij*, 107.
34. Ibid.
35. See Abū Ḥanīfah al-Nu'mān, *Al-'Ālim wa al-Muta'allim*, ed. Muhammad Zāhid al-Kawtharī (Cairo: Matba'at al-Anwār, 1949).

36. Qāḍī 'Abd al-Jabbār, *Sharḥ al-Uṣūl al-Khamsah* (Beirut, Lebanon: Dār Iḥyā' al-Turāth al-'Arabī, 2001), 87.
37. Izutsu, *The Concept*, 41–42.
38. Al-Malaṭī, *al-Tanbīh,* 49.
39. Ibid., 49–50.
40. Timani, *Modern Intellectual*, 30–31.
41. Ibid., 31.
42. A.J. Wensinck, *The Muslim Creed: Its Genesis and Historical Development*, v. 39 (London: Routledge, 2008), 40.
43. Al-Sābi'ī, *al-Khawārij wa al-Ḥaqīqa al-Ghā'ibah*, 121.
44. Wensinck, *The Muslim Creed*, 41.
45. Ibid.
46. Ibid.
47. Al-Sābi'ī, *al-Khawārij*, 121.
48. Ibid., 133.
49. Laṭīfah al-Bakkāy, *Ḥarakāt al-Khawārij* (Beirut, Lebanon: Dār al-Ṭalī'ah, 2001), 225–226.
50. Wensinck, *The Muslim Creed*, 42.
51. Ibid., 44.
52. Abū al-Ḥasan al-Ash'arī, *Maqālāt al-Islamiyyīn wa Ikhtilāf al-Muṣallīn*, vol. 1, ed. Muḥammad Muḥyī al-Dīn 'Abd al-Ḥamīd (Beirut, Lebanon: al-Maktabah al-'Aṣriyyah, 1990), 180.
53. Al-Bakkāy, *Ḥarakāt al-Khawārij*, 225–226.
54. Julius Wellhausen, *The Religio-Political Factions in Early Islam* (Oxford: North-Holland Publishing Company, 1975), 20–21.
55. Ibid.
56. Ibid.
57. Ibid.
58. Izutsu, *The Concept*, 17.
59. The Jahmite is a group affiliated with the Murji'ite school of theology. The group's founder is Ṣafwān b. Jahm, hence *al-jahmiyyah*, see al-Baghdādī, *al-Farq bayna al-Firaq*, 16–18.
60. Al-Malaṭī, *al-Tanbīh*, 98.
61. Ibid.
62. Ibid.
63. Ibid.
64. Wensinck, *The Muslim Creed,* 38.
65. Izutsu, *The Concept*, 44.
66. "Kitāb al-Irjā'," according to Joseph Givony, is an epistle (*Risālah*) that was preserved in *Kitāb al-Īmān* by Abū 'Abdallāh Muḥammad b. Yaḥya b. Abī 'Umar al-'Adanī (d. 856), Joseph Givony, "The Murji'a and the Theological School of Abu Hanifa: A Historical and Ideological Study," Ph.D. dissertation, University of Edinburgh, 1977, 16–25; see also Joseph Van Ess, "Das 'Kitab al-Irga'' des Hasan b. Muhammad b. al-Hanafiyya," *Arabica*, T. 21, Fasc. 1 (Feb. 1974): 20–52 and Daniel Lav, *Radical Islam and the Revival of Medieval Theology* (Cambridge: Cambridge University Press, 2012), 18.
67. Al-Baghdādī, *Al-Farq Bayna al-Firaq*, 49.
68. Lav, *Radical Islam*, 16.
69. Ibid.
70. Ibid.
71. Ibid.
72. Izutsu, *The Concept*, 45.
73. 'Abd al-Qāhir al-Baghdādī, *Firaq al-Shī'ah* (Istanbul: n.p., 1951), 6, quoted in Izutsu, *The Concept*, 45.
74. Izutsu, *The Concept*, 45.
75. Wilferd Madelung, *Religious Schools and Sects in Medieval Islam* (London: Variorum Reprints, 1985), 33.

76. Ibid.
77. Ibid.
78. Lav, *Radical Islam,* 24.
79. Al-Nuʻmān, *Al-ʻĀlim wa al-Mutaʻallim*, and *al-Fiqh al-Akbar* (n.c.: United Arab Emirates, Maktabat al-Furqān, 1999).
80. Joseph Schacht, "An Early Murci'ite Treatise: The Kitāb al-ʻĀlim wa'l-Muta'allim," *Oriens* Vol. 17 (Dec. 31, 1964): 97.
81. Lav, *Radical* Islam, 24.
82. Wilferd Madelung, "Early Sunni Doctrine Concerning Faith as Reflected in the 'Kitāb al-Īmān' of Abū 'Ubayd al-Qāsim b. Sallām (d. 224/839)," *Studia Islamica*, No. 32 (1970), 233.
83. Schacht, "An Early Murci'ite," 98.
84. Ibid., 101.
85. Ibid., 102.
86. Al-Nuʻmān, *Al-ʻĀlim wa al-Mutaʻallim*, 12.
87. Ibid., 13.
88. Ibid., 18.
89. Ibid., 19.
90. The Qur'an speaks of two angels (Munkar and Nakīr) questioning the soul and recording the answers while it sits in the grave (Q. 8:50), see Al-Nuʻmān, *Al-ʻĀlim wa al-Mutaʻallim,* 80.
91. Al-Nuʻmān, *Al-ʻĀlim wa al-Mutaʻallim,* 22–23.
92. Ibid.
93. Ibid.
94. Ibid.
95. Ibid., 27.
96. Ibid.
97. Madelung, "Early Sunni Doctrine," 234.
98. See, for example, J. Meric Pessagno, "The Murji'a, Iman and Abu 'Ubayd," *Journal of the American Oriental Society*, vol. 95, no. 3 (Jul.-Sep., 1975), 382–394.
99. Madelung, "Early Sunni Doctrine," 234.
100. Abū 'Ubayd al-Qāsim b. Sallām, *Kitāb al-Īman*, ed. Muḥammad Nāṣir al-Dīn al-Albānī (Beirut, Lebanon: al-Maktab al-Islāmī, 1983), 9–10.
101. Pessagno, "The Murji'a," 384.
102. Ibn Sallām, *Kitāb al-Īmān*, 10.
103. Ibid.
104. Ibid.
105. Ibn Sallām, *Kitāb al-Īmān*, 10–11.
106. Ibid., 11.
107. Ibid., 15.
108. Pessagno, "The Murji'a," 385.
109. Ibn Sallām, *Kitāb al-Īmān*, 24–25.
110. Ibid.
111. Ibid., 30.
112. Ibid.
113. Ibid., 40.
114. Pessagno, "The Murji'a," 385.
115. Ibn Sallām, *Kitāb al-Īmān*, 36.
116. Ibid. See the *ḥadīth* in *Ṣaḥīḥ Bukhārī*, Volume 008, Book 081, Hadith Number 763. Retrieved from www.hadithcollection.com. Accessed on December 9, 2013.
117. Madelung, "Early Sunni Doctrine," 248.
118. Abī al-Ḥusayn Muḥammad b. 'Alī Yaʻlā, *Tabaqāt al-Ḥanābilah*, vol. 1, ed. Muḥammad Ḥamid al-Faqī (Cairo: Matbaʻat al-Sunnah al-Muḥammadiyyah, n.d.), 311.
119. Ibn Sallām, *Kitāb al-Īmān*, 39.
120. Ibid., 38–39.
121. Ibid.
122. Ibid., 39.

123. Madelung, "Early Sunni Doctrine," 250.
124. Ibn Sallām, *Kitāb al-Īmān*, 43.
125. Ibid.
126. Ibid.
127. Ibid.
128. Ibid., 44.
129. Ibid.
130. Pessagno, "The Murji'a," 386.
131. Ibid.
132. Ibid., 387.
133. Ibid., 386–387.
134. Ibid.
135. Ibid., 393.
136. Ibid.
137. Ibid.
138. Ibid.
139. Ibid.
140. Ibid.
141. Abū Muḥammad 'Alī ibn Ḥazm, *al-Fiṣal fi al-Milal wa al-Ahwā' wa al-Niḥal*, vol. 3 (n.c.: Al-Maṭba'ah al-Adabiyyah, 1902), 204.
142. Muḥammad Al-Shahrastānī, *Kitāb al-Milal wa al-Niḥal*, part 1 (London: The Society for the Publication of Oriental Texts, n.d.), 84–85.
143. Ibid.
144. Ibid.
145. Izutsu, *The Concept*, 148–149.
146. Taqī al-Dīn Aḥmad ibn Taymiyyah, *Kitāb al-Īmān* (Bloomington, Ind.: Iman Publishing House, 1999), 153.
147. Ibid.
148. Ibid.
149. Al-Baghdādī, *al-Farq*, 168.
150. Ibid.
151. Izutsu, *The Concept*, 155.
152. Al-Baghdādī, *al-Farq*, 168.

Chapter Three

Takfīr in Later Islamic Theology

This chapter discusses the concept of *takfīr* in three medieval theological schools: the Mu'tazilites, the Ash'arites, and the Maturidites. The Mu'tazilites contributed to the debate by arguing for a third position between *kufr* and *īmān*. The Ash'arites, on the other hand, did not concern themselves with the meaning of *kufr*. Rather, they delved into the debate because they wanted first and foremost to highlight the meaning of *īmān*, and second they wanted to strengthen the traditional Sunni interpretation of *īmān*. Thus, they were interested in a debate that would define acts of *īmān* rather than acts of *kufr*. In contrast to the Mu'tazilite views on *kufr* as an act of free will, the Maturidites argued that *kufr* is God's creation. He wished it and has the foreknowledge that certain people choose to commit acts of *kufr*.

TAKFĪR IN THE MU'TAZILITE THOUGHT

The word *mu'tazilah* ("those who stand aloof") may be attributed to the school's middle position regarding the grave sinner.[1] Their position was in the middle between the Kharijites who declared a grave sinner a *kāfir* and the Murji'ites who taught that he remains a believer. Their reluctance to make a judgment on whether or not a grave sinner remains a member of the Islamic community was perhaps a factor in the rise of the Mu'tazilite school of theology. The general Mu'tazilite conception of *īmān* is the view that the acts of obedience are essential for belief and whoever neglects these acts is not a Believer.[2] The Mu'tazilites taught that *īmān* should be rejected if it contradicts reason and logic and that *īmān* is not only in performing the prayers but also in avoiding grave sins (*al-kabā'ir*). *Īmān* can be achieved, they argued, by committing acts of obedience. Without these qualifications, a man cannot be called a Believer.[3]

Al-Shahrastānī writes that the Mu'tazilites considered *īmān* a good trait upon which it gives its possessor the title "Believer," which is a title that earns its holder the praise.[4] The *fāsiq*, on the other hand, is someone who does not possess this good trait, and does not therefore deserve the same praise. He is not called Believer, al-Shahrastānī adds, and he is not absolutely a *kāfir*, for the *shahādah* (the testimony to God's oneness) and the good deeds are within him.[5] However, if he commits a grave sin and does not repent, he will be in Hell forever, for in the end time, according to the Qur'ān, the people are sorted into two groups: one in Heaven and the other in Hell (Q. 2:277; 4:55; 9:35). The punishment of the *fāsiqīn*, however, according to al-Shahrastānī, will be reduced and they will be elevated to a higher level above the *kuffār*.[6] This means that the Mu'tazilites do not deny the existence of some sort of belief in the heart of the *fāsiq*; yet they fall short of calling him a Believer.

Wāṣil b. 'Aṭā' (d. 748), to whom scholars attribute the rise of the Mu'tazilites,[7] says that *īmān* is a trait that if it exists in the heart, a man can be called a Believer. For Ibn 'Aṭā', the grave sinner is not a believer nor a *kāfir* nor a *munāfiq* but a *fāsiq*. The *fāsiq*, according to Ibn 'Aṭā', is called as such because he does not hold this praiseworthy trait of *īmān* and therefore does not deserve to be called a Believer. He is not called a Believer; at the same time, Ibn 'Aṭā' adds, he is not an absolute *kāfir* (*kāfir muṭlaq*), for the *taṣdīq* as well as the acts of obedience are in him and there is no ground to deny these traits.[8] What is interesting in Ibn 'Aṭā''s argument is his insistence on saying that a *fāsiq* should not be *called* a Believer as opposed to him saying he is *not* a Believer. One may argue that Ibn 'Aṭā' is not denying the *fāsiq* *īmān* so much as he is denying him the title "*mu'min*." For Ibn 'Aṭā', a *fāsiq* is a Believer because "he is not an absolute *kāfir*."[9] If he is not an absolute *kāfir*, then he is a Believer without calling him one.

The Mu'tazilite thinker Hishām al-Fuwaṭī says that *īmān*, which is the act of obedience, consists of two traits: 1) the *īmān* in God (*al-īmān billāh*), and 2) the *īmān* for God (*al-īmān lillāh*).[10] He argues that people should not say "*Īmān* in God" because not having *īmān in* God is not an act of *kufr* (*al- īmān billāh mā kāna tarkuh kafran billāh*). On the contrary, not having "*īmān for* God" (*īmān lillāh*) is an act of *kufr* as well as an act of *fisq*.[11] Sa'd al-Dīn Mas'ūd al-Taftazānī (d. 1390), the author of *Kashf al-Asrār*, a Persian commentary on the Qur'ān, writes: "The acts of obedience constitute a 'pillar' of the real essence of *īmān* in such a way that whoever neglects them is not a Believer."[12] Abū al-Qāsim Maḥmūd b. 'Umar al-Zamakhsharī (d. 1143), a Mu'tazilite scholar and Qur'ān commentator, says that verse 10:9 of the Qur'ān shows that the *īmān* that earns the person divine guidance, prosperity, and light on Judgment Day is the conditioned *īmān*, which is accompanied by good deeds (*al-'amal al-ṣāliḥ*). The *īmān* that is not accompanied by good deeds will not earn the person divine prosperity and light. The *īmān* that is accompanied by good deeds is a sure *īmān* free from any doubt.[13] Izutsu writes that emphasis on deeds led to the rise of a radical group among the Mu'tazilites known as the *Wa'īdiyyah* or the "Threateners."[14] The latter's central thesis is the view that one act of disobedience, Izutsu adds, makes a person a *kāfir* who will be destined for the Fire, and that the neglect of even one act of obedience cancels a man's *īmān.*[15]

Such views were not so popular among the Mu'tazilites, however. For instance, al-Shahrastānī refutes the *Wa'īdiyyah* views by arguing that their thesis goes against the majority of Qur'ānic verses, especially the message of Divine mercy.[16] He points out that the Qur'ān distinguishes between the word *īmān* and the word *'amal* and that God often addresses grave sinners by calling them Believers. Al-Shahrastānī opines that this Qur'ānic distinction between *īmān* and *'amal* renders *īmān* and *'amal* two different things and that *'amal* is not a fundamental "pillar" for *īmān.* For al-Shahrastānī, if the *Wa'īdiyyah* thesis were validated, then no person in the whole world would be a Believer except for the Prophets. It would also be inappropriate to give the name "Believer" on anyone until he has obtained in himself all the virtuous characteristics in actions and deeds without any exceptions.[17]

Moreover, al-Shahrastānī emphasizes that it would be absurd to affirm that even a single grave sin will earn its doer the status of *kufr* and will send the man to an eternal Hell although his punishment may be reduced (*takhfīf*) because of good works he may have done. Al-Shahrastānī says: "If all the acts of obedience are nullified by a single act of disobedience, how can punishment be alleviated and how can there be alleviation when the person is condemned to Hell forever? For someone who has been a *kāfir* for one hundred years, would not be unjust to send him to Hell forever? And how is it justifiable to impose eternal punishment on acts that are temporary or limited by time?"[18] Al-Shahrastānī, therefore, sees the Mu'tazilites of the *Wa'īdiyyah* doctrine as the antithesis of the Murji'ite position of postponement. The *Wa'īdiyyah* group, in the words of Izutsu, represents an extremist ideology within the Mu'tazilite School of theology, and not the whole of Mu'tazilism.[19] Al-Zamakhsharī, however, represents the general Mu'tazilite conception of the status of the grave sinner. He writes: "The true *īmān* is the belief in the truth and its affirmation by the tongue and the deeds. He who lacks the inner belief (*al-i'tiqād*)—although he affirms it by the tongue and the deeds—is a Hypocrite. He who lacks the verbal confession (*al-shahādah*) is a *Kāfir*, while he who lacks the doing (*al-'amal*) is a *Fāsiq*."[20] Thus, the Mu'tazilite thesis that the concept of *īmān* does not compromise actions makes *īmān* unaffected by acts of disobedience, even by grave sins. Hence, a believer never turns into a *kāfir* as long as he is a believer regardless of his actions.

The Mu'tazilite conception of *īmān* and *'amal* and their position on grave sin, according to Izutsu, "brings into the semantic field of *ummah*, or the

Muslim community a remarkable change in the distribution of concepts."[21] For instance, the Qur'ānic conception is based on a simple dichotomy: Believers vs. *kuffār*, and the text is not specific on the status of the *fāsiqīn*: Are they believers or *kuffār*? Since the Qur'ān remained silent on their status, the community, however, was to decide on where to situate them. The Kharijites, for instance, made them part of the *kuffār*, thus narrowing the realm of belief.[22] The Murji'ites, on the contrary, considered them believers, allotting them a place in the *ummah*, hence expanding the realm of belief.[23] The Mu'tazilites, on the other hand, made the *fāsiqīn* an independent group,[24] a position in the middle between the believers and the *kuffār*, that is, they are neither believers nor *kuffār*.

While the Mu'tazilites adduced Qur'ānic verses and *ḥadīths*[25] in support of their third, independent category for the *fāsiqīn*, their opponents, however, countered their claims and argued that in the Hereafter there will be only the Garden and the Fire and that the word *fāsiq* is synonymous with *kāfir*.[26] Although the Qur'ān does not offer a third place for the *fāsiqīn* in the Hereafter (the *kuffār* in Hell and the believers in Heaven), the Mu'tazilites strongly believed that the Qur'ān's reference to some people as *fāsiqīn* was an indication that they do not belong to the category of the *kuffār* and their actions are not classified as acts of *kufr*.[27]

Ibn Ḥazm argues that a Believer who has committed a grave sin loses his belief only in that particular point but not in its entirety.[28] In his own words, "Such a man is a Believer in regards to his good deeds, and a non-believer in regards to his evil acts."[29] For Ibn Ḥazm, therefore, "an act of disobedience does not lessen or impact the *taṣdīq* in the person who commits the act. He is a non-Believer in one of his works."[30] Ibn Ḥazm concludes that, "every *Kāfir* is an unjust disobedient *Fāsiq*, but not every unjust disobedient *Fāsiq* is a *Kāfir*. Some of them are Believers."[31]

Izutsu examines the thesis that classifies a *fāsiq* as neither a Believer nor a *kāfir*, but as a "hypocrite" *munāfiq*.[32] This thesis is similar to that of the Mu'tazilite; for the latter, a *fāsiq* is neither a Believer nor a *kāfir*, but belongs to a third category. Izutsu, however, tells us that some theologians, such as Ḥasan al-Baṣrī, branded a *fāsiq* as a hypocrite.[33] The latter is a person who professes belief in God and the Prophet but is not sincere in his profession (Q. 9:62; 9:64; 9:74). He professes belief while concealing *kufr* in his heart, which drives his actions. According to al-Ash'arī, the Ibāḍiyyah—a Kharijite sub-sect—termed the *fāsiq* "*munāfiq*,"[34] with the term occurring frequently in the Qur'ān: "The hypocrites [*al-munāfiqīna*] try to deceive God, but it is He who causes them to be deceived," (Q. 4:142) "wavering all the time between this and that, belonging neither to one side nor the other" (Q. 4:143). For the Ibāḍiyyah, the *munāfiq*—a name God Himself attached to certain people—is free from *shirk*: although not a Believer, he is a person fluctuating between *īmān* and *kufr*; he is a *kāfir* because he is against *tawḥīd* when he is in a state of unbelief.[35]

The concept that the grave sinner is a *munāfiq* was advanced by al-Ḥasan al-Baṣrī[36] (d. 728), one of the most important religious figures in early Islam whom the Muʻtazilites claim as one of their founders.[37] It was al-Baṣrī, Izutsu writes, rather than the Ibāḍiyyah, who put the argument forward that the grave sinner is a *munāfiq*. In al-Baṣrī's thought, work is important because each act is a manifestation of a pure heart, *qalb sāliḥ* or *qalb salīm*, and the pure heart is every believer's aim in this life.[38] Those who do not strive to attain *al-qalb al-salīm* are the *munāfiqīn*, the hypocrites. In other words, according to Izutsu, "a *Munāfiq* is a man whose concern is with the present world (*al-dunyah*) rather than with the world to come (*al-ākhirah*), who is not fearful of Divine chastisement in his life at every step . . . [who] does not take *īmān* seriously."[39] In *Ḥilyat al-Awliyā' wa Tabaqat al-Aṣfiyā'*, al-Baṣrī describes the believer as the one who, while committing good deeds and pious acts, fears that he will not be saved and the *munāfiq* as the one who commits evil acts and hopes that God will save him.[40] As for the *kāfir*, according to Izutsu, al-Baṣrī did not offer an opinion, believing that there is no hope for the *kuffār*. His main concern was with the *munāfiq* who, al-Baṣrī believed, should not be declared a *kāfir* and thus be expelled from the community.[41]

In reference to *ḥadīths* where the Prophet pointed out certain properties that make a person *munāfiq* and asserted that whoever had these properties in himself is a *munāfiq*, despite performing the fast and the prayers and insisting on being a Muslim, Ibn Ḥazm replies by arguing that the reference to *nifāq* in these particular *ḥadīths* is not the *nifāq* of *kufr*.[42] Ibn Ḥazm defines *nifāq* by saying that the term means "the hole of a field mouse" *nafaqa al-yarbu'*,[43] which is covered with dirt. Thus, a *munāfiq* is someone who hides something. That person, according to Ibn Ḥazm, is a *munāfiq* who covers something in the same way the Qur'ān in verse 57:20 calls the tiller a *kāfir* who covers the earth with seeds. In this sense, the terms *kāfir* and *munāfiq* are used only metaphorically.[44] Ibn Ḥazm adds that if a grave sinner is a *munāfiq*, then every human being becomes a *munāfiq* since no one, except the angels and the Prophets, is without a grave sin.[45] Therefore, committing a grave sin does not make anyone a *munāfiq*. Ibn Ḥazm contends that a *munāfiq* is a believing *fāsiq* (*mu'min fāsiq*) whose *īmān* is less (*nāqis*) than that of a believer who is not a *fāsiq*.[46]

In *Kitāb Sharḥ al-Uṣūl al-Khamsah*, the Muʻtazilites established the doctrine of the Divine Five Principles (*al-Uṣūl al-Khamsah*) and taught that whoever violates these principles is perhaps a *kāfir*, a *fāsiq*, or a wrongdoer.[47] The Five Principles are: 1) *Al-tawḥīd* (the oneness of God); 2) *al-'adl* (justice); 3) *al-wa'd wa al-wa'īd* (Divine Promise and Warning); 4) *al-manzilah bayna al-manzilatayn* (the Middle Position); and 5) *al-amr bi al-ma'rūf wa al-nahī 'an al-munkar* (commanding good and forbidding evil). He who violates *al-tawḥīd* (i.e., denies the things that God approves and approves the things that God denies) is a *kāfir*. He who violates *al-'adl* (i.e., committing oppression, persecuting the children of the *kuffār*, or causing disorder) is also a *kāfir*. He who violates *al-wa'd wa al-wa'īd* (i.e., denying that God promises repentance for the obedient and punishment for the disobedient) is a *kāfir*. He who says that God will not follow with His promises and threats because God is All-Merciful is a *kāfir*. He who says that God carries out on His promises and threats but He may do so under certain conditions and with exceptions is not a *kāfir* but is mistaken (*mukhṭi'*). He who goes against the middle position between *kufr* and belief and says that a grave sinner is not a believer is a *kāfir*. He who says that a grave sinner is a believer is a *fāsiq*. He who says that a grave sinner is neither a *kāfir* nor a believer but should be called a believer is *mukhṭi'*. He who denies that God has ordered the Muslims to command the good and prohibit evil is a *kāfir*, for he rejects a necessary principle in Islam. He who says, however, that God has ordered the Muslims to command good and prohibit evil only when a Muslim leader (*imām*) is present (*mashrūṭ bi wujūd al-imām*) is *mukhṭi'*.[48]

The Mu'tazilites taught that people are of two groups: those who receive God's grace and blessing and those who receive his punishment and wrath.[49] If a man is from the former group, he earns God's greater grace or another lesser grace. Those who receive God's greater grace are called prophets and messengers, while those who receive a lesser grace are called believers. The people who earn God's punishment and wrath are also of two groups: those who deserve the greater punishment and those who deserve a lesser punishment. Those who deserve the greater punishment are called *kuffār* or *mushrikūn* (polytheists), while those who deserve the lesser punishment are called *fāsiqūn* (neither believers nor *kuffār*).[50] Thus, the grave sinner (*ṣāḥib al-kabīrah*) is not called neither a believer, nor a *kāfir* or a *munāfiq*, but a *fāsiq* and therefore he does not qualify to be judged as a believer, a *kāfir*, or a *munāfiq*. He belongs to a status between the believer and the *kāfir* and his judgment should be somewhere between the two.[51]

For the Mu'tazilites, the grave sinner should not be declared a *munāfiq* because calling him as such will make him a *kāfir* who is subject to losing privileges allotted to him by the Muslim community, such as marriage to a Muslim woman, inheritance, and Islamic burial, among others.[52] The *munāfiq*, the Mu'tazilites taught, is someone who declares his *islām* while hiding his *kufr*. The grave sinner, on the other hand, is a believer who has committed an act of disobedience. The Mu'tazilites challenge fellow Muslims who taught that a grave sinner should be called a *munāfiq* but should not be treated as a *kāfir*. To these claims, the Mu'tazilites reply by arguing that if by calling the grave sinner *munāfiq* is meant to impose rulings pertaining to the *kuffār*, it is well established in Islam that the grave sinner is not subject to these rulings and he is not forbidden from marrying a Muslim woman, inheritance, and Islamic burial.[53] If calling the grave sinner a *munāfiq* is meant only to give the person the title *kāfir* without making him subject to the rulings of the *kuffār*, then their logic is absurd because the title *kāfir* is given to those who are subject to the rulings of the *kuffār*.[54]

In *K. Sharḥ al-Uṣūl al-Khamsah*, The Mu'tazilite theologian Qāḍī 'Abd al-Jabbār (d. 1025) argues against theologians who believed that a *kāfir* may become a believer although he chose *kufr* for himself. 'Abd al-Jabbār challenges this thesis by offering his views on the Islamic doctrine of *taklīf* (God does not impose on the people the things that they do not have the power to bear or that are beyond their capacity, Q. 2:286). 'Abd al-Jabbār contends that if *kufr* exists in the *kāfir* as an act of will, then it is impossible for that same person to become a believer since he chose for himself *kufr*, and it is absurd, he adds, that someone is both a *kāfir* and believer.[55] 'Abd al-Jabbār believed that God's *taklīf* (imposing *īmān* on the *kāfir*) is a *taklīf* beyond the capacity of the *kāfir* because the latter is capable of *kufr* only. Thus, his *taklīf* of *īmān*, which is something that he does not have the power to bear, according to 'Abd al-Jabbār, is evil and God does not commit evil.[56]

Moreover, in response to the theologians who argued that a *kāfir* is free to acquire *īmān*, 'Abd al-Jabbār taught that a *kāfir* is incapable of acquiring *īmān* because *takfīr* is the antithesis of *īmān*, and two opposites do not exist at the same time.[57] 'Abd al-Jabbār writes that if the theologians meant that a *kāfir* acquires *īmān* as a result of replacing *kufr* with belief, then they erred on this issue because they extended the choice to someone who is incapable of replacing one trait with another, that is, *kufr* with *īmān*. He challenges their theory by using the metaphor of the bird whose wings are clipped and is given the choice to fly. 'Abd al-Jabbār concludes by asking: How do theologians say that a *kāfir* is a person with the choice to acquire *īmān* when he chose for himself *kufr*? He was inclined toward *kufr* because ontologically he was incapable of acquiring belief.[58]

In the section on the Mu'tazilite middle position (*al-manzilah bayna al-manzilatayn*) of *K. al-Uṣūl al-Khamsah*, 'Abd al-Jabbār argues against the Murji'ites and the Kharijites for declaring a grave sinner a *mu'min* and a *kāfir*, respectively, and he justifies the Mu'tazilite argument for establishing an independent category for the grave sinner.[59] 'Abd al-Jabbār writes that it is religiously established that a grave sinner is not called a *mu'min* because, by committing a grave sin, that person has brought defamation and condemnation on himself.[60] It is also established, 'Abd al-Jabbār continues, that a *mu'min* is legally called as such because he has earned praise and loyalty for his belief. Based on these two established principles, it is not proper to call a grave sinner a *mu'min*.[61]

'Abd al-Jabbār argues for calling a grave sinner a *mu'min* with restrictions on the meaning of the term.[62] He writes, "although we objected to calling the grave sinner a *mu'min*, we do not object to calling him as such as long as we put restrictions on that name. It is proper to describe the grave sinner as believer in God and His Prophet, but the restrictions put on the title 'believer' makes it unbeneficial to its holder."[63]

'Abd al-Jabbār objects to the claims that *īmān* can be achieved by belief alone without works. He replies to the theologians who argue that God promised the Garden to those who believe in God and His Prophet.[64] 'Abd al-Jabbār contends that the promise has been made to those who believe in God and His Messenger on the condition that they commit acts of obedience and refrain from acts of disobedience. He justifies his argument by demonstrating that the Qur'ān praises the *mu'min* because of his works. If *īmān* is only by confessing belief in God and His Messenger, he writes, then a dumb man would not be called a *mu'min* for his inability to confess *īmān* by the tongue.[65] Thus, according to 'Abd al-Jabbār, a man cannot be called a *mu'min* unless he performs works that demonstrate belief, just like a person who is called a worshipper only at the time of performing the prayers.[66]

'Abd al-Jabbār refutes the Kharijite claims that a grave sinner is a *kāfir*.[67] He reiterates the Mu'tazilite belief that in principle the term *kāfir* is given to those who deserve God's Greater Punishment (*al-'iqāb al-'aẓīm*) and it justifies legal actions against the accuser such as denying him marriage, inheritance, and Islamic burial in Muslim cemeteries.[68] If it is established that only those who deserve God's Greater Punishment are declared *kuffār,* and if these punishments are not imposed on the grave sinner, according to 'Abd al-Jabbār, then the latter cannot be declared a *kāfir*.[69] In other words, a person cannot be called a *kāfir* if he is not subject to God's Greater Punishment, 'Abd al-Jabbār writes. He adds that the Prophet's companions and later Muslims in the early periods of Islam "agreed that the grave sinner is not subject to severing of ties with the community and therefore does not lose rights to marriage, inheritance, and Islamic burial in Muslim cemeteries."[70] 'Abd al-Jabbār uses an incident in early Islam to argue that a grave sinner is neither a Muslim nor a *kāfir*. He writes: "when a group of Muslims (proto-Kharijites) turned against 'Alī b. Abī Ṭālib, the latter called them grave sinners, or *bughāt*."[71] When 'Alī was asked, "Are they Muslims?" he replied, "If they were, we would not have fought them. They were our brothers yesterday and they turned against us today."[72] Thus, 'Alī neither called them Muslims nor *kuffār*. Abū al-Qāsim Maḥmūd ibn 'umar al-Zamakhsharī (d. 1143) writes that disobedience deserves blame and punishment unless it is canceled by acts of obedience or if the doer repents.[73] A major sin, he adds, cancels all acts of obedience, and whoever dies without committing acts of obedience, which, in turn, cancel all major sins, will die a *kāfir* and henceforth is condemned to eternal hellfire.[74]

The Mu'tazilite views on *takfīr* not only contributed to the theological debate but also compelled some theologians, most notably the Shāfi'i scholar

Abū al-Ḥasan al-Ash'arī (d. 936), to respond with views that challenged the group's thesis on *takfīr*. The Ash'arites emerged as a traditionist opposition movement to the Mu'tazilite views regarding the justice of God, free will, and the emphasis on reason as the sole source of religious truth.

TAKFĪR IN THE ASH'ARITE THOUGHT

The Ash'arite theological movement takes its name from its founder Abū al-Ḥasan al-Ash'arī (d. 935) "whose works marked a turning point in the history of Islamic theology."[75] Al-Ash'arī, a former Mu'tazilite, devoted himself to "strengthening attachment to Tradition while defending it by the rational arguments of *kalām*."[76] In the words of David Waines, "The purpose of the broad movement which flourished after al-Ash'arī was to challenge and modify the conclusions reached by the Mu'tazilites."[77] The following pages will discuss the Ash'arite theology of *takfīr* based on al-Ash'arī's two major works, *Kitāb al-Ibānah* and *Kitāb al-luma'*, and will demonstrate how this theology had emerged as a response to the Mu'tazilite views on *takfīr*.

Scholars have argued that al-Ash'arī's position on the concept of *takfīr* is quite obscure and difficult to define because al-Ash'arī himself did not intend to define the concept as much as to refute the Mu'tazilite views on *takfīr*. Since al-Ash'arī's intention of defining *takfīr* "is chiefly polemical," according to Izutsu, this perhaps led to some difficulty grasping the Ash'arite position regarding the concept of *īmān* and *takfīr*. Izutsu attributes this difficulty to the "divergent" accounts of the Ash'arite position on *īmān* and *takfīr* that have been handed down to us.[78] Walter C. Klein tells us that al-Ash'arī is "obscured by his work and its effects. . . . He left behind no confessions . . . virtually all his writings are polemical."[79] Although the Ash'arite position on *takfīr* and *īmān* was not intended to define these concepts, it did, however, strengthen the meaning of *īmān* as it is defined in the Sunni tradition.

In his *al-Ibānah 'an Uṣūl al-Diyānah*, al-Ash'arī dedicates two chapters to the topic of *īmān*: "On the Exposition of the Belief of the Deviators and Innovators" and "Concerning the Exposition of the Belief of the Followers of the Truth and the *Sunnah*."[80] In the first chapter, he begins by attacking the Mu'tazilites for deviating from the traditional Sunnite teachings about the nature of God; good and evil; the punishment of the grave; and the eternity of the Qur'ān. He rebukes the Mu'tazilites for interpreting the Qur'ān according to their opinions rather than following interpretations derived from Muḥammad's tradition, and he accuses them of opposing the traditions of the Prophet's Companions (*al-ṣaḥābah*) and "approximating the belief of the Magians."[81] Citing Muḥammad's words, he refers to the Mu'tazilites as "the Magians[82] of this community" because, according to him, "they have adopted the religion of the Magians and copied their tenets, and think that there are two creators, the one for good and the other for evil, just as the Magians think."[83] They are "deviators in their innovations," al-Ash'arī writes, "who dissent from the Book, and the *sunnah*, and that upon which the Prophet and his Companions take their stand and the Community have unanimously agreed."[84] It is important to ask that if al-Ash'arī believed that the Mu'tazilites deviated from orthodoxy, why did he then call them Magians and deviators and not *kuffār*, hypocrites, or *fāsiqūn* as the Kharijites and the Mu'tazilites did with other Muslims? The following discussion of al-Ash'arī's thought may provide some answers to this question.

In *K. al-Ibānah*, al-Ash'arī lays out the beliefs of the Sunnite tradition and refutes the "innovations" of the Mu'tazilites, the Kharijites, and the Murji'ites, among others. Al-Ash'arī begins his refutations by declaring that the belief "we hold and the religion we follow" are holding fast to the Qur'ān, to the *sunnah* of Muḥammad, to the traditions, and to what Aḥmad b. Ḥanbal[85] professed.[86] After affirming the beliefs of the Sunnite orthodoxy, he turns against the Mu'tazilites by declaring them *kuffār* for believing in the createdness of the Qur'ān. He writes: "We believe . . . that the Qur'ān is the uncreated Word of God, and that he who believes the Qur'ān is created is an infidel."[87] The *kuffār*, he contends, will be veiled from God when the believers see Him in Paradise, and that "we ought not to declare a single one of the people of the *qiblah* an infidel for a sin of which he is guilty, such as fornication or theft or the drinking of wine."[88] Here, al-Ash'arī makes it clear that a grave sin does not earn the believer the status of infidel. Rather, he asserts that committing "any of these mortal sins, such as fornication or theft or the like, presumptuously declaring it lawful and not acknowledging that it is forbidden, is an infidel."[89] In the above declarations, al-Ash'arī distinguishes between acts and belief. Committing a forbidden act is not a grave sin that leads to *kufr*; rather, believing that the act is not forbidden is a grave sin, which leads to *kufr*. It is an interesting and a novel thought to vindicate one person committing a mortal sin, yet convict another for failing to acknowledge the unlawfulness of that sinful act although he had never committed that sin. One may argue that al-Ash'arī had established a doctrine that puts more emphasis on belief and *taṣdīq* than on acts and deeds, although he affirms in his *al-Ibānah* that "faith consists of words and deeds, and is subject to increase and decrease."[90]

In *K. al-luma'*, al-Ash'arī begins by asking the question: What is faith in God? He answers by saying that faith is the belief (*taṣdīq*) in God.[91] On this, there is an agreement among those who speak the language of the Qur'ān. Al-Ash'arī asserts that the meaning of *īmān* in the language of the Qur'ān is the *taṣdīq* in God and that *taṣdīq* alone can be *īmān* because, as he argues, when the people say "so-and-so has faith in the punishment of the grave and intercession," they mean that he believes (*yuṣaddiq*) in that. Therefore, it must be that faith according to those who speak the language is *īmān*, which is belief or *taṣdīq*.[92]

The other question al-Ash'arī asks is: What about the sinner who is among the people of the *qiblah*, that is, "those who face Makkah when they pray," is he a believer? For al-Ash'arī, he is a believer by virtue of his faith, a sinner for his sins and grave misdeeds.[93] He adds that those who speak the language are in agreement that he who commits murder is a murderer, he who commits a sin is a sinner, he who disbelieves is a *kāfir*, he who believes (*taṣdīq*) is a believer (*muṣaddiq*), and he who has faith is a *mu'min* or believer.[94]

As for the status of the sinner (*fāsiq*), al-Ash'arī refutes the Mu'tazilite assertion that he is neither a believer nor a *kāfir*. He writes that if the sinner were neither a believer nor an unbeliever, he would then neither profess *tawḥīd* (the oneness of God) nor deny it. Therefore, since it cannot be that the sinner is neither a friend nor an enemy of God, al-Ash'arī writes, "it is impossible that the sinner is neither a believer nor a *kāfir*, as the Mu'tazilites claim."[95] He adds that if the sinner were a *mu'min*—confessing *tawḥīd*—before committing a sin, then committing adultery after confessing *tawḥīd* does not invalidate the name *īmān*, which has always been associated with him.[96]

Al-Ash'arī replies to the Mu'tazilite views regarding the grave sinner by asserting that prior to the advent of Wāṣil b. 'Aṭā', the Mu'tazilite chief, the Muslims were of two opinions: the Kharijites who declared a grave sinner a *kāfir*, and the "People of Rectitude" (*ahl al-istiqāmah*) who maintained that the grave sinner is a believer by virtue of his belief, a sinner by reason of his grave sin. No one has said, al-Ash'arī continues, that he is neither a *kāfir* nor a believer before Wāṣil b. 'Aṭā'. Al-Ash'arī maintains that there was a consensus among the Muslims that he who commits an act of obedience from among the people of prayer (*ahl al-ṣalāt*) must be either a believer or a *kāfir*. "One must conclude," he writes, "that Wāṣil's view is false."[97] Al-Ash'arī ends the section on *īmān* with the following:

> If it is said that he who has faith and committed a grave sin is neither a believer nor a *kāfir*, one should also say: he is a believer by virtue of his *īmān*, but one should not say he is a sinner by reason of his sin. Since this cannot be said, for there is no sin that does not belong to a sinner, their assertion is also impossible, for there is no faith that does not belong to a believer.[98]

In his chapter on *īmān*, it seems that al-Ash'arī is defending the views of the Sunnite orthodoxy against the views of the Mu'tazilites. Therefore, one may argue that the views on *īmān* as expressed by al-Ash'arī are not his *per se*; rather, they are the views of the established doctrine of the Sunnite orthodox tradition, and al-Ash'arī is only affirming and defending them against the unorthodoxy of the Mu'tazilites.

Early and modern Muslim theologians and scholars of Islamic theology were puzzled by al-Ash'arī's theory of *īmān* and argued that his position on *īmān* and *takfīr* is unclear since there are "divergent descriptions" of his position, which "is much more difficult to ascertain than it appears at first sight."[99] For instance, Ibn Ḥazm ascribes to al-Ash'arī the thesis that *īmān* is the "knowledge of (*ma'rifat*) God in the heart (*bi al-qalb*) alone even though a man manifests Judaism, Christianity, or any other way of *kufr* by his tongue or by his way of worship. If he knew (*'arifa*) God by his heart, then he is a Muslim of the people of Paradise (*min ahl al-jannah*)."[100]

Ibn Ḥazm adds that the Ash'arites asserted that in the language of the Qur'ān, *īmān* is the *taṣdīq* alone; *īmān* is *tawḥīd*. The act, however, cannot be called *tawḥīd* because it is not a form of *īmān*.[101] Izutsu challenges Ibn Ḥazm's views and contends that the attribution of this thesis to al-Ash'arī is wrong because al-Ash'arī "himself gives two completely different definitions of *īmān*" in *K. al-Ibānah* and *K. al-luma'*.[102] Izutsu argues that although al-Ash'arī explains the meaning of *īmān* by *taṣdīq*, he gives a different explanation of the latter term.[103] For instance, in *K. al-Ibānah*, al-Ash'arī states that faith "consists of words and deeds, and is subject to increase and decrease."[104] In this statement, it is clear that faith is saying *and* doing (*al- īmān qawl wa 'amal*). It is important to note that al-Ash'arī does not mention the word *taṣdīq* (i.e., knowing God by the heart) in *K. al-Ibānah* as he does in *K. al-luma'*. Izutsu points out this contradiction in the following:

> of the three basic elements of *īmān* he only mentions the second, i.e. "verbal confession" and the third, i.e. "good works," and completely omits the first. Examination of another definition of *īmān* given in *Luma'* makes it clear that his omission does not imply that he does not consider *taṣdīq* essential. On the contrary, he considers *taṣdīq* so important and essential that it needs no explicit mention.[105]

Taṣdīq is essential for al-Ash'arī, according to al-Shahrastānī:

> Al-Ash'arī said: *īmān* is *taṣdīq* by the heart. The saying by the tongue and the doing of the cardinal duties are supplementary. He who confesses (*ṣaddaqa*) by the heart to the oneness of God, acknowledges the prophets as truthful by the heart (*taṣdīqan bil qalb*) to what they have brought from God. The *īmān* of such a person is true *īmān*.[106]

Like al-Shahrastānī, al-Baghdādī, in his *K. Uṣūl al-Dīn*, emphasizes *taṣdīq* as essential for *īmān* without any mention of *īmān* as "saying and doing" (*qawl wa 'amal*). He writes: "Al-Ash'arī says that *īmān* is the *taṣdīq* in God and his Apostles and in what they brought from God. This *taṣdīq* is not truthful except by acknowledging Him (*bi ma'rifatihi*). *Kufr*, in his view, is the denying (*takdhīb*) of God and His Apostles."[107] In his *al-Farq Bayna al-Firaq*, al-Baghdādī says that the People of the *Sunnah* agree that knowledge (*ma'rifah*) and assent (*taṣdīq*) by the heart are the fundamentals of *īmān*. They differ, he adds, on the question whether verbal confession (*iqrār*) and the acts of obedience (*ṭā'āt*) are also essential to *īmān*.[108] Izutsu writes that the expression "they differ" suggests that "saying" and "doing" are not of primary significance to the concept, which makes *taṣdīq* more essential to *īmān*.[109]

What is *taṣdīq*, and how does al-Ash'arī define the term? According to al-Shahrastānī, al-Ash'arī defines *taṣdīq* as the knowledge of the existence of God, His Divinity, His eternity, and His attributes.[110] *Taṣdīq* is also a mental speech (*qawl fī al-nafs*) containing the knowledge, which is confessed by the tongue. This verbal confession is called *taṣdīq*. The "doing" of the basic religious duties, al-Shahrastānī adds, is also a kind of *taṣdīq* in a sense that it is the outward expression of *taṣdīq*.[111] Thus, the verbal confession (*shahādah*) and the religious duties are *taṣdīq* rather than *īmān*.

For al-Ash'arī, *taṣdīq* is the most fundamental part of *īmān*, and the *shahādah* is the least form of it. Al-Ash'arī recognized that the *shahādah* is mere words that some people use to hide their *kufr*.[112] Both the Qur'ān and the Prophet, al-Ash'arī asserts according to al-Shahrastānī, referred to such people as hypocrites (*munāfiqīn*) and the Qur'ān denied the existence of *īmān* in them.[113] The Qur'ān accuses them of lying and calls them liars (Q. 63:1). It is known for certain that *taṣdīq* by the heart is the most important foundation for *īmān*, while confession by the tongue is but a verbal expression of *taṣdīq*."[114] Al-Ash'arī believes that the *shahādah* is essential for revealing the *īmān*, which is in the heart of the people, and it is the "practical standard" by which the people are judged to be believers or *kuffār*. Thus, anyone who utters the *shahādah*, he contends, without denying what the Prophet has brought should not be called a *kāfir*. Anyone who opposes any of the Islamic fundamentals because of principles related to their schools of thought (*madhhab*), he writes, are not to be judged as *kuffār* but as committing deviations and innovations.[115]

Al-Ash'arī's definition of *īmān* and *kufr* is relatively inclusive in the sense that it narrowed the realm of *kufr* and widened the realm of belief. The latter could include those who committed acts of disobedience; the grave sinners; the hypocrites; and other Muslims deemed heretics because all are judged based on their inner belief, which only God has knowledge of. For this reason, al-Ash'arī was reluctant to call the Mu'tazilites *kuffār* since, according to his own definition of *īmān* and *taṣdīq*, they possess *taṣdīq* (i.e., they acknowledged the truth that the Prophet has brought from God). While the Kharijites and the Mu'tazilites held that faith increases and decreases (i.e., is quantitative) and the Murji'ites asserted that it is qualitative (not admitting either degree or measurement), the Ash'arites maintained, as Fazlur Rahman puts it, "that, with a genuine profession of faith, an irreducible amount of faith came into existence which was incapable of being eliminated by almost any external behavior, although it could increase or decrease beyond that point."[116] In the midst of the Ash'arite theological debate on the concept of *takfīr*, the Maturidites emerged and offered their response to the Ash'arite views as well as to the previous theological schools' theses on the subject.

TAKFĪR IN THE MATURIDITE THOUGHT

Maturidism (*al-Māturīdiyyah*)—a theological school contemporaneous with Ash'arism—contributed extensively to the concepts of *takfīr* and *īmān* in medieval Islam. Maturidism, which takes its name from its founder the Islamic scholar of jurisprudence and Qur'anic exegesis Muḥammad Abū Manṣūr al-Māturīdī (d. 944), differs from Ash'arism in some important theological issues. The views of the Maturidites on *takfīr* and *īmān* are expressed in *Kitāb Sharḥ al-Fiqh al-Akbar*,[117] attributed to al-Māturīdī, and in *Sharḥ al-'Aqā'id al-Nasafiyyah*, a work attributed to the Muslim Persian polymath Mas'ūd b. 'Umar al-Taftazānī (d. 1389 or 1390). Scholars have held that al-Taftazānī's *Sharḥ*, a commentary on 'Umar b. Muḥammad al-Nasafī's (d. 1142) *al-'Aqā'id al-Nasafiyyah*, is *the* most comprehensive work on the Maturidite theology.[118]

In *K. Sharḥ al-Fiqh al-Akbar*, al-Māturīdī writes that committing a grave sin does not make the person an unbeliever.[119] He is rather a *fāsiq* who has *īmān* in his heart. Al-Māturīdī separates between *īmān* and acts of disobedience, for the former has its place in the heart, whereas the latter is in the limbs. And because they are two different things and separate from one another, the acts of the limbs do not cancel the *īmān* in the heart. Al-Māturīdī cites verse 49:6 in the Qur'ān to support his claims: "If a *fāsiq* came to you with news, investigate."[120] Al-Māturīdī asserts that if the *fāsiq*, who committed a grave sin, is a *kāfir* because of his acts then the Qur'ān would not make him a witness. He also cites a *ḥadīth* where Muḥammad takes no action after a man has confessed an act of adultery to him. If adultery were an act of *kufr*, al-Māturīdī writes, Muḥammad would have declared him an apostate and hence killed him, or he would have asked him to come back to Islam.[121] Since Muḥammad did not take any action against the adulterer, this, according to al-Māturīdī, means that the *īmān*'s place is only in the heart.[122]

Another issue al-Māturīdī takes on is the nature of *īmān*: Is *īmān* God's creation, or is it within the power and responsibility of man? According to Izutsu, al-Māturīdī asserts that since "*īmān* consists in 'verbal confession' and 'assent,' *īmān* is a created thing . . . like fasting and ritual worship."[123]

The Maturidites are split on this issue. While the Maturidite school of Samarqand believes in the creation of *īmān*, the school of Bukhara contends that *īmān* has two aspects: One is God's creation, through "Divine Guidance" (*hidāyah*), meaning God's role and action, and the other is man's responsibility, through "being guided" (*ihtidā'*), which is man's action.[124]

If *īmān* is the creation of God, according to al-Māturīdī, then what are his views about *kufr*? Is *kufr* God's creation or man's responsibility? Al-Māturīdī objects to the Mu'tazilite claims that when a person commits an act of *kufr*, the act is committed against God's will and power for it is established that God has commended *īmān*. He asserts that this Mu'tazilite belief runs against God's nature, for it denies God's power over his creatures and makes the latter more powerful than their Creator by creating their own acts, that is, *kufr*. Al-Māturīdī reconciles this issue by contending that although everything is subject to God's power and will, it is not suitable to say that acts of disobedience and *kufr* are the will of God or are commended and desired by God. Al-Māturīdī makes an interesting point by saying that it is fitting to say that God created everything or God created *īmān* but it is inappropriate to say that God created *kufr* or other acts of evil.[125]

Supporters of the Mu'tazilite views offer their responses to some Qur'ānic verses that seem to support the anti-Mu'tazilite's thesis that God creates *kufr* in man. For instance, verse 5:60 says: "'Shall I tell you who deserves a worse punishment from God than [the one you wish upon] us: Those God distanced from Himself, was angry with, and condemned as apes and pigs, and those who worship idols: they are worse in rank and have strayed further from the right path." Also, verse 3:8 says: "Our Lord, do not let our hearts deviate after You have guided us." Al-Sharīf al-Murtaḍā (d. 1044 or 1045), a supporter of the Mu'tazilite views, refutes the claims that the above-mentioned verses show that it is God who creates *kufr* in man. Rather, he argues that although God created the *kāfir*, it does not necessarily follow that he created *kufr* in him. Al-Murtaḍā asserts that God creates man who is potentially a *kāfir*, and the potential to commit *kufr* is man's will and desire alone.[126] He adds that it is absurd to believe that God created *kufr* in the *kuffār* when the above verses condemn them for committing *kufr*.[127] Thus, al-Murtaḍā affirms that God does not directly create *kufr*; rather, He produces the situations for *kufr*, and it is man's choice when *kufr* is committed.

Ib Ḥazm summarizes the Mu'tazilite views on the creation of *kufr* by asserting that when a *kāfir* commits a *kufr*, his action is not willed or desired by God.[128] Ibn Ḥazm argues that God wills the existence of things, *kufr* being one of these things, but God does not wish that a man be a *kāfir*.[129] If a man commits a *kufr* because God wished it, then the *kāfir* should be rewarded, which is absurd, according to Ibn Ḥazm who affirms that God wishes the existence of *kufr* but commands His creatures against committing it.[130]

The Ash'arite position on the creation of *kufr* is somewhat similar to the Mu'tazilite views. Al-Ash'arī asserts that God has decreed *kufr* and determined its nature as something evil. He also affirms that acts of *kufr* were decreed and predetermined by God in a sense that He created them, but He has not commanded man to commit them.[131] For both the Mu'tazilites and the Ash'arites, *īmān* and *kufr* are both equally God's creation but one is commanded and the other is forbidden. God has decreed or commanded *kufr* to exist, but He has not commanded the people to commit *kufr*. Thus, "'decree' (*qaḍā'*)," according to Izutsu, "is to be differentiated from the 'object of decree' (*maqḍī*)."[132] The decree in the "former case is 'command' *amr*, whereas the 'decree' in the latter case is just announcing, informing, and writing down. The 'decree' itself is in both cases right."[133]

The Maturidite position on this issue is summarized by al-Māturīdī in his *Sharḥ al-Fiqh al-Akbar*.[134] He writes that God created *kufr* and wished it (*shā'a*). However, He commanded the *kāfir* to believe, though He did not wish it for him. As to the question whether God's wish (*mashī'at Allāh*) is satisfying or dissatisfying, al-Māturīdī says that it is satisfying. Then, he raises another question, if God wishes *kufr* and His wish is pleasing to Him, then why should He punish those who commit acts that please Him? Al-Māturīdī's answer is: God punishes man for doing what is not pleasing to Him.[135] He punishes the *kāfir* for his *kufr*, for *kufr* and acts of disobedience are not pleasing to God. But, if it is said that acts of disobedience and *kufr* are God's wish, is not His wish pleasing to Him? Al-Māturīdī answers: Yes, God's wish, desire, and will are pleasing to Him if a man commits acts of obedience, and they are equally displeasing if a man commits acts of disobedience. In this case, both acts are God's wish.[136]

Thus, for Maturidism, God wishes *īmān* on potential believer and *kufr* on potential *kāfir* but He does not command either one. God's wishes are not arbitrary; rather, He wishes *īmān* and *kufr* for people because of His eternal knowledge of their characters and the choices they will make in life. The Maturidite theological views argue that God created *kufr*, that He liked His creation of *kufr*, that He has the foreknowledge that certain people will choose *kufr*, but is displeased with people becoming *kuffār* by their choice of *kufr*.[137]

Some theologians challenged the views that the *kuffār* and the *fāsiqūn* are predestined to commit their *kufr* and *fisq*, which makes it absurd to expect from them *īmān* and obedience. In *Sharḥ al-'Aqā'id al-Nasafiyyah*, al-Taftazānī starts his discussion of *īmān* and *kufr* by arguing that God caused some people to commit *kufr* and *fisq*, not because God predestined their acts (*jabr*) but because *they* chose that path.[138] *Kufr* and *fisq*, according to al-Taftazānī, are committed by free will since God does not hold the doers responsible for acts they are incapable of performing.[139] "We know," al-Taftazānī contends, "that a thing may not be wanted but God ordains man to do it, and one thing may be wanted but God prevents him from doing it. These things happen for a reason, which is known to God alone."[140] Al-Taftazānī adds that when the scholars proved that man's power over his act (ability) precedes the act (responsibility occurs before the act), it then becomes necessary that it is the *kafir*'s burden to commit *īmān* (*mukallaf bi al-īmān*).[141]

Al-Taftazānī rejects the notion and deems it impossible that God imposes duties on people knowing that they have no power over their acts. He rather advocates the theological views that the people have the power over their acts because they possess the ability to perform these acts.[142] Possessing the ability (*al-qidrah*) allows them to choose their acts. For instance, he asserts that man has the *qidrah* to commit an act, and this *qidrah* can be used toward *īmān* or *kufr*. He rejects the views that some people have more *qidrah* to commit *kufr* than *īmān* and vice versa. He rather contends that the *kāfir* possesses the *qidrah* to fulfill his duties, that is, commit acts of *īmān*, but he may choose instead to use his *qidrah* to commit *kufr*.[143] In this case, al-Taftazānī writes, the *kāfir* lost his opportunity to commit *īmān* by his own free will, and therefore he is deserving of punishment and condemnation.[144] Al-Taftazānī supports his views on the notion that God does not burden the people with duties they cannot bear with the following Qur'ānic verse: "God does not burden [*lā yukallifu*] any soul with more than it can bear; each gains whatever good it has done, and suffers its bad. . . . 'Lord, do not burden us with more than we have strength to bear'" (Q. 2:286).[145]

In regard to the theological issue of the grave sin and whether the grave sinner is a *kāfir* or a *m'umin*, al-Taftazānī adopts the views that a grave sin does not make the believer a *kāfir*.[146] He starts by saying that there are different views on the definition of grave sin, or the acts that are defined as grave sins, but the most common acts, according to the Companions of Muḥammad, he tells us, are associating partners with God (*shirk)*, the killing of innocent life, committing adultery, fornicating, practicing magic, taking the possession of the orphans unlawfully, denying God, committing theft, drinking alcohol, and living off interest money.[147] He cites the twelfth-century theologian Nūr al-Dīn b. Abī Bakr al-Bukhārī who writes that, "each disobedient act added to the one above it is a lesser sin, but if it is added to the one below it, then it is a greater sin. And the greatest sin is to commit *kufr*, for there is no guilt greater than it."[148] Al-Taftazānī explains the definition above by saying that a grave sin (*al-kabīrah*) is a sin other than *kufr*.[149] Thus, a grave sin does not separate the believer from his *īmān* because, according to the views stated above, the *taṣdīq* remains in his heart, which is the true *īmān*. Here, al-Taftazānī juxtaposes the above views against the views of the Mu'tazilites who considered the grave sinner neither a *kāfir* nor a *mu'min* because, in their views, acts are part of *īmān*, and against the views of the Kharijites who declared the sinner, whether committing a grave sin or a lesser sin, a *kāfir*.[150]

For al-Taftazānī, *kufr* is a complicated issue, which can be defined in different ways. First, he asserts that true *īmān* is the *taṣdīq* by the heart, and believers cannot be described as non-believers unless they abandon or negate the *taṣdīq*.[151] In the event that a believer commits a grave sin due to desires or laziness but at the same time is conscious of God's punishment and reward and intends to repent, these acts are not considered negation of *taṣdīq*. Only acts that prove the negation of *taṣdīq* such as worshipping idols, throwing the Qur'ān in trash, or uttering words of *kufr* qualify the person as *kāfir*.[152] Here, al-Taftazānī makes distinctions between acts that are considered disobedient and acts that are deemed open *kufr*. A person committing the former remains a believer, but becomes a *kāfir* if he commits the latter.[153]

Al-Taftazānī offers a second definition of the grave sin based on the Qur'ān, which states that a person committing disobedient acts remains a believer: "You who believe [*āmanū*], fair retribution is prescribed for you in cases of murder . . . turn to God in sincere repentance" (Q. 2:178; 66:8). His third definition of the grave sin states that God refers to those who commit disobedient acts as believers. The term "believers" here, according to al-Taftazānī, refers to the Muslim community from Muḥammad and the early Muslims to "our day."[154] The term includes all the Muslims who have agreed to pray on the souls of *ahl al-qiblah* (the Muslims) and who have asked God to grant them His mercy despite "our knowledge of their disobedient acts and their failing to repent when they had the chance."[155] It was agreed, he says, that the same is not possible to non-believers (*dhālik lā yajūz li ghayr al-mu'min*).[156]

Al-Taftazānī shows that the views that the grave sinner is neither a *kāfir* nor a *munāfiq* nor a *mu'min* but a *fāsiq* are invalid because they run against the consensus of the scholars who reject the idea that the grave sinner is in a middle position between *kufr* and *īmān*.[157] Second, the views that the grave sinner is not a *mu'min* contradict the Qur'ānic teachings—"So, is someone who believes equal to someone who defies God? No, they are not equal" (Q. 32:18)—and the *ḥadīth*: "No one commits adultery while he is a believer."[158] Al-Taftazānī contends that although the above verse and *ḥadīth* equate *fisq* with *kufr*, what they actually mean is to highlight "the seriousness and severity of the grave sin and not to declare the *fāsiq* a *kāfir*."[159] Al-Taftazānī supports his views with Qur'ānic verses and *ḥadīths* that clearly indicate that the *fāsiqs* are believers because the community had agreed that the grave sinners have no *īmān*, but they are not *kuffār*. He cites *ḥadīth* reports stating that the early community (*al-ummah*) did not kill or punish the *fāsiqs* with apostasy laws, and they were buried in Muslim cemeteries. Moreover, he cites the Prophet Muḥammad who said that the *fāsiq* is a believer even though he has committed adultery or theft.[160] Al-Taftazānī asserts that despite the Qur'ānic and *ḥadīth* references to *fisq*, grave sin, and disobedience as *kufr*, the community has agreed that a grave sinner is not a *kāfir*.[161] It seems that al-Taftazānī distinguishes between grave sin and the ultimate or absolute sin (*al-kabīrah al-muṭlaqah*). For him the grave sin is the neglect of the religious duties, while the ultimate sin is the disbelief in God and his Prophet. He refers to the ultimate sin as the total *kufr*. The latter, he asserts, occurs when the person denies God and what the prophets have brought down from Him. He justifies declaring them *kuffār* because the Qur'ān speaks of their eternal punishment in the Hell Fire as well as it declares them unbelievers.[162]

Al-Taftazānī contends that the Prophet will intercede on behalf of the grave sinners but not the absolute *kuffār*: "So [Prophet] . . . ask for forgiveness for your sins and for the sins of believing men and women" (Q. 47:19). Those who were denied the Day of Judgment, "[n]o intercessor's plea will benefit them now" (Q. 74:48). The above verses differentiate between the sinners (the disobedient), who will be forgiven and removed from the eternal Hell Fire, and the *kuffār* (the deniers of Judgment Day), on whose behalf no intercession will occur. Al-Taftazānī argues that the verses above are a clear proof that intercession is inclusive to all sinners since the Qur'ān refers to them as *al-mu'minīna wa al-mu'mināti* (the believing men and believing women), whether the sin is minor or grave. Only those who deny the messages of the prophets are considered unbelievers for they will abide eternally in Hell.[163]

Al-Taftazānī's argument that the verses on intercession are inclusive to all sinners is a refutation of the Mu'tazilite belief that God forgives only those who commit a lesser sin but not those who commit a grave sin unless they repent.[164] Al-Taftazānī does not accept this Mu'tazilite conclusion, for he believes that the Qur'ān (9:72; 18:107; 99:7) is clear about the notion that as long as the grave sinner has the *taṣdīq* in his heart, he will not abide eternally in Hell. Al-Taftazānī concludes that eternal Hell Fire is the ultimate punishment, and if it is meant only for the *kuffār*, then condemning believers with grave sins runs against God's divine justice.[165]

The Maturidites also offered their views about the destiny of people in the afterlife. For instance, in the last two themes of his *K. al-Tawḥīd*, al-Māturīdī writes that a sinning believer is not condemned to eternal punishment in Hell; rather, at some point, he will be rewarded with admission to Paradise. His punishment for his sins is his entrance to Hell, but this will be for a limited period of time. Also, the Prophet is expected to intercede for sinning believers, according to al-Māturīdī. This belief goes against the Mu'tazilite and the Kharijite views that there is no hope for the Prophet's intercession on behalf of the sinners since all sinners automatically become *kuffār*.[166]

Medieval theological schools and sects engaged themselves in debates on the problem of *fiṭrah* (the natural religious beliefs of man) and provided different interpretations to this theological concept. For instance, the Mu'tazilites, al-Māturīdī tells us, argued that every human being is born in the *fiṭrah*, which they identify as *al-islām*.[167] It is not clear, however, what they meant by "*al-islām*" because the term in Arabic means both submission to one God or monotheism and the religious institutions that Muḥammad founded. Those who reject the *fiṭrah* commit *kufr*, according to the Mu'tazilites. They reject the *fiṭrah* not because of God's will but because of the choice they make.[168] Ibn Ḥazm maintains that children born to *kuffār* parents are Muslim and are destined to Paradise.[169] The *fiṭrah* for the Maturidites is a God-given quality,[170] "a kind of inborn quality," and "a good religious nature"[171] that leads the child to acknowledge the existence of God through the natural process of reasoning (*istidlāl*).[172] God's creatures, the Maturidites maintain, are created with the qualities of Islam (*ṣifāt al-islām*), not the qualities of disbelief (*ṣifāt al-kufr*).[173]

As for the concept of *taṣdīq* and its relationship to *īmān*, al-Taftazānī writes that *īmān* in the language is the *taṣdīq*, meaning that a person becomes a believer when he opens his heart to receive God's message and then accepts it and believes it to be true. He cites al-Ghazālī who says that *taṣdīq* is the receiving and accepting of the news. He also refers to *taṣdīq* using a Persian term: *geravidan*.[174] The latter, he writes, means the *taṣdīq* (being true and real) versus *taṣawwur* (untrue and unreal). When people receive news and information, they either take it as a truth or as *taṣawwur*. Here, al-Taftazānī tries to demonstrate that the term *kāfir* applies to those who have *taṣawwur*, that is, accepting what the Prophet has brought but not believing in it. The people who have *taṣawwur* rather than *taṣdīq* are the *kuffār* who believe in the Prophet's message and obey its teachings but at the same time hold to pagan practices and idol worship. For al-Taftazānī, this is *kufr* because it is a sign of denying the *tawḥīd*.[175] Those who have *taṣdīq* in God and His attributes but deny the doctrine of *tawḥīd* are called believers by name only, but in reality they are not for "Most of them will only believe in God while also joining others with Him" (Q. 12:106). Al-Taftazānī calls them *Mushrik Muṣaddiq*, or believers who associate partners with God.[176]

Since *taṣdīq* is proof of *īmān*, for al-Taftazānī, the confession by the tongue is secondary to *taṣdīq* because failing to confess by the tongue is not a sign of *kufr*. He writes that *taṣdīq* is the cornerstone of *īmān* that without which there is no *īmān*; however, the absence of confession by the tongue is not a proof of unbelief. Confession by the tongue is important only for the jurists and the scholars to determine the status of the person (i.e., Muslim or non-Muslim) and whether the *sharī'ah* applies on him or not. The confession of the tongue, he writes, is essential for the jurists to proscribe legal issues, while *taṣdīq* in the heart is only for God to know.[177] Al-Taftazānī asserts that he who has *taṣdīq* in the heart but does not confess by his tongue is a believer: "these are the people in whose hearts God has inscribed faith" (Q.

58:22) whose "hearts remain firm in faith" (Q. 16:106). But those who confess by the tongue yet do not have *taṣdīq* in the heart are the hypocrites,[178] according to al-Taftazānī, because "faith has not yet entered [their] hearts" (Q. 49:14).

On the views that *īmān* is only confession by the tongue without taking into consideration the *taṣdīq* in the heart, al-Taftazānī says that *īmān* is *taṣdīq* by the heart although the person never confesses by the tongue, and the proof is that the scholars never proclaimed that he who confesses by the tongue has a *taṣdīq* in his heart, or is a believer.[179] For this reason, he writes, it is right to deny the *īmān* of some of those who confess *īmān* by the tongue, for the Qur'ān says: "Some people say, 'We believe in God and the Last Day,' when really they do not believe" (Q. 2:8). Thus, for al-Taftazānī, confession by the tongue is not a sign of *īmān*.[180]

Al-Taftazānī also rejects the Karramite theological claim that *īmān* is by only uttering the two testimonies (no god but God and Muḥammad is His Messenger) as well as the claims of the theologians, the jurists, and the transmitters of *ḥadīths* who asserted that *īmān* is the *taṣdīq* by the heart as well as confession by the tongue and the fulfilling of religious duties. He writes that al-Nasafī rejected this definition of *īmān* because "the acts (the obligations) may increase or decrease but the *īmān* does not increase or decrease."[181] Unlike the Mu'tazilites who deny the *īmān* in those who neglect their religious duties, and unlike the Shāfi'ī school of law that believes that *īmān* remains even in a person who abandons these duties (considered a pillar of *īmān*), al-Taftazānī asserts that *īmān* remains strong in people despite the neglect of some of the religious duties. For him, the truth of *īmān* does not increase or decrease by the acts of disobedience, for the *taṣdīq* is beyond measurement, as is the case with the acts, but it becomes stronger with time as long as the person remains a *muṣaddiq*.[182] These views are affirmed by al-Māturīdī in his last part of *K. al-Tawḥīd*, where he addresses the characteristics of *īmān* and *kufr*. He asserts that *īmān* consists of affirmation with the heart (*taṣdīq*) and confession with the tongue.[183] Thus, deeds are excluded from being essential part of *īmān*, a belief that also runs against the Mu'tazilite and Kharijite views.

Al-Taftazānī argues that *īmān* and *islām* are one and the same for *islām* is the acceptance and *taṣdīq* of the message brought by the Prophet. He adds that it is against the *sharī'ah* to judge someone as being a *muslim* but not a *mu'min*, or a *mu'min* but not a *muslim*. "This is what we meant by their unity,"[184] he writes. The scholars, he adds, refused to separate the two, for denying one is denying the other. He cites a book entitled *al-Kifāyah*[185] that states that *īmān* is the *taṣdīq* in God and His message, and *islām* is the submission to and trust in His divinity.[186] Thus, submission and trust do not happen without the *taṣdīq* by the heart first. Therefore, *īmān* and *islām*, al-

Taftazānī asserts, are not two different things and one cannot be separated from the other.[187]

Al-Māturīdī contends that *islām* is the knowledge of God without the why (*ma'rifat Allāh bilā kayf*), meaning opening up the breast and accepting *islām*, while *īmān* is the knowledge of God in His divinity (i.e., acknowledging God's divinity) and its place is in the heart.[188] Al-Māturīdī writes that the heart is inside the breast and the knowledge of God is in the inner heart (*fu'ād*), which is inside the heart (*dākhil al-qalb*). The knowledge of God by *tawḥīd* (*ma'rifat Allāh bil wiḥdāniyyah*) (i.e., acknowledging God's unity) is in *al-sir*, the deepest part of the *fu'ād*.[189] Al-Māturīdī says that when *tawḥīd*, *ma'rifah*, *īmān*, and *islām*—which are neither one nor separate—unite, they become religion.[190]

Al-Māturīdī defines *īmān* as the *taṣdīq* in the word of God in the heart. *Īmān* is to believe in the entirety of the Qur'ān; denying one verse is equal to denying the whole Qur'ān, and this is an act of *kufr*,[191] he contends. On whether a person was a believer before acquiring *īmān*, al-Māturīdī refutes this notion. He objects to the Ash'arite views that Muḥammad's Companions were indeed believers prior to God's revelation. The Ash'arites contend that the first two caliphs, Abū Bakr and 'Umar, were indeed believers (*mu'minūn*) when they bowed down to their idols before the advent of Islam.[192] Al-Māturīdī rejects these views and argues that there was no *īmān* before God revealed the Qur'ān.[193]

The three schools discussed in this chapter have different takes on the meaning of *kufr*. The Mu'tazilites opposed the Kharijites' definition of *kufr* and insisted that a grave sinner is neither a believer nor a *kāfir* but still a member of the Muslim community, and that *kufr* and *īmān* are acts based on the human free will. The Ash'arites put less emphasis on acts and argued that belief in the heart matters the most. They believed that committing an act of disobedience is not *kufr*; however, believing that acts of disobedience are not *kufr* is *kufr*. The Maturidites took a different approach to the debate and emphasized that *kufr* is God's creation and that a *kāfir* may earn God's forgiveness if he rejects *kufr* after committing it.

NOTES

1. See, for example, Tāj al-Dīn Abū al-Fatḥ Muḥammad ibn 'Abd al-Karīm al-Shahrastānī, *Al-Milal wa al-Niḥal*, vol. 1, eds. Amīr 'Alī Muhannā and 'Alī Ḥasan Fā'ūr (Beirut, Lebanon: Dār al-Ma'rifah, 1996), 62.
2. Sulaymān S. 'Alam al-Dīn, *Al-Mu'tazilah* (Beirut, Lebanon: Nawfal, 2000), 266.
3. Ibid.
4. Al-Shahrastānī, *al-Milal*, 62.
5. Ibid.
6. Ibid.
7. Abū al-Ḥasan al-Ash'arī, *Maqālāt al-Islāmiyyīn*, vol. 1, ed. Muḥammad Muḥyī al-Dīn 'Abd al-Ḥamīd (Beirut, Lebanon: al-Maktabah al-'Aṣriyyah, 1990), 17.

8. 'Abd al-Raḥmān Badawī, *Madhāhib al-Islāmiyyīn* (Beirut, Lebanon: Dār al-'Ilm lil Malāyīn, 1997), 86.
9. Ibid.
10. Al-Ash'āri, *Maqālāt*, vol. 1, 330.
11. Ibid.
12. Quoted in Izutsu, *The Concept of Belief*, 160.
13. Abū al-Qāsim Maḥmūd b. 'Umar al-Zamakhsharī, *Tafsīr al-Kashāf*, 3rd ed., ed. Khalīl Ma'mūn Shīḥā (Beirut, Lebanon: Dār al-Ma'rifah, 2009), 457.
14. Izutsu, *The Concept of Belief*, 160.
15. Ibid.
16. Tāj al-Dīn Abū al-Fatḥ Muḥammad ibn 'Abd al-Karīm Al-Shahrastānī, *Nihāyat al-Aqdām fī 'Ilm al-Kalām*, ed. Alfred Guillaume (Cairo: Maktabat al-Thaqāfah al-Dīniyyah, 2009), 471.
17. Ibid.
18. Ibid., 472.
19. Izutsu, *The Concept of Belief*, 162.
20. Al-Zamakhsharī, *Tafsīr al-Kashāf*, 38.
21. Izutsu, *The Concept of Belief*, 100.
22. Al-Ash'arī, *Maqālāt*, vol. 1, 17.
23. Izutsu, *The Concept of Belief*, 100.
24. Al-Ash'arī, *Maqālāt*, vol. 1, 17.
25. Q. 32:18: "So, is someone who believes [*mu'minan*] equal to someone who defies God [*fāsiqan*]? No, they are not equal." The Prophet once said: "whoever dies without having worshipped anything besides Allah, will enter Paradise . . . Even if he had committed theft or committed illegal sexual intercourse" (*Ṣaḥīḥ Bukhārī* Volume 008, Book 076, Hadith Number 450).
26. Izutsu, *The Concept of Belief*, 49.
27. Ibid., 48–49.
28. Abū Muḥammad 'Alī ibn Aḥmad ibn Sa'īd ibn Ḥazm, *al-Fiṣal fī al-Milal wa al-Ahwā' wa al-Niḥal*, vol. 3 (Cairo, Egypt: Al-Maṭba'ah al-Adabiyyah, 1902), 233–234.
29. Ibid.
30. Ibid.
31. Ibid.
32. Izutsu, *The Concept of Belief*, 50.
33. Ibid.
34. Al-Ash'arī, *Maqālāt*, vol. 1, 183.
35. Ibid.
36. The school of al-Ḥasan al-Baṣrī was the first to study the philosophy of religion in early Islam. With the rise of the Mu'tazilites, the philosophy of religion became a popular field among many scholars and schools of thought in medieval Islam (see 'Alam al-Dīn, *al-Mu'tazilah*, 421).
37. Encyclopaedia Britannica. http://www.britannica.com/EBchecked/topic/256468/al-Hasan-al-Basri.
38. Izutsu, *The Concept of Belief*, 51.
39. Ibid., 52.
40. Abū Nu'aym al-Iṣfahānī, *Ḥilyat al-Awliyā' wa Tabaqat al-Aṣfiyā'*, vol. 2 (Beirut, Lebanon: Dār al-Fikr, 1996), 153.
41. Izutsu, *The Concept of Belief*, 52–53.
42. Abū Muḥammad 'Alī ibn Ḥazm, *al-Fiṣal fī al-Milal wa al-Ahwā' wa al-Niḥal*, vol. 3 (n.c.: Maktabat al-Salām al-'Ilmiyyah, 1930), 136.
43. Ibid.
44. Ibid.
45. Ibid., 136–137.
46. Ibid.
47. Qādī 'Abd al-Jabbār, *Sharḥ al-Uṣūl al-Khamsah* (Beirut, Lebanon: Dār Iḥyā' al-Turāth al-'Arabī, 2001), 77–78.

48. Ibid.
49. Ibid., 87.
50. Ibid.
51. Ibid.
52. Ibid., 88.
53. Ibid.
54. Ibid.
55. Ibid., 269.
56. Ibid., 270–271.
57. Ibid., 272–274.
58. Ibid.
59. Ibid., 474.
60. Ibid.
61. Ibid.
62. Ibid.
63. Ibid.
64. Ibid., 474–475.
65. Ibid.
66. Ibid.
67. Ibid., 481.
68. Ibid.
69. Ibid.
70. Ibid.
71. Ibid., 482.
72. Ibid.
73. Abū al-Qāsim Maḥmūd ibn 'umar al-Zamakhsharī, *al-Minhāj fī Uṣūl al-Dīn*, ed. Sabine Schmidtke (Beirut, Lebanon: Al-Dār al-'Arabiyyah lil 'Ulūm, 2007), 73.
74. Ibid.
75. David Waines, *An Introduction to Islam*, 2nd ed. (Cambridge: Cambridge University Press, 2004), 120.
76. Ibid.
77. Ibid., 121.
78. Izutsu, *The Concept of Belief*, 141.
79. Abū al-Ḥasan al-Ash'arī, *Al-Ibānah 'an Uṣūl ad-Diyānah*, trans. Walter C. Klein (New Haven, Conn.: American Oriental Society, 1940), 25.
80. Ibid.
81. Ibid., 47.
82. In the Qur'ān, Magians, or *al-Majūs*, are mentioned in verse 22:17.
83. Al-Ash'arī, *al-Ibānah*, 48.
84. Ibid., 49.
85. Aḥmad b. Ḥanbal (d. 855) was a traditionist who held that the Qur'ān and the *sunnah* are the groundwork of Islam.
86. Al-Ash'arī, *al-Ibānah*, 49.
87. Ibid., 51.
88. Ibid., 52.
89. Ibid.
90. Ibid., 53.
91. Abū al-Ḥasan al-Ash'arī, *Kitāb al-Luma'*, ed. Richard J. McCarthy, S. J. (Beirut, Lebanon: Al-Maṭba'ah al-Kāthūlīkiyyah, 1953), 75.
92. Ibid.
93. Ibid.
94. Ibid.
95. Ibid.
96. Ibid., 76.
97. Ibid.
98. Ibid.

99. Izutsu, *The Concept*, 140–141.
100. Ibn Ḥazm, *Kitāb al-Fiṣal,* 188.
101. Ibid., 189.
102. Izutsu, *The Concept*, 141.
103. Ibid.
104. Al-Ash'arī, *al-Ibānah*, 53.
105. Izutsu, *The Concept*, 141.
106. Al-Shahrastānī, *al-Milal wa al-Niḥal*, vol. 1, 114.
107. 'Abd al-Qāhir al-Baghdādī, *Kitāb Uṣūl al-Dīn* (Istanbul, Turkey: Maṭba'at al-Dawlah, 1928), 248.
108. 'Abd al-Qāhir al-Baghdādī, *al-Farq Bayna al-Firaq* (Cairo: n.p., 1910), 55.
109. Izutsu, *The Concept*, 143.
110. Al-Shahrastānī, *Nihāyat*, 468.
111. Ibid.
112. Al-Shahrastānī, *Nihāyat*, 469.
113. Ibid.
114. Ibid.
115. Ibid., 468–469.
116. Fazlur Rahman, *Islam,* 2nd ed. (Chicago: The University of Chicago Press, 1979), 93.
117. Muḥammad Abū Manṣūr al-Māturīdī, *Sharḥ al-Fiqh al-Akbar*, ed. 'Abd Allāh b. Ibrāhim al-Anṣārī (Ḥaydar Ābād al-Dakan, India: Dā'irat al-Ma'ārif al-Niẓāmiyyah, 1903). *Kitāb Sharḥ al-Fiqh al-Akbar* is al-Māturīdī's commentary on *al-Fiqh al-Akbar* for Abū Ḥanīfah al-Nu'mān b. Thābit al-Kūfī (d. 698).
118. Sa'd al-Dīn Masūd ibn 'umar ibn 'Abd Allāh al-Taftazānī, *Sharḥ al-'Aqā'id al-Nasafiyyah*, ed. Claude Salamah (Damascus, Syria: Manshūrāt Wizārat al-Thaqāfah wa al-Irshād al-Qawmī, 1974), 45.
119. Al-Māturīdī, *Sharḥ*, 7.
120. Ibid.
121. Ibid.
122. Ibid.
123. Izutsu, *The Concept*, 210.
124. Ibid., 211.
125. Al-Māturīdī, *Sharḥ*, 162.
126. Al-Sharīf al-Murtaḍā, *Āmālī al-Murtaḍā,* vol. 2 (Cairo: Dār Iḥyā' al-Kutub al-'Arabiyyah, 1954), 180–181.
127. Ibid.
128. Ibn Ḥazm, *Kitāb al-Fiṣal*, 142.
129. Ibid.
130. Ibid.
131. Abū al-Ḥasan al-Ash'arī, *al-Luma' fī al-Rad 'ala Ahl al-Zaygh wa al-Bida'*, ed. Ḥammūdah Ghurābah (Cairo, Egypt: Maṭba'at Miṣr, 1955), 82.
132. Izutsu, *The Concept* 221.
133. Ibid.
134. Al-Māturīdī, *Sharḥ*, 253–259.
135. Ibid.
136. Ibid.
137. Ibid.
138. Al-Taftazānī, *Sharḥ*, 84.
139. Ibid.
140. Ibid.
141. Ibid.
142. Ibid., 93–94.
143. Ibid.
144. Ibid.
145. Ibid.
146. Ibid., 115.

147. Ibid.
148. Ibid., 116.
149. Ibid.
150. Ibid.
151. Ibid., 116–117.
152. Ibid.
153. Ibid.
154. Ibid., 118.
155. Ibid.
156. Ibid.
157. Ibid., 119.
158. Ibid.
159. Ibid.
160. Ibid., 120.
161. Ibid., 121.
162. Ibid., 124–125.
163. Ibid.
164. Ibid., 127.
165. Ibid.
166. Quoted in Ulrich Rudolph, *Al-Maturidi and the Development of Sunni Theology in Samarqand* (Leiden and Boston: Brill, 2015), 309.
167. Al-Māturīdī, *Sharḥ*, 38.
168. Ibid.
169. Ibn Ḥazm, *Kitāb al-Fiṣal*, 130–131.
170. Al-Māturīdī, *Sharḥ*, 38.
171. Izutsu, *The Concept*, 227.
172. Al-Māturīdī, *Sharḥ*, 38.
173. Ibid.
174. The term *geravidan* means acceptance, obedience, and *īmān* in Muḥammad al-Tunjī, *al-Mu'jam al-Dhahabī* (Beirut, Lebanon: n.p., 1969), 501. It also means intention as well as *īmān* in S. Heim, *The English-Persian Dictionary* (Tehran, Iran: n.p., 1961), 701. Heim's translation of the term *geravidan* or *taṣdīq* gives it a new meaning: the intention. This shows that *īmān* is or requires the intention (al-Taftazānī, *Sharḥ*, 130).
175. Al-Taftazānī, *Sharḥ*, 130.
176. Ibid., 131.
177. Ibid., 132.
178. Ibid.
179. Ibid., 134.
180. Ibid.
181. Ibid., 135.
182. Ibid., 135–137.
183. Quoted in Rudolph, *Al-Maturidi*, 309.
184. Al-Taftazānī, *Sharḥ*, 141.
185. It is either *Kitāb al-Kifāyah fī al-Hidāyah* by Nūr al-Dīn al-Ṣābūnī (d. 580) or *Kitāb al-Bidāyah min al-Kifāyah*, ed., Fatḥallah Khalīf (n.c.: n.p., n.d.), 13.
186. Ibid.
187. Al-Taftazānī, *Sharḥ,* 140–141.
188. Al-Māturīdī, *Sharḥ*, 12–13.
189. Ibid.
190. Ibid.
191. Ibid., 16.
192. Ibid., 20.
193. Ibid.

Chapter Four

Takfīr in Medieval Islamic Thought

This chapter discusses the concept of *takfīr* in the thoughts of two prominent medieval Muslim theologians: Abū Ḥāmid Muḥammad b. Muḥammad al-Ṭūsī al-Ghazālī (d. 1111) and Taqī al-Dīn Aḥmad ibn Taymiyyah (d. 1328). These two thinkers wrote extensively on Islamic theology in general and the issue of *takfīr* in particular. In addition, al-Ghazālī's and Ibn Taymiyyah's works on Islamic theology and creed are considered in the Sunni world today to be among the most legitimate references for scholars, jurists, theologians, and students of Islamic studies. For this reason, their works have been selected for this project to explore their views on *takfīr* and how these views differ from the views expounded by classical and later theological schools.

AL-GHAZĀLĪ'S THEORY OF *TAKFĪR*

Al-Ghazālī, born in Ṭūs (Khurāsān) in 1058, was a scholar, a theologian, a jurist, and a Sufi (mystic) who has long been considered as one of the most prominent figures in Islam[1] and "best known of medieval Islam's religious intellectuals."[2] He earned the title "*Imām*" during his tenure as a professor in the Niẓāmiyyah College in Baghdad (1091–1095)[3] and is praised today as "one of the reformers of the religion."[4] His influence on Islamic thought and life is "unparalleled," according to many Sunni Muslim scholars.[5] He died in Ṭūs in 1111.

Al-Ghazālī's views on *takfīr* emerged as a reply to the theological schools that used the concept to denounce other Muslims who did not agree with their doctrines. Although al-Ghazālī denounced some Muslim philosophers as *kuffār*, he asserted that Muslims, especially theologians, were wrong to declare other Muslims infidel. He was against using the term lightly and believed that it was improper to declare Muslims infidel. Al-Ghazālī expressed his views on *takfīr* in several works such as *al-Munqidh min al-Ḍalāl*, *al-Iqtiṣād fī al-I'tiqād*, and *Fayṣal al-Tafriqah Bayna al-Islām wa al-Zandaqah.*[6] The following pages discuss in detail his views on *tafkīr*, point out how his definition of the concept differs from the theologians who preceded him, and highlight his contribution to the development of *takfīr* in Islamic thought.

In *Fayṣal al-Tafriqah*, al-Ghazālī begins by providing a definition of *kufr* and *īmān*. He writes that *kufr* is to deem anything that the Prophet brought to be a lie and *īmān* is to deem everything that the Prophet brought to be true.[7] He asserts that Jews and Christians are *kuffār* for denying the truth of Muḥammad, and so are the deists and the atheists (*al-dahriyyūn*) for denying the truth of "our Messenger" as well as the truths of all God's messengers.[8] He adds that *kufr* is a legal designation (*ḥukm shar'ī*), like slavery and freedom, and because it is a legal designation, it is justified to shed the blood of the *kāfir* and to pass a judgment stating that the *kāfir* will dwell in eternal Hellfire.[9] The legal designation of *kufr*, al-Ghazālī claims, is known based on a scriptural text (*naṣṣ*) or on analogy (*qiyās*).[10] He writes that there are explicit texts in regard to Jews and Christians as well as to deists, dualists, *zanādiqah* (heretics or "crypto-infidels"[11]), and atheists. They are all associationists (*mushrikūn*), for they deem the Prophet to be a liar. Every *kāfir*, he writes, deems the Prophet to be a liar, and everyone who deems the Prophet to be a liar is a *kāfir*.[12]

Al-Ghazālī writes that everyone who interprets a statement of the Lawgiver in accordance with the teachings of Islam has deemed the statement to be true. Deeming a statement to be a lie, on the other hand, is to deny its reality. And to deny its reality is to deem the Lawgiver to be a deceiver. This is absolute *kufr* and heresy (*zandaqah*).[13] He adds that it is inappropriate to brand as *kuffār* those who engage in figurative interpretation as long as they commit to the Rule of Figurative Interpretation (*Qānūn al-Ta'wīl*). Al-Ghazālī asks: "How is it proper to brand such figurative interpreters as *kuffār* when every Islamic party finds itself compelled to engage in figurative interpretation?"[14]

Those who rush to figurative interpretation, al-Ghazālī writes, using speculative presumptions without any definitive logical proofs should not be branded as *kuffār* in every instance.[15] Rather, one should look into their interpretation. If the figurative interpretation had no basis in fundamental principles, when it would not be proper to brand the interpreters as *kuffār*. Rather, they should be given the benefit of the doubt regarding interpretations that are not connected with the basic principles of creed. Those who alter the apparent meaning of a text without definitive logical proof must be branded as *kuffār*, that is, those who deny the resurrection of the body and physical punishment in the Hereafter on the basis of speculative presumptions without any definitive logical proof. Such people, he asserts, must absolutely be branded as *kuffār*.[16]

The spread of such belief is damaging to religion, he writes. Therefore, it is necessary to declare whoever is associated with it as *kāfir*; most Muslim philosophers associate themselves with this type of belief, he contends.[17] He writes that it is necessary to brand as a *kāfir* anyone among them who says that God knows nothing other than Himself or that God knows nothing except universals. This belief deems what the Prophet taught to be a lie, and it has nothing to do with levels of figurative interpretation. The evidence contained in the Qur'ān and the *ḥadīths* are too great to accept figurative interpretation.[18] Thus, for al-Ghazālī, the philosophers did not engage in figurative interpretation; rather, they denied what the Prophet taught to be true because they argued that believing in such teachings (i.e., resurrection of the body and the physical punishment in the Hereafter) is in the best interest of the people, for to believe that God watches over them generates hope and fear in their hearts.[19] Hence, the philosophers, in al-Ghazālī's view, deem it permissible for the Prophet to lie as long as it is beneficial to the masses.

These views of the philosophers, al-Ghazālī contends, constitute the first level of *zandaqah*. This is a stage between Mu'tazilism and absolute *zandaqah*. The approach of the Mu'tazilites, he tells us, is very close to that of the philosophers except for one issue: The Mu'tazilites do not deem it permissible for the Prophet to lie based on the excuses provided by the philosophers. Rather, they engage in figurative interpretation of the apparent meaning of the scripture whenever it appears to them to contradict some logical proof.[20] The philosopher, on the other hand, rejects the text if it contradicts logical proofs. For al-Ghazālī, it is absolute *zandaqah* to deny the resurrection of the body and to deny that the universe has a Creator.[21] Al-Ghazālī pronounced the philosophers as *kuffār* on three questions alone:

> One of them is the question of the world's pre-eternity and their statement that all substances are pre-eternal. The second is their statement that God's knowledge does not encompass the temporal particulars among individual [existents]. The third is their denial of the resurrection of bodies and their assembly at the [Day of Judgment]. These three doctrines do not agree with Islam in any respect. . . . This is manifest infidelity which none of the Islamic sects have believed.[22]

To support his views against the philosophers, al-Ghazālī cites the following *ḥadīth*: "My community will divide into more than seventy sects. All of them are destined to Heaven except for the *zanādiqah* [heretics]."[23]

Some modern and contemporary scholars have offered critical analysis studies of al-Ghazālī's *takfīr*. For instance, in a short book entitled *Qaḍiyat al-Takfīr 'inda al-Ghazālī*, Rifqī Zāhir, a professor of philosophy and Islamic studies at the prestigious al-Azhar University, argues that the "philosophers' views on the above-mentioned three questions do not lead to *kufr* even though these views are sometimes incorrect."[24] Zāhir criticizes al-Ghazālī for making a judgment against all philosophers based on his assessment of the most controversial philosophical views. Zāhir adds that al-Ghazālī's *The Incoherence of the Philosophers* deviated from its goal as a work to expose the un-Islamic views of the philosophers to a work that demonstrated al-Ghazālī's aggressive criticism and radical views on *takfīr*.[25] Zāhir argues that al-Ghazālī's accusations of *kufr* against the philosophers did not emerge out of a vacuum. Rather, they emerged in a time when 1) traditional Sunni orthodoxy prevailed as the dominant and correct belief and 2) there was a tendency to denounce "philosophical and rational sects" and accuse them of unbelief although they confess their Islam.[26]

Al-Ghazālī asserts that a full explanation of the grounds on which *kufr* is or is not declared requires a lengthy and detailed discussion covering all doctrines and schools of thought. This is a daunting task requiring several volumes to cover this explanation, he adds. As for the Advice (*al-waṣiyyah*), he writes, people should refrain from making judgments regarding *kufr* and *īmān* against the people who face Makkah in prayers (*ahl al-qiblah*), that is, the Muslims, as long as they testify that there is no god but God and Muḥammad is His Messenger, without contradicting this testimony. The contradiction, he writes, is to believe in the possibility that the Prophet committed lies, with or without an excuse. He affirms that branding people as *kuffār* is a dangerous matter (*fīhi khaṭar*). Remaining silent, however, entails no danger at all.[27]

As for the Law (*al-qānūn*), he contends, speculative matters (*al-naẓariyāt*) are of two types: One is connected with the fundamental principles of creed and the other with secondary issues. The fundamentals of *īmān* are three: the belief in God, His Prophet, and the Last Day. Anything else is secondary.[28] Based on this formula, al-Ghazālī asserts that there is no *takfīr* in secondary issues, with one exception: that a person rejects a religious tenet that was passed down from the Prophet. Even then, he contends, such a person should be deemed wrong, as is done with legal issues, or be condemned for innovation (*bid'ah*).[29] For al-Ghazālī, these acts are considered secondary issues and do not serve as a ground for condemning people as *kuffār*.

On the other hand, al-Ghazālī asserts that claims involving a secondary issue are considered acts of *kufr* if these claims affirm the possibility that the Prophet did not say the truth. One example he provides is the statement denying that the House at Makkah is the *Ka'bah*. This claim involves a secondary issue but it is an act of *kufr* because it contradicts what has been established on the authority of the Prophet via chains of transmission (*tawātur*).[30] Another example is the claim accusing Ā'ishah, Muḥammad's wife, of indecency. Those who made this claim have committed *kufr*, al-

Ghazālī writes, because they contradicted the Qur'ān's affirmation of her innocence.[31] Thus, for al-Ghazālī, these are claims that involve secondary issues, but they are acts of *kufr* because they contradict the Qur'ān, they deem the Prophet to have lied, or they deny *tawātur*. He adds that a person may deny *tawātur* by the tongue but cannot ignore it in the heart. For al-Ghazālī, on the other hand, to deny the truth of an isolated (*āḥād*) *ḥadīth* is not an act of *kufr* since *āḥād ḥadīths* are not transmitted or based on a unanimous consensus.[32]

As to the three fundamental principles (God, the Prophet, and the Last Day) and the *ḥadīths* that are not open to figurative interpretation, al-Ghazālī writes, contradicting them is a pure lie that constitutes an act of *kufr*.[33] However, what is open to figurative interpretation is constituted a *bid'ah* only if the interpretation is revealed to the masses, which they cannot understand, and causes a threat to religion, such as the Mu'tazilites' negation of the beatific vision, which is a *bid'ah* but not *kufr*. Regarding matters that appear to be a threat to religion, al-Ghazālī adds, determining their legal status is subject to the scholars' scrutiny. They are either declared acts of *kufr* or a *bid'ah*.[34]

Matters that appear to be a threat to the religion, according to al-Ghazālī, are the claims made by some who refer to themselves as Sufis who assert that they have reached a state between themselves and God, they are no longer obligated to pray, and drinking wine and other forms of disobedience are permitted to them.[35] Al-Ghazālī renders their killing licit "for executing one person of them is better than killing a hundred *kāfir*, because the harm they bring to the religion is greater than that caused by the *kuffār*."[36] They are more dangerous to the religion than the *kuffār* because they are masked infidels with the ability to deceive people, whereas the exposed unbelief of the *kuffār* discourages people from listening to their ideas.[37]

Al-Ghazālī opposes the notion that the branding of *kufr* or its negation must be based on certainty in every case. He contends that *takfīr* is a legal designation (*ḥukm shar'ī*) that refers to a person's losing property, shedding his blood, and condemning him to eternal Hellfire. The basis upon which *takfīr* is established, he writes, is the same upon which the rest of the rules are established.[38] Sometimes they are established based on certainty; sometimes on a majority of probabilities; and sometimes on sources that one finds himself hesitant to fully commit to (*yataraddadu fīh*). Whenever one finds himself hesitant to make a decision, he writes, it is more appropriate to refrain from declaring *takfīr*. The rush to *takfīr*, he adds, is the habit of those who are overtaken by ignorance.[39]

Al-Ghazālī declares the *kufr* of those who contradict a *mutawātir* (mass transmitted) text while claiming that they are simply providing a figurative interpretation of it. They have committed *kufr*, al-Ghazālī explains, because their figurative interpretations have no basis in the language. An example he provides is the doctrine of some Batinites (*al-bāṭiniyyah*)[40] who say that God is one in the sense that He gives oneness and creates, and knowing in the sense that He gives knowledge to others and creates it, and He exists in the sense that He brings others into existence. For al-Ghazālī, this is a "plain *kufr* (*kufrun ṣaraḥ*)."[41]

Al-Ghazālī establishes a formula in which suspected instances of *kufr* can be investigated to determine whether they are truly acts of *kufr* or not. Investigating suspected instances of *kufr* includes: 1) Investigating whether a sacred or legal text (*naṣṣ*), whose apparent meaning has been altered, is open to figurative interpretation; 2) Investigating whether a doctrinal statement has been established based on a *naṣṣ* that is deemed an isolated report (*ḥadīth āḥād*), a report based on chains of transmission (*tawātur*), or one whose validity was established based on a mere consensus (*ijmā*'); 3) Investigating whether the author of some doctrinal statements had access to a *naṣṣ* through *tawātur* or consensus; 4) Investigating whether a doctrinal statement has been established based on logic when the statement contradicts the apparent meaning of a *naṣṣ*; and 5) Investigating whether the public disclosure of a doctrinal statement causes serious harm to religion or not. For al-Ghazālī, such instances serve as grounds for the accusations of *kufr*.[42]

Investigating suspected instances of *kufr* involves an examination of the text of a doctrine (a mode of examination similar to that of source criticism) and an examination of the legal status and implications of certain points of that doctrine (a mode of examination arriving at a legal verdict).[43] For instance, in his *K. al-Mustaẓhirī*, al-Ghazālī examines the textual basis and legal status of Ismā'īlī claims on the Imamate. In the first examination, al-Ghazālī investigates whether the assertion that the Imamate, 'Alī's succession of Muḥammad, is *ḥadīth tawātur*. Al-Ghazālī argues that there is no evidence to establish the *tawātur* of such claims. By doing so, he establishes that the Ismā'īlī claims for the Imamate are not in line with the history of the Muslim community.[44] Moreover, al-Ghazālī writes that the Ismā'īlī denial of a bodily resurrection and a physical garden and a fire in the afterlife contradicts the message brought by the Prophet. Thus, their doctrine of the afterlife, according to al-Ghazālī, makes the Prophet a liar (*takdhīb*), an act that warrants the charge of *takfīr*.[45] For al-Ghazālī, the Ismā'īlī claims are "of deviation and transgression" and charging them of *takfīr* is a judgment arrived at after a "thorough investigation and not an arbitrary opinion."[46] By asserting that examination of suspected instances of *kufr* is a prerequisite to reaching a judgment on *kufr*, al-Ghazālī demonstrates that he was, in the words of Farouk Mitha, "keen to formulate a theory of unbelief, the value of which would not only safeguard *takfīr* from being misused but would also serve as a key construct in supporting the idea of an orthodoxy."[47] F. Griffel, echoing Mitha, speaks of al-Ghazālī's political action to establish a criteria or "rule of law" against *kufr* "as a development from earlier meaning of *kufr*."[48]

Al-Ghazālī strongly criticizes some speculative theologians for declaring common Muslims as *kuffār*. These "most extreme and excessive theologians," he writes, alleged that whoever does not know speculative theology (*al-kalām*) as they know it and does not know the sources of religion that they have established is a *kāfir*.[49] These theologians, al-Ghazālī claims, have narrowed God's all-encompassing mercy on the people and made Paradise exclusive to a small group of speculative theologians. Al-Ghazālī attacks these theologians by saying that anyone who believes that the way to *īmān* is through speculative theology has committed an innovation (*bid'ah*).[50] He tells us that *īmān* is a light that God sends to the hearts of His servants as a gift and gratuity from Him. He provides incidents where people declared their *islām* in the presence of Prophet for seeing the prophecy on his face. For al-Ghazālī, *īmān* came to the hearts of these people without occupying themselves with speculative theology. Rather, the light of *īmān* appeared in their hearts like a white flash, and it increased in intensity as a result of their recitation of the Qur'ān and the purification of their hearts.[51]

Al-Ghazālī concludes his *Fayṣal al-Tafriqah* by saying that the plain truth is that whoever firmly believes what the Messenger brought, including the Qur'ān, is a Believer (*mu'min*), even if he does not know the proofs of his belief. By contrast, the *īmān* that results from the proofs of speculative theology is extremely weak and subject to collapse when facing the simplest sophism. The well-established *īmān*, he adds, is the *īmān* of the common people that develops in their hearts from childhood due to their exposure to religious experiences.[52]

Al-Ghazālī writes that God's mercy will encompass not only Muslims but also many bygone communities like the Christians of Byzantium and the Turks. These communities will receive God's mercy because they did not come in contact with the message of Islam. Al-Ghazālī divides these communities into three categories: 1) a group that never heard of Muḥammad, therefore, they are excused; 2) a group that lived in territories in close proximity to Muslims, had contacts with the Muslims, and had heard of Muḥammad, his character, and the miracles associated with him. These are the *kuffār* and the atheists; and 3) a group whose status falls between these two poles. These people had heard of Muḥammad's name but not of his character and attributes. In al-Ghazālī's opinion, this group, like the first, is excused for not having *īmān*.[53]

In *Fayṣal al-Tafriqah*, al-Ghazālī does not aim to establish a definition of *takfīr*, nor does he concern himself with refuting the established theological doctrines in regard to the concept. Rather, al-Ghazālī attempts to highlight his dissatisfaction with the theological schools for their (theological) intolerance of other Muslims and to express his frustration at the liberal theologies of his time that conflated "rationalistic deism with Islam"[54] and exposed the religion to a great danger. He brings to light new ideas to help mend the rift between theological schools. More importantly, he aims to assert that theology is not the monopoly of one group or another but it is an enterprise of learning and reflecting. For al-Ghazālī, theological orthodoxy as well as loose theology or "theological *laissez-faire*,"[55] as Sherman Jackson calls it, is dangerous to Islam. Thus, in his works, al-Ghazālī tries to establish a reconciliatory theology that tolerates and accepts doctrines that are otherwise condemned and attacked and, at the same time, "defend the *community* against a veritable cyclone of charges and counter-charges of Unbelief."[56]

Al-Ghazālī finds it improper for one group of theologians to enjoy a monopoly over the truth (i.e., one group declaring the *kufr* of their adversaries while denying them this privilege). He had also some issues with this free, unrestricted application of the definition of *kufr*. His aim was to redefine the grounds for declaring *kufr*, and while doing that he attempted to narrow the application of the definition. While narrowing the realm of *kufr* to those who reject the truthfulness of the Prophet, he stretched it enough to include those who masked their unbelief, or *al-zanādiqah* (heretics) as he calls them. For al-Ghazālī, *al-zanādiqah* (including some Muslim philosophers) masked their unbelief, that is, their denial of the Prophet's truthfulness, with their application of the figurative interpretation of the revelation. Thus, one may conclude that the way al-Ghazālī narrows and then stretches the realm of *kufr* to bring some in (i.e., unbelieving Muslim philosophers) and to keep others outside [i.e., the Mu'tazilites and believing philosophers like Abū al-Ḥasan Muḥammad b. Yūsuf al-'Āmirī (d. 992)],[57] he provided a new definition of *kufr* that is "both precise and restrictive."[58]

For al-Ghazālī, *kufr* is a status conferred upon Muslims who violate basic theological doctrines such as the fundamentals (*uṣūl*) of Islam (i.e., belief in God, Muḥammad, and the Last Hour and doctrines based on consensus (*ijmā'*), or on the authority of Muḥammad via reliable transmissions (*tawātur*). The question remains: Why does al-Ghazālī limit *kufr* to the violation of these doctrines? Al-Ghazālī provides a new and interesting approach to the concept of *kufr*. For instance, unlike the theological schools that deemed the figurative interpretations of Muḥammad's statements invalid because they contradicted their views, he believed that both the literal and figurative interpretations of Muḥammad's statements are valid truths. What matters for al-Ghazālī is not how people understand the statements of Muḥammad but how they react to these statements. For instance, al-Ghazālī insists that charges of *kufr* are given to those who deem the Prophet to have uttered lies in teaching such beliefs. This is *kufr*, he contends, because accusing the Prophet of delivering lies is dangerous to the religion and the community. The accusation that Muḥammad lied is equal to denying the truth that Muḥammad brought. Thus, denying the truth is a reactionary doctrine against revelation, which is an act of *kufr*, as opposed to a figurative interpretation that affirms the authenticity of the revelation but with a different understanding of its doctrine. Now, to declare *kufr* against those who have different understanding of the revelation is considered "theological intolerance."[59]

Like theological intolerance, loose theology or "theological *laissez-faire*" must be shunned, according to al-Ghazālī. Loose theology seems to be not of a great concern to him since he admits that some of the Muslims who fit in this category, such as the philosophers (*al-falāsifa*), are believers in God. In the words of Sherman Jackson, al-Ghazālī brings up the discussion of *laissez-faire* theology not to make a judgment against the philosophers but rather his aim is to demonstrate the "limits of theological tolerance" in Islam.[60] For al-Ghazālī, *kufr* is to reject the truthfulness of Muḥammad. This means that no person can be accused of *kufr* based on moral or religious behavior. Thus, according to al-Ghazālī's definition of *kufr*, "the modern Muslim tendency to employ '*kāfir* (Unbeliever)' as a moral, ethnic, cultural, or even civilizational delineator," Jackson argues, "shows itself to be a patent misuse of the category."[61] Likewise, to refrain from declaring the unbelief of those who commit acts of *kufr* only to preserve their status as believers can be also a distortion of Islamic doctrine.

Al-Ghazālī criticizes some theologians for misunderstanding the nature of *takfīr* itself. He tells us that these theologians base *takfīr* solely on Reason. For al-Ghazālī, *takfīr* is a legal issue based on Revelation. He supports his argument by demonstrating that a judgment on *takfīr* has legal implications. Because of these legal implications, al-Ghazālī concludes that *takfīr* is essentially a legal problem dealt with according to Divine Law. Basing a judgment on Divine Law, al-Ghazālī asserts, becomes a matter of jurisprudence.[62]

Al-Ghazālī is against a loose theology of *takfīr*. His theory is that the question of *takfīr* should come up only when the fundamentals of Islam (*uṣūl al-dīn*) are violated. For al-Ghazālī, a mere violation of moral or religious behavior is not a ground for *takfīr*. These views, however, were challenged almost two centuries later when Ibn Taymiyyah propagated a more radical and loose interpretation of *takfīr*.

IBN TAYMIYYAH'S THEORY OF *TAKFĪR*

Ibn Taymiyyah was born in 1263 in Ḥarrān, near Mosul in northern Iraq, to a family known for its learning. His father, 'Abd al-Ḥalīm ibn Taymiyyah (d. 1284), was a distinguished scholar of *ḥadīths* and his grandfather, Majd al-Dīn ibn Taymiyyah (d. 1255), was the most prominent jurist in the Ḥanbalī school of law. Ibn Taymiyyah studied the Qur'ān, *ḥadīths*, jurisprudence, and the Arabic language at the feet of more than 200 scholars. At the age of 22, he succeeded his father as a lecturer of Qur'ānic and *ḥadīth* studies. The political situation of his time, which was rife with turmoil, chaos, and disorder, shaped his views on Islam in general and on Islamic and other religious sects in particular.[63] The following paragraphs explore his views on *takfīr* and *īmān* as well as his replies to the theological schools in regard to these two concepts.

In his *Kitāb al-Īmān*, Ibn Taymiyyah starts his discussion of *īmān* and *kufr* by supporting the views that some acts of disobedience can be categorized as unbelief but others cannot. Three kinds of disobedience are mentioned: unbelief (*kufr*), sin (*fusuq*), and rebellion (*'iṣyān*). The *kuffār*, the criminals, and the sinful, Ibn Taymiyyah writes, will be questioned on the Day of Judgment. The *kāfir* does not praise God for the good things that happen to him; therefore, he will punished.[64] God gives guidance only to the believers, according to Ibn Taymiyyah, whereas the *kuffār* are left in the dark. For instance, God made the good things permissible to eat only for the believers: "You who believe, eat the good things We have provided for you" (Q. 2:172). As for the *kuffār*, "God does not allow, permit, prohibit, or even excuse anything for them to eat or not to eat."[65] For Ibn Taymiyyah, God gives permission regarding eating only to the believers, and he does not give them permission unless they believe.[66] Ibn Taymiyyah interprets the Qur'ānic verse 2:172 to mean that the *kuffār*'s "possessions do not belong to them legally . . . to own something legally one must have the ability and permission of the Prophet . . . this permission comes only under one condition: *īmān*."[67]

Based on his own interpretation of the Qur'ānic verse 2:172, Ibn Taymiyyah provides legal justification for the believers to confiscate the possessions of the *kuffār*, when captured by force, and to make these possessions theirs legally because God, he writes, makes the spoils permissible for them and not others. God makes the *kuffār*'s possessions legal to "those who deserve it . . . (to the believers who worship Him and seek His sustenance through worship)."[68] Not only do the possessions of the *kuffār* become the property of the believers but their captives also, whom they are allowed to buy from each other because "this falls in the category of claiming permissible things."[69]

In a chapter on the meaning of the words *kufr* and *nifāq*, Ibn Taymiyyah writes that when the term *kufr* is cited in the Qur'ān and the *Sunnah* by itself, it implies *nifāq* and unbelief, as indicated in the following Qurānic verses: "The deeds of anyone who rejects faith will come to nothing, and in the Hereafter he will be one of the losers" (Q. 5:5) and "Anyone who does not believe in God, His Angels, His Scriptures, His Messengers, and the Last Day has gone far, far astray" (Q. 4:136). For Ibn Taymiyyah, *kufr* implies both hypocrisy (i.e., rejecting belief and pretending to behave as a Muslim) and unbelief, wherein "the unbeliever rejects belief, openly and explicitly."[70] Ibn Taymiyyah cites the ninth-century Muslim scholar Sahl ibn 'Abd Allāh al-Tustarī who said: "*Īmān* is confession (*qawl*), deed (*'amal*), intention (*niyyah*), and *sharī'ah* (*sunnah*). If *īmān* is confession without deed, it will be *kufr*; if *īmān* is both confession and deed without intention, it will be hypocrisy (*nifāq*); if *īmān* is confession, deed, and intention without *sunnah*, it will be invention (*bid'ah*)."[71]

Ibn Taymiyyah makes a distinction not only between *kufr* and *nifāq* but also between *kufr* and polytheism (*shirk*). For instance, he writes that the meaning of the word *al-mushrikūn,* the polytheists, applies only to the People of the Book (Jews and Christians) and to other communities. He refers to verse 22:17 to justify his claims: "As for the believers, those who follow the Jewish faith, the Sabians, the Christians, the Magians, and the idolaters, God will judge between them on the Day of Resurrection." Ibn Taymiyyah believes that the word "idolaters" ("*mushrikūn*") may also refer to the Jews and Christians since the Qur'ān places those who disbelieve among the People of the Book in the same category with the *mushrikūn* (Q. 98:1; 98:6).

Ibn Taymiyyah explains the meanings of the terms disobedience (*al-ma'ṣiyah*), sin, and *kufr*. He writes that if the word *al-ma'ṣiyah* is used "in an absolute sense with reference" to God and the Prophet, then *kufr* and sin are implied for whoever disobeys God and His Messenger, for them is Hellfire (Q. 72:23). In fact, the Qur'ān makes a reference to the ancient people of 'Ād who rejected the signs of God and disobeyed His Messengers (Q. 11:59). Ibn Taymiyyah's reasoning is that disobedient people are considered *kuffār* because they disobeyed God's Messengers. He writes that the above verse "generalizes the use of the term, disobedience, beyond that of the people of 'Ād to all Messengers."[72] By this reasoning, Ibn Taymiyyah extended the definition of *kufr* to include acts of mere disobedience such as loud wailing, beating the cheeks, uncovering the hair, and tearing of the clothes in grief. For Ibn Taymiyyah, "Disobeying Allah in any just matter means disobeying Him in all of what he commands [i.e., Islamic legislation and guidelines]."[73]

Ibn Taymiyyah does not distinguish between *ma'ṣiyah*, *kufr*, and *fusuq*. He writes that disobeying God and His Apostle (*ma'ṣiyat Allāh wa rasūlahu)* is *kufr* and *fusuq*[74] and that the Qur'an makes no distinction between *ma'ṣiyah*, *kufr*, and *fusuq*: "He has made disbelief [*kufr*], mischief [*fusuq*], and disobedience [*'iṣyān*] hateful to you" (49:7). Ibn Taymiyyah adds: "It is well known that the *fāsiq* is disobedient as well."[75] It is important to note that Ibn Taymiyyah places *ma'ṣiyah* and *fusuq* in the category of lesser *kufr*. He writes that the Muslims are of two groups: the believers and the hypocrites; the latter are in the hellfire. The *fāsiq*, however, is a believer who has committed a minor guilt or invention. Committing guilt or invention does not make the person a *kāfir* in the heart (*kāfir fī al-bāṭin*), unless he is a *munāfiq*.[76] He adds: "whoever has an *īmān* [believe] in the Messenger and in what he brought although he committed some wrongdoings is not basically a *kāfir* (*laysa bikāfir aṣlan*)."[77]

Ibn Taymiyyah refutes the Murji'ite claims that a *munāfiq* (hypocrite) is a *mu'min* because he declared his *īmān* openly and that he who declares his *īmān* openly is a true *mu'min.*[78] He writes that the Murji'ites say the *munāfiq* is a *mu'min* because *īmān* is the confession by the tongue (*al-qawl al-ẓāhir*); others call him a Muslim because of the apparent submission (*al-istislām al-ẓāhir*).[79] Ibn Taymiyyah argues that a *munāfiq* is not a *mu'min*; those who called him a *mu'min* have committed an error.[80] He writes: "He who has knowledge of God and *taṣdīq* in his heart yet rejects God's Messenger, like what the Jews and others did, God has called them *kuffār* and never called them *mu'minīn* although they adhered to some forms of *īmān*."[81]

Ibn Taymiyyah offers his replies to the Mu'tazilites who believed that a Muslim who has committed a grave sin is neither a *kāfir* nor a believer but in a position in the middle. According to Muḥammad Ḥarbī, the creation of the middle position is problematic for Ibn Taymiyyah. The latter believes that the Mu'tazilites committed a grave mistake because the middle position defies logic since a thing is against reason to be both in existence and not in existence.[82] For Ibn Taymiyyah, Ḥarbī adds, the middle position indicates that those who have committed a grave sin are neither *kuffār* nor believers, which is impossible. Ibn Taymiyyah argues that the Mu'tazilites would have had a better argument if they said that the grave sinner is neither a total believer nor a *kāfir*.[83] For Ibn Taymiyyah, the grave sinners are disobedient believers who are wrongly condemned to the eternal fire in Hell. He questions: "Since the Mu'tazilites refused to declare them *kuffār*, then how do they condemn them to Hell?" Their judgment, he contends, has no basis in both the Qur'ān and the *ḥadīth* and it contradicts God's justice and mercy. For Ibn Taymiyyah, therefore, there must be either *kufr* or belief. If there is not *kufr*, then there must be some amount of belief.[84]

Ibn Taymiyyah also offers his reply to the Mu'tazilite belief regarding the Prophet's intercession on behalf of the grave sinners. The Mu'tazilites believe that God will not extend intercession to the grave sinners; rather, He will condemn them to the eternal fire in Hell. The Mu'tazilites support their views with Qur'ānic verses that clearly state that on the Day of Judgment "no soul will stand in place of another, no intercession will be accepted for it, nor any ransom" (Q. 2:48; 2:123; 74:48). Ibn Taymiyyah challenges their belief and their understanding of these verses by citing the beliefs of the Sunnis (*ahl-al-Sunnah*), who argue that the verses on Qur'ānic intercession are twofold: First, they refer to a different category of people: the polytheists not the grave sinners. Ibn Taymiyyah writes: "intercession does not help the polytheists (*al-mushrikīn*) because God denied it to them in verses 74: 41–48: 'What drove you to the Scorching Fire?' . . . and they will answer, 'We did not pray; we did not feed the poor; we indulged with others [in mocking the believers]; we denied the Day of Judgement until the Certain End came upon us.' No intercessor's plea will benefit them now." Second, Ibn Taymiyyah adds, the Qur'ānic verses on intercession are meant to affirm that God "will deny intercession not only to the polytheists but also to Muslim, Jewish, and

Christian heretics who erroneously believed that God's creatures have the power to intercede on their behalf without His permission."[85]

Ibn Taymiyyah refutes the Ash'arite belief that knowledge of God (*ma'rifat Allāh*) alone is belief. For the Ash'arites, the meaning of belief is the *taṣdīq* (assent) in the heart. Ibn Taymiyyah challenges this belief by arguing that a mere Knowledge of God is not *taṣdīq*. Knowledge of God without having true hope and fear—hope for eternal reward and fear of eternal hell—is not *īmān*.[86] For Ibn Taymiyyah, he who believes in heaven without having hope for it and in hell without having fear of it is not a believer. He writes: "Iblīs, *the devil*, is not called a believer in Allāh even if he believes in the existence of Allāh and His Oneness. . . . Also, Pharaoh . . . is not called a believer although he has the knowledge that Allāh sent Moses as a Prophet and that He has sent the revelation."[87] Here, Ibn Taymiyyah makes a distinction between *taṣdīq* and belief. He writes that he "who believes in his heart but falsifies that belief with his own tongue is called in the speech of the Arabs a *muṣaddiq*, confirmer of belief."[88] However, he adds, "if he believes in something and acts contrary to that belief, they [the Arabs] would not say: he believes in it."[89] *Taṣdīq* without work, Ibn Taymiyyah writes, is a lesser *taṣdīq*, which is not *taṣdīq* (*nāqis kalma'dūm*).[90] Ibn Taymiyyah concludes that, "there is not one single piece of evidence in the speech of the Arabs to support [the Ash'arite] claim that the word *muṣaddiq* is synonymous with the word 'believer.'"[91]

Ibn Taymiyyah turns to the Murji'ites and offers his criticism of their theory on *īmān* and *kufr*. He writes that the Murji'ites committed two basic errors. The first error is that they believed that "Īmān, *belief*, was a mere *taṣdīq*, *assent*, and *'ilm*, *knowledge*, without work, state, action, will, love and humility in the heart."[92] For Ibn Taymiyyah, this is one of the Murji'ites' gravest errors, for the works of the heart are part of the duties and obligations that God and the Prophet have laid out.[93] Ibn Taymiyyah writes:

> There is the essence of *īmān* and there are its consequences. *Kufr* is opposed to *īmān* in both respects. The essence of *īmān* is belief (*taṣdīq*) and confession (*iqrār*), and the actions of the heart or the body are its consequences. The opposite of *iqrār* and *taṣdīq* which is the essence of *īmān* is the rejection (*kufr*) of God and what He has said, and the refusal to believe (*taṣdīq*) in Him and in His words. The opposite of *īmān* which is action and not confession is the *kufr* on account of which one does not go out of the *millah* [community] but a *kufr* which renders the actions worthless.[94]

The second basic error the Murji'ites ever committed, according to Ibn Taymiyyah, was the belief that if the Legislator (i.e., the Prophet) judges someone to be a *kāfir* who will abide forever in Hell, "then the reason would be the lack of knowledge and taṣdīq, *assent*, in his heart."[95] In this matter, Ibn Taymiyyah writes, the Murji'ites "contradicted common sense, reason, and law and all that people considered right through their natural disposition."[96] Ibn Taymiyyah challenges this belief by claiming that those who denied that God sent down the Messengers knew in their hearts that these Messengers were right but they disbelieved them out of their love for the religion they followed. Ibn Taymiyyah refers again to Iblīs who has *taṣdīq* in his heart but is "considered among the most blasphemous of all people who really knew" that he was wrong and that the Messengers were right.[97]

Ibn Taymiyyah refutes the views that *īmān* is *taṣdīq* by pointing out that if *īmān* is *taṣdīq* then *takdhīb* (reflection of the truth) is *kufr*, which is not the case.[98] He contends that the opposite of *taṣdīq* is *takdhīb*, while *takdhīb* is not the opposite of *īmān*. Ibn Taymiyyah explains his theory by saying that it is known in the language to say to someone who brings a report: "you are truthful or a liar (*ṣadaqta aw kadhabta*). We say: *ṣaddaqnāhu* (we regard him as truthful) rather than *āmannā lahu* (we have belief in him) for the opposite of *īmān* is *kufr*."[99] Ibn Taymiyyah writes that the opposite of *takdhīb* is not *kufr* and *kufr* does not necessarily involve *takdhīb*.[100] Ibn Taymiyyah adds that *kufr* can sometimes be *takdhīb* and it can also be *taṣdīq* with refutations of the truth, opposition, and antagonism. For Ibn Taymiyyah, a mere *taṣdīq* is not *īmān* and *taṣdīq* without submission is *kufr*.[101]

Ibn Taymiyyah declares many Islamic sects and groups as *kuffār* for deviating from Islamic teachings and beliefs. For instance, he declares the *kufr* of the Mu'tazilites for believing in the creation of the Qur'ān. He writes: "Whoever says that the Qur'ān is created is *kafir*."[102] He declares the Jahmites, a Mu'tazilite sub-sect, *kuffār* for denying God's divine attributes.[103] Ibn Taymiyyah believes that these attributes are necessary to God, and to deny what is necessary to God is equal to denying God Himself. Also, he refutes the Jahmite belief that God is everywhere and not in a particular place. He contends that this idea was never entertained by the Prophet or by any one of his Companions, and that no one denied the notion that God is separated from this world.[104] Thus, for Ibn Taymiyyah, the Mu'tazilites and the Jahmites are ignorant and unbelievers (*ḍāllūn wa kāfirūn*) for they separated themselves from the Islamic community (*jamā'āt al-muslimīn*).[105]

Other groups that earned Ibn Taymiyyah's *takfīr* are some Sufis, Muslim esoteric sects (*al-bāṭiniyyah*), and the philosophers.[106] Some Sufis are declared *kuffār* for having some beliefs that, for Ibn Taymiyyah, are considered a deviation from Islam.[107] He in particular directs his attacks against the Sufi mystic Abū 'Abd Allāh Muḥammad ibn 'Alī ibn Muḥammad ibn 'Arabī (d. 1240), also known as "the Greatest Master" or (*al-shaykh al-akbar*),[108] and declares him a *kāfir* for believing in the Unity of Being,[109] or *waḥdat al-wujūd*.[110] Ibn Taymiyyah condemns these Sufis for denying the doctrine of the Promise and the Threat (*al-wa'd wa al-wa'īd*) and for not distinguishing between the right that God has commended and the wrong that He has prohibited. Ibn Taymiyyah writes that Ibn 'Arabī and the Sufis who follow his teachings belong to the religion of the polytheists (*dīn al-mushrikīn*). Ibn Taymiyyah issued a judgment saying that Ibn 'Arabī committed a clear *kufr*,[111] putting "the Greatest Master amid the cohort of 'heretics' and 'grave sinners.'"[112]

Some contemporary scholars offered their criticisms of Ibn Taymiyyah's accusations of *kufr* against some Sufi groups in general and Ibn 'Arabī in particular. For instance, Muḥammad Ḥarbī, in a book entitled *Ibn Taymiyyah wa Mawqifihi min Ahammi al-Firaqi wa al-Diyānāti fī 'Aṣrihi*, says that Ibn Taymiyyah's accusations of plain *kufr* (*al-kufr al-ṣarḥ*) against Ibn 'Arabī can be justified in a time that witnessed the rise of groups that, for Ibn Taymiyyah, promoted doctrines that deviated from the path of Islam.[113]

Due to the popularity of the deviant schools, Ibn Taymiyyah became more aggressive in exposing their un-Islamic doctrines, writes Ḥarbī. However, despite their deviant doctrines, Ḥarbī adds, it was inappropriate for a figure like Ibn Taymiyyah to rush and declare Ibn 'Arabī a *kāfir*. Ḥarbī does not say that people like Ibn 'Arabī should be tolerated but he argues that to declare someone a *kāfir* based on a book he wrote may be unfair. Ḥarbī writes that Ibn Taymiyyah tolerated Ibn 'Arabī until he read his book *Fuṣūṣ al-Ḥikam*. Ḥarbī argues that Ibn Taymiyyah committed injustice against Ibn 'Arabī "by judging him based on a book that may have been written by someone else."[114]

Ibn Taymiyyah also condemns the Batinite sects as heretics and *kuffār*. Ibn Taymiyyah accuses them of being hypocrites for they declare that Islam can be known both esoterically (*bāṭinan*) and exoterically (*ẓāhiran*) and that only the Prophets and *imāms* are qualified to know the religion esoterically.[115] He believes that their doctrines and practices are contradictory to the Qur'ān and the *ḥadīths*. Ibn Taymiyyah declared them both *kuffār* and hypocrites because "they claim to be believers while hiding their unbelief."[116]

Ibn Taymiyyah's accusations of *kufr* were directed not only at "heretic" Muslims, but also were extended to Sunni, orthodox Muslims such as the Mongols. During the time of Ibn Taymiyyah, the Mongols converted to Islam and ruled their portion of the Islamic world with mixed legal system (i.e., *sharī'ah* and other imported laws). Ibn Taymiyyah—believing that the Mongol rulers threatened the purity of Islam with their mixed legal regime—issued a religious ruling, or *fatwā*, suggesting *jihād* against them.[117] When Muslim scholars voiced their objections to waging *jihād* against believing Muslims, Ibn Taymiyyah declares them *kuffār* "who must be fought on a more consistent basis than other obvious infidels."[118] Some scholars argue that Ibn Taymiyyah introduced a new formula for assessing and measuring the beliefs of believing Muslims. For instance, Ibn Taymiyyah's judgment of the Mongol's Islamicness "was set alongside those characteristic of the Mongol code of honor."[119] Here, Ibn Taymiyyah identified and defined the character of believing Muslims and the quality of their belief.[120] Traditionally, David Cook writes, making a judgment on the character of other Muslims who confessed the *shahādah* (the belief in God and His Messenger) "was a novel conclusion."[121]

Ibn Taymiyyah also attacked Muslim philosophers and accused them of committing acts of *kufr*.[122] The issues that Ibn Taymiyyah had with the philosophers are their belief that God did not will the creation of the world. In other words, the world emanated from Him without His Will. He also attacked them for believing that God knows only universals, but not particulars, for believing in the eternity of the world, and for believing that the philosophers are more capable of understanding the Truth than the Messengers.[123] Ibn Taymiyyah replies to the philosophers by accusing them of *kufr*. They have committed a greater *kufr*, he claims, because they know the truth about God, the world, and His Messengers but they insist on their beliefs.[124] According to Tawfīq al-Ṭawīl: "It is hard to ignore Ibn Taymiyyah's hostility to the philosophers and philosophy, and this hostility is obvious in his works where he prohibited the discussion of logic. His style of argument and criticism was very aggressive."[125]

Like many Islamic sects and groups, the Jews were not spared Ibn Taymiyyah's accusations of committing unbelief. The Jews, according to Ibn Taymiyyah, committed "stubborn kufr" (*kufr 'inād*) because they rejected Muhammad's prophecy although they had knowledge of it.[126] *Kufr 'inād*, according to Ibn Taymiyyah, is when the *kāfir* has knowledge of God but rejects what is important in the *īmān*.[127] Ibn Taymiyyah writes that God has called the Jews and the Christians to believe in the message of the Qur'ān and also to believe that Muḥammad is the Messenger of God (5:81; 4:65) but "they do not have in their hearts the true *īmān*."[128] Ibn Taymiyyah interprets verse 4:65—"they will not be true believers until they let you decide between them in all matters of dispute"—to mean that God intended for the Jews and the Christians to obey Muḥammad as a Messenger sent to them and failing to do so is an act of *kufr*. For Ibn Taymiyyah, believing in Muhammad's prophecy is an essential condition for true *īmān*.[129]

One would question why did Ibn Taymiyyah not take into account the Qur'ānic verse 2:62, which brands the Jews and Christians as believers, and why did he not make the effort to reconcile the apparent contradictions in these verses? Contemporary scholars of Ibn Taymiyyah, like Muḥammad Ḥarbī, have argued that the accusations of *takfīr* against Jews, Christians, and others emanated not from Ibn Taymiyyah's desires to attack these groups but out of his concerns for Islam's purity *vis-à-vis* the various deviant religious doctrines of his time.[130] "Ibn Taymiyya's ideology was shaped and nurtured not only by his teachers and the doctrines he studied, by also by the political situation of his time, which was characterized by turmoil, chaos, and disorder."[131] Abdul Hakim I. al-Matroudi tells us that Ibn Taymiyyah "found himself assuming the role of political reformer working in several spheres."[132] Thus, the political situation of his time and the rise of various religious doctrines may have contributed to the development of *takfīr* in Ibn Taymiyyah's thought.

Both al-Ghazālī and Ibn Taymiyyah are considered two of the most prominent figures in Islam who contributed tremendously to Islamic theology and jurisprudence. Their contribution, however, was not limited to offering theories on who is a believer and who is not. Rather, they took the concept of *takfīr* to a new level and applied it to sects and groups of their times. Classical theologians, in contrast, discussed *takfīr* as a concept and a theory and were very careful not to cross the boundaries to the practical application of the theory. Their reluctance to do so may be attributed to two reasons: 1) they realized the danger resulting from accusations of *takfīr*, and 2) in their time, Islam was at its zenith in intellectual and cultural productivity. Thus, they did not see threats from doctrines deviant to Islam. That changed later during the times of al-Ghazālī and Ibn Taymiyyah who saw in *takfīr* a means for protecting Islam against deviations in a time of intellectual and cultural decline. This trend that started with al-Ghazālī and Ibn Taymiyyah continued in later centuries all the way to the twenty-first century, where *takfīr* has become the norm. The question remains: To what extent is the rise of *takfīr*-ist groups in modern time related to the continued decline of the Islamic civilization?

NOTES

1. Yazeed Said, *Ghazali's Politics in Context* (London and New York: Routledge, 2013), 10.
2. Abū Ḥāmid Muḥammad b. Muḥammad al-Ghazālī, *The Incoherence of the Philosophers*, trans. Michael E. Marmura (Provo, Utah: Brigham Young University Press, 2000), xvi.
3. 'Abd al-Amīr al-A'sam, *al-Faylasūf al-Ghazālī*, 2nd ed. (Beirut, Lebanon: Dār al-Andalus, 1981), 30–47.
4. Abū Ḥāmid Muḥammad b. Muḥammad al-Ghazālī, *al-Munqidh min al-Ḍalāl*, eds. Jamīl Ṣalībā and Kāmil 'Ayyād (Beirut, Lebanon: Dār al-Andalus, 1981), 12.
5. Yūsuf al-Qaradāwī, *al-Imām al-Ghazālī* (Beirut, Lebanon: Mū'asasat al-Risālah, 1994), ii.
6. Abū Ḥāmid Muḥammad b. Muḥammad al-Ghazālī, *Fayṣal al-Tafriqah Bayna al-Islām wa al-Zandaqah* (n.c.: Dār Iḥyā' al-Kutub al-'Arabiyyah, 1961), *Kitāb al-Iqtiṣād fī al-I'tiqād* (Beirut, Lebanon: Dār al-Kutub al-'Ilmiyyah, 1988), and *al-Munqidh min al-Ḍalāl*.
7. Al-Ghazālī, *Fayṣal*, 1961, 134.
8. Ibid.
9. Ibid.
10. Quoted in Sherman A. Jackson, *On the Boundaries of Theological Tolerance in Islam* (Oxford: Oxford University Press, 2002), 55–59.
11. Ibid.
12. Al-Ghazālī, *Fayṣal*, 1961, 174.
13. Ibid., 184.
14. Ibid.
15. Ibid., 190–191.
16. Ibid.
17. Ibid., 192.
18. Ibid.

19. Ibid.
20. Ibid., 192–193.
21. Ibid.
22. Al-Ghazālī, *The Incoherence*, 226.
23. Al-Ghazālī, *Fayṣal*, 1961, 193.
24. Rifqī Zāhir, *Qaḍiyat al-Takfīr 'inda al-Ghazālī* (Cairo, Egypt: Tawzī' Maktabat al-Azhar, 1977), 64.
25. Ibid.
26. Ibid., 64–65.
27. Al-Ghazālī, *Fayṣal*, 1961, 195.
28. Ibid.
29. Ibid.
30. Ibid., 196.
31. Ibid.
32. Ibid.
33. Ibid., 196–197.
34. Ibid.
35. Ibid., 197.
36. Ibid.
37. Ibid.
38. Ibid.
39. Ibid.
40. For the Batinite doctrines and beliefs, see 'Abd al-Qāhir al-Baghdādī, *al-Farq Bayna al-Firaq*, 213–237.
41. Al-Ghazālī, *Fayṣal*, 1961, 198.
42. Ibid., 199–201.
43. Farouk Mitha, *Al-Ghazali and the Ismailis: A Debate on Reason and Authority in Medieval Islam* (London and New York: I.B. Tauris, 2001), 65–66.
44. Ibid., 69.
45. Ibid.
46. Ibid., 67.
47. Ibid., 67–68.
48. F. Griffel, *al-Ghazali's Philosophical Theology* (Oxford: Oxford University Press, 2009), 104.
49. Al-Ghazālī, *Fayṣal*, 1961, 202.
50. Ibid.
51. Ibid.
52. Ibid., 204.
53. Abū Ḥāmid Muḥammad b. Muḥammad al-Ghazālī, *Fayṣal al-Tafriqah Bayna al-Islām wa al-Zandaqah*, ed. Maḥmūd Bījū (n.c.: n.p., 1993), 84.
54. Jackson, *On the Boundaries*, 5.
55. Ibid.
56. Ibid., 40.
57. Al-Shahrastānī ranks al-'Āmirī as one of the major Muslim philosophers alongside Abū-'Alī al-Ḥusayn ibn-'Abd Allāh ibn sīnā (d. 1037) and Abū Naṣr al-Fārābī (d. 951). His *Kitāb al-I'lām bi Manāqib al-Islām* satisfied al-Ghazālī's "decisive criterion for Islam"; see Jackson, *On the Boundaries*, 59.
58. Jackson, *On the Boundaries*, 58.
59. Ibid., 5.
60. Ibid., 7.
61. Ibid.
62. Al-Ghazālī, *Kitāb al-Iqtiṣād*, 155–156.
63. See Muḥammad 'Abdul-Ḥaqq Anṣārī, trans., *Ibn Taymiyyah Expounds on Islam: Selected Writings of Shaykh al-Islam Taqi ad-Din Ibn Taymiyyah on Islamic Faith, Life, and Society* (Riyadh, Saudi Arabia: The General Administration of Culture and Publication, 2000), xxv; Abdul Hakim I. Al-Matroudi, *The Ḥanbali School of Law and Ibn Taymiyyah: Conflict or*

Conciliation (London: Routledge, 2006), 16; Hussam S. Timani, "The Islamic Context of Global *Jihadism*: Why Ibn Taymiyya (d. 1328) Matters," in Lori J. Underwood, ed., *The Root of All Evil? Religious Perspectives on Terrorism* (New York: Peter Lang Publishing, 2013), 45–46.

64. Ibn Taymiyyah, *Kitāb al-Īmān*, *Book of Faith*, trans. Salman Hassan al-Ani and Shadia Ahmad Tel (Bloomington, Indiana: Iman Publishing House, 1999), 56–57.
65. Ibid., 60.
66. Ibid.
67. Ibid.
68. Ibid., 61.
69. Ibid.
70. Ibid., 66.
71. Ibn Taymiyyah, *Majmū' Fatāwā*, vol. 7 (Al-Madinah, Saudia Arabia: Mujamma' al-Malik Fahd, 2004), 171.
72. Ibn Taymiyyah, *Kitāb al-Īmān*, 72.
73. Ibid., 74.
74. Ibn Taymiyyah, *Majmū'*, vol. 7, 59.
75. Ibid., 61.
76. Ibid., 217.
77. Ibid.
78. Ibid., 141.
79. Ibid.
80. Ibid.
81. Ibid.
82. Quoted in Muḥammad Ḥarbī, *Ibn Taymiyyah wa Mawqifihi min Ahammi al-Firaqi wa al-Diyānāti fī 'Aṣrihi* (Beirut, Lebanon: 'Ālam al-Kutub, 1987), 132–133.
83. Ibid.
84. Quoted in ibid.
85. Quoted in ibid., 143.
86. Ibn Taymiyyah, *K. al-Īmān*, 137.
87. Ibid.
88. Ibid., 138.
89. Ibid.
90. Ibn Taymiyyah, *Majmū'*, vol. 7, 131.
91. Ibn Taymiyyah, *K. al-Īmān*, 138.
92. Ibid., 200.
93. Ibid.
94. Anṣārī, *Ibn Taymiyyah*, 323.
95. Ibn Taymiyyah, *K. al-Īmān*, 137.
96. Ibid.
97. Ibid., 201.
98. Ibn Taymiyyah, *Majmū'*, vol. 7, 292.
99. Ibid.
100. Ibid.
101. Ibid.
102. Ibid., 619.
103. Ibid.
104. Quoted in Ḥarbī, *Ibn Taymiyyah*, 65–67, 138.
105. Quoted in ibid.
106. Ibn Taymiyyah, *Majmū'*, vol. 7, 586–597.
107. Ibid.
108. Syafaatun Almirzanah, *When Mystic Masters Meet: Towards a New Matrix for Christian-Muslim Dialogue* (New York: Blue Dome Press, 2011), 8.
109. Ibn Taymiyyah, *Majmū'*, vol. 7, 593–595.

110. The concept of the Unity of Being is taken to mean "that there is only one Being, and that all existence is nothing but the manifestation of that One Being . . . nothing can exist outside of the One" (Almirzanah, *When Mystic Masters*, 168–171).
111. Quoted in Ḥarbī, *Ibn Taymiyyah*, 194.
112. Alexander D. Knysh, *Ibn 'Arabi in the Later Islamic Tradition: The Making of a Polemical Image in Medieval Islam* (Albany: State University of New York Press, 1999), 89.
113. Quoted in Ḥarbī, *Ibn Taymiyyah*, 194
114. Ibid.
115. Ḥarbī, *Ibn Taymiyyah*, 236.
116. Ibid., 268.
117. Timani, "The Islamic Context of Global Jihadism," 39.
118. Ibid.
119. John Kelsay, *Arguing the Just War in Islam* (Cambridge, Mass.: Harvard University Press, 2007), 122.
120. David Cook, *Understanding Jihad* (Berkeley: University of California Press, 2005), 65.
121. Ibid.
122. Ibn Taymiyyah, *Majmū'*, vol. 7, 585–597.
123. Ibid.
124. Ibid.
125. Tawfīq al-Ṭawīl, *Qiṣṣat al-Nizā' Bayna al-Dīn wa al-Falsafah* (n.c.: Maṭba'at al-I'timād, n.d.), 123–124.
126. Ibn Taymiyyah, *Majmū'*, vol. 7, 146.
127. Ibid.
128. Ibid., 150.
129. Ibid.
130. Ḥarbī, *Ibn Taymiyyah*, 311.
131. Timani, "The Islamic Context," 46.
132. Al-Matroudi, *The Ḥanbali School of Law,* 17.

Conclusion

This book has explored the meaning, origin, and types of *takfīr* in Islam and demonstrated that although the term is mentioned in the revealed text, neither the Qur'ān nor Muḥammad used the term to indicate faithlessness. The term was used in early Islam, and in particular during the time of the Prophet Muḥammad, to describe those who committed acts that some people believed were deviations from the path of Islam. As chapter 1 has demonstrated, the Qur'ānic concept of *takfīr* was applied to certain acts only to point out the danger posed by these actions not to the religion but to the individual and the community. Thus, in early Islam, the concept was never used to mean unfaithfulness.

In early and later Islamic theology, *takfīr* gained momentum as a field of study. Theologians became entangled in extensive debates on *takfīr* not because they were interested in defining the concept but because they were concerned about defining the meaning of *īmān*: What constitutes *īmān*, and how is it measured? Through their debates on the meaning of *īmān*, the concept of *takfīr* had to be defined. With the theologians being engaged in this theological discourse, the meaning of *takfīr* was defined based on the school's understanding of *īmān*. Early and later Muslim theologians looked at *takfīr* as a theory worth studying and exploring and paid little attention to the practical application of it. This would explain the reasons for the various interpretations that these schools of theology provided starting with the Kharijites in the seventh century to the Maturidites centuries later. This book has attempted to demonstrate how *takfīr* has evolved throughout the centuries from being a term used to condemn (and warn against) unacceptable actions to a term used on individuals and communities to strip them of their belief, dignity, and linkage to the Divine (i.e., declaring them unbelievers).

Theologians have long debated the acts that are grounds for *takfīr*. Some have argued that either inward or outward belief or *īmān* should be enough to make the individual a believer; some have insisted on outward demonstrations as observing the prayers, paying alms-tax, and fasting, as an indication of belief. Others were satisfied with inner belief only, which means refraining from making a judgment against the person, believing that God will be the final judge because only God knows what is in the heart of the individual. These differing views on what makes a person a *kāfir* are but an indication that *takfīr* was never a key theological issue in early Islam, specifically at the time of Muḥammad and the Rightly Guided Caliphs (632–661). As this book has attempted to demonstrate, neither the Qur'ān nor the Prophet Muḥammad used *takfīr* as a means of declaring individuals unbelievers. Historical, political, social, and religious circumstances subsequent to the first two centuries of Islam contributed to the development of *takfīr* because the emerging theological schools and religio-political movements of the time utilized the precept against others to affirm the legitimacy of their own religious and political thoughts.

The development of *takfīr* in classical and medieval Islamic thought can be divided into three historical eras: 1) the formative era; 2) the classical era; and 3) the medieval era. In the formative era (i.e., the era of the coming down of the Revelation and the preaching of the Prophet Muḥammad), the term *takfīr* appeared in Qur'ānic verses and in *ḥadīth* reports as a method only to distinguish those who accepted God and His Messenger from those who did not. In Qur'ānic and *ḥadīth* contexts, terms like *takfīr*, *kufr*, and *kuffār* were used to describe behavior, identify ideologies, and distinguish two sets of communities (believers vs. unbelievers). During the time of Muḥammad and after his death, Muslims who committed acts of disobedience were not branded as *kuffār*. For example, when the first caliph Abū Bakr (r. 632–634) assumed the leadership of the Muslim community after the Prophet, a group of Muslims refused to pay the prescribed almsgiving (*zakāt*). Abū Bakr and everyone else identified them at that time as *ahl al-riddah* (the people who turned away). The term "*riddah*" was applied to them but not "*kuffār*." They went in history as *ahl al-riddah* not as *al-kuffār*. One may conclude that *kufr* was a term applied to those who have not openly accepted Islam, that is, members of the Muslim community. This societal attitude to the concept of *takfīr* remained unchanged for decades until the rise of the Kharijites in 657–658. With the rise of the Kharijites, *takfīr* took a new meaning and became a term associated with belief (*īmān*) rather than membership in the community.

In the classical era, the theological schools subsequent to the Kharijites defined membership in the community based on the *īmān* of the individual. Classical theologians tried to define *īmān* as a way to protect Islam from deviations. The rise of the Kharijite ideology in the seventh century, especially their novel idea of declaring a Muslim ruler infidel (*kāfir*), not only challenged the theological status quo, but it was also seen as a deviation from the Qur'ān and the *sunnah*. Thus, this deviant Kharijite thought on *takfīr* prompted religious thinkers and jurists to respond to the Kharijites with their own definition of *īmān*. In this era, the theologians' debate on the concept of *kufr* was to defend *īmān* and Islam against deviations. Yet, their dealings with *kufr* were confined to the theoretical level only: what makes a person a *kāfir*, and what makes him a believer.

In the third era (medieval Islam), the concept of *takfīr* was transformed from being a topic to be debated and studied to a tool used by theologians and jurists not only to defend their own theological, religious, and political views but also to silence theological and religious sects and groups that were considered deviant to Islam. This can be seen in al-Ghazālī's and Ibn Taymiyyah's attacks against Muslim philosophers, the Mongols, Sufis, Shī'ah sects, and others. The question remains: To what extent have al-Ghazālī's and Ibn Taymiyyah's theological discourses on *takfīr* impacted the thoughts of modern and contemporary Muslim groups and set a precedent for them to use *takfīr* to recruit, make political gains, and eliminate rivalries? For example, the second half of the twentieth century witnessed the rise of *takfīr*-ist Muslim groups, like the Egyptian *Jamā'at al-Takfīr wa al-Hijrah*, that declared Egyptian society a *mujtama' kāfir* (unbelieving society) and made the *hijrah* (emigration) away from those declared infidels. As more groups have adopted this ideology, most notably al-Qā'ida, the Islamic State of Iraq and the Levant (ISIL), Boko Haram, and other splinter groups, further research is needed to explore how the theological discourse on *takfīr* in medieval Islam has made an impact on twentieth- and twenty-first-century groups.

Bibliography

'Abd al-Bāqī, Muḥammad Fu'ād. *Al-Mu'jam al-Mufahras li Alfāẓ al-Qur'ān al-Karīm*. n.c.: Dār al-Fikr, n.d.

'Abd al-Jabbār, Qādī. *Sharḥ al-Uṣūl al-Khamsa*. Beirut, Lebanon: Dār Iḥyā' al-Turāth al-'Arabī, 2001.

Abdel Haleem, M. A. S. *The Qur'an*. Oxford: Oxford University Press, 2005.

'Alam al-Dīn, Sulaymān S. *Al-Mu'tazilah*. Beirut, Lebanon: Nawfal, 2000.

Al-Albānī, Muḥammad Nāsir al-Dīn. *Fitnat al-Takfīr*. n.c.: Dār ibn Khazīmah, 1997.

'Alī, 'Abdullah Yūsuf. *The Meaning of the Holy Qur'ān*, 11th ed. Beltsville, Md.: Amana Publications, 2008.

Almirzanah, Syafaatun. *When Mystic Masters Meet: Towards a New Matrix for Christian Muslim Dialogue*. New York: Blue Dome Press, 2011.

Al-'Āmilī, Mālik Muṣṭafā Wahbī. *Ẓāhirat al-Takfīr fī al-Fikr al-Islāmī*. Beirut, Lebanon: Dār al-Hādī, 2007.

Anṣārī, Muḥammad 'Abdul-Ḥaqq, trans. *Ibn Taymiyyah Expounds on Islam: Selected Writings of Shaykh al-Islam Taqi ad-Din Ibn Taymiyyah on Islamic Faith, Life, and Society*. Riyadh, Saudi Arabia: The General Administration of Culture and Publication, 2000.

Al-A'sam, 'Abd al-Amīr. *Al-Faylasūf al-Ghazālī*, 2nd ed. Beirut, Lebanon: Dār al- Andalus, 1981.

Al-Ash'arī, Abū al-Ḥasan. *Maqālāt al-Islāmiyyīn*, ed. Muḥammad Muḥyī al-Dīn 'Abd al-Ḥamīd. Beirut, Lebanon: al-Maktabah al-'Aṣriyyah, 1990.

———. *Al-Ibānah 'an Uṣūl ad-Diyānah*, trans. Walter C. Klein. New Haven, Conn.: American Oriental Society, 1940.

———. *Kitāb al-Luma'*, ed. Richard J. McCarthy, S. J. Beirut, Lebanon: Al-Maṭba'ah al-Kāthūlīkiyyah, 1953.

———. *Al-Luma' fī al-Rad 'ala Ahl al-Zaygh wa al-Bida'*, ed. Ḥammūdah Ghurābah. Cairo, Egypt: Maṭba'at Miṣr, 1955.

Aydin, Mahmut. "Islam in a World of Diverse Faiths: A Muslim View." *Islam and Inter-Faith Relations*, eds. Perry Schmidt-Leukel and Lloyd Ridgeon. London: SCM Press, 2007.

Badawī, Abd al-Raḥmān. *Madhāhib al-Islāmiyyīn*. Beirut, Lebanon: Dār al-'Ilm lil Malāyīn, 1997.

Al-Badr, 'Abd al-Razzāq b. 'Abd al-Muḥsin. *Ziyādat al-Īmān wa Nuqṣānuh*. Riyadh, Saudi Arabia: Dār Kunūz Ishbīlyah, 2006.

Al-Baghawī, Abū Muḥammad Ḥusayn b. Mas'ūd. *Tafsīr al-Baghawī*, vol. 4, ed. Muḥammad 'Abd Allāh al-Nimr, et al. Riyadh, Saudi Arabia: Dār Ṭaybah, 1989.

Al-Baghdādī, 'Abd al-Qāhir. *Al-Farq Bayna al-Firaq*. Beirut: Dār al-Kutub al-'Ilmiyyah, n.d.

———. *Firaq al-Shī'ah*. Istanbul: n.p., 1951.

———. *Kitāb Uṣūl al-Dīn*. Istanbul, Turkey: Maṭba'at al-Dawlah, 1928.
———. *Al-Farq Bayna al-Firaq*. Cairo: n.p., 1910.
Al-Bakkāy, Laṭīfah. *Ḥarakāt al-Khawārij*. Beirut, Lebanon: Dār al-Ṭalī'ah, 2001.
Al-Balādhurī, Aḥmad b. Yaḥyā. *Ansāb al-Ashrāf*, vol. 5, eds. Suhayl Zakkār and Riyad Zarkalī. N.c.: Dār al-Fikr, 1997.
Block, C. Jonn. "Philoponian Monophysitism in South Arabia at the Advent of Islam with Implications for the English Translation of '*Thalatha*' in Qur'ān 4. 171 and 5. 73." *Journal of Islamic Studies* 23:1 (2012).
Al-Bustī, Ismā'īl b. Aḥmad. *Kitāb al-Baḥth 'an Adillat al-Takfīr wa al-Tafsīq*. Köln., Germany: Al-Kamel Verlag, 2009.
Cook, David. *Understanding Jihad*. Berkeley, Calif.: University of California Press, 2005.
Encyclopaedia Britannica. http://www.britannica.com/EBchecked/topic/256468/al-Hasan-al-Basri.
Al-Fayrūz Abādī, Majd al-Dīn Muḥammad b. Ya'qūb. *Al-Qāmūs al-Muḥīṭ*, ed. Muḥammad Na'īm al-'Arqasūsī, 8th ed. Beirut, Lebanon: Mū'asasat al-Risālah, 2005.
Fiqh-us-Sunnah. www.hadithcollection.com.
Al-Ghazālī, Abū Ḥāmid Muḥammad. *Qawā'id al-'Aqā'id*, ed. Mūsa Muḥammad 'Alī. Beirut, Lebanon: 'Ālam al-Kutub, 1985.
———. *Kitāb al-Iqtiṣād fī al-I'tiqād*. Beirut, Lebanon: Dār al-Kutub al-'Ilmiyyah, 1988.
———. *Fayṣal al-Tafriqah Bayna al-Islām wa al-Zandaqah*. N.c.: Dār Ihyā' al-Kutub al-'Arabiyyah, 1961.
———. *Fayṣal al-Tafriqah Bayna al-Islām wa al-Zandaqah*, ed. Maḥmūd Bījū. N.c.: n.p., 1993.
———. *Al-Munqidh min al-Ḍalāl*, eds. Jamīl Ṣalībā and Kāmil 'Ayyād. Beirut, Lebanon: Dār al-Andalus, 1981.
———. *The Incoherence of the Philosophers*, trans. Michael E. Marmura. Provo, Utah: Brigham Young University Press, 2000.
Givony, Joseph. "The Murji'a and the Theological School of Abu Hanifa: A Historical and Ideological Study." Ph.D. dissertation, University of Edinburgh, 1977.
Goodman, Lenn E. *Islamic Humanism*. Oxford: Oxford University Press, 2003.
Griffel, F. *Al-Ghazali's Philosophical Theology*. Oxford: Oxford University Press, 2009.
Ḥalīmah, 'Abd al-Mun'im Muṣtafā. *Qawā'id fī al-Takfīr*. Amman, Jordan, Dār al-Bashīr, 1994.
Ḥarbī, Muḥammad. *Ibn Taymiyyah wa Mawqifihi min Ahammi al-Firaqi wa al-Diyānāti fī 'Aṣrihi*. Beirut, Lebanon: 'Ālam al-Kutub, 1987.
Heim, S. *The English-Persian Dictionary*. Tehran, Iran: n.p., 1961.
Al-Ḥumṣī, Muḥammad Ḥasan, ed. *Al-Qur'ān al-Karīm*. Beirut, Lebanon: Dār al-Rashīd, n.d.
Ibn 'Abd al-Wahhāb, Muḥammad. *Al-Rasā'il al-Shakhsiyyah*. https://d1.islamhouse.com/data/ar/ih_books/single2/ar_Personal_Message.pdf.
———. *Majmū'at al-Tawḥīd*. N.c.: Dār Ihyā' al-Turāth, n.d.
Ibn al-Ḥajjāj, Muslim, *Ṣaḥīḥ Muslim*, vol. 1, trans. Nasiruddin al-Khattab. Riyadh, Saudi Arabia: Darussalam, 2007.
Ibn 'Alī Ya'Lā, Abī al-Ḥusayn Muḥammad. *Tabaqāt al-Ḥanābilah*, vol. 1, ed. Muḥammad Ḥāmid al-Faqī. Cairo: Maṭba'at al-Sunnah al-Muḥammadiyyah, n.d.
Ibn Bāz, 'Abd al-'Azīz b. 'Abd Allāh. *Majmū' Fatāwā*, vol. 9. Riyadh, Saudi Arabia: Al-Buḥūth al-'Ilmiyyah lil Iftā', 1999.
Ibn Ḥazm, Abū Muḥammad 'Alī ibn Aḥmad ibn Sa'īd. *Al-Iḥkām fī Uṣūl al-Aḥkām*. Beirut, Lebanon: Dār al-Āfāq al-Jadidah, 1983.
———. *Al-Fiṣal fī al-Milal wa al-Ahwā' wa al-Niḥal*. Beirut, Lebanon: Dār al-Ṣādir, n.d.
———. *Al-Fiṣal fī al-Milal wa al-Ahwā' wa al-Niḥal*. Cairo, Egypt: Al-Maṭba'ah al-Adabiyyah, 1902.
———. *Al-Fiṣal fī al-Milal wa al-Ahwā' wa al-Niḥal*. N.c.: Maktabat al-Salām al-'Ilmiyyah, 1930.
Ibn Kathīr, Ismā'īl. *Tafsīr al-Qur'ān*, ed. Sāmī b. Muḥammad al-Salāmah, 2nd ed. Riyadh, Saudi Arabia: Dār Ṭaybah, 1999.
Ibn Manẓūr, Jamāl al-Dīn. *Lisān al-'Arab*, v. 5. Cairo, Egypt: Dār al-Ma'ārif, n.d.

Ibn Sallām, Abū 'Ubayd al-Qāsim. *Kitāb al-Īmān*, ed. Muḥammad Nāṣir al-Dīn al-Albānī. Beirut, Lebanon: Al-Maktab al-Islāmī, 1983.

Ibn Taymiyyah, Taqī al-Dīn Aḥmad. *Majmū' Fatāwā*. Al-Madina, Saudi Arabia: Mujamma' al-Malik Fahd, 2004.

———. *Kitāb al-Īmān, Book of Faith*, trans. Salman Ḥassan al-Ani and Shadia Ahmad Tel. Bloomington, Ind.: Iman Publishing House, 1999.

Al-Iṣfahānī, Abū Nu'aym. *Ḥilyat al-Awliyā' wa Ṭabaqāt al-Aṣfiyā'*. Beirut, Lebanon: Dār al-Fikr, 1996.

Al-Iyād, Al-Qādī. *Kitāb al-Shifā bi Ta'rīf Ḥuqūq al-Muṣṭafā*, trans. Aisha Abdarrahman Bewley. Islamicsystem.blogspot.com/2012/09/imam-al-qadi-iyad-judgements-concerning.html.

Izutsu, Toshihiko. *The Concept of Belief in Islamic Theology*. New York: Books for Libraries, 1980.

Jackson, Sherman A. *On the Boundaries of Theological Tolerance in Islam*. Oxford: Oxford University Press, 2002.

Al-Jabbār, Qādī 'Abd. *Sharḥ al-Uṣūl al-Khamsah*. Beirut, Lebanon: Dār Iḥyā' al-Turāth al-'Arabī, 2001.

Kāmil, 'Umar 'Abd Allāh. *Al-Khawārij al-Judud: Al-Mutaṭariffūn*. Cairo, Egypt: Maktabat al-Turāth al-Islamī, 1998.

Kelsay, John. *Arguing the Just War in Islam*. Cambridge, Mass.: Harvard University Press, 2007.

Khalīf, Fatḥallah, ed. *Kitāb al-Bidāyah min al-Kifāyah*. N.c.: n.p., n.d.

Khatab, Sayed. *The Political Thought of Sayyid Qutb: The Theory of Jahiliyyah*. London and New York: Routledge, 2006.

Al-Khatīb, Ḍuḥā. *Ḍawābit al-Takfīr*. Beirut, Lebanon: Dār al-Burāq, 2007.

Knysh, Alexander D. *Ibn 'Arabi in the Later Islamic Tradition: The Making of a Polemical Image in Medieval Islam*. Albany: State University of New York Press, 1999.

Al-Kulaynī, Muḥammad Ibn Ya'qūb. *Al-Kāfī*. Beirut, Lebanon: Manshūrāt al-Fajr, 2007.

Lav, Daniel. *Radical Islam and the Revival of Medieval Theology*. Cambridge: Cambridge University Press, 2012.

Madelung, Wilferd. *Religious Schools and Sects in Medieval Islam*. London: Variorum Reprints, 1985.

———. "Early Sunni Doctrine Concerning Faith as Reflected in the 'Kitāb al-Īmān' of Abū 'Ubayd al-Qāsim b. Sallām (d. 224/839)." *Studia Islamica*, No. 32 (1970).

Al-Malaṭī, Muḥammad b. Aḥmad. *Al-Tanbīh wa-al-Radd 'alā Ahl al-Ahwā' wa al-Bida'*, ed. Muḥammad Zāhid b. al-Ḥasan al-Kawtharī. Beirut, Lebanon: Maktabat al-Ma'ārif, 1968.

Maliks Muwatta. www.hadithcollection.com.

Al-Matroudi, Abdul Hakim I. *The Ḥanbalī School of Law and Ibn Taymiyyah: Conflict or Conciliation*. London: Routledge, 2006.

Al-Māturīdī, Muḥammad Abū Manṣūr. *Sharḥ al-Fiqh al-Akbar*, ed. 'Abd Allāh b. Ibrāhim al-Anṣārī. Ḥaydar Ābād al-Dakan, India: Dā'irat al-Ma'ārif al-Niẓāmiyyah, 1903.

Mitha, Farouk. *Al-Ghazali and the Ismailis: A Debate on Reason and Authority in Medieval Islam*. London and New York: I.B. Tauris, 2001.

Al-Murtaḍā, al-Sharīf. *Āmālī al-Murtaḍā*. Cairo: Dār Iḥyā' al-Kutub al-'Arabiyyah, 1954.

Al-Najjār, 'Āmir. *Al-Khawārij: 'Aqīdatan wa Fikran wa Falsafatan*. Beirut, Lebanon: 'Ālam al-Kutub, 1986.

Al-Nawawī, Abū Zakariyyah Yaḥyā ibn Sharaf. *Sharḥ Ṣaḥīḥ Muslim*. Cairo: Al-Maṭba'a al-Maṣriyyah, 1929.

Al-Nu'mān, Abū Ḥanīfah. *Al-Sharḥ al-Muyassar*, ed. Muḥammad b. 'Abd al-Raḥmān al-Khamīs. Ajman, UAE: Maktabat al-Furqān, n.d.

———. *Al-'Ālim wa al-Muta'allim*, ed. Muḥammad Zāhid al-Kawtharī. Cairo: Matba'at al-Anwār, 1949.

———. *Al-'Ālim wa al-Muta'allim*. Cairo, Egypt: Matba'at al-Anwār, 1989.

———. *Al-Fiqh al-Akbar*. N.c.: United Arab Emirates, Maktabat al-Furqān, 1999.

Pessagno, J. Meric. "The Murji'a, Iman and Abu 'Ubayd." *Journal of the American Oriental Society*, vol. 95, no. 3 (Jul.-Sep., 1975).

Al-Qaradāwī, Yūsuf. *Al-Imām al-Ghazālī*. Beirut, Lebanon: Mū'asasat al-Risālah, 1994.

Rahman, Fazlur. *Islam,* 2nd ed. Chicago: The University of Chicago Press, 1979.
Al-Rāzī, Fakhr al-Dīn. *Tafsīr al-Rāzī*. Beirut, Lebanon: Dār al-Fikr, 1981.
Riḍā, Aḥmad. *Matn al-Lugha*. Beirut, Lebanon: Dār al-Ḥayāt, 1959.
———. *Mu'jam Matn al-Lugha*, vol. 5. Beirut, Lebanon: Dār al-Ḥayāt, 1960.
———. *Mu'jam Matn al-Lugha*, vol. 2. Beirut, Lebanon: Al-Ḥayāt, 1985.
Rudolph, Ulrich. *Al-Maturidi and the Development of Sunni Theology in Samarqand*. Leiden and Boston: Brill, 2015.
Al-Sābi'ī, Nāsir b. Sulaymān b. Sa'īd. *Al-Khawārij wa al-Ḥaqīqah al-Ghā'ibah*, 2nd ed. Muscat, Oman: Maktabat al-Jīl al-Wa'īd, 2003.
———. *Al-Khawārij wa al-Ḥaqīqah al-Ghā'ibah*. Beirut, Lebanon: Dār al-Muntaẓar, 2000.
Ṣaḥīḥ Bukhārī. www.hadithcollection.com.
Ṣaḥīḥ Muslim. www.hadithcollection.com.
Al-Sāhimī, 'Abd al-Salām b. Salīm. *Fikr al-Takfīr Qadīman wa Ḥadīthan*. Cairo, Egypt: Dār al-Imām Aḥmad, 2005.
Said, Yazeed. *Ghazali's Politics in Context*. London and New York: Routledge, 2013.
Salhab, Fāyiz. *Al-Takfīr wa al-Takfīr al-Muḍād: Dirāsah*. Damascus, Syria: Dār al-Farqad, 2005.
Al-Sāmarrā'ī, Nu'mān 'Abd al-Razzāq. *Al-Takfīr fī al-Qur'ān wa al-Sunna Qadīman wa Ḥadīthan*. Riyadh, Saudi Arabia: Markaz al-Malik Fayṣal lil Buḥūth wa al-Dirāsāt al-Islāmiyyah, 2007.
Schacht, Joseph. "An Early Murci'ite Treatise: The Kitāb al-'Ālim wa'l-Muta'allim." *Oriens* Vol. 17 (Dec. 31, 1964).
Al-Shābandar, Ghālib Ḥasan. *Lā Takfīr fī al-Qur'ān*. Beirut, Lebanon: Ḍār al-Hudā, 2009.
Al-Shāfi'ī, Muḥammad ibn Idrīs. *Kitāb al-Umm*, vol. 1, ed. Rif'at Fawzī 'Abd al-Muṭalib. Cairo, Egypt: Dār al-Wafā', 2001.
Al-Shahrastānī, Tāj al-Dīn. *Al-Milal wa al-Niḥal*, eds. 'Āmir 'Alī Muhannā and 'Alī Ḥasan Fā'ūr. Beirut, Lebanon: Dār al-Ma'rifah, 1996.
———. *Nihāyat al-Aqdām fī 'Ilm al-Kalām*, ed. Alfred Guillaume. Cairo: Maktabat al-Thaqāfah al-Dīniyyah, 2009.
———. *Kitāb al-Milal wa al-Niḥal*. London: The Society for the Publication of Oriental Texts, n.d.
Sunnahonline. sunnahonline.com/library/beliefs-and-methodology/87-types-of-kufr-disbelief.
Al-Ṭabarī, Abū Ja'far Muḥammad ibn Jarīr. *Jāmi' al-Bayān 'an Ta'wīl āy al-Qur'ān*, eds. Bashār 'Awaḍ Ma'rūf and 'Iṣām Fāris al-Ḥirstānī. Beirut, Lebanon: Mu'asasat al-Risālah, 1994.
———. *Tafsīr al-Qur'ān*, 3rd ed. Beirut, Lebanon: Dār al-Ma'rifah, 1978.
———. *Ta'rīkh al-Umam wa al-Mulūk*, vol. 3. Beirut, Lebanon: Dār al-Kutub al-'Ilmiyyah, 1997.
Al-Taftazānī, Sa'd al-Dīn Masūd ibn 'Umar ibn 'Abd Allāh. *Sharḥ al-'Aqā'id al-Nasafiyyah*, ed. Claude Salamah. Damascus, Syria: Manshūrāt Wizārat al-Thaqāfah wa al-Irshād al-Qawmī, 1974.
Al-Ṭawīl, Tawfīq. *Qiṣṣat al-Nizā' Bayna al-Dīn wa al-Falsafah*. N.c.: Maṭba'at al-I'timād, n.d.
Timani, Hussam S. *Modern Intellectual Readings of the Kharijites*. New York: Peter Lang Publishing, 2008.
———. "Death and Dying in Islam." In *Ultimate Journey: Death and Dying in the World's Major Religions*, ed. Steven J. Rosen. Westport, Conn.: Praeger, 2008.
———. "The Islamic Context of Global *Jihadism*: Why Ibn Taymiyya (d. 1328) Matters." In Lori J. Underwood, ed., *The Root of All Evil? Religious Perspectives on Terrorism*. New York: Peter Lang Publishing, 2013.
Al-Tunjī, Muḥammad. *Al-Mu'jam al-Dhahabī*. Beirut, Lebanon: n.p., 1969.
'Ubaydāt, Maḥmūd Salīm. *Qaḍiyyat al-Īmān wa al-Takfīr fī ārā' al-Muslimīn*. Amman, Jordan: Dār al-Bashīr, 1996.
Van Ess, Joseph. "Das 'Kitab al-Irga'' des Hasan b. Muhammad b. al-Hanafiyya." *Arabica*, T. 21, Fasc. 1 (Feb. 1974).
Volpi, Frederic, ed. *Political Islam: A Reader*. New York: Routledge, 2011.

Waines, David. *An Introduction to Islam*, 2nd ed. Cambridge: Cambridge University Press, 2004.

Wellhausen, Julius. *The Religio-Political Factions in Early Islam*. Oxford: North-Holland Publishing Company, 1975.

Wensinck, A. J. *The Muslim Creed: Its Genesis and Historical Development*, v. 39. London: Routledge, 2008.

Zāhir, Rifqī. *Qaḍiyyat al-Takfīr 'inda al-Ghazālī*. Cairo, Egypt: Maktabat al-Azhar, 1977.

Al-Zamakhsharī, Abū al-Qāsim Maḥmūd b. 'Umar. *Tafsīr al-Kashāf*, 3rd ed., ed. Khalīl Ma'mūn Shīḥā. Beirut, Lebanon: Dār al-Ma'rifah, 2009.

———. *Al-Minhāj fī Uṣūl al-Dīn*, ed. Sabine Schmidtke. Beirut, Lebanon: Al-Dār al-'Arabiyyah lil 'Ulūm, 2007.

Al-Zarkashī, Badr al-Dīn. *Al-Manthūr fī al-Qawā'id*, vol. 2, ed. Muḥammad Ḥasan Ismā'īl. Beirut, Lebanon: Dār al-Kutub al-'Ilmiyyah, 2000.

Index

About the Author

Hussam S. Timani is an associate professor of philosophy and religion and co-director of the Middle East and North Africa Studies Program at Christopher Newport University. He teaches courses on world religions, Islam, and interfaith studies. Dr. Timani earned his BA in Political Science from the University of California, Irvine, and his PhD in Islamic Studies from the University of California, Los Angeles. He is the author of *Modern Intellectual Readings of the Kharijites* (2008) and co-editor of *Strangers in this World: Multireligious Reflections on Immigration* (2015). He has also authored more than fifteen book chapters and journal/encyclopedic articles on classical and modern Islamic religious thought, comparative theology, and interfaith studies. Dr. Timani is a frequent speaker on Islam and interfaith dialogue and leads community seminars on scriptural reasoning and interfaith dialogue. He is a recipient of the 2017 *National Association for the Advancement of Colored People (NAACP) Award* for leadership and service as interfaith guidepost both in the Hampton Roads community and the Commonwealth of Virginia and the 2009 *Rumi Forum* Education Award for service, leadership, and dedication to the cause of dialogue, peace, tolerance, community service, and understanding.

www.ingramcontent.com/pod-product-compliance
Ingram Content Group UK Ltd.
Pitfield, Milton Keynes, MK11 3LW, UK
UKHW041122071026
15587UKWH00009B/79

9 780739 194256